Ford Ranger & Mazda B-series Pick-ups Automotive Repair Manual

by Eric Jorgensen, Alan Ahlstrand and John H Haynes
Member of the Guild of Motoring Writers

Models covered:
All Ford Ranger models - 1993 through 2000
All Mazda B2300, B3000 and B4000
pick-ups - 1994 through 2000

(1D9 - 36071)   ABCDEF

Haynes Publishing Group
Sparkford Nr Yeovil
Somerset BA22 7JJ England

Haynes North America, Inc
861 Lawrence Drive
Newbury Park
California 91320 USA

About this manual

Its purpose

The purpose of this manual is to help you get the best value from your vehicle. It can do so in several ways. It can help you decide what work must be done, even if you choose to have it done by a dealer service department or a repair shop; it provides information and procedures for routine maintenance and servicing; and it offers diagnostic and repair procedures to follow when trouble occurs.

We hope you use the manual to tackle the work yourself. For many simpler jobs, doing it yourself may be quicker than arranging an appointment to get the vehicle into a shop and making the trips to leave it and pick it up. More importantly, a lot of money can be saved by avoiding the expense the shop must pass on to you to cover its labor and overhead costs. An added benefit is the sense of satisfaction and accomplishment that you feel after doing the job yourself.

Using the manual

The manual is divided into Chapters. Each Chapter is divided into numbered Sections, which are headed in bold type between horizontal lines. Each Section consists of consecutively numbered paragraphs.

At the beginning of each numbered Section you will be referred to any illustrations which apply to the procedures in that Section. The reference numbers used in illustration captions pinpoint the pertinent Section and the Step within that Section. That is, illustration 3.2 means the illustration refers to Section 3 and Step (or paragraph) 2 within that Section.

Procedures, once described in the text, are not normally repeated. When it's necessary to refer to another Chapter, the reference will be given as Chapter and Section number. Cross references given without use of the word "Chapter" apply to Sections and/or paragraphs in the same Chapter. For example, "see Section 8" means in the same Chapter.

References to the left or right side of the vehicle assume you are sitting in the driver's seat, facing forward.

Even though we have prepared this manual with extreme care, neither the publisher nor the author can accept responsibility for any errors in, or omissions from, the information given.

NOTE

A **Note** provides information necessary to properly complete a procedure or information which will make the procedure easier to understand.

CAUTION

A **Caution** provides a special procedure or special steps which must be taken while completing the procedure where the Caution is found. Not heeding a Caution can result in damage to the assembly being worked on.

WARNING

A **Warning** provides a special procedure or special steps which must be taken while completing the procedure where the Warning is found. Not heeding a Warning can result in personal injury.

Acknowledgements

We are grateful to the Ford Motor Company for assistance with technical information and certain illustrations.

© Haynes North America, Inc. 1996, 1998, 1999, 2000
With permission from J.H. Haynes & Co. Ltd.

A book in the Haynes Automotive Repair Manual Series

Printed in the U.S.A.

All rights reserved. No part of this book may be reproduced or transmitted in any form or by any means, electronic or mechanical, including photocopying, recording or by any information storage or retrieval system, without permission in writing from the copyright holder.

ISBN 1 56392 391 2

Library of Congress Catalog Card Number 00-104970

While every attempt is made to ensure that the information in this manual is correct, no liability can be accepted by the authors or publishers for loss, damage or injury caused by any errors in, or omissions from, the information given.

Contents

Introductory pages
About this manual	0-2
Introduction to the Ford Ranger, Mazda B2300, B3000 and B4000 pick-ups	0-4
Vehicle identification numbers	0-5
Buying parts	0-6
Maintenance techniques, tools and working facilities	0-6
Jacking and towing	0-12
Booster battery (jump) starting	0-13
Automotive chemicals and lubricants	0-14
Conversion factors	0-15
Safety first!	0-16
Troubleshooting	0-17

Chapter 1
Tune-up and routine maintenance — 1-1

Chapter 2 Part A
Four-cylinder engines — 2A-1

Chapter 2 Part B
3.0L V6 engine — 2B-1

Chapter 2 Part C
4.0L V6 engine — 2C-1

Chapter 2 Part D
General engine overhaul procedures — 2D-1

Chapter 3
Cooling, heating and air conditioning systems — 3-1

Chapter 4
Fuel and exhaust systems — 4-1

Chapter 5
Engine electrical systems — 5-1

Chapter 6
Emissions and engine control systems — 6-1

Chapter 7 Part A
Manual transmission — 7A-1

Chapter 7 Part B
Automatic transmission — 7B-1

Chapter 7 Part C
Transfer case — 7C-1

Chapter 8
Clutch and drivetrain — 8-1

Chapter 9
Brakes — 9-1

Chapter 10
Suspension and steering systems — 10-1

Chapter 11
Body — 11-1

Chapter 12
Chassis electrical system — 12-1

Wiring diagrams — 12-23

Index — IND-1

Photographer, author, and Haynes mechanic with 1995 Ford Ranger Pick-up

Introduction to the Ford Ranger and Mazda B-series pick-ups

The models covered by this manual are available in a variety of trim options.

Engine options include the inline four, 3.0L V6 and 4.0L V6.

Chassis layout is conventional, with the engine mounted at the front and the power being transmitted through either a manual or automatic transmission to a driveshaft and solid rear axle on 2WD models. On 4WD models a transfer case transmits power to the front axle by way of a driveshaft. Transmissions used are a five-speed overdrive manual and four-speed overdrive automatic.

2WD models use twin I-beam front suspension with coil springs and radius arms. 4WD models use a similar independent front suspension with a two-piece front driveaxle assembly, coil springs and radius arms. Both types use semi-elliptical leaf springs at the rear.

All models are equipped with power assisted front disc and rear drum brakes. Rear anti-lock brakes are used on some models; others are equipped with four-wheel anti-lock brakes.

Vehicle identification numbers

Vehicle identification numbers

Modifications are a continuing and unpublicized process in vehicle manufacturing. Since spare parts manuals and lists are compiled on a numerical basis, the individual vehicle numbers are necessary to correctly identify the component required.

Vehicle Identification Number (VIN)

This very important identification number is located on a plate attached to the left side dashboard just inside the windshield **(see illustration)**. The VIN also appears on the Vehicle Certificate of Title and Registration. It contains information such as where and when the vehicle was manufactured, engine type, the model year and the body style.

Vehicle Certification Label

The Vehicle Certification label (VC label, also referred to as the Truck Safety Compliance Certification label) is attached to the end of the left (driver's side) door **(see illustration)**. The upper half of the label contains the name of the manufacturer, the month and year of production, the Gross Vehicle Weight Rating (GVWR), the Gross Axle Weight Rating (GAWR) and the certification statement.

The VC label also contains the VIN number, which is used for warranty identification of the vehicle, and provides such information as manufacturer, type of restraint system, body type, engine, transmission, model year and vehicle serial number.

Engine numbers

Labels containing the engine code, calibration and serial numbers, as well as plant name, can be found on the valve cover, as well as stamped on the engine itself **(see illustration)**.

Calibration label

This label **(see illustration)**, found on the driver's door pillar, contains information that will be required when purchasing fuel or emissions components

Manual transmission numbers

The manual transmission identification number and serial numbers can be found on a tag on the left or right side of the transmission main case **(see illustration)**.

Automatic transmission numbers

Tags with the automatic transmission serial number, build date and other information are attached with a bolt, usually at the extension housing **(see illustration)**.

Vehicle Emissions Control Information (VECI) label

This label is found under the hood (see Chapter 6).

The Vehicle Identification Number (VIN) is visible through the driver's side of the windshield

The Vehicle Certification label, also known as the Truck Safety Compliance Certification label is located on the left front door pillar

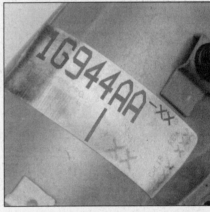

An engine identification label can usually be found on the valve cover

The calibration label is located on the driver's door pillar

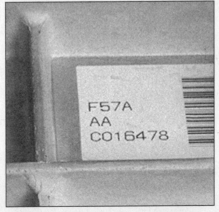

The manual transmission identification label is mounted on the side of the transmission case

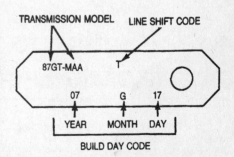

The automatic transmission identification label is attached with a bolt, usually to the extension housing

Buying parts

Replacement parts are available from many sources, which generally fall into one of two categories - authorized dealer parts departments and independent retail auto parts stores. Our advice concerning these parts is as follows:

Retail auto parts stores: Good auto parts stores will stock frequently needed components which wear out relatively fast, such as clutch components, exhaust systems, brake parts, tune-up parts, etc. These stores often supply new or reconditioned parts on an exchange basis, which can save a considerable amount of money. Discount auto parts stores are often very good places to buy materials and parts needed for general vehicle maintenance such as oil, grease, filters, spark plugs, belts, touch-up paint, bulbs, etc. They also usually sell tools and general accessories, have convenient hours, charge lower prices and can often be found not far from home.

Authorized dealer parts department: This is the best source for parts which are unique to the vehicle and not generally available elsewhere (such as major engine parts, transmission parts, trim pieces, etc.).

Warranty information: If the vehicle is still covered under warranty, be sure that any replacement parts purchased - regardless of the source - do not invalidate the warranty!

To be sure of obtaining the correct parts, have engine and chassis numbers available and, if possible, take the old parts along for positive identification.

Maintenance techniques, tools and working facilities

Maintenance techniques

There are a number of techniques involved in maintenance and repair that will be referred to throughout this manual. Application of these techniques will enable the home mechanic to be more efficient, better organized and capable of performing the various tasks properly, which will ensure that the repair job is thorough and complete.

Fasteners

Fasteners are nuts, bolts, studs and screws used to hold two or more parts together. There are a few things to keep in mind when working with fasteners. Almost all of them use a locking device of some type, either a lockwasher, locknut, locking tab or thread adhesive. All threaded fasteners should be clean and straight, with undamaged threads and undamaged corners on the hex head where the wrench fits. Develop the habit of replacing all damaged nuts and bolts with new ones. Special locknuts with nylon or fiber inserts can only be used once. If they are removed, they lose their locking ability and must be replaced with new ones.

Rusted nuts and bolts should be treated with a penetrating fluid to ease removal and prevent breakage. Some mechanics use turpentine in a spout-type oil can, which works quite well. After applying the rust penetrant, let it work for a few minutes before trying to loosen the nut or bolt. Badly rusted fasteners may have to be chiseled or sawed off or removed with a special nut breaker, available at tool stores.

If a bolt or stud breaks off in an assembly, it can be drilled and removed with a special tool commonly available for this purpose. Most automotive machine shops can perform this task, as well as other repair procedures, such as the repair of threaded holes that have been stripped out.

Flat washers and lockwashers, when removed from an assembly, should always be replaced exactly as removed. Replace any damaged washers with new ones. Never use a lockwasher on any soft metal surface (such as aluminum), thin sheet metal or plastic.

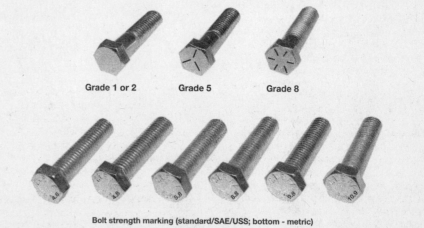

Maintenance techniques, tools and working facilities

Fastener sizes

For a number of reasons, automobile manufacturers are making wider and wider use of metric fasteners. Therefore, it is important to be able to tell the difference between standard (sometimes called U.S. or SAE) and metric hardware, since they cannot be interchanged.

All bolts, whether standard or metric, are sized according to diameter, thread pitch and length. For example, a standard 1/2 - 13 x 1 bolt is 1/2 inch in diameter, has 13 threads per inch and is 1 inch long. An M12 - 1.75 x 25 metric bolt is 12 mm in diameter, has a thread pitch of 1.75 mm (the distance between threads) and is 25 mm long. The two bolts are nearly identical, and easily confused, but they are not interchangeable.

In addition to the differences in diameter, thread pitch and length, metric and standard bolts can also be distinguished by examining the bolt heads. To begin with, the distance across the flats on a standard bolt head is measured in inches, while the same dimension on a metric bolt is sized in millimeters (the same is true for nuts). As a result, a standard wrench should not be used on a metric bolt and a metric wrench should not be used on a standard bolt. Also, most standard bolts have slashes radiating out from the center of the head to denote the grade or strength of the bolt, which is an indication of the amount of torque that can be applied to it. The greater the number of slashes, the greater the strength of the bolt. Grades 0 through 5 are commonly used on automobiles. Metric bolts have a property class (grade) number, rather than a slash, molded into their heads to indicate bolt strength. In this case, the higher the number, the stronger the bolt. Property class numbers 8.8, 9.8 and 10.9 are commonly used on automobiles.

Strength markings can also be used to distinguish standard hex nuts from metric hex nuts. Many standard nuts have dots stamped into one side, while metric nuts are marked with a number. The greater the number of dots, or the higher the number, the greater the strength of the nut.

Metric studs are also marked on their ends according to property class (grade). Larger studs are numbered (the same as metric bolts), while smaller studs carry a geometric code to denote grade.

It should be noted that many fasteners, especially Grades 0 through 2, have no distinguishing marks on them. When such is the case, the only way to determine whether it is standard or metric is to measure the thread pitch or compare it to a known fastener of the same size.

Standard fasteners are often referred to as SAE, as opposed to metric. However, it should be noted that SAE technically refers to a non-metric fine thread fastener only. Coarse thread non-metric fasteners are referred to as USS sizes.

Since fasteners of the same size (both standard and metric) may have different strength ratings, be sure to reinstall any bolts, studs or nuts removed from your vehicle in their original locations. Also, when replacing a fastener with a new one, make sure that the new one has a strength rating equal to or greater than the original.

Tightening sequences and procedures

Most threaded fasteners should be tightened to a specific torque value (torque is the twisting force applied to a threaded component such as a nut or bolt). Overtightening the fastener can weaken it and cause it to break, while undertightening can cause it to eventually come loose. Bolts, screws and studs, depending on the material they are made of and their thread diameters, have specific torque values, many of which are noted in the Specifications at the beginning of each Chapter. Be sure to follow the torque recommendations closely. For fasteners not assigned a specific torque, a general torque value chart is presented here as a guide. These torque values are for dry (unlubricated) fasteners threaded into steel or cast iron (not aluminum). As was previously mentioned, the size and grade of a fastener determine the amount of torque that can safely be applied to it. The figures listed here are approximate for Grade 2 and Grade 3 fasteners. Higher grades can tolerate higher torque values.

Fasteners laid out in a pattern, such as cylinder head bolts, oil pan bolts, differential cover bolts, etc., must be loosened or tightened in sequence to avoid warping the com-

Metric thread sizes	Ft-lbs	Nm
M-6	6 to 9	9 to 12
M-8	14 to 21	19 to 28
M-10	28 to 40	38 to 54
M-12	50 to 71	68 to 96
M-14	80 to 140	109 to 154

Pipe thread sizes		
1/8	5 to 8	7 to 10
1/4	12 to 18	17 to 24
3/8	2 to 33	30 to 44
1/2	25 to 35	34 to 47

U.S. thread sizes		
1/4 - 20	6 to 9	9 to 12
5/16 - 18	12 to 18	17 to 24
5/16 - 24	14 to 20	19 to 27
3/8 - 16	22 to 32	30 to 43
3/8 - 24	27 to 38	37 to 51
7/16 - 14	40 to 55	55 to 74
7/16 - 20	40 to 60	55 to 81
1/2 - 13	55 to 80	75 to 108

Standard (SAE and USS) bolt dimension/grade marks
- G   Grade marks (bolt strength)
- L   Length (in inches)
- T   Tread pitch (number of threads per inch)
- D   Nominal diameter (in inches)

Metric bolt dimensions/grade marks
- P   Property class (bolt strength)
- L   Length (in millimeters)
- T   Thread pitch (distance between threads in millimeters)
- D   Diameter

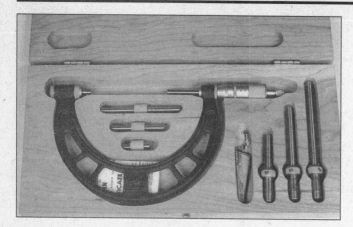

Micrometer set

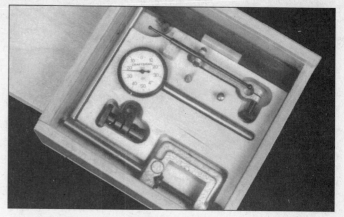

Dial indicator set

ponent. This sequence will normally be shown in the appropriate Chapter. If a specific pattern is not given, the following procedures can be used to prevent warping.

Initially, the bolts or nuts should be assembled finger-tight only. Next, they should be tightened one full turn each, in a criss-cross or diagonal pattern. After each one has been tightened one full turn, return to the first one and tighten them all one-half turn, following the same pattern. Finally, tighten each of them one-quarter turn at a time until each fastener has been tightened to the proper torque. To loosen and remove the fasteners, the procedure would be reversed.

Component disassembly

Component disassembly should be done with care and purpose to help ensure that the parts go back together properly. Always keep track of the sequence in which parts are removed. Make note of special characteristics or marks on parts that can be installed more than one way, such as a grooved thrust washer on a shaft. It is a good idea to lay the disassembled parts out on a clean surface in the order that they were removed. It may also be helpful to make sketches or take instant photos of components before removal.

When removing fasteners from a component, keep track of their locations. Sometimes threading a bolt back in a part, or putting the washers and nut back on a stud, can prevent mix-ups later. If nuts and bolts cannot be returned to their original locations, they should be kept in a compartmented box or a series of small boxes. A cupcake or muffin tin is ideal for this purpose, since each cavity can hold the bolts and nuts from a particular area (i.e. oil pan bolts, valve cover bolts, engine mount bolts, etc.). A pan of this type is especially helpful when working on assemblies with very small parts, such as the carburetor, alternator, valve train or interior dash and trim pieces. The cavities can be marked with paint or tape to identify the contents.

Whenever wiring looms, harnesses or connectors are separated, it is a good idea to identify the two halves with numbered pieces of masking tape so they can be easily reconnected.

Gasket sealing surfaces

Throughout any vehicle, gaskets are used to seal the mating surfaces between two parts and keep lubricants, fluids, vacuum or pressure contained in an assembly.

Many times these gaskets are coated with a liquid or paste-type gasket sealing compound before assembly. Age, heat and pressure can sometimes cause the two parts to stick together so tightly that they are very difficult to separate. Often, the assembly can be loosened by striking it with a soft-face hammer near the mating surfaces. A regular hammer can be used if a block of wood is placed between the hammer and the part. Do not hammer on cast parts or parts that could be easily damaged. With any particularly stubborn part, always recheck to make sure that every fastener has been removed.

Avoid using a screwdriver or bar to pry apart an assembly, as they can easily mar the gasket sealing surfaces of the parts, which must remain smooth. If prying is absolutely necessary, use an old broom handle, but keep in mind that extra clean up will be necessary if the wood splinters.

After the parts are separated, the old gasket must be carefully scraped off and the gasket surfaces cleaned. Stubborn gasket material can be soaked with rust penetrant or treated with a special chemical to soften it so it can be easily scraped off. A scraper can be fashioned from a piece of copper tubing by flattening and sharpening one end. Copper is recommended because it is usually softer than the surfaces to be scraped, which reduces the chance of gouging the part. Some gaskets can be removed with a wire brush, but regardless of the method used, the mating surfaces must be left clean and smooth. If for some reason the gasket surface is gouged, then a gasket sealer thick enough to fill scratches will have to be used during reassembly of the components. For most applications, a non-drying (or semi-drying) gasket sealer should be used.

Hose removal tips

Warning: *If the vehicle is equipped with air conditioning, do not disconnect any of the A/C hoses without first having the system depressurized by a dealer service department or a service station.*

Hose removal precautions closely parallel gasket removal precautions. Avoid scratching or gouging the surface that the hose mates against or the connection may leak. This is especially true for radiator hoses. Because of various chemical reactions, the rubber in hoses can bond itself to the metal spigot that the hose fits over. To remove a hose, first loosen the hose clamps that secure it to the spigot. Then, with slip-joint pliers, grab the hose at the clamp and rotate it around the spigot. Work it back and forth until it is completely free, then pull it off. Silicone or other lubricants will ease removal if they can be applied between the hose and the outside of the spigot. Apply the same lubricant to the inside of the hose and the outside of the spigot to simplify installation.

As a last resort (and if the hose is to be replaced with a new one anyway), the rubber can be slit with a knife and the hose peeled from the spigot. If this must be done, be careful that the metal connection is not damaged.

If a hose clamp is broken or damaged, do not reuse it. Wire-type clamps usually weaken with age, so it is a good idea to replace them with screw-type clamps whenever a hose is removed.

Tools

A selection of good tools is a basic requirement for anyone who plans to maintain and repair his or her own vehicle. For the owner who has few tools, the initial investment might seem high, but when compared to the spiraling costs of professional auto maintenance and repair, it is a wise one.

To help the owner decide which tools are needed to perform the tasks detailed in this manual, the following tool lists are offered: *Maintenance and minor repair, Repair/overhaul* and *Special*.

The newcomer to practical mechanics

Maintenance techniques, tools and working facilities

Dial caliper

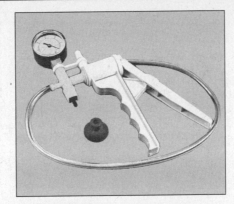
Hand-operated vacuum pump

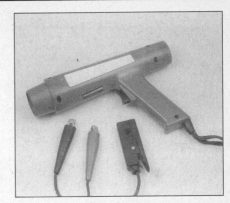
Timing light

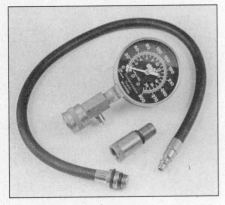
Compression gauge with spark plug hole adapter

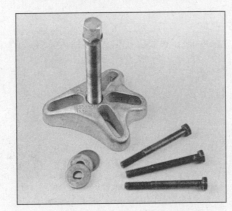
Damper/steering wheel puller

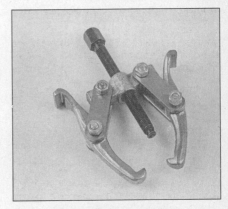
General purpose puller

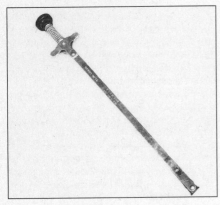
Hydraulic lifter removal tool

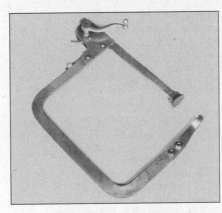
Valve spring compressor

Valve spring compressor

Ridge reamer

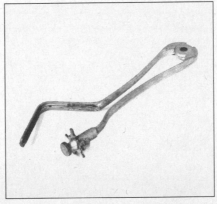
Piston ring groove cleaning tool

Ring removal/installation tool

Maintenance techniques, tools and working facilities

Ring compressor

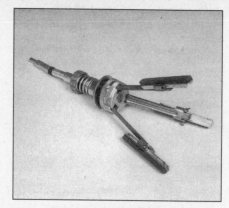
Cylinder hone

Brake hold-down spring tool

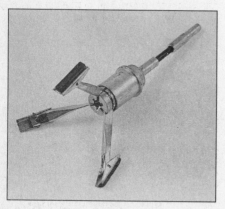
Brake cylinder hone

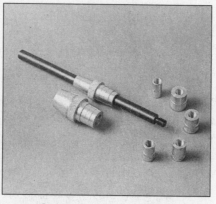
Clutch plate alignment tool

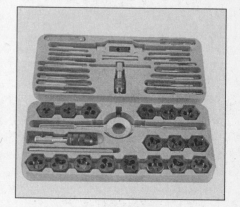
Tap and die set

should start off with the *maintenance and minor repair* tool kit, which is adequate for the simpler jobs performed on a vehicle. Then, as confidence and experience grow, the owner can tackle more difficult tasks, buying additional tools as they are needed. Eventually the basic kit will be expanded into the *repair and overhaul* tool set. Over a period of time, the experienced do-it-yourselfer will assemble a tool set complete enough for most repair and overhaul procedures and will add tools from the special category when it is felt that the expense is justified by the frequency of use.

Maintenance and minor repair tool kit

The tools in this list should be considered the minimum required for performance of routine maintenance, servicing and minor repair work. We recommend the purchase of combination wrenches (box-end and open-end combined in one wrench). While more expensive than open end wrenches, they offer the advantages of both types of wrench.

Combination wrench set (1/4-inch to 1 inch or 6 mm to 19 mm)
Adjustable wrench, 8 inch
Spark plug wrench with rubber insert
Spark plug gap adjusting tool
Feeler gauge set
Brake bleeder wrench
Standard screwdriver (5/16-inch x 6 inch)
Phillips screwdriver (No. 2 x 6 inch)
Combination pliers - 6 inch
Hacksaw and assortment of blades
Tire pressure gauge
Grease gun
Oil can
Fine emery cloth
Wire brush
Battery post and cable cleaning tool
Oil filter wrench
Funnel (medium size)
Safety goggles
Jackstands (2)
Drain pan

Note: *If basic tune-ups are going to be part of routine maintenance, it will be necessary to purchase a good quality stroboscopic timing light and combination tachometer/dwell meter. Although they are included in the list of special tools, it is mentioned here because they are absolutely necessary for tuning most vehicles properly.*

Repair and overhaul tool set

These tools are essential for anyone who plans to perform major repairs and are in addition to those in the maintenance and minor repair tool kit. Included is a comprehensive set of sockets which, though expensive, are invaluable because of their versatility, especially when various extensions and drives are available. We recommend the 1/2-inch drive over the 3/8-inch drive. Although the larger drive is bulky and more expensive, it has the capacity of accepting a very wide range of large sockets. Ideally, however, the mechanic should have a 3/8-inch drive set and a 1/2-inch drive set.

Socket set(s)
Reversible ratchet
Extension - 10 inch
Universal joint
Torque wrench (same size drive as sockets)
Ball peen hammer - 8 ounce
Soft-face hammer (plastic/rubber)
Standard screwdriver (1/4-inch x 6 inch)
Standard screwdriver (stubby - 5/16-inch)
Phillips screwdriver (No. 3 x 8 inch)
Phillips screwdriver (stubby - No. 2)
Pliers - vise grip
Pliers - lineman's
Pliers - needle nose
Pliers - snap-ring (internal and external)
Cold chisel - 1/2-inch
Scribe
Scraper (made from flattened copper tubing)
Centerpunch
Pin punches (1/16, 1/8, 3/16-inch)
Steel rule/straightedge - 12 inch
Allen wrench set (1/8 to 3/8-inch or 4 mm to 10 mm)
A selection of files
Wire brush (large)
Jackstands (second set)
Jack (scissor or hydraulic type)

Maintenance techniques, tools and working facilities

Note: *Another tool which is often useful is an electric drill with a chuck capacity of 3/8-inch and a set of good quality drill bits.*

Special tools

The tools in this list include those which are not used regularly, are expensive to buy, or which need to be used in accordance with their manufacturer's instructions. Unless these tools will be used frequently, it is not very economical to purchase many of them. A consideration would be to split the cost and use between yourself and a friend or friends. In addition, most of these tools can be obtained from a tool rental shop on a temporary basis.

This list primarily contains only those tools and instruments widely available to the public, and not those special tools produced by the vehicle manufacturer for distribution to dealer service departments. Occasionally, references to the manufacturer's special tools are included in the text of this manual. Generally, an alternative method of doing the job without the special tool is offered. However, sometimes there is no alternative to their use. Where this is the case, and the tool cannot be purchased or borrowed, the work should be turned over to the dealer service department or an automotive repair shop.

*Valve spring compressor
Piston ring groove cleaning tool
Piston ring compressor
Piston ring installation tool
Cylinder compression gauge
Cylinder ridge reamer
Cylinder surfacing hone
Cylinder bore gauge
Micrometers and/or dial calipers
Hydraulic lifter removal tool
Balljoint separator
Universal-type puller
Impact screwdriver
Dial indicator set
Stroboscopic timing light (inductive pick-up)
Hand operated vacuum/pressure pump
Tachometer/dwell meter
Universal electrical multimeter
Cable hoist
Brake spring removal and installation tools
Floor jack*

Buying tools

For the do-it-yourselfer who is just starting to get involved in vehicle maintenance and repair, there are a number of options available when purchasing tools. If maintenance and minor repair is the extent of the work to be done, the purchase of individual tools is satisfactory. If, on the other hand, extensive work is planned, it would be a good idea to purchase a modest tool set from one of the large retail chain stores. A set can usually be bought at a substantial savings over the individual tool prices, and they often come with a tool box. As additional tools are needed, add-on sets, individual tools and a larger tool box can be purchased to expand the tool selection. Building a tool set gradually allows the cost of the tools to be spread over a longer period of time and gives the mechanic the freedom to choose only those tools that will actually be used.

Tool stores will often be the only source of some of the special tools that are needed, but regardless of where tools are bought, try to avoid cheap ones, especially when buying screwdrivers and sockets, because they won't last very long. The expense involved in replacing cheap tools will eventually be greater than the initial cost of quality tools.

Care and maintenance of tools

Good tools are expensive, so it makes sense to treat them with respect. Keep them clean and in usable condition and store them properly when not in use. Always wipe off any dirt, grease or metal chips before putting them away. Never leave tools lying around in the work area. Upon completion of a job, always check closely under the hood for tools that may have been left there so they won't get lost during a test drive.

Some tools, such as screwdrivers, pliers, wrenches and sockets, can be hung on a panel mounted on the garage or workshop wall, while others should be kept in a tool box or tray. Measuring instruments, gauges, meters, etc. must be carefully stored where they cannot be damaged by weather or impact from other tools.

When tools are used with care and stored properly, they will last a very long time. Even with the best of care, though, tools will wear out if used frequently. When a tool is damaged or worn out, replace it. Subsequent jobs will be safer and more enjoyable if you do.

How to repair damaged threads

Sometimes, the internal threads of a nut or bolt hole can become stripped, usually from overtightening. Stripping threads is an all-too-common occurrence, especially when working with aluminum parts, because aluminum is so soft that it easily strips out.

Usually, external or internal threads are only partially stripped. After they've been cleaned up with a tap or die, they'll still work. Sometimes, however, threads are badly damaged. When this happens, you've got three choices:

1) *Drill and tap the hole to the next suitable oversize and install a larger diameter bolt, screw or stud.*
2) *Drill and tap the hole to accept a threaded plug, then drill and tap the plug to the original screw size. You can also buy a plug already threaded to the original size. Then you simply drill a hole to the specified size, then run the threaded plug into the hole with a bolt and jam nut. Once the plug is fully seated, remove the jam nut and bolt.*
3) *The third method uses a patented thread repair kit like Heli-Coil or Slimsert. These easy-to-use kits are designed to repair damaged threads in straight-through holes and blind holes. Both are available as kits which can handle a variety of sizes and thread patterns. Drill the hole, then tap it with the special included tap. Install the Heli-Coil and the hole is back to its original diameter and thread pitch.*

Regardless of which method you use, be sure to proceed calmly and carefully. A little impatience or carelessness during one of these relatively simple procedures can ruin your whole day's work and cost you a bundle if you wreck an expensive part.

Working facilities

Not to be overlooked when discussing tools is the workshop. If anything more than routine maintenance is to be carried out, some sort of suitable work area is essential.

It is understood, and appreciated, that many home mechanics do not have a good workshop or garage available, and end up removing an engine or doing major repairs outside. It is recommended, however, that the overhaul or repair be completed under the cover of a roof.

A clean, flat workbench or table of comfortable working height is an absolute necessity. The workbench should be equipped with a vise that has a jaw opening of at least four inches.

As mentioned previously, some clean, dry storage space is also required for tools, as well as the lubricants, fluids, cleaning solvents, etc. which soon become necessary.

Sometimes waste oil and fluids, drained from the engine or cooling system during normal maintenance or repairs, present a disposal problem. To avoid pouring them on the ground or into a sewage system, pour the used fluids into large containers, seal them with caps and take them to an authorized disposal site or recycling center. Plastic jugs, such as old antifreeze containers, are ideal for this purpose.

Always keep a supply of old newspapers and clean rags available. Old towels are excellent for mopping up spills. Many mechanics use rolls of paper towels for most work because they are readily available and disposable. To help keep the area under the vehicle clean, a large cardboard box can be cut open and flattened to protect the garage or shop floor.

Whenever working over a painted surface, such as when leaning over a fender to service something under the hood, always cover it with an old blanket or bedspread to protect the finish. Vinyl covered pads, made especially for this purpose, are available at auto parts stores.

Jacking and towing

Jacking

Warning: *On vehicles equipped with a limited slip differential, do not run the engine with the vehicle on a jack.*

When lifting the vehicle with either the supplied jack or a floor jack, block the wheel that is diagonally opposite the one being lifted to prevent the vehicle from moving.

Warning: *If equipped with an under-chassis mounted spare tire, remove the tire or tire carrier from the rack before the vehicle is raised in order to avoid sudden weight release from the chassis.*

When using the vehicle jack, the manufacturer recommends positioning the jack under the rear axle housing or under the front shock absorber **(see illustration)**.

If you're using a floor jack to raise the vehicle, place the jack under the outer end of the control arm (front) or under the outer end of the axle housing (rear) **(see illustrations)**.

Caution: *Don't raise the rear of the vehicle with a floor jack positioned under the differential. The cover could become distorted or cracked, resulting in lubricant leakage.*

When raising the vehicle for service, always place jackstands under the frame rails. NEVER work under the vehicle when it is supported only by a jack!

Towing

When the vehicle is towed with the rear wheels raised, the steering wheel must be clamped in the straight ahead position with a special device designed for use during towing. **Warning:** *Don't use the steering lock to keep the front wheels pointed straight ahead. It's not strong enough for this purpose.*

The vehicle can be towed with all four wheels on the ground, provided the following conditions are met (depending on drivetrain):

2WD models with an automatic transmission

a) Parking brake released
b) Transmission in Neutral (N)
c) Maximum speed of 35 mph*
d) Maximum distance of 50 miles*

*If this speed or distance will be exceeded, disconnect the driveshaft at the differential and support it securely under the vehicle.

2WD models with a manual transmission

a) Parking brake released
b) Transmission in Neutral (N)
c) Maximum speed of 55 mph

4WD models with a manual shift transfer case

a) Parking brake released
b) Transmission in Neutral (N)
c) Transfer case in Neutral (N)
d) Manual locking hubs in the FREE position
e) Maximum speed of 55 mph

4WD with an electric shift transfer case

a) Parking brake released
b) Transmission in Neutral (N)
c) Transfer case placed in the 2H position
d) Hubs unlocked
e) If the vehicle is equipped with an automatic transmission, limit speed to 35 mph and distance to 50 miles*

*If this speed or distance will be exceeded, disconnect the driveshaft at the differential and support it securely under the vehicle.

Equipment specifically designed for towing should be used. It should be attached to the frame members of the vehicle, not the bumpers or brackets.

Safety is a major consideration when towing and all applicable state and local laws must be obeyed. A safety chain system must be used at all times. Remember that power steering and power brakes will not work with the engine off.

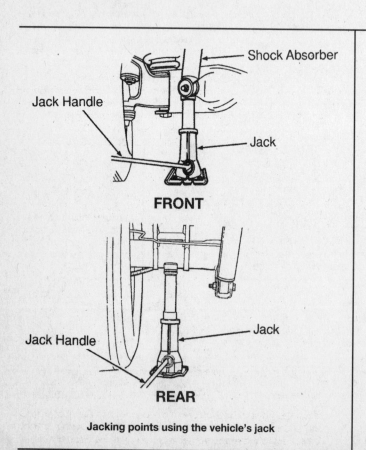

Jacking points using the vehicle's jack

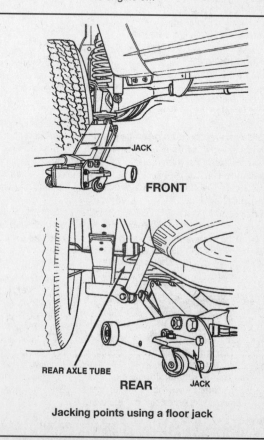

Jacking points using a floor jack

Booster battery (jump) starting

Observe these precautions when using a booster battery to start a vehicle:

a) Before connecting the booster battery, make sure the ignition switch is in the Off position.
b) Turn off the lights, heater and other electrical loads.
c) Your eyes should be shielded. Safety goggles are a good idea.
d) Make sure the booster battery is the same voltage as the dead one in the vehicle.
e) The two vehicles MUST NOT TOUCH each other!
f) Make sure the transaxle is in Neutral (manual) or Park (automatic).
g) If the booster battery is not a maintenance-free type, remove the vent caps and lay a cloth over the vent holes.

Connect the red jumper cable to the positive (+) terminals of each battery **(see illustration)**.

Connect one end of the black jumper cable to the negative (-) terminal of the booster battery. The other end of this cable should be connected to a good ground on the vehicle to be started, such as a bolt or bracket on the body.

Start the engine using the booster battery, then, with the engine running at idle speed, disconnect the jumper cables in the reverse order of connection.

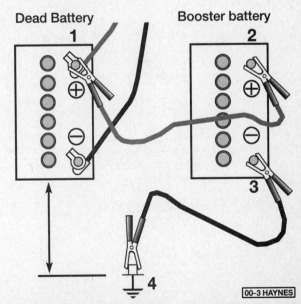

Make the booster battery cable connections in the numerical order shown (note that the negative cable of the booster battery is NOT attached to the negative terminal of the dead battery)

Automotive chemicals and lubricants

A number of automotive chemicals and lubricants are available for use during vehicle maintenance and repair. They include a wide variety of products ranging from cleaning solvents and degreasers to lubricants and protective sprays for rubber, plastic and vinyl.

Cleaners

Carburetor cleaner and choke cleaner is a strong solvent for gum, varnish and carbon. Most carburetor cleaners leave a dry-type lubricant film which will not harden or gum up. Because of this film it is not recommended for use on electrical components.

Brake system cleaner is used to remove grease and brake fluid from the brake system, where clean surfaces are absolutely necessary. It leaves no residue and often eliminates brake squeal caused by contaminants.

Electrical cleaner removes oxidation, corrosion and carbon deposits from electrical contacts, restoring full current flow. It can also be used to clean spark plugs, carburetor jets, voltage regulators and other parts where an oil-free surface is desired.

Demoisturants remove water and moisture from electrical components such as alternators, voltage regulators, electrical connectors and fuse blocks. They are non-conductive, non-corrosive and non-flammable.

Degreasers are heavy-duty solvents used to remove grease from the outside of the engine and from chassis components. They can be sprayed or brushed on and, depending on the type, are rinsed off either with water or solvent.

Lubricants

Motor oil is the lubricant formulated for use in engines. It normally contains a wide variety of additives to prevent corrosion and reduce foaming and wear. Motor oil comes in various weights (viscosity ratings) from 0 to 50. The recommended weight of the oil depends on the season, temperature and the demands on the engine. Light oil is used in cold climates and under light load conditions. Heavy oil is used in hot climates and where high loads are encountered. Multi-viscosity oils are designed to have characteristics of both light and heavy oils and are available in a number of weights from 5W-20 to 20W-50.

Gear oil is designed to be used in differentials, manual transmissions and other areas where high-temperature lubrication is required.

Chassis and wheel bearing grease is a heavy grease used where increased loads and friction are encountered, such as for wheel bearings, balljoints, tie-rod ends and universal joints.

High-temperature wheel bearing grease is designed to withstand the extreme temperatures encountered by wheel bearings in disc brake equipped vehicles. It usually contains molybdenum disulfide (moly), which is a dry-type lubricant.

White grease is a heavy grease for metal-to-metal applications where water is a problem. White grease stays soft under both low and high temperatures (usually from -100 to +190-degrees F), and will not wash off or dilute in the presence of water.

Assembly lube is a special extreme pressure lubricant, usually containing moly, used to lubricate high-load parts (such as main and rod bearings and cam lobes) for initial start-up of a new engine. The assembly lube lubricates the parts without being squeezed out or washed away until the engine oiling system begins to function.

Silicone lubricants are used to protect rubber, plastic, vinyl and nylon parts.

Graphite lubricants are used where oils cannot be used due to contamination problems, such as in locks. The dry graphite will lubricate metal parts while remaining uncontaminated by dirt, water, oil or acids. It is electrically conductive and will not foul electrical contacts in locks such as the ignition switch.

Moly penetrants loosen and lubricate frozen, rusted and corroded fasteners and prevent future rusting or freezing.

Heat-sink grease is a special electrically non-conductive grease that is used for mounting electronic ignition modules where it is essential that heat is transferred away from the module.

Sealants

RTV sealant is one of the most widely used gasket compounds. Made from silicone, RTV is air curing, it seals, bonds, waterproofs, fills surface irregularities, remains flexible, doesn't shrink, is relatively easy to remove, and is used as a supplementary sealer with almost all low and medium temperature gaskets.

Anaerobic sealant is much like RTV in that it can be used either to seal gaskets or to form gaskets by itself. It remains flexible, is solvent resistant and fills surface imperfections. The difference between an anaerobic sealant and an RTV-type sealant is in the curing. RTV cures when exposed to air, while an anaerobic sealant cures only in the absence of air. This means that an anaerobic sealant cures only after the assembly of parts, sealing them together.

Thread and pipe sealant is used for sealing hydraulic and pneumatic fittings and vacuum lines. It is usually made from a Teflon compound, and comes in a spray, a paint-on liquid and as a wrap-around tape.

Chemicals

Anti-seize compound prevents seizing, galling, cold welding, rust and corrosion in fasteners. High-temperature ant-seize, usually made with copper and graphite lubricants, is used for exhaust system and exhaust manifold bolts.

Anaerobic locking compounds are used to keep fasteners from vibrating or working loose and cure only after installation, in the absence of air. Medium strength locking compound is used for small nuts, bolts and screws that may be removed later. High-strength locking compound is for large nuts, bolts and studs which aren't removed on a regular basis.

Oil additives range from viscosity index improvers to chemical treatments that claim to reduce internal engine friction. It should be noted that most oil manufacturers caution against using additives with their oils.

Gas additives perform several functions, depending on their chemical makeup. They usually contain solvents that help dissolve gum and varnish that build up on carburetor, fuel injection and intake parts. They also serve to break down carbon deposits that form on the inside surfaces of the combustion chambers. Some additives contain upper cylinder lubricants for valves and piston rings, and others contain chemicals to remove condensation from the gas tank.

Miscellaneous

Brake fluid is specially formulated hydraulic fluid that can withstand the heat and pressure encountered in brake systems. Care must be taken so this fluid does not come in contact with painted surfaces or plastics. An opened container should always be resealed to prevent contamination by water or dirt.

Weatherstrip adhesive is used to bond weatherstripping around doors, windows and trunk lids. It is sometimes used to attach trim pieces.

Undercoating is a petroleum-based, tar-like substance that is designed to protect metal surfaces on the underside of the vehicle from corrosion. It also acts as a sound-deadening agent by insulating the bottom of the vehicle.

Waxes and polishes are used to help protect painted and plated surfaces from the weather. Different types of paint may require the use of different types of wax and polish. Some polishes utilize a chemical or abrasive cleaner to help remove the top layer of oxidized (dull) paint on older vehicles. In recent years many non-wax polishes that contain a wide variety of chemicals such as polymers and silicones have been introduced. These non-wax polishes are usually easier to apply and last longer than conventional waxes and polishes.

Conversion factors

Length (distance)
Inches (in) X 25.4 = Millimetres (mm) X 0.0394 = Inches (in)
Feet (ft) X 0.305 = Metres (m) X 3.281 = Feet (ft)
Miles X 1.609 = Kilometres (km) X 0.621 = Miles

Volume (capacity)
Cubic inches (cu in; in^3) X 16.387 = Cubic centimetres (cc; cm^3) X 0.061 = Cubic inches (cu in; in^3)
Imperial pints (Imp pt) X 0.568 = Litres (l) X 1.76 = Imperial pints (Imp pt)
Imperial quarts (Imp qt) X 1.137 = Litres (l) X 0.88 = Imperial quarts (Imp qt)
Imperial quarts (Imp qt) X 1.201 = US quarts (US qt) X 0.833 = Imperial quarts (Imp qt)
US quarts (US qt) X 0.946 = Litres (l) X 1.057 = US quarts (US qt)
Imperial gallons (Imp gal) X 4.546 = Litres (l) X 0.22 = Imperial gallons (Imp gal)
Imperial gallons (Imp gal) X 1.201 = US gallons (US gal) X 0.833 = Imperial gallons (Imp gal)
US gallons (US gal) X 3.785 = Litres (l) X 0.264 = US gallons (US gal)

Mass (weight)
Ounces (oz) X 28.35 = Grams (g) X 0.035 Ounces (oz)
Pounds (lb) X 0.454 = Kilograms (kg) X 2.205 = Pounds (lb)

Force
Ounces-force (ozf; oz) X 0.278 = Newtons (N) X 3.6 = Ounces-force (ozf; oz)
Pounds-force (lbf; lb) X 4.448 = Newtons (N) X 0.225 = Pounds-force (lbf; lb)
Newtons (N) X 0.1 = Kilograms-force (kgf; kg) X 9.81 = Newtons (N)

Pressure
Pounds-force per square inch (psi; lbf/in^2; lb/in^2) X 0.070 = Kilograms-force per square centimetre (kgf/cm^2; kg/cm^2) X 14.223 = Pounds-force per square inch (psi; lbf/in^2; lb/in^2)
Pounds-force per square inch (psi; lbf/in^2; lb/in^2) X 0.068 = Atmospheres (atm) X 14.696 = Pounds-force per square inch (psi; lbf/in^2; lb/in^2)
Pounds-force per square inch (psi; lbf/in^2; lb/in^2) X 0.069 = Bars X 14.5 = Pounds-force per square inch (psi; lbf/in^2; lb/in^2)
Pounds-force per square inch (psi; lbf/in^2; lb/in^2) X 6.895 = Kilopascals (kPa) X 0.145 = Pounds-force per square inch (psi; lbf/in^2; lb/in^2)
Kilopascals (kPa) X 0.01 = Kilograms-force per square centimetre (kgf/cm^2; kg/cm^2) X 98.1 = Kilopascals (kPa)

Torque (moment of force)
Pounds-force inches (lbf in; lb in) X 1.152 = Kilograms-force centimetre (kgf cm; kg cm) X 0.868 = Pounds-force inches (lbf in; lb in)
Pounds-force inches (lbf in; lb in) X 0.113 = Newton metres (Nm) X 8.85 = Pounds-force inches (lbf in; lb in)
Pounds-force inches (lbf in; lb in) X 0.083 = Pounds-force feet (lbf ft; lb ft) X 12 = Pounds-force inches (lbf in; lb in)
Pounds-force feet (lbf ft; lb ft) X 0.138 = Kilograms-force metres (kgf m; kg m) X 7.233 = Pounds-force feet (lbf ft; lb ft)
Pounds-force feet (lbf ft; lb ft) X 1.356 = Newton metres (Nm) X 0.738 = Pounds-force feet (lbf ft; lb ft)
Newton metres (Nm) X 0.102 = Kilograms-force metres (kgf m; kg m) X 9.804 = Newton metres (Nm)

Power
Horsepower (hp) X 745.7 = Watts (W) X 0.0013 = Horsepower (hp)

Velocity (speed)
Miles per hour (miles/hr; mph) X 1.609 = Kilometres per hour (km/hr; kph) X 0.621 = Miles per hour (miles/hr; mph)

Fuel consumption*
Miles per gallon, Imperial (mpg) X 0.354 = Kilometres per litre (km/l) X 2.825 = Miles per gallon, Imperial (mpg)
Miles per gallon, US (mpg) X 0.425 = Kilometres per litre (km/l) X 2.352 = Miles per gallon, US (mpg)

Temperature
Degrees Fahrenheit = (°C x 1.8) + 32   Degrees Celsius (Degrees Centigrade; °C) = (°F - 32) x 0.56

*It is common practice to convert from miles per gallon (mpg) to litres/100 kilometres (l/100km), where mpg (Imperial) x l/100 km = 282 and mpg (US) x l/100 km = 235

Safety first!

Regardless of how enthusiastic you may be about getting on with the job at hand, take the time to ensure that your safety is not jeopardized. A moment's lack of attention can result in an accident, as can failure to observe certain simple safety precautions. The possibility of an accident will always exist, and the following points should not be considered a comprehensive list of all dangers. Rather, they are intended to make you aware of the risks and to encourage a safety conscious approach to all work you carry out on your vehicle.

Essential DOs and DON'Ts

DON'T rely on a jack when working under the vehicle. Always use approved jackstands to support the weight of the vehicle and place them under the recommended lift or support points.

DON'T attempt to loosen extremely tight fasteners (i.e. wheel lug nuts) while the vehicle is on a jack - it may fall.

DON'T start the engine without first making sure that the transmission is in Neutral (or Park where applicable) and the parking brake is set.

DON'T remove the radiator cap from a hot cooling system - let it cool or cover it with a cloth and release the pressure gradually.

DON'T attempt to drain the engine oil until you are sure it has cooled to the point that it will not burn you.

DON'T touch any part of the engine or exhaust system until it has cooled sufficiently to avoid burns.

DON'T siphon toxic liquids such as gasoline, antifreeze and brake fluid by mouth, or allow them to remain on your skin.

DON'T inhale brake lining dust - it is potentially hazardous (see *Asbestos* below).

DON'T allow spilled oil or grease to remain on the floor - wipe it up before someone slips on it.

DON'T use loose fitting wrenches or other tools which may slip and cause injury.

DON'T push on wrenches when loosening or tightening nuts or bolts. Always try to pull the wrench toward you. If the situation calls for pushing the wrench away, push with an open hand to avoid scraped knuckles if the wrench should slip.

DON'T attempt to lift a heavy component alone - get someone to help you.

DON'T rush or take unsafe shortcuts to finish a job.

DON'T allow children or animals in or around the vehicle while you are working on it.

DO wear eye protection when using power tools such as a drill, sander, bench grinder, etc. and when working under a vehicle.

DO keep loose clothing and long hair well out of the way of moving parts.

DO make sure that any hoist used has a safe working load rating adequate for the job.

DO get someone to check on you periodically when working alone on a vehicle.

DO carry out work in a logical sequence and make sure that everything is correctly assembled and tightened.

DO keep chemicals and fluids tightly capped and out of the reach of children and pets.

DO remember that your vehicle's safety affects that of yourself and others. If in doubt on any point, get professional advice.

Asbestos

Certain friction, insulating, sealing, and other products - such as brake linings, brake bands, clutch linings, torque converters, gaskets, etc. - may contain asbestos. Extreme care must be taken to avoid inhalation of dust from such products, since it is hazardous to health. If in doubt, assume that they do contain asbestos.

Fire

Remember at all times that gasoline is highly flammable. Never smoke or have any kind of open flame around when working on a vehicle. But the risk does not end there. A spark caused by an electrical short circuit, by two metal surfaces contacting each other, or even by static electricity built up in your body under certain conditions, can ignite gasoline vapors, which in a confined space are highly explosive. Do not, under any circumstances, use gasoline for cleaning parts. Use an approved safety solvent.

Always disconnect the battery ground (-) cable at the battery before working on any part of the fuel system or electrical system. Never risk spilling fuel on a hot engine or exhaust component. It is strongly recommended that a fire extinguisher suitable for use on fuel and electrical fires be kept handy in the garage or workshop at all times. Never try to extinguish a fuel or electrical fire with water.

Fumes

Certain fumes are highly toxic and can quickly cause unconsciousness and even death if inhaled to any extent. Gasoline vapor falls into this category, as do the vapors from some cleaning solvents. Any draining or pouring of such volatile fluids should be done in a well ventilated area.

When using cleaning fluids and solvents, read the instructions on the container carefully. Never use materials from unmarked containers.

Never run the engine in an enclosed space, such as a garage. Exhaust fumes contain carbon monoxide, which is extremely poisonous. If you need to run the engine, always do so in the open air, or at least have the rear of the vehicle outside the work area.

If you are fortunate enough to have the use of an inspection pit, never drain or pour gasoline and never run the engine while the vehicle is over the pit. The fumes, being heavier than air, will concentrate in the pit with possibly lethal results.

The battery

Never create a spark or allow a bare light bulb near a battery. They normally give off a certain amount of hydrogen gas, which is highly explosive.

Always disconnect the battery ground (-) cable at the battery before working on the fuel or electrical systems.

If possible, loosen the filler caps or cover when charging the battery from an external source (this does not apply to sealed or maintenance-free batteries). Do not charge at an excessive rate or the battery may burst.

Take care when adding water to a non maintenance-free battery and when carrying a battery. The electrolyte, even when diluted, is very corrosive and should not be allowed to contact clothing or skin.

Always wear eye protection when cleaning the battery to prevent the caustic deposits from entering your eyes.

Household current

When using an electric power tool, inspection light, etc., which operates on household current, always make sure that the tool is correctly connected to its plug and that, where necessary, it is properly grounded. Do not use such items in damp conditions and, again, do not create a spark or apply excessive heat in the vicinity of fuel or fuel vapor.

Secondary ignition system voltage

A severe electric shock can result from touching certain parts of the ignition system (such as the spark plug wires) when the engine is running or being cranked, particularly if components are damp or the insulation is defective. In the case of an electronic ignition system, the secondary system voltage is much higher and could prove fatal.

Troubleshooting

Contents

Symptom	Section
Engine	
Engine backfires	13
Engine diesels (continues to run) after switching off	15
Engine hard to start when cold	4
Engine hard to start when hot	5
Engine lacks power	12
Engine lopes while idling or idles erratically	8
Engine misses at idle speed	9
Engine misses throughout driving speed range	10
Engine rotates but will not start	2
Engine stalls	11
Engine starts but stops immediately	7
Engine will not rotate when attempting to start	1
Pinging or knocking engine sounds during acceleration or uphill	14
Starter motor noisy or excessively rough in engagement	6
Starter motor operates without rotating engine	3
Engine electrical system	
Battery will not hold a charge	16
Ignition light fails to come on when key is turned on	18
Ignition light fails to go out	17
Fuel system	
Excessive fuel consumption	19
Fuel leakage and/or fuel odor	20
Cooling system	
Coolant loss	25
External coolant leakage	23
Internal coolant leakage	24
Overcooling	22
Overheating	21
Poor coolant circulation	26
Clutch	
Clutch pedal stays on floor when disengaged	32
Clutch slips (engine speed increases with no increase in vehicle speed)	28
Fails to release (pedal pressed to the floor - shift lever does not move freely in and out of Reverse)	27
Grabbing (chattering) as clutch is engaged	29
Squeal or rumble with clutch fully disengaged (pedal depressed)	31
Squeal or rumble with clutch fully engaged (pedal released)	30
Manual transmission	
Difficulty in engaging gears	37
Noisy in all gear	34

Symptom	Section
Noisy in Neutral with engine running	33
Noisy in one particular gear	35
Oil leakage	38
Slips out of high gear	36
Automatic transmission	
Fluid leakage	42
General shift mechanism problems	39
Transmission slips, shifts rough, is noisy or has no drive in forward or reverse gears	41
Transmission will not downshift with accelerator pedal pressed to the floor	40
Transfer case	
Lubricant leaks from the vent or output shaft seals	46
Noisy or jumps out of four-wheel drive Low range	45
Transfer case is difficult to shift into the desired range	43
Transfer case noisy in all gears	44
Driveshaft	
Knock or clunk when the transmission is under initial load (just after transmission is put into gear)	48
Metallic grinding sound consistent with vehicle speed	49
Oil leak at front of driveshaft	47
Vibration	50
Axles	
Noise	51
Oil leakage	53
Vibration	52
Brakes	
Brake pedal feels spongy when depressed	57
Brake pedal pulsates during brake application	60
Excessive brake pedal travel	56
Excessive effort required to stop vehicle	58
Noise (high-pitched squeal with the brakes applied)	55
Pedal travels to the floor with little resistance	59
Vehicle pulls to one side during braking	54
Suspension and steering systems	
Excessive pitching and/or rolling around corners or during braking	63
Excessive play in steering	65
Excessive tire wear (not specific to one area)	67
Excessive tire wear on inside edge	69
Excessive tire wear on outside edge	68
Excessively stiff steering	64
Lack of power assistance	66
Shimmy, shake or vibration	62
Tire tread worn in one place	70
Vehicle pulls to one side	61

This section provides an easy reference guide to the more common problems which may occur during the operation of your vehicle. These problems and possible causes are grouped under various components or systems; i.e. Engine, Cooling System, etc., and also refer to the Chapter and/or Section which deals with the problem.

Remember that successful troubleshooting is not a mysterious black art practiced only by professional mechanics. It's simply the result of a bit of knowledge combined with an intelligent, systematic approach to the problem. Always work by a process of elimination, starting with the simplest solution and working through to the most complex - and never overlook the obvious. Anyone can forget to fill the gas tank or leave the lights on overnight, so don't assume that you are above such oversights.

Finally, always get clear in your mind why a problem has occurred and take steps to ensure that it doesn't happen again. If the electrical system fails because of a poor connection, check all other connections in the system to make sure that they don't fail as well. If a particular fuse continues to blow,

Troubleshooting

find out why - don't just go on replacing fuses. Remember, failure of a small component can often be indicative of potential failure or incorrect functioning of a more important component or system.

Engine

1 Engine will not rotate when attempting to start

1 Battery terminal connections loose or corroded. Check the cable terminals at the battery. Tighten the cable or remove corrosion as necessary.
2 Battery discharged or faulty. If the cable connections are clean and tight on the battery posts, turn the key to the On position and switch on the headlights and/or windshield wipers. If they fail to function, the battery is discharged.
3 Automatic transmission not completely engaged in Park or Neutral or clutch pedal not completely depressed.
4 Broken, loose or disconnected wiring in the starting circuit. Inspect all wiring and connectors at the battery, starter solenoid and ignition switch.
5 Starter motor pinion jammed in flywheel ring gear. If manual transmission, place transmission in gear and rock the vehicle to manually turn the engine. Remove starter and inspect pinion and flywheel at earliest convenience (Chapter 5).
6 Starter solenoid faulty (Chapter 5).
7 Starter motor faulty (Chapter 5).
8 Ignition switch faulty (Chapter 12).

2 Engine rotates but will not start

1 Fuel tank empty.
2 Fault in the fuel injection system (Chapter 4).
3 Battery discharged (engine rotates slowly). Check the operation of electrical components as described in the previous Section.
4 Battery terminal connections loose or corroded (see previous Section).
5 Fuel pump faulty (Chapter 4).
6 Excessive moisture on, or damage to, ignition components (see Chapter 5).
7 Worn, faulty or incorrectly gapped spark plugs (Chapter 1).
8 Broken, loose or disconnected wiring in the starting circuit (see previous Section).
9 Broken, loose or disconnected wires at the ignition coil pack or faulty coil pack (Chapter 5).
10 Broken, loose or damaged timing belt (Chapter 2A).

3 Starter motor operates without rotating engine

1 Starter pinion sticking. Remove the starter (Chapter 5) and inspect.
2 Starter pinion or flywheel teeth worn or broken. Remove the flywheel/driveplate access cover and inspect.

4 Engine hard to start when cold

1 Battery discharged or low. Check as described in Section 1.
2 Fault in the fuel or electrical systems (Chapters 4 and 5).

5 Engine hard to start when hot

1 Air filter clogged (Chapter 1).
2 Fault in the fuel or electrical systems (Chapters 4 and 5).
3 Fuel not reaching the injectors (see Chapter 4).

6 Starter motor noisy or excessively rough in engagement

1 Pinion or flywheel gear teeth worn or broken. Remove the cover at the rear of the engine (if equipped) and inspect.
2 Starter motor mounting bolts loose or missing.

7 Engine starts but stops immediately

1 Loose or faulty electrical connections at distributor, coil or alternator.
2 Fault in the fuel or electrical systems (Chapters 4 and 5).
3 Vacuum leak at the gasket surfaces of the intake manifold or throttle body. Make sure all mounting bolts/nuts are tightened securely and all vacuum hoses connected to the manifold are positioned properly and in good condition.

8 Engine lopes while idling or idles erratically

1 Vacuum leakage. Check the mounting bolts/nuts at the throttle body and intake manifold for tightness. Make sure all vacuum hoses are connected and in good condition. Use a stethoscope or a length of fuel hose held against your ear to listen for vacuum leaks while the engine is running. A hissing sound will be heard. A soapy water solution will also detect leaks.
2 Fault in the fuel or electrical systems (Chapters 4 and 5).
3 Plugged PCV valve (see Chapters 1 and 6).
4 Air filter clogged (Chapter 1).
5 Fuel pump not delivering sufficient fuel to the fuel injectors (see Chapter 4).
6 Leaking head gasket. Perform a compression check (Chapter 2).
7 Camshaft lobes worn (Chapter 2).

9 Engine misses at idle speed

1 Spark plugs worn, fouled or not gapped properly (Chapter 1).
2 Fault in the fuel or electrical systems (Chapters 4 and 5).
3 Faulty spark plug wires (Chapter 1).
4 Vacuum leaks at intake or hose connections. Check as described in Section 8.
5 Uneven or low cylinder compression. Check compression as described in Chapter 1.

10 Engine misses throughout driving speed range

1 Fuel filter clogged and/or impurities in the fuel system (Chapter 1).
2 Faulty or incorrectly gapped spark plugs (Chapter 1).
3 Fault in the fuel or electrical systems (Chapters 4 and 5).
4 Defective spark plug wires (Chapter 1).
5 Faulty emissions system components (Chapter 6).
6 Low or uneven cylinder compression pressures. Remove the spark plugs and test the compression with a gauge (Chapter 2).
7 Weak or faulty ignition system (Chapter 5).
8 Vacuum leaks at the throttle body, intake manifold or vacuum hoses (see Section 8).

11 Engine stalls

1 Idle speed incorrect. Refer to the VECI label.
2 Fuel filter clogged and/or water and impurities in the fuel system (Chapter 1).
3 Fault in the fuel system or sensors (Chapters 4 and 6).
4 Faulty emissions system components (Chapter 6).
5 Faulty or incorrectly gapped spark plugs (Chapter 1). Also check the spark plug wires (Chapter 1).
6 Vacuum leak at the throttle body, intake manifold or vacuum hoses. Check as described in Section 8.

12 Engine lacks power

1 Fault in the fuel or electrical systems (Chapters 4 and 5).
2 Faulty or incorrectly gapped spark plugs (Chapter 1).
3 Faulty coil (Chapter 5).
4 Brakes binding (Chapter 1).
5 Automatic transmission fluid level incorrect (Chapter 1).
6 Clutch slipping (Chapter 8).
7 Fuel filter clogged and/or impurities in the fuel system (Chapter 1).
8 Emissions control system not functioning properly (Chapter 6).

Troubleshooting

9   Use of substandard fuel. Fill the tank with the proper octane fuel.
10 Low or uneven cylinder compression pressures. Test with a compression tester, which will detect leaking valves and/or a blown head gasket (Chapter 2).

13 Engine backfires

1   Emissions system not functioning properly (Chapter 6).
2   Fault in the fuel or electrical systems (Chapters 4 and 5).
3   Faulty secondary ignition system (cracked spark plug insulator or faulty plug wires) (Chapters 1 and 5).
4   Fuel injection system malfunction (Chapter 4).
5   Vacuum leak at the throttle body, intake manifold or vacuum hoses. Check as described in Section 8.
6   Valves sticking (Chapter 2).
7   Crossed plug wires (Chapter 1).

14 Pinging or knocking engine sounds during acceleration or uphill

1   Incorrect grade of fuel. Fill the tank with fuel of the proper octane rating.
2   Fault in the fuel or electrical systems (Chapters 4 and 5).
3   Improper spark plugs. Check the plug type against the VECI label located in the engine compartment. Also check the plugs and wires for damage (Chapter 1).
4   Faulty emissions system (Chapter 6).
5   Vacuum leak. Check as described in Section 9.

15 Engine diesels (continues to run) after switching off

1   Idle speed too high. Refer to Chapter 1.
2   Fault in the fuel or electrical systems (Chapters 4 and 5).
3   Excessive engine operating temperature. Probable causes of this are a low coolant level (see Chapter 1), malfunctioning thermostat, clogged radiator or faulty water pump (see Chapter 3).

Engine electrical system

16 Battery will not hold a charge

1   Alternator drivebelt defective or not adjusted properly (Chapter 1).
2   Electrolyte level low or battery discharged (Chapter 1).
3   Battery terminals loose or corroded (Chapter 1).
4   Alternator not charging properly (Chapter 5).

5   Loose, broken or faulty wiring in the charging circuit (Chapter 5).
6   Short in the vehicle wiring causing a continuous drain on the battery (refer to Chapter 12 and the Wiring Diagrams).
7   Battery defective internally.

17 Ignition light fails to go out

1   Fault in the alternator or charging circuit (Chapter 5).
2   Alternator drivebelt defective or not properly adjusted (Chapter 1).

18 Ignition light fails to come on when key is turned on

1   Instrument cluster warning light bulb defective (Chapter 12).
2   Alternator faulty (Chapter 5).
3   Fault in the instrument cluster printed circuit, dashboard wiring or bulb holder (Chapter 12).

Fuel system

19 Excessive fuel consumption

1   Dirty or clogged air filter element (Chapter 1).
2   Emissions system not functioning properly (Chapter 6).
3   Fault in the fuel or electrical systems (Chapters 4 and 5).
4   Fuel injection system malfunction (Chapter 4).
5   Low tire pressure or incorrect tire size (Chapter 1).

20 Fuel leakage and/or fuel odor

1   Leak in a fuel feed or vent line (Chapter 4).
2   Tank overfilled. Fill only to automatic shut-off.
3   Evaporative emissions system canister clogged (Chapter 6).
4   Vapor leaks from system lines (Chapter 4).
5   Fuel injection system malfunction (Chapter 4).

Cooling system

21 Overheating

1   Insufficient coolant in the system (Chapter 1).
2   Water pump drivebelt defective or not adjusted properly (Chapter 1).
3   Radiator core blocked or radiator grille dirty and restricted (see Chapter 3).
4   Thermostat faulty (Chapter 3).
5   Fan blades broken or cracked (Chapter 3).

6   Radiator cap not maintaining proper pressure. Have the cap pressure tested by gas station or repair shop.

22 Overcooling

1   Thermostat faulty (Chapter 3).
2   Inaccurate temperature gauge (Chapter 12).

23 External coolant leakage

1   Deteriorated or damaged hoses or loose clamps. Replace hoses and/or tighten the clamps at the hose connections (Chapter 1).
2   Water pump seals defective. If this is the case, water will drip from the weep hole in the water pump body (Chapter 3).
3   Leakage from the radiator core or side tank(s). This will require the radiator to be professionally repaired (see Chapter 3 for removal procedures).
4   Engine drain plug leaking (Chapter 1) or water jacket core plugs leaking (see Chapter 2).

24 Internal coolant leakage

Note: *Internal coolant leaks can usually be detected by examining the oil. Check the dipstick and inside of the valve cover for water deposits and an oil consistency like that of a milkshake.*

1   Leaking cylinder head gasket. Have the cooling system pressure tested.
2   Cracked cylinder bore or cylinder head. Dismantle the engine and inspect (Chapter 2).
3   Intake manifold gasket leaking (Chapters 2B and 2C).

25 Coolant loss

1   Too much coolant in the system (Chapter 1).
2   Coolant boiling away due to overheating (see Section 15).
3   External or internal leakage (see Sections 23 and 24).
4   Faulty radiator cap. Have the cap pressure tested.

26 Poor coolant circulation

1   Inoperative water pump. A quick test is to pinch the top radiator hose closed with your hand while the engine is idling, then let it loose. You should feel the surge of coolant if the pump is working properly (see Chapter 1).
2   Restriction in the cooling system. Drain, flush and refill the system (Chapter 1). If necessary, remove the radiator (Chapter 3) and have it reverse flushed.
3   Water pump drivebelt defective or not adjusted properly (Chapter 1).
4   Thermostat sticking (Chapter 3).

Troubleshooting

Clutch

27 Fails to release (pedal pressed to the floor - shift lever does not move freely in and out of Reverse)

1. Leak in the clutch hydraulic system. Check the master cylinder, slave cylinder and lines (Chapter 8).
2. Clutch plate warped or damaged (Chapter 8).

28 Clutch slips (engine speed increases with no increase in vehicle speed)

1. Clutch plate oil soaked or lining worn. Remove clutch (Chapter 8) and inspect.
2. Clutch plate not seated. It may take 30 or 40 normal starts for a new one to seat.
3. Pressure plate worn (Chapter 8).

29 Grabbing (chattering) as clutch is engaged

1. Oil on clutch plate lining. Remove (Chapter 8) and inspect. Correct any leakage source.
2. Worn or loose engine or transmission mounts. These units move slightly when the clutch is released. Inspect the mounts and bolts (Chapter 2).
3. Worn splines on clutch plate hub. Remove the clutch components (Chapter 8) and inspect.
4. Warped pressure plate or flywheel. Remove the clutch components and inspect.

30 Squeal or rumble with clutch fully engaged (pedal released)

1. Release bearing binding on transmission bearing retainer. Remove clutch components (Chapter 8) and check bearing. Remove any burrs or nicks; clean and relubricate bearing retainer before installing.

31 Squeal or rumble with clutch fully disengaged (pedal depressed)

1. Worn, defective or broken release bearing (Chapter 8).
2. Worn or broken pressure plate springs (or diaphragm fingers) (Chapter 8).

32 Clutch pedal stays on floor when disengaged

1. Linkage or release bearing binding. Inspect the linkage or remove the clutch components as necessary.

2. Make sure proper pedal stop (bumper) is installed.

Manual transmission

Note: *All the following references are in Chapter 7, unless noted.*

33 Noisy in Neutral with engine running

1. Input shaft bearing worn.
2. Damaged main drive gear bearing.
3. Worn countershaft bearings.
4. Worn or damaged countershaft endplay shims.

34 Noisy in all gears

1. Any of the above causes, and/or:
2. Insufficient lubricant (see the checking procedures in Chapter 1).

35 Noisy in one particular gear

1. Worn, damaged or chipped gear teeth for that particular gear.
2. Worn or damaged synchronizer for that particular gear.

36 Slips out of high gear

1. Transmission loose on clutch housing.
2. Dirt between the transmission case and engine or misalignment of the transmission (Chapter 7).

37 Difficulty in engaging gears

1. Clutch not releasing completely (see Chapter 8).
2. Loose, damaged or out-of-adjustment shift linkage. Make a thorough inspection, replacing parts as necessary (Chapter 7).

38 Oil leakage

1. Excessive amount of lubricant in the transmission (see Chapter 1 for correct checking procedures). Drain lubricant as required.
2. Driveaxle oil seal or speedometer oil seal in need of replacement (Chapter 7).

Automatic transmission

Note: *Due to the complexity of the automatic transmission, it's difficult for the home mechanic to properly diagnose and service this component. For problems other than the following, the vehicle should be taken to a dealer service department or a transmission shop.*

39 General shift mechanism problems

1. Chapter 7 deals with checking and adjusting the shift linkage on automatic transmissions. Common problems which may be attributed to poorly adjusted linkage are:
   a) *Engine starting in gears other than Park or Neutral.*
   b) *Indicator on shifter pointing to a gear other than the one actually being selected.*
   c) *Vehicle moves when in Park.*
2. Refer to Chapter 7 to adjust the linkage.

40 Transmission will not downshift with accelerator pedal pressed to the floor

Kickdown cable misadjusted.

41 Transmission slips, shifts rough, is noisy or has no drive in forward or reverse gears

1. There are many probable causes for the above problems, but the home mechanic should be concerned with only one possibility - fluid level.
2. Before taking the vehicle to a repair shop, check the level and condition of the fluid as described in Chapter 1. Correct fluid level as necessary or change the fluid and filter if needed. If the problem persists, have a professional diagnose the probable cause.
3. If the transmission shifts late and the shifts are harsh, suspect a faulty vacuum diaphragm (Chapter 7).

42 Fluid leakage

1. Automatic transmission fluid is a deep red color. Fluid leaks should not be confused with engine oil, which can easily be blown by air flow to the transmission.
2. To pinpoint a leak, first remove all built-up dirt and grime from around the transmission. Degreasing agents and/or steam cleaning will achieve this. With the underside clean, drive the vehicle at low speeds so air flow will not blow the leak far from its source. Raise the vehicle and determine where the leak is coming from. Common areas of leakage are:
   a) *Pan: Tighten the mounting bolts and/or replace the pan gasket as necessary (see Chapter 7).*
   b) *Filler pipe: Replace the rubber seal where the pipe enters the transmission case.*
   c) *Transmission oil lines: Tighten the connectors where the lines enter the transmission case and/or replace the lines.*
   d) *Vent pipe: Transmission overfilled and/or water in fluid (see checking procedures, Chapter 1).*

Troubleshooting 0-21

e) **Speedometer connector:** Replace the O-ring where the speedometer cable enters the transmission case (Chapter 7).

Transfer case

43 Transfer case is difficult to shift into the desired range

1   Speed may be too great to permit engagement. Stop the vehicle and shift into the desired range.
2   Shift linkage loose, bent or binding. Check the linkage for damage or wear and replace or lubricate as necessary (Chapter 7).
3   If the vehicle has been driven on a paved surface for some time, the driveline torque can make shifting difficult. Stop and shift into two-wheel drive on paved or hard surfaces.
4   Insufficient or incorrect grade of lubricant. Drain and refill the transfer case with the specified lubricant. (Chapter 1).
5   Worn or damaged internal components. Disassembly and overhaul of the transfer case may be necessary (Chapter 7).

44 Transfer case noisy in all gears

Insufficient or incorrect grade of lubricant. Drain and refill (Chapter 1).

45 Noisy or jumps out of four-wheel drive Low range

1   Transfer case not fully engaged. Stop the vehicle, shift into Neutral and then engage 4L.
2   Shift linkage loose, worn or binding. Tighten, repair or lubricate linkage as necessary.
3   Shift fork cracked, inserts worn or fork binding on the rail. Disassemble and repair as necessary (Chapter 7).

46 Lubricant leaks from the vent or output shaft seals

1   Transfer case is overfilled. Drain to the proper level (Chapter 1).
2   Vent is clogged or jammed closed. Clear or replace the vent.
3   Output shaft seal incorrectly installed or damaged. Replace the seal and check contact surfaces for nicks and scoring.

Driveshaft

47 Oil leak at seal end of driveshaft

Defective transmission or transfer case oil seal. See Chapter 7 for replacement procedures. While this is done, check the splined yoke for burrs or a rough condition which may be damaging the seal. Burrs can be removed with crocus cloth or a fine whetstone.

48 Knock or clunk when the transmission is under initial load (just after transmission is put into gear)

1   Loose or disconnected rear suspension components. Check all mounting bolts, nuts and bushings (see Chapter 10).
2   Loose driveshaft bolts. Inspect all bolts and nuts and tighten them to the specified torque.
3   Worn or damaged universal joint bearings. Check for wear (see Chapter 8).

49 Metallic grinding sound consistent with vehicle speed.

Pronounced wear in the universal joint bearings. Check as described in Chapter 8.

50 Vibration

Note: *Before assuming that the driveshaft is at fault, make sure the tires are perfectly balanced and perform the following test.*
1   Install a tachometer inside the vehicle to monitor engine speed as the vehicle is driven. Drive the vehicle and note the engine speed at which the vibration (roughness) is most pronounced. Now shift the transmission to a different gear and bring the engine speed to the same point.
2   If the vibration occurs at the same engine speed (rpm) regardless of which gear the transmission is in, the driveshaft is NOT at fault since the driveshaft speed varies.
3   If the vibration decreases or is eliminated when the transmission is in a different gear at the same engine speed, refer to the following probable causes.
4   Bent or dented driveshaft. Inspect and replace as necessary (see Chapter 8).
5   Undercoating or built-up dirt, etc. on the driveshaft. Clean the shaft thoroughly and recheck.
6   Worn universal joint bearings. Remove and inspect (see Chapter 8).
7   Driveshaft and/or companion flange out of balance. Check for missing weights on the shaft. Remove the driveshaft (see Chapter 8) and reinstall 180-degrees from original position, then retest. Have the driveshaft professionally balanced if the problem persists.

Axles

51 Noise

1   Road noise. No corrective procedures available.
2   Tire noise. Inspect tires and check tire pressures (Chapter 1).
3   Rear wheel bearings loose, worn or damaged (Chapter 8).

52 Vibration

See probable causes under Driveshaft. Proceed under the guidelines listed for the driveshaft. If the problem persists, check the rear wheel bearings by raising the rear of the vehicle and spinning the rear wheels by hand. Listen for evidence of rough (noisy) bearings. Remove and inspect (see Chapter 8).

53 Oil leakage

1   Pinion seal damaged (see Chapter 8).
2   Axleshaft oil seals damaged (see Chapter 8).
3   Differential inspection cover leaking. Tighten the bolts or replace the gasket as required (see Chapters 1 and 8).

Brakes

Note: *Before assuming that a brake problem exists, make sure that the tires are in good condition and inflated properly (see Chapter 1), that the front end alignment is correct and that the vehicle is not loaded with weight in an unequal manner.*

54 Vehicle pulls to one side during braking

1   Defective, damaged or oil contaminated disc brake pads or shoes on one side. Inspect as described in Chapter 9.
2   Excessive wear of brake shoe or pad material or drum/disc on one side. Inspect and correct as necessary.
3   Loose or disconnected front suspension components. Inspect and tighten all bolts to the specified torque (Chapter 10).
4   Defective drum brake or caliper assembly. Remove the drum or caliper and inspect for a stuck piston or other damage (Chapter 9).
5   Inadequate lubrication of front brake caliper slide rails. Remove caliper and lubricate slide rails (Chapter 9).

55 Noise (high-pitched squeal with the brakes applied)

1   Disc brake pads worn out. The noise comes from the wear sensor rubbing against the disc (does not apply to all vehicles) or the actual pad backing plate itself if the material is completely worn away. Replace the pads with new ones immediately (Chapter 9). If the pad material has worn completely away, the brake discs should be inspected for damage

as described in Chapter 9.
2   Missing or damaged brake pad insulators (disc brakes). Replace pad insulators (see Chapter 9).
3   Linings contaminated with dirt or grease. Replace pads or shoes.
4   Incorrect linings. Replace with correct linings.

56 Excessive brake pedal travel

1   Partial brake system failure. Inspect the entire system (Chapter 9) and correct as required.
2   Insufficient fluid in the master cylinder. Check (Chapter 1), add fluid and bleed the system if necessary (Chapter 9).
3   Rear brakes not adjusting properly. Make a series of starts and stops while the vehicle is in Reverse. If this does not correct the situation, remove the drums and inspect the self-adjusters (Chapter 9).

57 Brake pedal feels spongy when depressed

1   Air in the hydraulic lines. Bleed the brake system (Chapter 9).
2   Faulty flexible hoses. Inspect all system hoses and lines. Replace parts as necessary.
3   Master cylinder mounting bolts/nuts loose.
4   Master cylinder defective (Chapter 9).

58 Excessive effort required to stop vehicle

1   Power brake booster not operating properly (Chapter 9).
2   Excessively worn linings or pads. Inspect and replace if necessary (Chapter 9).
3   One or more caliper pistons or wheel cylinders seized or sticking. Inspect and rebuild as required (Chapter 9).
4   Brake linings or pads contaminated with oil or grease. Inspect and replace as required (Chapter 9).
5   New pads or shoes installed and not yet seated. It will take a while for the new material to seat against the drum (or rotor).

59 Pedal travels to the floor with little resistance

1   Little or no fluid in the master cylinder reservoir caused by leaking wheel cylinder(s), leaking caliper piston(s), loose, damaged or disconnected brake lines. Inspect the entire system and correct as necessary.
2   Worn master cylinder seals (Chapter 9).

60 Brake pedal pulsates during brake application

1   Caliper improperly installed. Remove and inspect (Chapter 9).
2   Disc or drum defective. Remove (Chapter 9) and check for excessive lateral runout and parallelism. Have the disc or drum resurfaced or replace it with a new one.

Suspension and steering systems

61 Vehicle pulls to one side

1   Tire pressures uneven (Chapter 1).
2   Defective tire (Chapter 1).
3   Excessive wear in suspension or steering components (Chapter 10).
4   Front end in need of alignment.
5   Front brakes dragging. Inspect the brakes as described in Chapter 9.

62 Shimmy, shake or vibration

1   Tire or wheel out-of-balance or out-of-round. Have professionally balanced.
2   Loose, worn or out-of-adjustment rear wheel bearings (Chapter 1).
3   Shock absorbers and/or suspension components worn or damaged (Chapter 10).

63 Excessive pitching and/or rolling around corners or during braking

1   Defective shock absorbers. Replace as a set (Chapter 10).
2   Broken or weak springs and/or suspension components. Inspect as described in Chapter 10.

64 Excessively stiff steering

1   Lack of fluid in power steering fluid reservoir (Chapter 1).
2   Incorrect tire pressures (Chapter 1).
3   Lack of lubrication at steering joints (see Chapter 1).
4   Front end out of alignment.
5   Lack of power assistance (see Section 62).

65 Excessive play in steering

1   Loose front wheel bearings (Chapters 1 and 10).
2   Excessive wear in suspension or steering components (Chapter 10).
3   Steering gearbox damaged or out of adjustment (Chapter 10).

66 Lack of power assistance

1   Steering pump drivebelt faulty or not adjusted properly (Chapter 1).
2   Fluid level low (Chapter 1).
3   Hoses or lines restricted. Inspect and replace parts as necessary.
4   Air in power steering system. Bleed the system (Chapter 10).

67 Excessive tire wear (not specific to one area)

1   Incorrect tire pressures (Chapter 1).
2   Tires out-of-balance. Have professionally balanced.
3   Wheels damaged. Inspect and replace as necessary.
4   Suspension or steering components excessively worn (Chapter 10).

68 Excessive tire wear on outside edge

1   Inflation pressures incorrect (Chapter 1).
2   Excessive speed in turns.
3   Front end alignment incorrect (excessive toe-in). Have professionally aligned.
4   Suspension arm bent or twisted (Chapter 10).

69 Excessive tire wear on inside edge

1   Inflation pressures incorrect (Chapter 1).
2   Front end alignment incorrect (toe-out). Have professionally aligned.
3   Loose or damaged steering components (Chapter 10).

70 Tire tread worn in one place

1   Tires out-of-balance.
2   Damaged or buckled wheel. Inspect and replace if necessary.
3   Defective tire (Chapter 1).

Chapter 1
Tune-up and routine maintenance

Contents

Section		Section	
Air filter replacement	30	Fuel filter replacement	31
Automatic transmission fluid check/change	9	Fuel system check	21
Automatic transmission shift linkage lubrication	15	Introduction	2
Battery check, maintenance and charging	13	Maintenance schedule	1
Brake system check	25	Manual transmission lubricant level check and change	17
Chassis lubrication	16	Positive Crankcase Ventilation (PCV) valve check and replacement	20
CHECK ENGINE light	See Chapter 6	Power steering fluid level check	8
Cooling system check	22	Safety checks	6
Cooling system servicing (draining, flushing and refilling)	29	Spark plugs, wires, distributor cap and rotor check and replacement	14
Differential lubricant level check and change	18	Steering and suspension check	24
Drivebelt check, adjustment and replacement	12	Tire and tire pressure checks	5
Driveshaft and driveaxle yoke lubrication (4WD models)	32	Tire rotation	10
Engine oil and filter change	7	Transfer case lubricant level check and change	19
Exhaust system check	23	Tune-up general information	3
Fluid level checks	4	Underhood hose check and replacement	11
Front hub lock, spindle bearing and wheel bearing maintenance (4WD models)	27	Windshield wiper blade check and replacement	28
Front wheel bearing check, repack and adjustment (2WD models)	26		

Specifications

Recommended lubricants and fluids

Note: *Listed here are manufacturer recommendations at the time this manual was written. Manufacturers occasionally upgrade their fluid and lubricant specifications, so check with your local auto parts store for current recommendations.*

Engine oil	
Type	API grade SG
Viscosity	See accompanying chart
Power steering fluid type	MERCON automatic transmission fluid
Brake and clutch fluid type	DOT 3 heavy duty brake fluid
Automatic transmission fluid type	MERCON automatic transmission fluid
Manual transmission lubricant type	MERCON automatic transmission fluid
Transfer case lubricant type	MERCON automatic transmission fluid
Coolant type	50/50 mixture of ethylene glycol-based antifreeze and water
Front wheel bearing grease	NLGI No. 2 lithium base grease containing polyethylene and molybdenum disulfide
Automatic locking front hubs	NLGI No. 2 lithium base grease containing polyethylene and molybdenum disulfide
Front spindle and thrust bearings (4WD)	NLGI No. 2 lithium base grease containing polyethylene and molybdenum disulfide
Caliper slide rail grease	Disc brake caliper slide rail grease
Chassis grease	NLGI No. 2 lithium base grease containing polyethylene and molybdenum disulfide
Differential lubricant*	API GL-5 SAE 90 Hypoid gear lubricant

*Traction-Lok axles add 4 oz. of friction modifier (Ford part no. E0AZ-19580-AA, or equivalent) when oil is changed.

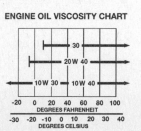

ENGINE OIL VISCOSITY CHART

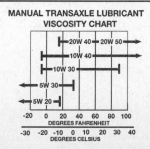

MANUAL TRANSAXLE LUBRICANT VISCOSITY CHART

Engine oil viscosity chart - for best fuel economy and cold-weather starting, select the lowest recommended viscosity rating for the operating temperature range

1-2 Chapter 1 Tune-up and routine maintenance

Capacities (approximate)

Engine oil (includes one quart for filter change)
   Four-cylinder and 4.0L V6 engines .... 5 qts
   3.0L V6 engine .... 4.5 qts
Cooling system*
   Without air conditioning .... 7.8 qts
   With air conditioning .... 8.6 qts
Transfer case (Warner 13-54) .... 2.5 pints
Front axle differential .... 3.5 pints
Rear axle differential .... 5.0 pints
Automatic transmission (drain and refill) .... 3.0 qts**
Manual transmission .... 5.6 pints

*Capacity may vary +/- 15% due to equipment variations. Most service refills take only 80% of listed capacity because some coolant remains in the engine.

**Add more fluid as needed, 1/2 pint at a time, to bring it up to the correct level on the dipstick. Do not overfill. After overhaul, approximately 5 to 6 more quarts of fluid will be required.

Cylinder and coil terminal locations - four-cylinder engine

Brakes

Disc brake pad thickness (minimum) .... See Chapter 9
Drum brake shoe lining thickness (minimum) .... See Chapter 9

Ignition system

Firing order
   Four-cylinder engine 1-3-4-2
   V6 engines 1-4-2-5-3-6
Ignition timing .... Not adjustable
Spark plug type and gap
   Four-cylinder engines
     Type
       1993 through 1995 .... Motorcraft AWSF32PP
       1996 and 1997 .... Motorcraft AWSF32F
       1998 and later .... Motorcraft AWSTF32F
     Gap .... 0.042 to 0.046 inch
   3.0L V6 engine
     Type
       1993 through 1999 .... Motorcraft AWSF32PP
       2000 .... Motorcraft AGSF22PP
     Gap .... 0.042 to 0.046 inch
   4.0L V6 engine
     Type
       1993 through 1997 .... Motorcraft AWSF42P
       1998 and later .... Motorcraft AGSF22PP
     Gap .... 0.052 to 0.056 inch
Idle speed .... Non adjustable

Cylinder location and distributor rotation - 1993 and 1994 3.0L V6 engine

Cylinder and coil terminal locations - 1995 and later 3.0L V6 engine

Torque specifications

                                                                                                  Ft-lbs (unless otherwise indicated)

Wheel lug nuts 100*
Front hub adjusting nut (2WD models)
   Step 1 .... 17 to 25
   Step 2 .... Back off 1/2-turn
   Step 3 .... 18 to 20 in-lbs
Front hub adjusting nut (4WD models)
   Manual locking hubs
     Step 1 35
     Step 2 .... Back off 1/4-turn
     Step 3 .... 16 in-lbs
     Step 4 (outer locknut) 150
     Endplay .... 0 to 0.003 inch
     Hub turning torque (maximum) .... 25 in-lbs
   Automatic locking hubs
     Step 1 35
     Step 2 .... Back-off 1/4-turn
     Step 3 .... 16 in-lbs
     Endplay .... 0 to 0.003 inch
     Hub turning torque (maximum) .... 25 in-lbs
Automatic transmission pan bolts .... 96 to 120 in-lbs
Spark plugs .... 7 to 15

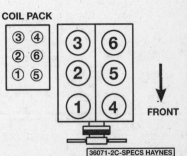

Cylinder and coil terminal locations - 4.0L V6 engine

* Whenever the lug nuts are removed, recheck lug nut torque after 500 miles.

Chapter 1 Tune-up and routine maintenance

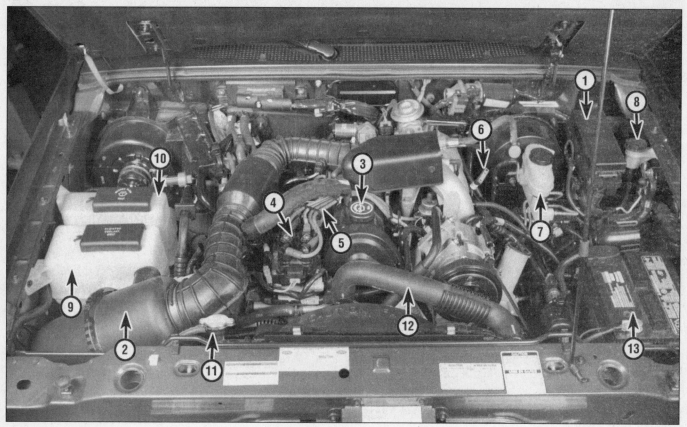

Typical four-cylinder engine compartment component locations

1. Power distribution (fuse and relay) box
2. Air cleaner housing
3. Engine oil filler cap
4. Ignition coil pack
5. Ignition wires
6. Engine oil dipstick
7. Brake master cylinder reservoir
8. Clutch fluid reservoir
9. Engine coolant reservoir
10. Windshield washer reservoir
11. Radiator cap
12. Upper radiator hose
13. Battery

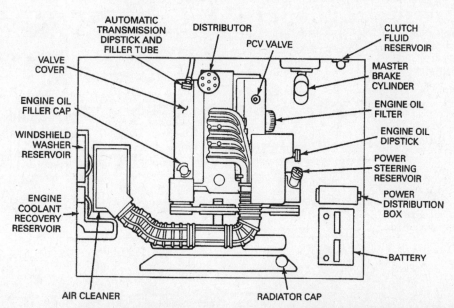

Typical 1993 and 1994 3.0L V6 engine compartment component locations

Chapter 1 Tune-up and routine maintenance

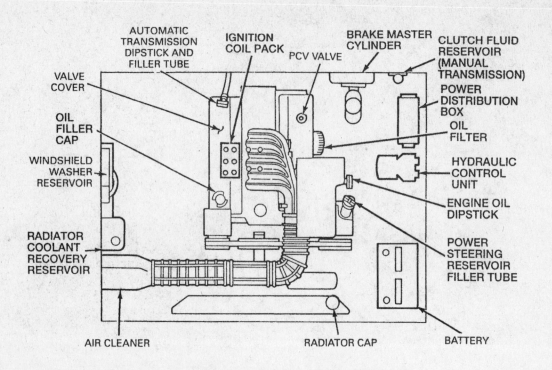

Typical 1995 and later 3.0L V6 engine compartment component locations

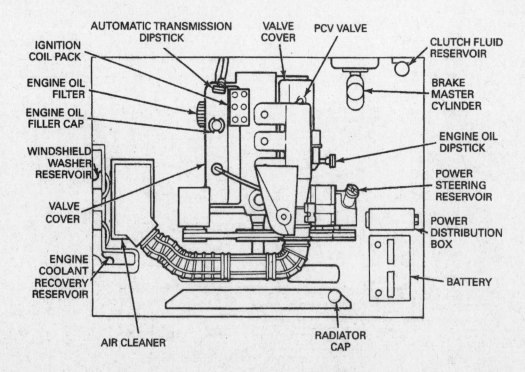

Typical 4.0L V6 engine compartment component locations

Chapter 1 Tune-up and routine maintenance

Typical engine compartment underside components

1. Radiator drain fitting (obscured by lower support)
2. Lower radiator hose
3. Grease fittings

Typical rear underside components

1. Rear shock absorber
2. Leaf spring assembly
3. Muffler
4. Fuel tank
5. Rear differential

1 Ford Ranger/Mazda B-series Maintenance schedule

The following maintenance intervals are based on the assumption that the vehicle owner will be doing the maintenance or service work, as opposed to having a dealer service department do the work. Although the time/mileage intervals are loosely based on factory recommendations, most have been shortened to ensure, for example, that such items as lubricants and fluids are checked/changed at intervals that promote maximum engine/driveline service life. Also, subject to the preference of the individual owner interested in keeping his or her vehicle in peak condition at all times, and with the vehicle's ultimate resale in mind, many of the maintenance procedures may be performed more often than recommended in the following schedule. We encourage such owner initiative.

Because off-road use necessitates more frequent maintenance, a separate schedule is included for vehicles used off road.

When the vehicle is new it should be serviced initially by a factory authorized dealer service department to protect the factory warranty. In many cases the initial maintenance check is done at no cost to the owner (check with your dealer service department for more information).

Every 250 miles or weekly, whichever comes first

Check the engine oil level (Section 4)
Check the engine coolant level (Section 4)
Check the brake fluid level (Section 4)
Check the clutch fluid level (Section 4)
Check the windshield washer fluid level (Section 4)
Check the tires and tire pressures (Section 5)
Perform owner safety checks (Section 6)

Every 3,000 miles or 3 months, whichever comes first

All items listed above, plus . . .
Change the engine oil and oil filter (Section 7)
Check the power steering fluid level (Section 8)
Check the automatic transmission fluid level (Section 9)
Rotate the tires (Section 10)
Check the air filter (Section 30)

Every 6,000 miles or 6 months, whichever comes first

All the items listed above, plus . . .
Inspect/replace the underhood hoses (Section 11)
Check the drivebelt (Section 12)
Check/service the battery (Section 13)
Inspect/lubricate automatic transmission shift linkage (Section 15)
Lubricate the chassis (Section 16)

Every 12,000 miles or 12 months, whichever comes first

All items listed above, plus . . .
Check/replenish the manual transmission lubricant (Section 17)
Check the differential lubricant level (Section 18)
Check the transfer case lubricant level (Section 19)
Check/replace the PCV valve (Section 20)
Check the fuel system (Section 21)
Inspect the cooling system (Section 22)
Inspect the exhaust system (Section 23)
Inspect the steering and suspension components (Section 24)
Inspect the brakes and lubricate caliper friction points (Section 25)
Check/lubricate the front wheel bearings (2WD) (Section 26)
Check hub lock, spindle and front wheel bearing lubrication (4WD) (Section 27)
Inspect/replace the windshield wiper blades (Section 28)

Every 24,000 miles or 24 months, whichever comes first

All items listed above plus . . .
Change the automatic transmission fluid and filter (Section 9)
Service the cooling system (drain, flush and refill) (Section 22)
Check/replace the spark plug wires (Section 14)

Every 30,000 miles or 30 months, whichever comes first

Replace the spark plugs (1993 2.3L engine)
Replace the air filter (Section 30)
Check/replace the fuel filter (Section 31)
Change the differential lubricant (Section 18)
Change the transfer case lubricant (Section 19)
Lubricate the driveshaft slip yokes (Section 32)
Lubricate right front driveaxle slip yoke (4WD) (Section 32)

Every 60,000 miles or 60 months, whichever comes first

Replace the spark plugs (all except 1993 2.3L engine)

Off-road operation

If the vehicle is driven off road, perform the following maintenance items every 1,000 miles. If the vehicle is driven in mud or water, perform the items daily.

Inspect the brakes (Section 25)
Inspect the front wheel bearings (2WD models) (Section 26)
Check hub lock, spindle and front wheel bearing lubrication (4WD) (Section 27)
Inspect exhaust system (Section 23)
Lubricate driveshaft slip yoke and U-joints (if equipped with grease fittings) (Sections 16 and 32)

Chapter 1 Tune-up and routine maintenance

2 Introduction

This Chapter is designed to help the home mechanic maintain his or her vehicle with the goals of maximum performance, economy, safety and reliability in mind.

Included is a master maintenance schedule (page 1-6), followed by procedures dealing specifically with each item on the schedule. Visual checks, adjustments, component replacement and other helpful items are included. Refer to the accompanying illustrations of the engine compartment and the underside of the vehicle for the locations of various components. Servicing the vehicle, in accordance with the mileage/time maintenance schedule and the step-by-step procedures will result in a planned maintenance program that should produce a long and reliable service life. Keep in mind that it is a comprehensive plan, so maintaining some items but not others at specified intervals will not produce the same results.

As you service the vehicle, you will discover that many of the procedures can - and should - be grouped together because of the nature of the particular procedure you're performing or because of the close proximity of two otherwise unrelated components to one another.

For example, if the vehicle is raised for chassis lubrication, you should inspect the exhaust, suspension, steering and fuel systems while you're under the vehicle. When you're rotating the tires, it makes good sense to check the brakes since the wheels are already removed. Finally, let's suppose you have to borrow or rent a torque wrench. Even if you only need it to tighten the spark plugs, you might as well check the torque of as many critical fasteners as time allows.

The first step in this maintenance program is to prepare yourself before the actual work begins. Read through all the procedures you're planning to do, then gather up all the parts and tools needed. If it looks like you might run into problems during a particular job, seek advice from a mechanic or an experienced do-it-yourselfer.

3 Tune-up general information

The term tune-up is used in this manual to represent a combination of individual operations rather than one specific procedure.

If, from the time the vehicle is new, the routine maintenance schedule is followed closely and frequent checks are made of fluid levels and high wear items, as suggested throughout this manual, the engine will be kept in relatively good running condition and the need for additional work will be minimized.

More likely than not, however, there will be times when the engine is running poorly due to a lack of regular maintenance. This is even more likely if a used vehicle, which has not received regular and frequent maintenance checks, is purchased. In such cases, an engine tune-up will be needed outside of the regular maintenance intervals.

The first step in any tune-up or diagnostic procedure to help correct a poor running engine is a cylinder compression check. A compression check (see Chapter 2, Part D) will help determine the condition of internal engine components and should be used as a guide for tune-up and repair procedures. If, for instance, a compression check indicates serious internal engine wear, a conventional tune-up will not improve the performance of the engine and would be a waste of time and money. Because of its importance, the compression check should be done by someone with the right equipment and the knowledge to use it properly.

The following procedures are those most often needed to bring as generally poor running engine back into a proper state of tune.

Minor tune-up
Check all engine related fluids
   (see Section 4)
Check all underhood hoses
   (see Section 11)
Check and adjust the drivebelt
   (see Section 12)
Clean, inspect and test the battery
   (see Section 13)
Replace the spark plugs (see Section 14)
Inspect the spark plug wires
   (see Section 14)
Check the PCV valve (see Section 20)
Check the cooling system
   (see Section 22)
Check the air filter (see Section 30)

Major tune-up
All items listed under minor tune-up, plus . . .
Replace the spark plug wires
   (see Section 14)
Check the fuel system (see Chapter 4)
Check the ignition system (see Chapter 5)
Check the charging system
   (see Chapter 5)

4.4a Remove the engine oil dipstick, wipe it clean, then reinsert it all the way before withdrawing it for an accurate oil level check

4 Fluid level checks (every 250 miles or weekly, whichever comes first)

Refer to illustrations 4.4a, 4.4b, 4.6, 4.9 and 4.15

Note: *The following are fluid level checks to be done on a 250 mile or weekly basis. Additional fluid level checks can be found in specific maintenance procedures which follow. Regardless of intervals, be alert to fluid leaks under the vehicle which would indicate a fault to be corrected immediately.*

1 Fluids are an essential part of the lubrication, cooling, brake and windshield washer systems. Because the fluids gradually become depleted and/or contaminated during normal operation of the vehicle, they must be periodically replenished. See *Recommended lubricants and fluids* at the beginning of this Chapter before adding fluid to any of the following components. **Note:** *The vehicle must be on level ground when fluid levels are checked.*

Engine oil

2 Engine oil is checked with a dipstick, which is located on the side of the engine (refer to the underhood illustration at the front of this Chapter for dipstick location). The dipstick extends through a metal tube down into the oil pan.

3 The engine oil should be checked before the vehicle has been driven, or about 15 minutes after the engine has been shut off. If the oil is checked immediately after driving the vehicle, some of the oil will remain in the upper part of the engine, resulting in an inaccurate reading on the dipstick.

4 Pull the dipstick out of the tube **(see illustration)** and wipe all of the oil away from the end with a clean rag or paper towel. Insert the clean dipstick all the way back into the tube and pull it out again. Note the oil at the end of the dipstick. At its highest point, the oil should be above the ADD mark, in the SAFE range **(see illustration)**.

5 It takes one quart of oil to raise the level from the ADD mark to the FULL or MAX mark on the dipstick. Do not allow the level to drop

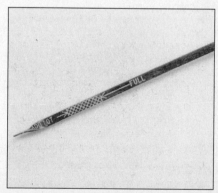

4.4b The oil level should appear between the ADD 1 QT and SAFE or FULL marks; don't overfill the engine with oil

4.6 To add oil, remove the filler cap from the valve cover or filler neck

4.9 The coolant recovery reservoir is combined with the windshield washer fluid reservoir (there are separate compartments for the two fluids)

4.15 Clutch fluid is contained in a separate reservoir (right arrow) next to the brake master cylinder (left arrow) - clean the rubber-lined cap(s) before returning it to the reservoir

below the ADD mark or oil starvation may cause engine damage. Conversely, overfilling the engine (adding oil above the FULL or MAX mark) may cause oil fouled spark plugs, oil leaks or oil seal failures.

6   To add oil, remove the filler cap located on the valve cover or filler neck **(see illustration)**. After adding oil, wait a few minutes to allow the level to stabilize, then pull the dipstick out and check the level again. Add more oil if required. Install the filler cap and tighten it by hand only.

7   Checking the oil level is an important preventive maintenance step. A consistently low oil level indicates oil leakage through damaged seals, defective gaskets or past worn rings or valve guides. The condition of the oil should also be noted. If the oil looks milky in color or has water droplets in it, the cylinder head gasket(s) may be blown or the head(s) or block may be cracked. The engine should be repaired immediately. Whenever you check the oil level, slide your thumb and index finger up the dipstick before wiping off the oil. If you see small dirt or metal particles clinging to the dipstick, the oil should be changed (see Section 4).

Engine coolant

Warning: *Do not allow antifreeze to come in contact with your skin or painted surfaces of the vehicle. Rinse off spills immediately with plenty of water. Antifreeze is highly toxic if ingested. Never leave antifreeze lying around in an open container or in puddles on the floor; children and pets are attracted by its sweet smell and may drink it. Check with local authorities about disposing of used antifreeze. Many communities have collection centers which will see that antifreeze is disposed of safely.*

8   All vehicles covered by this manual are equipped with a pressurized coolant recovery system. A white plastic coolant reservoir located at the front of the engine compartment is connected by a hose to the radiator filler neck. If the engine overheats, coolant escapes through a valve in the radiator cap and travels through the hose into the reservoir. As the engine cools, the coolant is automatically drawn back into the cooling system to maintain the correct level.

9   The coolant level in the reservoir **(see illustration)** should be checked regularly. **Warning:** *Do not remove the radiator cap to check the coolant level when the engine is warm!* The level in the reservoir varies with the temperature of the engine. When the engine is cold, the coolant level should be at or slightly above the COLD FULL mark on the reservoir. Once the engine has warmed up, the level should be at or near the FULL HOT mark. If it isn't, allow the engine to cool, then remove the cap from the reservoir and add a 50/50 mixture of ethylene glycol based antifreeze and water. Don't use rust inhibitors or additives.

10   Drive the vehicle and recheck the coolant level. If only a small amount of coolant is required to bring the system up to the proper level, water can be used. However, repeated additions of water will dilute the antifreeze and water solution. In order to maintain the proper ratio of antifreeze and water, always top up the coolant level with the correct mixture. An empty plastic milk jug or bleach bottle makes an excellent container for mixing coolant.

11   If the coolant level drops consistently, there may be a leak in the system. Inspect the radiator, hoses, filler cap, drain plugs and water pump (see Section 22). If no leaks are noted, have the radiator cap pressure tested by a service station.

12   If you have to remove the radiator cap, wait until the engine has cooled completely, then wrap a thick cloth around the cap and turn it to the first stop. If coolant or steam escapes, let the engine cool down longer, then remove the cap.

13   Check the condition of the coolant as well. It should be relatively clear. If it's brown or rust colored, the system should be drained, flushed and refilled. Even if the coolant appears to be normal, the corrosion inhibitors wear out, so it must be replaced at the specified intervals.

Brake and clutch fluid

Warning: *Brake fluid can harm your eyes and damage painted surfaces, so use extreme caution when handling or pouring it. Do not use brake fluid that has been standing open or is more than one year old. Brake fluid absorbs moisture from the air, which can cause a dangerous loss of brake effectiveness. Use only the specified type of brake fluid. Mixing different types (such as DOT 3 or 4 and DOT 5) can cause brake failure.*

14   The brake master cylinder is mounted at the left (driver's side) rear corner of the engine compartment. The clutch fluid reservoir (used on models with manual transmissions) is mounted adjacent to it.

15   To check the clutch fluid level, observe the level through the translucent reservoir. The level should be at or near the step molded into the reservoir. If the level is low, remove the reservoir cap to add the specified fluid **(see illustration)**.

16   The brake fluid level is checked by looking through the plastic reservoir mounted on the master cylinder. The fluid level should be between the MAX and MIN lines on the reservoir **(see illustration 4.15)**. If the fluid level is low, wipe the top of the reservoir and the cap with a clean rag to prevent contamination of the system as the cap is unscrewed. Top up with the recommended brake fluid, but do not overfill.

17   While the reservoir cap is off, check the master cylinder reservoir for contamination. If rust deposits, dirt particles or water droplets are present, the system should be drained and refilled by a dealer service department or repair shop.

18   After filling the reservoir to the proper level, make sure the cap is seated to prevent fluid leakage and/or contamination.

19   The fluid level in the master cylinder will drop slightly as the disc brake pads wear. A very low level may indicate worn brake pads. Check for wear (see Section 25).

20   If the brake fluid level drops consistently, check the entire system for leaks immediately. Examine all brake lines, hoses and connections, along with the calipers, wheel cylinders and master cylinder (see Section 25).

Chapter 1 Tune-up and routine maintenance

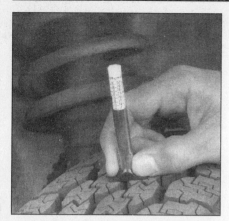

5.2 Use a tire tread depth indicator to monitor tire wear - they are available at auto parts stores and service stations and cost very little

21   When checking the fluid level, if you discover one or both reservoirs empty or nearly empty, the brake or clutch hydraulic system should be checked for leaks and bled (see Chapters 8 and 9).

Windshield washer fluid

22   Fluid for the windshield washer system is stored in a plastic reservoir in the engine compartment. The windshield washer reservoir is combined with the coolant reservoir (there are separate compartments for the two different fluids) **(see illustration 4.9)**.

23   In milder climates, plain water can be used in the reservoir, but it should be kept no more than 2/3 full to allow for expansion if the water freezes. In colder climates, use windshield washer system antifreeze, available at any auto parts store, to lower the freezing point of the fluid. This comes in concentrated or pre-mixed form. If you purchase concentrated antifreeze, mix the antifreeze with water in accordance with the manufacturer's directions on the container. **Caution:** *Do not use cooling system antifreeze - it will damage the vehicle's paint.*

5 Tire and tire pressure checks (every 250 miles or weekly, whichever comes first)

Refer to illustrations 5.2, 5.3, 5.4a, 5.4b and 5.8

1   Periodic inspection of the tires may save you the inconvenience of being stranded with a flat tire. It can also provide you with vital information regarding possible problems in the steering and suspension systems before major damage occurs.

2   Tires are equipped with 1/2-inch wide bands that will appear when tread depth reaches 1/16-inch, which is when the tires are worn out. Tread wear can be monitored with a simple, inexpensive device known as a tread depth indicator **(see illustration)**.

UNDERINFLATION

CUPPING

OVERINFLATION

INCORRECT TOE-IN OR EXTREME CAMBER

Cupping may be caused by:
- Underinflation and/or mechanical irregularities such as out-of-balance condition of wheel and/or tire, and bent or damaged wheel.
- Loose or worn steering tie-rod or steering idler arm.
- Loose, damaged or worn front suspension parts.

FEATHERING DUE TO MISALIGNMENT

5.3 This chart will help you determine the condition of the tires, the probable cause(s) of abnormal wear and the corrective action necessary

3   Note any abnormal tire wear **(see illustration)**. Tread pattern irregularities such as cupping, flat spots and more wear on one side that the other are indications of front end alignment and/or balance problems. If any of these conditions are noted, take the vehicle to a tire shop or service station to correct the problem.

4   Look closely for cuts, punctures and embedded nails or tacks. Sometimes a tire will hold air pressure for a short time or leak down very slowly after a nail has embedded itself in the tread. If a slow leak persists, check the valve stem core to make sure it is tight **(see illustration)**. Examine the tread for an object that may have embedded itself in t the tire or for a "plug" that may have begun to leak (radial tire punctures are repaired with a plug that is installed in the puncture). If a puncture is suspected, it can be easily verified by spraying a solution of soapy water onto the puncture **(see illustration)**. The soapy solution will bubble if there is a leak. Unless the puncture is unusually large, a tire shop or service station can usually repair the tire.

5   Carefully inspect the inner sidewall of each tire for evidence of brake fluid leakage. If you see any, inspect the brakes immediately.

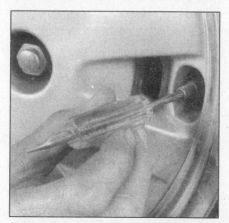

5.4a If a tire loses air on a steady basis, check the valve stem core first to make sure it's snug (special inexpensive wrenches are commonly available at auto parts stores)

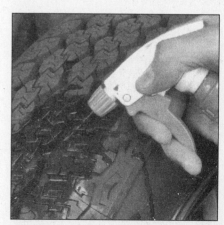

5.4b If the valve stem core is tight, raise the corner of the vehicle with the low tire and spray a soapy water solution onto the tread as the tire is turned slowly - leaks will cause small bubbles to appear

Chapter 1 Tune-up and routine maintenance

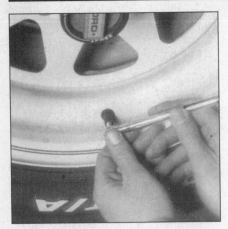

5.8 To extend the life of the tires, check the air pressure at least once a week with an accurate gauge (don't forget the spare!)

6 Correct air pressure adds miles to the lifespan of the tires, improves mileage and enhances overall ride quality. Tire pressure cannot be accurately estimated by looking at a tire, especially if it's a radial. A tire pressure gauge is essential. Keep an accurate gauge in the glove compartment. The pressure gauges attached to the nozzles of air hoses at gas stations are often inaccurate.

7 Always check tire pressure when the tires are cold. Cold, in this case, means the vehicle has not been driven over a mile in the three hours preceding a tire pressure check. A pressure rise of four to eight pounds is not uncommon once the tires are warm.

8 Unscrew the valve cap protruding from the wheel or hubcap and push the gauge firmly onto the valve stem **(see illustration)**. Note the reading on the gauge and compare the figure to the recommended tire pressure shown in your owner's manual or on the tire placard on the passenger side door or door pillar. Be sure to reinstall the valve cap to keep dirt and moisture out of the valve stem mechanism. Check all four tires and, if necessary, add enough air to bring them to the recommended pressure.

9 Don't forget to keep the spare tire inflated to the specified pressure (refer to your owner's manual or the placard attached to the door pillar). Note that the pressure recommended for temporary (mini) spare tires is higher than for the tires on the vehicle.

6 Safety checks (every 250 miles or weekly, whichever comes first)

1 Most of these checks can be easily performed while the vehicle is being driven, simply by paying attention to the specified items. The checks are intended to make the vehicle owner aware of potential safety problems before they occur.

2 Check the seat belts for wear, fraying and cuts. Make sure the buckles latch securely and that the automatic retractors function correctly. Do not try to repair seat belts; always replace them if any problems are found.

3 Make sure the ignition key cannot be removed when the transmission is in any gear other than Park (automatic) or Reverse (manual). Make sure the steering column locks when the key is removed from the ignition. It may be necessary to rotate the steering wheel slightly to lock the steering column.

4 Check the parking brake. The easiest way to do this is to park on a steep hill, set the parking brake and see if it keeps the vehicle from rolling.

5 If equipped with an automatic transmission, also check the Park mechanism. Place the transmission in Park, release the parking brake and note whether the transmission holds the vehicle from rolling. If the vehicle rolls while in Park, the transmission should be taken to a qualified shop for repairs.

6 If equipped with an automatic transmission, see if the shift indicator shows the proper gear. If it doesn't, refer to Chapter 7B for linkage adjustment procedures.

7 If equipped with an automatic transmission, make sure the vehicle starts only in Park or Neutral. If it starts in any other gear, refer to Chapter 7B for switch adjustment procedures.

8 If equipped with a manual transmission, the starter should operate only when the clutch pedal is pressed to the floor. If the starter operates when it shouldn't, refer to Chapter 8 for clutch/starter interlock switch service.

9 Make sure the brakes do not pull to one side while stopping. The brake pedal should feel firm, but excessive effort should not be required to stop the vehicle. If the pedal sinks too low, if you have to pump it more than once to get a firm pedal, or if pedal effort is too high, refer to Chapter 9 for repair procedures. A squealing sound from the front brakes may be caused by the pad wear indicators. Refer to Chapter 9 for pad replacement procedures.

10 Rearview mirrors should be clean and undamaged. They should hold their position when adjusted.

11 Sun visors should hold their position when adjusted. They should remain securely out of the way when lifted off the windshield.

12 Make sure the defroster blows heated air onto the windshield. If it doesn't, refer to Chapter 3 for heating system service.

13 The horn should sound with a clearly audible tone every time it is operated. If not, refer to Chapter 12.

14 Make sure the windows are clean and undamaged.

15 Turn on the lights, then walk around the car and make sure they work. Check headlights in both the high beam and low beam positions. Check the turn indicators for one side of the vehicle, then for the other side. If possible, have an assistant watch the brake lights while you push the pedal. If no assistant is available, the brake lights can be checked by backing up to a wall or garage

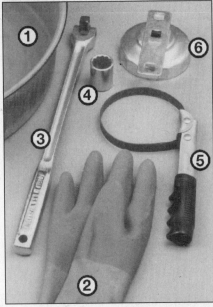

7.2 These tools are required when changing the engine oil and filter

1 **Drain pan** - It should be fairly shallow in depth, but wide to prevent spills
2 **Rubber gloves** - When removing the drain plug and filter, you will get oil on your hands (the gloves will prevent burns)
3 **Breaker bar** - Sometimes the oil drain plug is tight and a long breaker bar is needed to loosen it
4 **Socket** - To be used with the breaker bar or a ratchet (must be the correct size to fit the drain plug - six-point preferred)
5 **Filter wrench** - This is a metal band-type wrench, which requires clearance around the filter to be effective
6 **Filter wrench** - This type fits on the bottom of the filter and can be turned with a ratchet or breaker bar (different size wrenches are available for different types of filters)

door, then pressing the pedal. There should be two distinct patches of red light when the brake pedal is pressed.

16 Make sure locks operate smoothly when the key is turned. Lubricate locks if necessary with Ford Lock Lubricant or an equivalent product. Make sure all latches hold securely.

7 Engine oil and filter change (every 3,000 miles or 3 months, whichever comes first)

Refer to illustrations 7.2, 7.7 and 7.16

1 Frequent oil changes are the most important preventive maintenance procedures that can be done by the home mechanic. As engine oil ages, it becomes diluted and contaminated, which leads to premature engine wear.

Chapter 1 Tune-up and routine maintenance

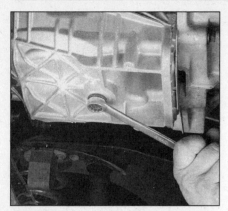

7.7 Remove the oil pan drain plug with a box-end wrench or socket - an open-end or adjustable wrench shouldn't be used, since it may round off the corners of the drain plug

7.16 Lubricate the oil filter gasket with clean engine oil before installing the filter on the engine

8.2 On all models, the power steering fluid reservoir is located on the left side of the engine, just below the air conditioning compressor (if equipped) - four-cylinder engine shown, V6 engines similar

2 Make sure that you have all the necessary tools before you begin this procedure **(see illustration)**. You should also have plenty of rags or newspapers handy for mopping up oil spills.

3 Start the engine and allow it to reach normal operating temperature - oil and sludge will flow more easily when warm. If new oil, a filter or tools are needed, use the vehicle to go get them and warm up the engine oil at the same time. Park on a level surface and shut off the engine when it's warmed up. Remove the oil filler cap from the valve cover.

4 Access to the oil drain plug and filter will be improved if the vehicle can be lifted on a hoist, driven onto ramps or supported by jackstands. **Warning:** *DO NOT work under a vehicle supported only by a bumper, hydraulic or scissors-type jack - always use jackstands!*

5 Raise the vehicle and support it on jackstands. Make sure it is safely supported!

6 If you haven't changed the oil on this vehicle before, get under it and locate the drain plug and the oil filter. The exhaust components will be hot as you work, so note how they are routed to avoid touching them.

7 Being careful not to touch the hot exhaust components, position a drain pan under the plug in the bottom of the engine. Clean the area around the plug, then remove the plug **(see illustration)**. It's a good idea to wear an old glove while unscrewing the plug the final few turns to avoid being scalded by hot oil. It will also help to hold the drain plug against the threads as you unscrew it, then pull it away from the drain hole suddenly. This will place your arm out of the way of the hot oil, as well as reducing the chances of dropping the drain plug into the drain pan.

8 It may be necessary to move the drain pan slightly as oil flow slows to a trickle. Inspect the old oil for the presence of metal particles.

9 After all the oil has drained, wipe off the drain plug with a clean rag. Any small metal particles clinging to the plug would immediately contaminate the new oil.

10 Reinstall the plug and tighten it securely, but don't strip the threads.
11 Move the drain pan into position under the oil filter.
12 Loosen the oil filter by turning it counterclockwise with a filter wrench. Any standard filter wrench will work.
13 Sometimes the oil filter is screwed on so tightly that it can't be loosened. If it is, punch a metal bar or long screwdriver directly through it, as close to the engine as possible, and use it as a T-bar to turn the filter. Be prepared for oil to spurt out of the canister as it's punctured.
14 Once the filter is loose, use your hands to unscrew it from the block. Just as the filter is detached from the block, immediately tilt the open end up to prevent oil inside the filter from spilling out.
15 Using a clean rag, wipe off the mounting surface on the block. Also, make sure that none of the old gasket remains stuck to the mounting surface. It can be removed with a scraper if necessary.
16 Compare the old filter with the new one to make sure they are the same type. Smear some engine oil on the rubber gasket of the new filter and screw it into place **(see illustration)**. Overtightening the filter will damage the gasket, so don't use a filter wrench. Most filter manufacturers recommend tightening the filter by hand only. Normally, they should be tightened 3/4-turn after the gasket contacts the block, but be sure to follow the directions on the filter or container.
17 Remove all tools and materials from under the vehicle, being careful not to spill the oil in the drain pan, then lower the vehicle.
18 Add new oil to the engine through the oil filler cap in the rocker arm cover. Use a funnel to prevent oil from spilling onto the top of the engine. Pour four quarts of fresh oil into the engine. Wait a few minutes to allow the oil to drain into the pan, then check the level on the dipstick (see Section 4 if necessary). If the oil level is in the SAFE range, install the filler cap.
19 Start the engine and run it for about a minute. While the engine is running, look under the vehicle and check for leaks at the oil pan drain plug and around the oil filter. If either one is leaking, stop the engine and tighten the plug or filter slightly.
20 Wait a few minutes, then recheck the level on the dipstick. Add oil as necessary to bring the level into the SAFE range.
21 During the first few trips after an oil change, make it a point to check frequently for leaks and proper oil level.
22 The old oil drained from the engine cannot be reused in its present state and should be discarded. Oil reclamation centers, auto repair shops and gas stations will normally accept the oil, which can be recycled. After the oil has cooled, it can be drained into a container (plastic jugs, bottles, milk cartons, etc.) for transport to a disposal site.

8 Power steering fluid level check (every 3,000 miles or 3 months, whichever comes first)

Refer to illustrations 8.2 and 8.5

1 Check the power steering fluid level periodically to avoid steering system problems, such as damage to the pump. **Caution:** *DO NOT hold the steering wheel against either stop (extreme left or right turn) for more than five seconds. If you do, the power steering pump could be damaged.*
2 The power steering pump, located at the left front corner of the engine on all models, is equipped with a twist-off cap with an integral fluid level dipstick **(see illustration)**.
3 Park the vehicle on level ground and apply the parking brake.
4 Run the engine until it has reached normal operating temperature. With the engine at idle, turn the steering wheel back-and-forth several times to get any air out of the steering system. Shut the engine off, remove the cap by turning it counterclockwise, wipe the dipstick clean and reinstall the cap (make sure it is seated).
5 Remove the cap again and note the fluid

1-12 Chapter 1 Tune-up and routine maintenance

8.5 Check the power steering fluid level with the engine at normal operating temperature - fluid should not go above the Full Hot mark

9.5 Make sure the area around the transmission dipstick is clean (arrow), then pull it out of the tube - fluid level is checked with the engine idling

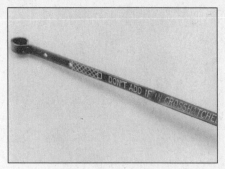

9.6 Follow the directions stamped on the transmission dipstick to get an accurate reading

level. It must be between the two lines designating the FULL HOT range **(see illustration)** (be sure to use the proper temperature range on the dipstick when checking the fluid level - the FULL COLD lines on the reverse side of the dipstick are only usable when the engine is cold).

6   Add small amounts of fluid until the level is correct. **Caution:** *Do not overfill the pump. If too much fluid is added, remove the excess with a clean syringe or suction pump.*

7   Check the power steering hoses and connections for leaks and wear (see Section 11).

8   Check the condition and tension of the serpentine drivebelt (see Section 12).

9   Automatic transmission fluid check/change (every 3,000 miles or 3 months, whichever comes first)

Fluid level check (every 3,000 miles or 3 months)

Refer to illustrations 9.5 and 9.6
Caution: *The use of transmission fluid other than the type listed in this Chapter's Specifications could result in transmission malfunctions or failure.*

1   The automatic transmission fluid should be carefully maintained. Low fluid level can lead to slipping or loss of drive, while overfilling can cause foaming and loss of fluid. Either condition can cause transmission damage.

2   Since transmission fluid expands as it heats up, the fluid level should only be checked when the transmission is warm (at normal operating temperature). If the vehicle has just been driven over 20 miles (32 km), the transmission can be considered warm. **Caution:** *If the vehicle has just been driven for a long time at high speed or in city traffic, in hot weather, or if it has been pulling a*

trailer, an accurate fluid level reading cannot be obtained. Allow the transmission to cool down for about 30 minutes. You can also check the transmission fluid level when the transmission is cold. If the vehicle has not been driven for over five hours and the fluid is about room temperature (70 to 95-degrees F), the transmission is cold. However, the fluid level is normally checked with the transmission warm to ensure accurate results.

3   Immediately after driving the vehicle, park it on a level surface, set the parking brake and start the engine. While the engine is idling, depress the brake pedal and move the selector lever through all the gear ranges, beginning and ending in Park.

4   Locate the automatic transmission dipstick tube in the engine compartment (see the illustration at the front of this chapter for dipstick location).

5   With the engine still idling, pull the dipstick away from the tube **(see illustration)**, wipe it off with a clean rag, push it all the way back into the tube and withdraw it again, then note the fluid level.

6   If the transmission is cold, the level should be in the room temperature range on the dipstick (between the two circles); if it's warm, the fluid level should be in the operating temperature range (between the two lines) **(see illustration)**. If the level is low, add the specified automatic transmission fluid through the dipstick tube - use a clean funnel to prevent spills.

7   Add just enough of the recommended fluid to fill the transmission to the proper level. It takes about one pint to raise the level from the low mark to the high mark when the fluid is hot, so add the fluid a little at a time and keep checking the level until it's correct.

8   The condition of the fluid should also be checked along with the level. If the fluid is black or a dark reddish-brown color, or if it smells burned, it should be changed (see below). If you are in doubt about its condition, purchase some new fluid and compare the two for color and smell.

Fluid and filter change (every 24,000 miles or 24 months)

Refer to illustrations 9.14, 9.16 and 9.19

9   At the specified intervals, the transmission fluid should be drained and replaced. Since the fluid will remain hot long after driving, perform this procedure only after the engine has cooled down completely.

10   Before beginning work, purchase the specified transmission fluid (see *Recommended lubricants and fluids* at the front of this Chapter), a new filter and gasket. Never reuse the old filter or gasket!

11   Other tools necessary for this job include jackstands to support the vehicle in a raised position, a drain pan capable of holding at least eight quarts, newspapers and clean rags.

12   Raise the vehicle and support it securely on jackstands. DO NOT crawl under the vehicle when it is supported only by a jack! Place the drain pan beneath the transmission.

13   If you're working on a vehicle with a 4.0L engine, pry the transmission heat shield loose from the edge of the transmission fluid pan. It isn't necessary to remove the heat shield from the vehicle.

14   With the drain pan in place, remove the mounting bolts from the front and sides of the transmission fluid pan **(see illustration)**.

15   Loosen the rear pan bolts approximately four turns.

9.14 Remove the bolts around the front and sides of the transmission pan . . .

Chapter 1 Tune-up and routine maintenance

9.16 ... and let the pan hang down so the fluid can drain

9.19 Once the pan is removed the filter is accessible

16 Carefully pry the transmission pan loose with a screwdriver. Let the pan hang down so the fluid can drain **(see illustration)**. Don't damage the pan or transmission gasket surfaces or leaks could develop.
17 Remove the remaining bolts, pan and gasket. Carefully clean the gasket surface of the transmission to remove all traces of the old gasket and sealant.
18 Drain the fluid from the transmission pan, clean it with solvent and dry it with compressed air.
19 Remove the filter from the mount inside the transmission **(see illustration)**.
20 Install a new filter and gasket. Tighten the mounting bolt securely.
21 Make sure the gasket surface on the transmission pan is clean, then install a new gasket. Put the pan in place against the transmission and install the bolts. Working around the pan, tighten each bolt a little at a time until the torque listed in this Chapter's Specifications is reached. Don't overtighten the bolts!
22 Lower the vehicle and add automatic transmission fluid through the filler tube. **Caution:** *Refer to the Specifications at the front of this chapter for the correct amount and type of transmission fluid. Use of the wrong type or the wrong amount can cause transmission damage.*
23 With the transmission in Park and the parking brake set, run the engine at a fast idle, but don't race it.
24 Move the gear selector through each range and back to Park. Check the fluid level. Add fluid if needed to reach the correct level.
25 Check under the vehicle for leaks after the first few trips.

10 Tire rotation (every 3,000 miles or 3 months, whichever comes first)

Refer to illustration 10.2

1 The tires should be rotated at the specified intervals and whenever uneven wear is noticed. Since the vehicle will be raised and the tires checked anyway, check the brakes also (see Section 25).
2 It is recommended that the tires be rotated in a specific pattern **(see illustration)**.
3 Refer to the information in *Jacking and towing* at the front of this manual for the proper procedure to follow when raising the vehicle and changing a tire. If the brakes must be checked, don't apply the parking brake as stated.
4 The vehicle must be raised on a hoist or supported on jackstands to get all four tires off the ground. Make sure the vehicle is safely supported!
5 After the rotation procedure is finished, check and adjust the tire pressures as necessary and be sure to check the lug nut tightness.

11 Underhood hose check and replacement (every 6,000 miles or 6 months, whichever comes first)

Caution: *Replacement of air conditioning hoses must be left to a dealer service department or air conditioning shop that has the equipment to depressurize the system safely.*

Never disconnect air conditioning hoses or components until the system has been depressurized.

General

1 High temperatures under the hood can cause deterioration of the rubber and plastic hoses used for engine, accessory and emission systems operation. Periodic inspection should be made for cracks, loose clamps, material hardening and leaks.
2 Information specific to the cooling system can be found in Section 22.
3 Most (but not all) hoses are secured to the fitting with clamps. Where clamps are used, check to be sure they haven't lost their tension, allowing the hose to leak. If clamps aren't used, make sure the hose has not expanded and/or hardened where it slips over the fitting, allowing it to leak.

PCV system hose

4 To reduce hydrocarbon emissions, crankcase blow-by gas is vented through the PCV valve to the intake manifold via a rubber hose on most models. The blow-by gases mix with incoming air in the intake manifold before being burned in the combustion chambers.
5 Check the PCV hose for cracks, leaks and other damage. Disconnect it from the valve cover and the intake manifold and check the inside for obstructions. If it's clogged, clean it out with solvent.

Vacuum hoses

Refer to illustration 11.6

6 It's quite common for vacuum hoses, especially those in the emissions system, to be color coded or identified by colored stripes molded into them. Various systems require hoses with different wall thicknesses, collapse resistance and temperature resistance. When replacing hoses, be sure the new ones are made of the same material. A number of hoses connect to vacuum fittings on the intake manifold **(see illustration)**.
7 Often the only effective way to check a hose is to remove it completely from the vehicle. If more than one hose is removed, be

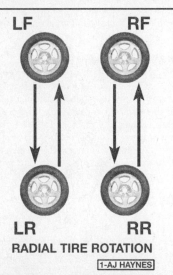

10.2 Tire rotation diagram

11.6 Vacuum hoses are connected to fittings on the intake manifold

Chapter 1 Tune-up and routine maintenance

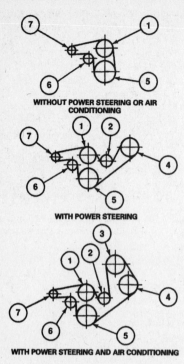

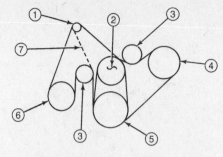

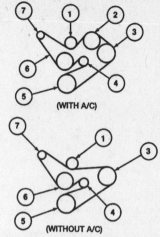

12.2a Serpentine belt routing details - four-cylinder engine models

1. Water pump
2. Idler pulley
3. Air conditioning compressor
4. Power steering pump
5. Crankshaft pulley
6. Belt tensioner
7. Alternator

12.2b Serpentine belt routing details - 1993 through 1995 3.0L engine models

1. Alternator
2. Water pump
3. Idler pulley
4. Power steering pump
5. Crankshaft pulley
6. Air conditioning compressor
7. Without air conditioning

12.2c Serpentine belt routing details - 1996 3.0L engine models

1. Idler pulley
2. Air conditioning compressor
3. Power steering pump
4. Belt tensioner
5. Crankshaft pulley
6. Water pump
7. Alternator

sure to label the hoses and fittings to ensure correct installation.

8 When checking vacuum hoses, be sure to include any plastic T-fittings in the check. Inspect the fittings for cracks and the hose where it fits over each fitting for distortion, which could cause leakage.

9 A small piece of vacuum hose can be used as a stethoscope to detect vacuum leaks. Hold one end of the hose to your ear and probe around vacuum hoses and fittings, listening for the "hissing" sound characteristic of a vacuum leak. **Warning:** *When probing with the vacuum hose stethoscope, be careful not to come into contact with moving engine components such as the drivebelt, cooling fan, etc.*

Fuel hoses

Warning: *There are certain precautions which must be taken when servicing or inspecting fuel system components. Work in a well ventilated area and do not allow open flames (cigarettes, appliance pilot lights, etc.) or bare light bulbs near the work area. Mop up any spills immediately and do not store fuel-soaked rags where they could ignite.*

10 The fuel lines are usually under pressure, so if any fuel lines are to be disconnected be prepared to catch spilled fuel.

Warning: *These vehicles are equipped with fuel injection; you must relieve the fuel system pressure before servicing the fuel lines. Refer to Chapter 4 for the fuel system pressure relief procedure.*

11 Check all rubber fuel lines for deterioration and chafing. Check especially for cracks in areas where the hose bends and just before fittings, such as where a hose attaches to the fuel pump, fuel filter and carburetor or fuel injection system.

12 High quality fuel line, specifically designed for fuel-injection systems, must be used for fuel line replacement. Never, under any circumstances, use unreinforced vacuum line, clear plastic tubing or water hose for fuel lines.

13 Spring-type clamps are commonly used on fuel lines. These clamps often lose their tension over a period of time, and can be "sprung" during removal. Replace all spring-type clamps with screw clamps whenever a hose is replaced.

Metal lines

14 Sections of metal line are often used for fuel line between the fuel pump and carburetor or fuel injection system. Check carefully to make sure the line isn't bent, crimped or cracked.

15 If a section of metal fuel line must be replaced, use seamless steel tubing only, since copper and aluminum tubing do not have the strength necessary to withstand the vibration caused by the engine.

16 Check the metal brake lines where they enter the master cylinder and brake proportioning unit (if used) for cracks in the lines and loose fittings. Any sign of brake fluid leakage calls for an immediate thorough inspection of the brake system.

Nylon fuel lines

17 Nylon fuel lines are used at several points in fuel injection systems. These lines require special materials and methods for repair. Refer to Chapter 4 for details.

Power steering hoses

18 Check the power steering hoses for leaks, loose connections and worn clamps. Tighten loose connections. Worn clamps or leaky hoses should be replaced.

12 Drivebelt check, adjustment and replacement (every 6,000 miles or 6 months, whichever comes first)

Refer to illustrations 12.2a through 12.2e, 12.3, 12.4a, 12.4b, 12.5a, 12.5b and 12.6

1 The accessory drivebelt is located at the front of the engine. The single serpentine belt drives the water pump, alternator, power steering pump and air conditioning compressor. The condition and tension of the drivebelt is critical to the operation of the engine and accessories. Excessive tension causes bearing wear, while insufficient tension produces slippage, noise, component vibration and belt failure. Because of their composition and the high stress to which they are subjected, drivebelts stretch and continue to deteriorate as they get older. As a result, they must be periodically checked. The serpentine belt has an automatic tensioner and requires no adjustment for the life of the belt.

Check

2 These vehicles use a single V-ribbed belt to drive all of the accessories. This is also known as a "serpentine" belt because of the winding path it follows between various drive, accessory and idler pulleys **(see illustrations)**.

3 With the engine off, open the hood and

Chapter 1 Tune-up and routine maintenance   1-15

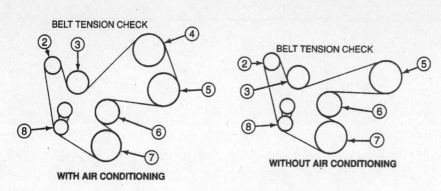

12.2d Serpentine belt routing details - 4.0L engine models

2 Alternator
3 Idler pulley
4 Air conditioning compressor
5 Power steering pump
6 Water pump
7 Crankshaft pulley
8 Tensioner

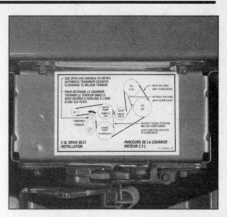

12.2e The belt's pathway is illustrated on a decal on the radiator support

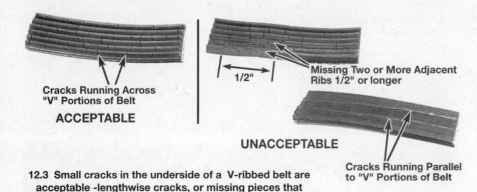

12.3 Small cracks in the underside of a V-ribbed belt are acceptable - lengthwise cracks, or missing pieces that cause the belt to make noise, are cause for replacement

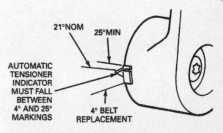

12.4a On four-cylinder engines, the belt tensioner indicator must fall between the 4-degree and 25-degree markings on the tensioner if not, the belt must be replaced

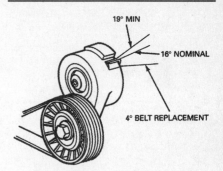

12.4b On 3.0L V6 engines, the belt tensioner indicator must fall between the 4-degree and 19-degree markings on the tensioner - if not, the belt must be replaced. On 4.0L V6 engines, there is no tensioner specification; if the belt is noisy, slips or seems loose, replace it

locate the drivebelt at the front of the engine. With a flashlight, check the belt for separation of the rubber plies from each side of the core, a severed core, separation of the ribs from the rubber and torn or worn ribs. Also check for fraying and glazing, which gives the belt a shiny appearance. Cracks in the rib side of V-ribbed belts are acceptable, as are small chunks missing from the ribs. If a V-ribbed belt has lost chunks bigger than 1/2-inch (13 mm) from two adjacent ribs, or if the missing chunks cause belt noise, the belt should be replaced **(see illustration)**. Both sides of the belt should be inspected, which means you'll have to twist it to check the underside. Use your fingers to feel the belt where you can't see it. If any of the above conditions are evident, replace the belt as described below.

Adjustment

4   Tension is set by an automatic tensioner. Manual adjustment is not required **(see illustrations)**.

Replacement

5   To replace a serpentine belt, rotate the tensioner counterclockwise (2.3L and 4.0L engines) or clockwise (3.0L engine) to lift it off the belt **(see illustrations)**. Slip the belt off the pulleys.

12.5a To loosen the belt, attach a 3/8-inch drive ratchet . . .

12.5b . . . and rotate it against the spring force in the direction shown to move the tensioner off the belt

Chapter 1 Tune-up and routine maintenance

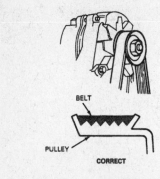

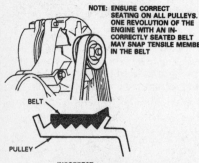

12.6 V-ribbed drivebelts must be centered on the pulleys, not offset

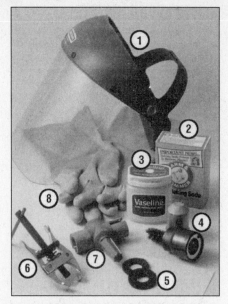

13.1 Tools and materials required for battery maintenance

1. **Face shield/safety goggles** - When removing corrosion with a brush, the acidic particles can easily fly up into your eyes
2. **Baking soda** - A solution of baking soda and water can be used to neutralize corrosion
3. **Petroleum jelly** - A layer of this on the battery posts will help prevent corrosion
4. **Battery post/cable cleaner** - This wire brush cleaning tool will remove all traces of corrosion from the battery posts and cable clamps
5. **Treated felt washers** - Placing one of these on each post, directly under the cable clamps, will help prevent corrosion
6. **Puller** - Sometimes the cable clamps are very difficult to pull off the posts, even after the nut/bolt has been completely loosened. This tool pulls the clamp straight up and off the post without damage.
7. **Battery post/cable cleaner** - Here is another cleaning tool which is a slightly different version of number 4 above, but it does the same thing
8. **Rubber gloves** - Another safety item to consider when servicing the battery; remember that's acid inside the battery!

13.4 If equipped with a maintenance-type battery, remove the cell caps to check the water level in the battery - if the level is low, add distilled water only

13.8a Battery terminal corrosion usually appears as light, fluffy powder

6 Hold the tensioner in the released position. Install a new belt and make sure it is routed correctly **(see illustrations 12.2a through 12.2e)**. Be sure the ribs of the new belt engage the pulley ribs correctly **(see illustration)**. Release the tensioner.

13 Battery check, maintenance and charging (every 6,000 miles or 6 months, whichever comes first)

Check and maintenance

Refer to illustrations 13.1, 13.4, 13.8a, 13.8b, 13.8c and 13.8d

Warning: Certain precautions must be followed when checking and servicing the battery. Hydrogen gas, which is highly flammable, is always present in the battery cells, so keep lighted tobacco and all other flames and sparks away from it. The electrolyte inside the battery is actually dilute sulfuric acid, which will cause injury if splashed on your skin or in your eyes. It will also ruin clothes and painted surfaces. When removing the battery cables, always detach the negative cable first and hook it up last!

1 Battery maintenance is an important procedure which will help ensure that you are not stranded because of a dead battery. Several tools are required for this procedure **(see illustration)**.

2 Before servicing the battery, always turn the engine and all accessories off and disconnect the cable from the negative terminal of the battery.

3 A sealed (sometimes called maintenance free) battery is standard equipment. The cell caps cannot be removed, no electrolyte checks are required and water cannot be added to the cells. However, if an aftermarket battery has been installed and it is a type that requires regular maintenance, the following procedures can be used.

4 If equipped with a maintenance-type battery, check the electrolyte level in each of the battery cells **(see illustration)**. It must be above the plates. There's usually a split-ring indicator in each cell to indicate the correct level. If the level is low, add distilled water only, then install the cell caps. **Caution:** Overfilling the cells may cause electrolyte to spill over during periods of heavy charging, causing corrosion and damage to nearby components.

5 If the positive terminal and cable clamp on your vehicle's battery is equipped with a rubber protector, make sure that it's not torn or damaged. It should completely cover the terminal.

6 The external condition of the battery should be checked periodically. Look for damage such as a cracked case.

7 Check the tightness of the battery cable clamps to ensure good electrical connections and inspect the entire length of each cable, looking for cracked or abraded insulation and frayed conductors.

8 If corrosion (visible as white, fluffy deposits) is evident, remove the cables from the terminals, clean them with a battery brush and reinstall them **(see illustrations)**. Corrosion can be kept to a minimum by installing specially treated washers available at auto parts stores or by applying a layer of petroleum jelly or grease to the terminals and cable clamps after they are assembled.

9 Make sure that the battery carrier is in good condition and that the hold-down clamp bolt is tight. If the battery is removed (see Chapter 5 for the removal and installa-

Chapter 1 Tune-up and routine maintenance   1-17

13.8b Removing the cable from a battery post with a wrench - sometimes special battery pliers are required for this procedure if corrosion has caused deterioration of the nut hex (always remove the ground cable first and hook it up last!)

13.8c Regardless of the type of tool used on the battery posts, a clean, shiny surface should be the result

13.8d When cleaning the cable clamps, all corrosion must be removed (the inside of the clamp is tapered to match the taper on the post, so don't remove too much material)

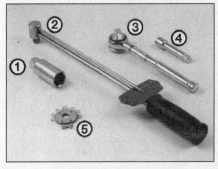

14.2 Tools required for changing spark plugs

1 **Spark plug socket -** *This will have special padding inside to protect the spark plug's ceramic insulator*
2 **Torque wrench -** *Although not mandatory, using this tool is the best way to ensure the plugs are tightened properly*
3 **Ratchet -** *Standard hand tool to fit the spark plug socket*
4 **Extension -** *Depending on model and accessories, you may need special extensions and universal joints to reach one or more of the plugs*
5 **Spark plug gap gauge -** *This gauge for checking the gap comes in a variety of styles. Make sure the gap for your engine is included.*

tion procedure), make sure that no parts remain in the bottom of the carrier when it's reinstalled. When reinstalling the hold-down clamp, don't overtighten the bolt.
10 Corrosion on the carrier, battery case and surrounding areas can be removed with a solution of water and baking soda. Apply the mixture with a small brush, let it work, then rinse it off with plenty of clean water.
11 Any metal parts of the vehicle damaged by corrosion should be coated with a zinc-based primer, then painted.
12 Additional information on the battery and jump starting can be found in Chapter 5 and the front of this manual.

Charging

13 Remove all of the cell caps (if equipped) and cover the holes with a clean cloth to prevent spattering electrolyte. Disconnect the negative battery cable and hook the battery charger leads to the battery posts (positive to positive, negative to negative), then plug in the charger. Make sure it is set at 12-volts if it has a selector switch.
14 If you're using a charger with a rate higher than two amps, check the battery regularly during charging to make sure it doesn't overheat. If you're using a trickle charger, you can safely let the battery charge overnight after you've checked it regularly for the first couple of hours.
15 If the battery has removable cell caps, measure the specific gravity with a hydrometer every hour during the last few hours of the charging cycle. Hydrometers are available inexpensively from auto parts stores - follow the instructions that come with the hydrometer. Consider the battery charged when there's no change in the specific gravity reading for two hours and the electrolyte in the cells is gassing (bubbling) freely. The specific gravity reading from each cell should be very close to the others. If not, the battery probably has a bad cell(s).
16 Some batteries with sealed tops have built-in hydrometers on the top that indicate the state of charge by the color displayed in the hydrometer window. Normally, a bright-colored hydrometer indicates a full charge and a dark hydrometer indicates the battery still needs charging. Check the battery manufacturer's instructions to be sure you know what the colors mean.
17 If the battery has a sealed top and no built-in hydrometer, you can hook up a digital voltmeter across the battery terminals to check the charge. A fully charged battery should read 12.6-volts or higher.
18 Further information on the battery and jump starting can be found in Chapter 5 and at the front of this manual.

14 Spark plugs, wires, distributor cap and rotor check and replacement (every 30,000 miles or 30 months, whichever comes first)

Spark plugs

Refer to illustrations 14.2, 14.5a, 14.5b, 14.6 and 14.10.

1 The spark plugs are located on the sides of the engine. **Note:** *Four cylinder models are equipped with two spark plugs per cylinder, for a total of eight spark plugs. The spark plugs on the intake manifold side of the engine are very difficult to access, requiring removal of intake system components and/or several special tools (long extensions, swivel socket, etc.). Refer to Chapter 4 for additional information on the intake system.* The maximum replacement interval for spark plugs specified by Ford is as follows:

a) 1993 2.3L engine - 30,000 miles
b) 1993 3.0L and 4.0L engines - 60,000 miles
c) 1994 2.3L and 4.0L engines - 60,000 miles
d) 1994 3.0L engine - 100,000 miles*
e) 1995 and later four-cylinder engines - 60,000 miles
f) 1995 and later 3.0L and 4.0L engines - 100,000 miles*

* While Ford specifies a 100,000-mile interval for the 1994 and later 3.0L engine and 1995 and later 4.0L engine, we recommend a 60,000-mile interval for these engines, since fouling and erosion can often cause misfiring before 100,000 miles.

2 In most cases, the tools necessary for spark plug replacement include a spark plug socket which fits into a ratchet (spark plug sockets are padded inside to prevent damage to the porcelain insulators on the new plugs and to hold the plugs in the socket during

removal and installation), various extensions and a gap gauge to check and adjust the gaps on the new plugs **(see illustration)**. A special plug wire removal tool is available for separating the wire boots from the spark plugs, but it isn't absolutely necessary. A torque wrench should be used to tighten the new plugs.

3   The best approach when replacing the spark plugs is to purchase the new ones in advance, adjust them to the proper gap and replace the plugs one at a time. When buying the new spark plugs, be sure to obtain the correct type for your particular engine. This information can be found in this Chapter's Specifications on the *Vehicle Emission Control Information* label located under the hood and in the vehicle owner's manual. If differences exist between the plug specified on the emissions label and in the owner's manual or in this Chapter's Specifications-cations, assume the emissions label is correct.

4   Allow the engine to cool completely before attempting to remove any of the plugs. While you are waiting for the engine to cool, check the new plugs for defects and adjust the gaps.

5   The gap is checked by inserting the proper thickness gauge between the electrodes at the tip of the plug **(see illustration)**. The gap between the plugs should be the same as the one specified on the *Vehicle Emissions Control Information* label or in this Chapter's Specifications. The gauge wire should just slide between the electrodes with a slight amount of drag. If the gap is incorrect, use the adjuster on the gauge body to bend the curved side electrode slightly until the specified gap is obtained **(see illustration)**. If the side electrode is not exactly over the center electrode, bend it with the adjuster until it is. Check for cracks in the porcelain insulator (if any are found, the plug should not be used).

6   With the engine cool, remove the spark plug wire from one spark plug. Pull only on the boot at the end of the wire - do not pull on the wire. A plug wire removal tool should be used if available **(see illustration)**.

7   If compressed air is available, use it to blow any dirt or foreign material away from the spark plug hole. A common bicycle pump

14.5a Spark plug manufacturers recommend using a wire type gauge when checking the gap - if the wire does not slide between the electrodes with a slight drag, adjustment is required

14.5b To change the gap, bend the *side* electrode only, as indicated by the arrows, and be very careful not to crack or chip the porcelain insulator surrounding the center electrode

will also work. The idea here is to eliminate the possibility of debris falling into the cylinder as the spark plug is removed.

8   Place the spark plug socket over the plug and remove it from the engine by turning in a counterclockwise direction.

9   Compare the spark plug to those shown in the spark plug condition chart on the inside of the back cover to get an indication of the general running condition of the engine.

10   Thread one of the new plugs into the hole until you can no longer turn it with your fingers, then tighten it with a torque wrench (if available) or the ratchet. It's a good idea to slip a short length of rubber hose over the end of the plug to use as a tool to thread it into place, particularly if the cylinder head is made of aluminum **(see illustration)**. The hose will grip the plug well enough to turn it, but will start to slip if the plug begins to cross-thread in the hole - this will prevent damaged threads and the accompanying repair costs.

11   Before pushing the spark plug wire onto the end of the plug, inspect it following the procedures outlined below.

12   Attach the plug wire to the new spark plug, again using a twisting motion on the boot until it is seated on the spark plug.

13   Repeat the procedure for the remaining spark plugs, replacing them one at a time to prevent mixing up the spark plug wires.

Spark plug wires

Refer to illustrations 14.16a, 14.16b, 14.16c and 14.21

Note: *Every time a spark plug wire is detached from a spark plug or the coil, silicone dielectric compound (a white grease available at auto parts stores) should be applied to the inside of each boot before reconnection. Use a small standard screwdriver to coat the entire inside surface of each boot with a thin layer of the compound.*

14   The spark plug wires should be checked and, if necessary, replaced at the same time new spark plugs are installed.

15   The easiest way to identify bad wires is to make a visual check while the engine is running. In a dark, well-ventilated garage, start the engine and look at each plug wire. Be careful not to come into contact with any moving engine parts. If there is a break in the wire, you will see arcing or a small spark at the damaged area. If arcing is noticed, make a note to obtain new wires.

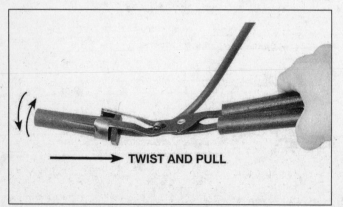

14.6 When removing the spark plug wires, pull only on the boot and twist it back-and-forth

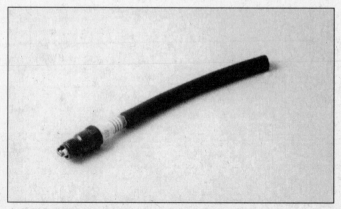

14.10 A length of 3/8-inch ID rubber hose will save time and prevent damaged threads when installing the spark plugs

Chapter 1 Tune-up and routine maintenance

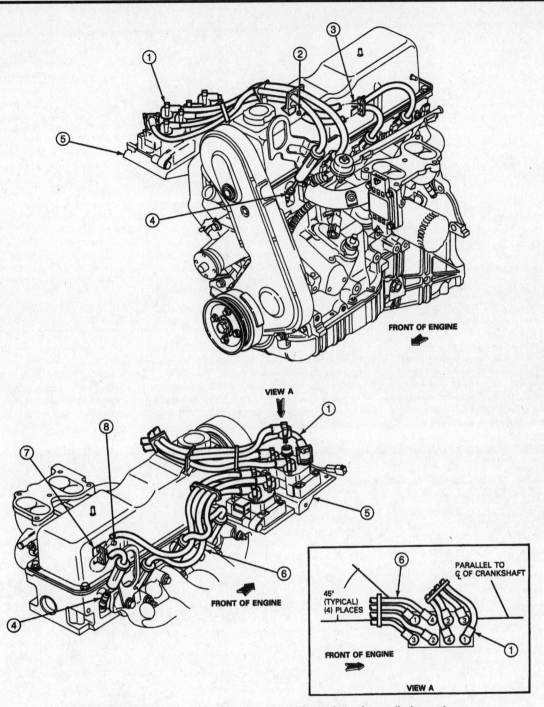

14.16a Coil terminal and spark plug wire connection points - four-cylinder engine

1. Coil wire and bracket assembly
2. T-stud
3. Wire retainer
4. Spark plug
5. Ignition coil and bracket
6. Right ignition wire and bracket assembly (indicated by letter R)
7. Retainer
8. T-stud

16 The spark plug wires should be inspected one at a time, beginning with the spark plug for the number one cylinder, to prevent confusion. Clearly label each original plug wire with a piece of tape marked with the correct number. The plug wires must be reinstalled in the correct order to ensure proper engine operation (see illustrations).

17 Disconnect the spark plug wire from the first spark plug. A removal tool can be used (see illustration 14.6), or you can grab the wire boot, twist it slightly and pull the wire free. Do not pull on the wire itself, only on the rubber boot.

18 Push the wire and boot back onto the end of the spark plug. It should fit snugly. If it doesn't, detach the wire and boot once more and use a pair of pliers to carefully crimp the metal connector inside the wire boot until it does.

19 Using a clean rag, wipe the entire length

Chapter 1 Tune-up and routine maintenance

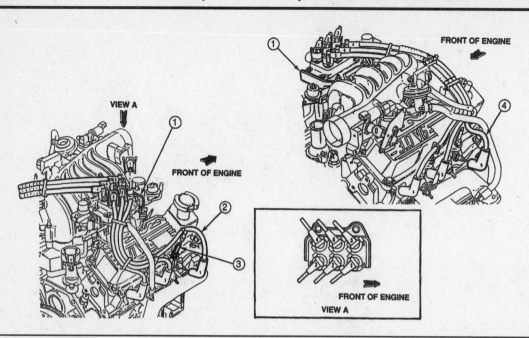

14.16b Coil terminal and spark plug wire connection points - 1995 and later 3.0L V6 engine (for connection points on 1993 and 1994 3.0L engines with distributor ignition, see the diagram in this Chapter's Specifications)

1. Ignition coil
2. Spark plug wire
3. Valve cover bolt or stud
4. Spark plug wire

of the wire to remove built-up dirt and grease.
20   Once the wire is clean, check for burns, cracks and other damage. Do not bend the wire sharply or you might break the conductor.
21   Disconnect the wire from the ignition coil pack or distributor. Squeeze the locking tabs together and pull the end of the wire out of the terminal (see illustration). Check for corrosion and a tight fit. Replace the wire in the coil pack.
22   Inspect each of the remaining spark plug wires, making sure that each one is securely fastened at the coil pack or distributor and spark plug when the check is complete.
23   If new spark plug wires are required, purchase a set for your specific engine model. Pre-cut wire sets with the boots already installed are available. Remove and replace the wires one at a time to avoid mix-ups in the firing order.

14.16c Coil terminal and spark plug wire connection points - 4.0L engine

1. Four-wire clip
2. No. 4 plug wire
3. No. 6 plug wire
4. No. 5 plug wire
5. No. 1 plug wire
6. No. 2 plug wire
7. No. 3 plug wire
8. Clip
9. Wire separator
10. Clip
11. Wire separator

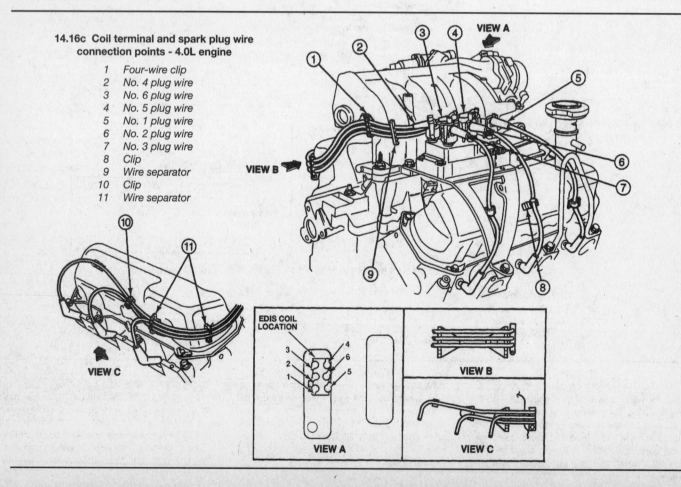

Chapter 1 Tune-up and routine maintenance

1-21

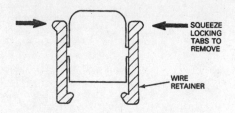

14.21 Squeeze the locking tabs together, twist and pull to detach the spark plug wires from the coil pack

Distributor cap and rotor (1993 and 1994 3.0L engine only)

Refer to illustrations 14.25 and 14.28

24 The distributor cap is located at the rear of the engine. Unscrew the two screws - one on either side of the lower part of the distributor cap. Pull up on the cap, with the wires attached, to separate it from the distributor, then position it to one side.

25 The rotor is now visible on the end of the distributor shaft. Check it carefully for cracks and carbon tracks. Make sure the center terminal spring tension is adequate and look for corrosion and wear on the rotor tip **(see illustration)**. If in doubt about its condition, replace it with a new one.

26 If replacement is required, detach the rotor from the shaft and install a new one. The rotor is press fit on the shaft and can be pried or pulled off.

27 The rotor is indexed to the shaft with a locating boss so it can only be installed one way. It has an internal key that must line up with a slot in the end of the shaft (or vice versa).

28 Check the distributor cap for carbon tracks, cracks and other damage. Closely examine the terminals on the inside of the cap for excessive corrosion and damage **(see illustration)**. Slight deposits are normal. Again, if in doubt about the condition of the cap, replace it with a new one. Be sure to apply a small dab of silicone dielectric grease to each terminal before installing the cap.

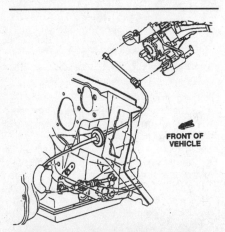

15.3b Automatic transmission shift cable details - 1995 and later models

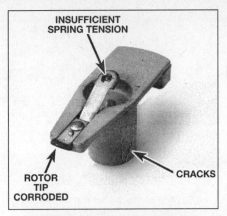

14.25 Check the rotor carefully for cracks, carbon tracks, corrosion and a bent or worn spring

Also, make sure the carbon brush (center terminal) is correctly installed in the cap - a wide gap between the brush and rotor will result in rotor burn-through and/or damage to the distributor cap.

29 To replace the cap, simply separate it from the distributor and transfer the spark plug wires, one at a time, to the new cap. Be very careful not to mix up the wires!

30 Reattach the cap to the distributor, making sure you align the square alignment locator on the cap with the lug on the distributor. Tighten the screws to hold it in place.

15 Automatic transmission shift linkage lubrication (every 6,000 miles or 6 months, whichever comes first)

Refer to illustrations 15.3a, 15.3b, 15.3c and 15.3d

1 Open the hood and locate the shift cable pivot point on the steering column. Also locate the cable end and pivot point on the transmission lever (raise the vehicle and support it securely on jackstands, if necessary).

2 Clean the cable ends and pivot points.

3 Carefully pry the cable ends off the pivot

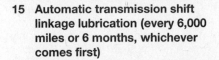

15.3c The lubrication points are accessible between the transmission and the exhaust pipe on the driver's side of the vehicle

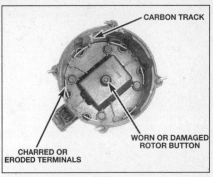

14.28 Cracks or carbon tracks inside the distributor cap will cause the engine to run poorly, so check the cap carefully (1993 and 1994 3.0L engines only)

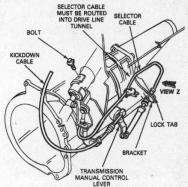

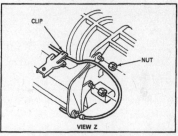

15.3a Automatic transmission shift cable details -1993 and 1994 models

points and lubricate the pivot points with multi-purpose grease **(see illustrations)**. Reconnect the cable(s).

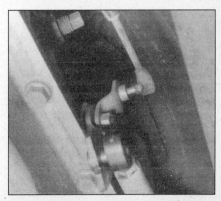

15.3d Carefully pry the kickdown cable end off the ballstud to lubricate the stud

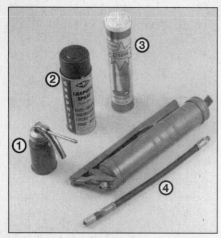

16.1 Materials required for chassis and body lubrication 13p6

1. *Engine oil* - Light engine oil in a can like this can be used for door and hood hinges
2. *Graphite spray* - Used to lubricate lock cylinders
3. *Grease* - Grease, in a variety of types and weights, is available for use in a grease gun. Check the Specifications for your requirements
4. *Grease gun* - A common grease gun, shown here with a detachable hose and nozzle, is needed for chassis lubrication. After use, clean it thoroughly!

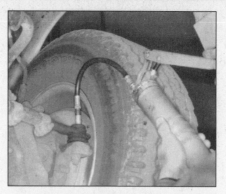

16.6a Lubricate the tie-rod ends (one on each side of the vehicle) . . .

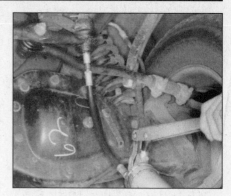

16.6b . . . the Pitman arm connection to the steering linkage . . .

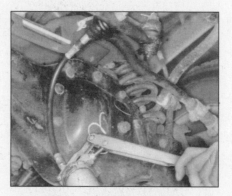

16.6c . . . and the steering linkage cross rod

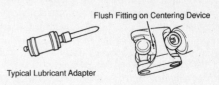

16.7 An adapter is needed for the flush fitting on the U-joint centering device

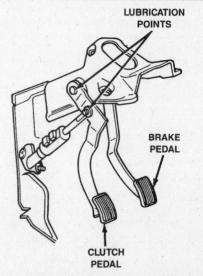

16.13 Clutch linkage lubrication points

16 Chassis lubrication (every 6,000 miles or 6 months, whichever comes first)

Refer to illustrations 16.1, 16.6a, 16.6b, 16.6c, 16.7, 16.13, 16.14a and 16.14b

1 Refer to *Recommended lubricants and fluids* at the front of this Chapter to obtain the necessary grease, etc. You'll also need a grease gun **(see illustration)**. Occasionally plugs will be installed rather than grease fittings. If so, grease fittings will have to be purchased and installed.

2 Look under the vehicle and locate the grease fittings or plugs in the tie-rod ends. If there are plugs, remove them and buy grease fittings, which will thread into the component. A dealer or auto parts store will be able to supply the correct fittings. Straight, as well as angled, fittings are available.

3 For easier access under the vehicle, raise it with a jack and place jackstands under the frame. Make sure the vehicle is safely supported - DO NOT crawl under the vehicle when it is supported only by the jack! If the wheels are to be removed at this interval for tire rotation or brake inspection, loosen the lug nuts slightly while the vehicle is still on the ground.

4 Before beginning, force a little grease out of the nozzle to remove any dirt from the end of the gun. Wipe the nozzle clean with a rag.

5 With the grease gun and plenty of clean rags, crawl under the vehicle.

6 Wipe the tie-rod end grease fitting nipple clean and push the nozzle firmly over it. Squeeze the trigger on the grease gun to force grease into the component **(see illustrations)**. They should be lubricated until the rubber seal is firm to the touch. Don't pump too much grease into the fitting as it could rupture the seal. If grease escapes around the grease gun nozzle, the nipple is clogged or the nozzle is not completely seated on the fitting. Resecure the gun nozzle to the fitting and try again. If necessary, replace the fitting with a new one.

7 Wipe the excess grease from the components and the grease fitting. Repeat the procedure for the remaining fitting(s). **Note:** *Original equipment universal joints don't have grease fittings, but they may have been replaced with U-joints that do have grease fittings. Some models are equipped with a flush grease fitting on the driveshaft centering device* **(see illustration)**. *If this is the case, lubricate the fittings with a grease gun at the interval listed in the Maintenance schedule. You'll need an adapter to lubricate the flush fitting.*

8 Open the hood and smear a little chassis grease on the hood latch mechanism. Have an assistant pull the hood release lever from inside the vehicle as you lubricate the cable at the latch.

9 Lubricate all the hinges (door, hood, etc.) with engine oil to keep them in proper working order.

10 The key lock cylinders can be lubricated with spray graphite or silicone lubricant, which is available at auto parts stores.

11 Lubricate the door weatherstripping with silicone spray. This will reduce chafing and retard wear.

12 Lubricate the parking brake linkage. Note that two different types of grease are required. Use multi-purpose grease on the linkage, adjuster assembly and connectors; use speedometer cable lubricant on parts of the cable that touch other parts of the vehicle. Lubricate the cable twice, once with the

16.14a Remove the throttle linkage protective cove (2.3L engine shown)

16.14b After the cover is removed, carefully pry the linkage from the stud, grease the stud and reconnect the linkage

17.1 Manual transmission fill (upper) and drain (lower) plugs

parking brake set and once with it released.
13   On manual transmission equipped models, lubricate the clutch linkage pivot points **(see illustration)**.
14   Remove the protective cover from the throttle lever ballstud **(see illustration)**. Carefully unsnap the throttle linkage from the ballstud **(see illustration)**. Lubricate the ballstud with multi-purpose grease, then reconnect the linkage and install the protective cover.

17 Manual transmission lubricant level check and change (every 12,000 miles or 12 months, whichever comes first)

Refer to illustration 17.1
Note: *The transmission lubricant level and quality should not deteriorate under normal driving conditions. However, it's recommended that you check the level occasionally. The most convenient time would be when the vehicle is raised for another reason, such as an engine oil change.*

Level check

1   The transmission has a check/fill plug which must be removed to check the lubricant level **(see illustration)**. If the vehicle is raised to gain access to the plug, be sure to support it safely on jackstands - DO NOT crawl under a vehicle which is supported only by a jack!
2   Remove the plug from the transmission and use your little finger to reach inside the housing and feel the lubricant level. It should be at or very near the bottom of the plug hole.
3   If it isn't, add the recommended lubricant through the plug hole with a syringe or squeeze bottle.
4   Install and tighten the plug securely and check for leaks after the first few miles of driving.

Lubricant change

5   Manual transmission lubricant does not normally need changing during the life of the vehicle, but if you wish to do so, place a drain pan beneath the drain plug **(see illustration 17.1)**. Remove the filler plug, then remove the drain plug and let the lubricant drain into a pan. Let the lubricant drain for 10 minutes or more, then reinstall the drain plug and tighten securely.
6   Fill the transmission to the bottom of the filler plug hole with the recommended lubricant.
7   Install and tighten the filler plug securely and check for leaks after the first few miles of driving.

18 Differential lubricant level check and change (every 12,000 miles or 12 months, whichever comes first)

Refer to illustrations 18.2a, 18.2b, 18.3, 18.7, 18.8, 18.11a, 18.11b and 18.12
Note: *This procedure applies to both the front and rear differential. Ford recommends that differential lubricant be changed only at 100,000-mile intervals - and not checked between intervals - unless a leak is suspected or the differential has been submerged in water. Before starting this procedure, check for oil leaks at the differential pinion seal and cover gasket.*

Level check

1   The differential has a check/fill plug which must be removed to check the lubri-

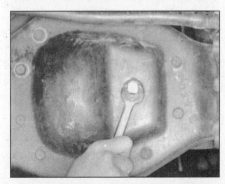

18.2b The check/fill plug for the front differential on 4WD models is located in the axle housing - use an open-end wrench to loosen it

18.2a The check/fill plug for the rear differential is located in the differential housing, facing the front of the vehicle

cant level. If the vehicle is raised to gain access to the plug, be sure to support it safely on jackstands - DO NOT crawl under the vehicle when it's supported only by the jack!
2   Remove the lubricant check/fill plug from the differential **(see illustrations)**. Use a 3/8-inch drive ratchet and short extension without a socket to unscrew the plug from the rear differential; use an open end wrench to unscrew the plug from the front differential.
3   Use your little finger as a dipstick to make sure the lubricant level is even with the bottom of the plug hole **(see illustration)**. If not, use a syringe to add the recommended

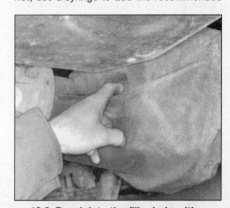

18.3 Reach into the filler hole with a finger to check the lubricant level

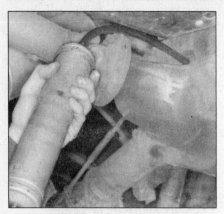

18.7 A suction pump can be used to remove the old lubricant from the differential - it's the only method for differentials that don't have cover or a drain plug

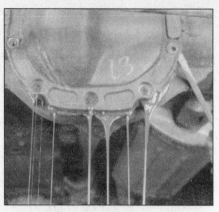

18.8 The rear differential can be drained by removing the cover if you don't have a suction pump

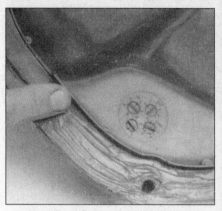

18.11a If you drain the differential by removing the cover, apply a thin film of RTV sealant to the differential cover just before installation . . .

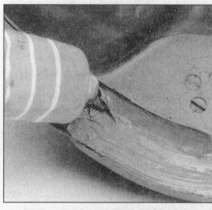

18.11b . . . then apply a thick bead all around the inside edge of the cover

lubricant until it just starts to run out of the opening. On some models a tag is located in the area of the plug which gives information regarding lubricant type, particularly on models equipped with a limited slip differential.

4   Install the plug and tighten it securely.

Lubricant change

5   If it is necessary to change the differential lubricant, remove the check/fill plug **(see illustration 18.2a or 18.2b)**, then drain the differential. Some differentials can be drained by removing the drain plug, while on some rear differentials it's necessary to remove the cover plate on the differential housing. As an alternative, a hand suction pump can be used to remove the differential lubricant through the filler hole. This is the only way to drain front axles which don't have a drain plug. If you remove the cover plate, obtain a tube of silicone sealant to be used when reinstalling the differential cover.

6   If equipped with a drain plug, remove the plug and allow the differential lubricant to drain completely. After the lubricant has drained, install the plug and tighten it securely.

7   If a suction pump is being used, insert the flexible hose **(see illustration)**. Work the hose down to the bottom of the differential housing and pump the lubricant out.

8   If the differential is being drained by removing the cover plate, remove all of the bolts except the two near the top. Loosen the remaining two bolts and use them to keep the cover loosely attached. Allow the lubricant to drain into the pan, then completely remove the cover **(see illustration)**.

9   Using a lint-free rag, clean the inside of the cover and the accessible areas of the differential housing. As this is done, check for chipped gears and metal particles in the lubricant, indicating that the differential should be more thoroughly inspected and/or repaired.

10   Clean all old gasket material from the cover and differential housing.

11   Apply a thin film of RTV sealant to the cover mating surface, then run a thick bead all the way around inside the cover bolt holes **(see illustrations)**.

12   Place the cover on the differential housing and install the bolts. Tighten the bolts securely in a criss-cross pattern **(see illustration)**. Don't overtighten them or the cover may be distorted and leaks may develop.

13   On all models, use a hand pump, syringe or funnel to fill the differential housing with the specified lubricant until it's level with the bottom of the plug hole.

14   Install the check/fill plug and tighten it securely.

19 Transfer case lubricant level check and change (every 12,000 miles or 12 months, whichever comes first)

Refer to illustrations 19.2, 19.6 and 19.7
Note: *This procedure applies to 4WD models only.*

Level check

1   The transfer case has a check/fill plug which must be removed to check the lubricant level. If the vehicle is raised to gain access to the plug, be sure to support it safely on jackstands - DO NOT crawl under a vehicle which is supported only by a jack!

2   Remove the transfer case damper (if equipped) to gain access to the fill plug **(see illustration)**.

3   Remove the plug from the transfer case and use your little finger to reach inside the

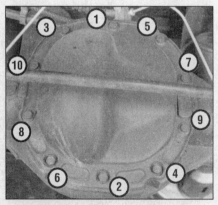

18.12 Tighten the cover bolts in a criss-cross pattern - don't overtighten the bolts, or the cover may be distorted, causing it to leak

19.2 If the vehicle has a transfer case damper, unbolt it to gain access to the fill and drain plugs

Chapter 1 Tune-up and routine maintenance

19.6 If the vehicle's skid plate will obstruct removal of the drain plug, unbolt it from the frame rail to gain access to the plug

housing and feel the lubricant level. It should be at or very near the bottom of the plug hole.
4   If it isn't, add the recommended lubricant through the plug hole with a syringe or squeeze bottle.
5   Install and tighten the plug securely and check for leaks after the first few miles of driving.

Lubricant change

6   To change the lubricant, first note whether the transfer case skid plate will be in the way when the lubricant is drained. If it will be, remove it **(see illustration)**.
7   Remove the fill plug first. This will speed draining. Remove the drain plug **(see illustration)** and let the lubricant drain.
8   Clean the drain plug threads, then rein-

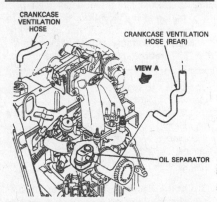

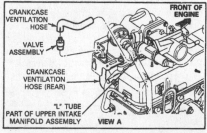

20.1a The PCV valve on four-cylinder engines is mounted in the oil separator on the left side of the engine

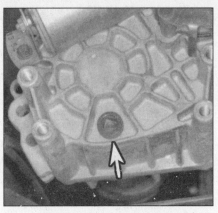

19.7 Location of the transfer case drain plug

stall the plug after the lubricant has finished draining.
9   Fill the transaxle to the bottom of the filler plug threads with the recommended lubricant listed in this Chapter's Specifications.
10 Install the filler plug and tighten it securely.

20 Positive Crankcase Ventilation (PCV) valve check and replacement (every 12,000 miles or 12 months, whichever comes first)

Refer to illustrations 20.1a, 20.1b and 20.2
Note: *To maintain the efficiency of the PCV system, clean the hoses and check the PCV valve at the intervals recommended in the maintenance schedule. For additional information on the PCV system, refer to Chapter 6.*
1   Locate the PCV valve **(see illustrations)**.
2   To check the valve, first pull it out of the grommet in the oil separator (2.3L) or valve cover (3.0L and 4.0L). Shake the valve **(see illustration)**. It should rattle, indicating that it

20.1b On 3.0L and 4.0L engines, the PCV valve is mounted in the valve cover near the brake booster (arrow)

is not clogged with deposits. If the valve does not rattle, replace it with a new one. If it does rattle, reinstall it.
3   Start the engine and allow it to idle, then disconnect the PCV hose. If vacuum is felt, the PCV valve system is working properly (see Chapter 6 for additional PCV system information).
4   If no vacuum is felt, the oil filler cap, hoses or valve cover gasket may be leaking or the PCV valve may be bad. Check for vacuum leaks at the valve, filler cap and all hoses.
5   Pull straight up on the valve to remove it. Check the rubber grommet for cracks and distortion. If it's damaged, replace it.
6   If the valve is clogged, the hose is also probably plugged. Remove the hose and clean with solvent.
7   After cleaning the hose, inspect it for damage, wear and deterioration. Make sure it fits snugly on the fittings.
8   If necessary, install a new PCV valve.
Note: *The elbow (if equipped) is not part of the PCV valve. A new valve will not include the elbow. The original must be transferred to the new valve. If a new elbow is purchased, it may be necessary to soak it in warm water for up to an hour to slip it onto the new valve. Do not attempt to force the elbow onto the valve or it will break.*
9   Install the clean PCV system hose. Make sure that the PCV valve and hose are secure.

21 Fuel system check (every 12,000 miles or 12 months, whichever comes first)

Warning: *Gasoline is extremely flammable, so take extra precautions when you work on any part of the fuel system. Don't smoke or allow open flames or bare light bulbs near the work area, and don't work in a garage where a natural gas-type appliance (such as a water heater or a clothes dryer) with a pilot light is present. Since gasoline is carcinogenic, wear latex gloves when there's a possibility of*

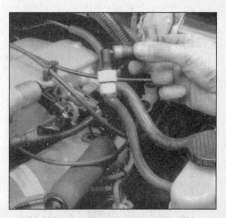

20.2 Shake the valve to test it - if it doesn't rattle, the valve is stuck and should be replaced

22.3 The radiator cap seals and the sealing surfaces in the radiator filler neck should be checked for built-up corrosion - the radiator cap should be replaced if the seals are brittle or deteriorated

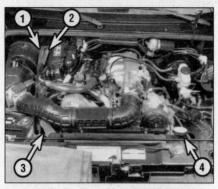

22.4a Radiator and heater hose locations (4.0L engine models) - the lower radiator hose, not shown, is connected to the bottom of the radiator

1 Heater return
2 Heater supply
3 Upper radiator hose
4 Coolant recovery hose

Check for a chafed area that could fail prematurely.

Check for a soft area indicating the hose has deteriorated inside.

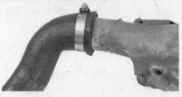

Overtightening the clamp on a hardened hose will damage the hose and cause a leak.

Check each hose for swelling and oil-soaked ends. Cracks and breaks can be located by squeezing the hose.

being exposed to fuel, and, if you spill any fuel on your skin, rinse it off immediately with soap and water. Mop up any spills immediately and do not store fuel-soaked rags where they could ignite. When you perform any kind of work on the fuel system, wear safety glasses and have a Class B type fire extinguisher on hand.

1 If you smell gasoline while driving or after the vehicle has been sitting in the sun, inspect the fuel system immediately.
2 Remove the fuel filler cap and inspect it for damage and corrosion. The gasket should have an unbroken sealing imprint. If the gasket is damaged or corroded, install a new cap.
3 Inspect the fuel feed and return lines for cracks. Make sure that the connections between the fuel lines and the fuel injection system and between the fuel lines and the in-line fuel filter are tight. **Warning:** *The fuel system pressure must be relieved before servicing fuel system components. The fuel system pressure relief procedure is outlined in Chapter 4.*
4 Since some components of the fuel system - the fuel tank and some of the fuel feed and return lines, for example - are underneath the vehicle, they can be inspected more easily with the vehicle raised on a hoist. If that's not possible, raise the vehicle and support it on jackstands.
5 With the vehicle raised and safely supported, inspect the gas tank and filler neck for punctures, cracks or other damage. The connection between the filler neck and the tank is particularly critical. Sometimes a rubber filler neck will leak because of loose clamps or deteriorated rubber. Inspect all fuel tank mounting brackets and straps to be sure the tank is securely attached to the vehicle. **Warning:** *Do not, under any circumstances, try to repair a fuel tank (except rubber components). A welding torch or any open flame can easily cause fuel vapors inside the tank to explode.*
6 Carefully check all rubber hoses and metal or nylon lines leading away from the fuel tank. Check for loose connections, deteriorated hoses, crimped lines and other dam-

age. Repair or replace damaged sections as necessary (see Chapter 4).

22 Cooling system check (every 12,000 miles or 12 months, whichever comes first)

Refer to illustrations 22.3, 22.4a and 22.4b

1 Many major engine failures can be attributed to a faulty cooling system. If the vehicle is equipped with an automatic transmission, the cooling system also plays an important role in prolonging transmission life because it cools the fluid.
2 The engine should be cold for the cooling system check, so perform the following procedure before the vehicle is driven for the day or after it has been shut off for at least three hours.
3 Remove the radiator cap **(see illustration)** and clean it thoroughly, inside and out, with clean water. Also clean the filler neck on the radiator. The presence of rust or corrosion in the filler neck means the coolant should be changed (see Section 29). The coolant inside the radiator should be relatively clean and transparent. If it's rust colored, drain the system and refill with new coolant.
4 Carefully check the radiator hoses and smaller diameter heater hoses **(see illustrations)**. Inspect each coolant hose along its entire length, replacing any hose which is cracked, swollen or deteriorated. Cracks will show up better if the hose is squeezed. Pay close attention to hose clamps that secure the hoses to cooling system components. Hose clamps can pinch and puncture hoses, resulting in coolant leaks.
5 Make sure all hose connections are tight. A leak in the cooling system will usually show up as white or rust colored deposits on the area adjoining the leak. If wire-type clamps are used on the hoses, it may be a good idea to replace them with screw-type clamps.
6 Clean the front of the radiator and air conditioning condenser with compressed air, if available, or a soft brush. Remove all bugs, leaves, etc. embedded in the carburetor fins. Be extremely careful not to damage the cooling fins or cut your fingers on them.
7 If the coolant level has been dropping consistently and no leaks are detectable, have the radiator cap and cooling system pressure checked at a service station.

23 Exhaust system check (every 12,000 miles or 12 months, whichever comes first)

Refer to illustrations 23.4a, 23.4b, 23.4c and 23.4d

1 With the engine cold (at least three hours after the vehicle has been driven), check the complete exhaust system from the

Chapter 1 Tune-up and routine maintenance

23.4a A catalytic converter is mounted beside the transmission - on automatic transmission models, the heat shield should be securely mounted in its clips

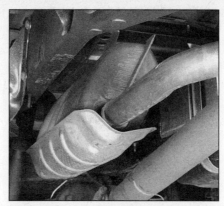

23.4b Also check the heat shield beneath the muffler

23.4c The catalytic converter and muffler should be securely attached to their brackets and the rubber mounts should be in good condition

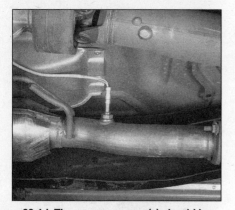

23.4d The oxygen sensor(s) should be securely attached and the insulation on the sensor wires should be in good condition

24.10a To check the suspension balljoints, try to move the lower edge of each front tire in-and-out while watching/feeling for movement at the top of the tire . . .

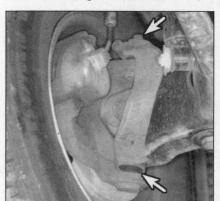

24.10b . . . if there's movement, repeat the test and look for looseness at the balljoints

engine to end of the tailpipe. Ideally, the inspection should be done with the vehicle on a hoist to permit unrestricted access. If a hoist isn't available, raise the vehicle and support it securely on jackstands.
2 Check the exhaust pipes and connections for evidence of leaks, severe corrosion and damage. Make sure that all brackets and hangers are in good condition and are tight.
3 At the same time, inspect the underside of the body for holes, corrosion, open seams, etc. which may allow exhaust gases to enter the passenger compartment. Seal all body openings with silicone or body putty.
4 Rattles and other noises can often be traced to the exhaust system, especially the mounts, hangers and heat shields. Try to move the pipes, muffler and catalytic converter **(see illustrations)**. If the components can come in contact with the body or suspension parts, secure the exhaust system with new mounts.
5 Check the running condition of the engine by inspecting inside the end of the tailpipe. The exhaust deposits here are an indication of engine state-of-tune. If the pipe is black and sooty or coated with white deposits, the engine may need a tune-up, including a thorough fuel system inspection.

24 Steering and suspension check (every 12,000 miles or 12 months, whichever comes first)

Refer to illustrations 24.10a, 24.10b, 24.11a and 24.11b
Note: *The steering linkage and suspension components should be checked periodically. Worn or damaged suspension and steering linkage components can result in excessive and abnormal tire wear, poor ride quality and vehicle handling and reduced fuel economy. For detailed illustrations of the steering and suspension components, refer to Chapter 10.*

Shock absorber check

1 Park the vehicle on level ground, turn the engine off and set the parking brake. Check the tire pressures.
2 Push down at one corner of the vehicle, then release it while noting the movement of the body. It should stop moving and come to rest in a level position with one or two bounces.
3 If the vehicle continues to move up-and-down or if it fails to return to its original position, a worn or weak shock absorber is prob-

ably the reason.
4 Repeat the above check at each of the three remaining corners of the vehicle.
5 Raise the vehicle and support it on jackstands.
6 Check the shock absorbers for evidence of fluid leakage. A light film of fluid is no cause for concern. Make sure that any fluid noted is from the shocks and not from any other source. If leakage is noted, replace the shocks as a set.
7 Check the shock absorbers to be sure that they are securely mounted and undamaged. Check the upper mounts for damage and wear. If damage or wear is noted, replace the shock absorbers as a set.
8 If the shock absorbers must be replaced, refer to Chapter 10 for the procedure.

Steering and suspension check

9 Visually inspect the steering system components for damage and distortion (see Chapter 10). Look for leaks and damaged seals, boots and fittings.
10 Clean the lower end of the steering knuckle. Have an assistant grasp the lower edge of the tire and move the wheel in-and-out **(see illustration)** while you look for

Chapter 1 Tune-up and routine maintenance

24.11a To check the steering gear and idler arm mounts and tie-rod connections for play, grasp each front tire like this and try to move it back-and-forth . . .

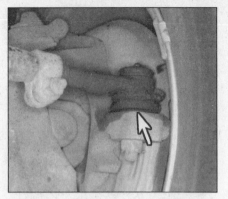

24.11b . . . if play is noted, check the steering gear mounts and make sure that they're tight; look for looseness at the tie-rod ends (shown) and the steering linkage connections

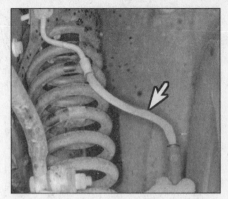

25.7a The brake hoses at the front of the vehicle (arrow) should be inspected and replaced if they show any defects

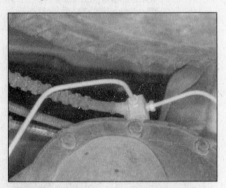

25.7b The rear brake hose meets the metal brake lines at a junction block on the axle housing - they should also be inspected and replaced if they show any defects

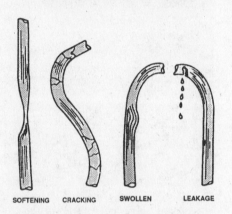

25.7c Typical brake hose defects

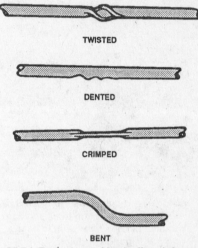

25.7d Typical metal brake line defects

movement at the steering knuckle-to-axle arm balljoints **(see illustration)**. If there is any movement, the balljoint(s) must be replaced.

11 Grasp each front tire at the front and rear edges, push in at the front, pull out at the rear and feel for play in the steering linkage **(see illustrations)**. If any freeplay is noted, check the steering gear mounts and the tie-rod balljoints for looseness. If the steering gear mounts are loose, tighten them. If the tie-rods are loose, the balljoints may be worn (check to make sure the nuts are tight). Additional steering and suspension system illustrations can be found in Chapter 10.

Front wheel bearing check

12 Refer to Section 26 (2WD) or Section 27 (4WD) for the wheel bearing check, repack and adjustment procedure.

25 Brake system check (every 12,000 miles or 12 months, whichever comes first)

Refer to illustrations 25.7a, 25.7b, 25.7c, 25.7d, 25.11 and 25.17

Warning: *Dust produced by lining wear and deposited on brake components may contain asbestos, which is hazardous to your health.*

DO NOT blow it out with compressed air and DO NOT inhale it! DO NOT use gasoline or solvents to remove the dust. Brake system cleaner should be used to flush the dust into a drain pan. After the brake components are wiped with a damp rag, dispose of the contaminated rag(s) and brake cleaner in a covered and labeled container. Try to use non-asbestos replacement parts whenever possible.

Note: *In addition to the specified intervals, the brake system should inspected each time the wheels are removed or a malfunction is indicated. Because of the obvious safety considerations, the following brake system checks are some of the most important maintenance procedures you can perform on your vehicle.*

Symptoms of brake system problems

1 The disc brakes have built-in wear indicators which should make a high-pitched squealing or scraping noise when they're worn to the replacement point. When you hear this noise, replace the pads immediately or expensive damage to the brake discs could result.

2 Any of the following symptoms could indicate a potential brake system defect. The vehicle pulls to one side when the brake pedal is depressed, the brakes make squealing or dragging noises when applied, brake travel is excessive, the pedal pulsates and brake fluid leaks are noted (usually on the inner side of the tire or wheel). If any of these conditions are noted, inspect the brake system immediately. If the BRAKE light on the dash is on, a hydraulic system leak is indicated, which could be at the master cylinder, wheel cylinder, caliper, hose or line. Hydraulic leaks must be repaired immediately (see Chapter 9). If the antilock brake system light on the dash is on, verify the brakes are functioning correctly, then carefully drive the vehicle to a dealer service department for repairs.

Brake lines and hoses

Note: *Steel tubing is used throughout the brake system, with the exception of flexible, reinforced hoses at the front wheels and as connectors at the rear axle. Periodic inspection of these lines is very important.*

3 Park the vehicle on level ground and turn the engine off.

4 Remove the wheel covers. Loosen, but do not remove, the lug nuts on all four wheels.

Chapter 1 Tune-up and routine maintenance

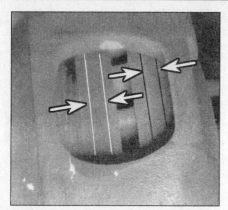

25.11 The lining thickness of the front disc brake pads (arrows) can be checked through the caliper inspection hole

25.17 Carefully peel back the rubber boot on each end of the wheel cylinder - if the exposed area is covered with brake fluid, or if fluid runs out, the wheel cylinder must be overhauled or replaced

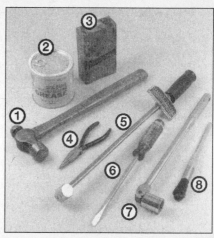

26.1 Tools and materials needed for front wheel bearing maintenance

1. **Hammer** - A common hammer will do just fine
2. **Grease** - High-temperature grease that is formulated specially for front wheel bearings should be used
3. **Wood block** - If you have a scrap piece of 2 x 4, it can be used to drive the new seal into the hub
4. **Needle-nose pliers** - Used to straighten and remove the cotter pin in the spindle
5. **Torque wrench** - This is very important in this procedure; if the bearing is too tight, the wheel won't turn freely - if it's too loose, the wheel will "wobble" on the spindle. Either way, it could mean extensive damage.
6. **Screwdriver** - Used to remove the seal from the hub (a long screwdriver would be preferred)
7. **Socket/breaker bar** - Needed to loosen the nut on the spindle if it's extremely tight
8. **Brush** - Together with some clean solvent, this will be used to remove old grease from the hub and spindle

5 Raise the vehicle and support it securely on jackstands (see *Jacking and towing* at the front of this book, or refer to your owner's manual, if necessary).

6 Remove the wheels.

7 Check all brake lines and hoses for cracks, chafing of the outer cover, leaks, blisters and distortion. Check the brake hoses at front and rear of the vehicle for softening, cracks, bulging, or wear from rubbing on other components **(see illustrations)**. Check all threaded fittings for leaks and make sure the brake hose mounting bolts and clips are secure.

8 If leaks or damage are discovered, they must be fixed immediately. Refer to Chapter 9 for detailed brake system repair procedures.

Disc brakes

9 If it hasn't already been done, raise the vehicle and support it securely on jackstands. Remove the front wheels.

10 The disc brake calipers, which contain the pads, are now visible. Each caliper has an outer and an inner pad - all pads should be checked.

11 Note the pad thickness by looking through the inspection hole in the caliper **(see illustration)**. If the lining material is 1/8-inch thick or less, or if it is tapered from end-to-end, the pads should be replaced (see Chapter 9). Keep in mind that the lining material is riveted or bonded to a metal plate or shoe - the metal portion is not included in this measurement.

12 Check the condition of the brake disc. Look for score marks, deep scratches and overheated areas (they will appear blue or discolored). If damage or wear is noted, the disc can be removed and resurfaced by an automotive machine shop or replaced with a new one. Refer to Chapter 9 for more detailed inspection and repair procedures.

13 Remove the calipers without disconnecting the brake hoses (see Chapter 9). Lubricate the caliper slide rails and the inner pad slots on the steering knuckles with the special caliper slide grease listed in this Chapter's Specifications.

Drum brakes

14 Refer to Chapter 9 and remove the rear brake drums.

15 Note the thickness of the lining material on the rear brake shoes and look for signs of contamination by brake fluid or grease. If the lining material is within 1/16-inch of the recessed rivets or metal shoes, replace the brake shoes with new ones. The shoes should also be replaced if they are cracked, glazed (shiny lining surfaces), or contaminated with brake fluid or grease. See Chapter 9 for the replacement procedure.

16 Check the shoe return and hold-down springs and the adjusting mechanism to make sure they are installed correctly and in good condition. Deteriorated or distorted springs, if not replaced, could allow the linings to drag and wear prematurely.

17 Check the wheel cylinders for leakage by carefully peeling back the rubber boots **(see illustration)**. Slight moisture behind the boots is acceptable. If brake fluid is noted behind the boots or if it runs out of the wheel cylinder, the wheel cylinders must be overhauled or replaced (see Chapter 9).

18 Check the drums for cracks, score marks, deep scratches and hard spots, which will appear as small discolored areas. If imperfections cannot be removed with emery cloth, the drums must be resurfaced by an automotive machine shop (see Chapter 9 for more detailed information).

19 Refer to Chapter 9 and install the brake drums.

20 Install the wheels, but don't lower the vehicle yet.

Parking brake

Note: *The parking brake cable and linkage should be periodically lubricated (see Section 16). This maintenance procedure helps prevent the parking brake cable adjuster or the linkage from binding and adversely affecting the operation or adjustment of the parking brake.*

21 The easiest, and perhaps most obvious, method of checking the parking brake is to park the vehicle on a steep hill with the parking brake set and the transmission in Neutral. If the parking brake doesn't prevent the vehicle from rolling, refer to Chapter 9 and adjust it.

26 Front wheel bearing check, repack and adjustment (2WD models) (every 12,000 miles or 12 months, whichever comes first)

Refer to illustrations 26.1, 26.7, 26.9, 26.15, 26.16 and 26.23

1 In most cases the front wheel bearings will not need servicing until the brake pads are changed. However, the bearings should be checked whenever the front of the vehicle is raised for any reason. Several items, including a torque wrench and special

1-30   Chapter 1   Tune-up and routine maintenance

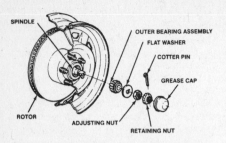

26.7 An exploded view of the front wheel bearing components (2WD models)

26.9 Pull out on the hub to dislodge the outer bearing

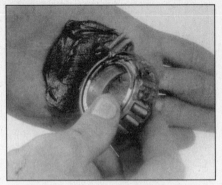

26.15 Pack each wheel bearing by working the grease into the rollers from the back side

26.16 Apply a thin coat of grease to the spindle, particularly where the seal rides

grease, are required for this procedure **(see illustration)**.
2   With the vehicle securely supported on jackstands, spin each wheel and check for noise, rolling resistance and freeplay.
3   Move the wheel in-and-out on the spindle **(see illustration 24.11a)**. If there's any noticeable movement, the bearings should be checked and then repacked with grease or replaced if necessary.
4   Remove the wheel.
5   Remove the brake caliper (see Chapter 9) and hang it out of the way on a piece of wire. **Warning:** *DO NOT allow the brake caliper to hang by the rubber hose!*
6   Pry the grease cap out of the hub with a screwdriver or hammer and chisel.
7   Straighten the bent ends of the cotter pin, then pull the cotter pin out of the retaining nut and spindle **(see illustration)**. Discard the cotter pin and use a new one during reassembly.
8   Remove the retainer, the adjusting nut and flat washer from the end of the spindle.
9   Pull the hub assembly out slightly, then push it back into its original position. This should force the outer bearing off the spindle enough so it can be removed **(see illustration)**.
10   Pull the hub off the spindle.
11   Use a screwdriver to pry the grease seal out of the rear of the hub. As this is done, note how the seal is installed.
12   Remove the inner wheel bearing from the hub.
13   Use solvent to remove all traces of old grease from the bearings, hub and spindle. A small brush may prove helpful; however make sure no bristles from the brush embed themselves inside the bearing rollers. Allow the parts to air dry.
14   Carefully inspect the bearings for cracks, heat discoloration, worn rollers, etc. Check the bearing races inside the hub for wear and damage. If the bearing races are defective, the hubs should be taken to a machine shop with the facilities to remove the old races and press new ones in. Note that the bearings and races come as matched sets and new bearings should never be installed on old races.
15   Use high-temperature front wheel bearing grease to pack the bearings. Work the grease completely into the bearings, forcing it between the rollers, cone and cage from the back side **(see illustration)**.
16   Apply a thin coat of grease to the spindle at the outer bearing seat, inner bearing seat, shoulder and seal seat **(see illustration)**.
17   Put a small quantity of grease inboard of each bearing race inside the hub. Using your finger, form a dam at these points to provide extra grease availability and to keep thinned grease from flowing out of the bearing.
18   Place the grease-packed inner bearing into the rear of the hub and put a little more grease outward of the bearing.
19   Place a new seal over the inner bearing and tap the seal evenly into place with a hammer and block of wood until it's flush with the hub.
20   Carefully place the hub assembly onto the spindle and push the grease-packed outer bearing into position.
21   Install the flat washer and adjusting nut. Tighten the nut only slightly.
22   Spin the hub in a forward direction to seat the bearings and remove any grease or burrs which could cause excessive bearing play later.
23   While spinning the wheel, tighten the adjusting nut to the specified torque (Step 1 in this Chapter's Specifications) **(see illustration)**.
24   Loosen the nut 1/2-turn, no more.
25   Tighten the nut to the specified torque (Step 3 in this Chapter's Specifications). Install a new cotter pin through the hole in the spindle and retainer nut. If the holes don't line up, don't turn the nut. Instead, remove the retainer and try it in a different position. The notches in the retainer are offset for this purpose. Keep trying the retainer in different positions until the holes line up.
26   Bend the ends of the cotter pin until they're flat against the nut. Cut off any extra length which could interfere with the grease cap.
27   Install the grease cap, tapping it into place with a hammer.
28   Install the caliper (see Chapter 9).
29   Install the wheel and lug nuts. Tighten the lug nuts to the torque listed in this Chapter's Specifications.
30   Check the bearings in the manner described earlier in this Section.
31   Lower the vehicle.

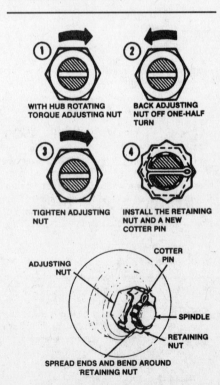

26.23 Wheel bearing adjustment procedure (2WD models)

Chapter 1 Tune-up and routine maintenance

27.1 The tool on the left is used for automatic locking hubs; the tool on the right is used for manual locking hubs

27.7 Make alignment marks on the locking hub and wheel hub, then take the locking hub off the wheel studs

27 Front hub lock, spindle bearing and wheel bearing maintenance (4WD models) (every 12,000 miles or 12 months, whichever comes first)

Removal (manual and automatic hubs)

Refer to illustrations 27.1 and 27.7

1 In most cases the front wheel bearings will not need servicing until the brake pads are changed. However, the bearings should be checked whenever the front of the vehicle is raised for any reason. Several items, including a torque wrench and special grease, are required for this procedure **(see illustration 26.1)**. In addition to these tools you'll need a four-pronged spindle nut spanner wrench for manual locking hubs or a 2-3/8 inch hex locknut wrench for automatic locking hubs **(see illustration)**. These may be available at four-wheel drive shops or auto parts stores. You'll also need a pair of snap-ring pliers.

2 With the vehicle securely supported on jackstands, spin each wheel and check for noise, rolling resistance and freeplay.

3 Move the wheel in-and-out on the spindle **(see illustration 24.11a)**. If there's any noticeable movement, the bearings should be checked and then repacked with grease or replaced if necessary.

4 The lubrication of hub locks, as well as spindle needle and thrust bearings on vehicles so equipped, should be checked at the intervals specified in the maintenance schedule.

5 Remove the wheel.

6 Remove the brake caliper (see Chapter 9) and hang it out of the way on a piece of wire. **Warning:** *DO NOT allow the brake caliper to hang by the hose!*

7 Remove the retaining washers (if equipped) from the wheel studs. Make alignment marks on the locking hub and wheel hub, then take the locking hub off **(see illustration)**.

Manual locking hubs

Refer to illustrations 27.8, 27.9, 27.10, 27.16 and 27.21

8 Using snap-ring pliers, carefully expand the snap-ring just enough to remove it from the end of the spindle shaft **(see illustration)**.

9 Remove the axleshaft spacer(s) **(see illustration)**.

10 Remove the outer wheel bearing locknut with a four-prong spanner wrench, available at

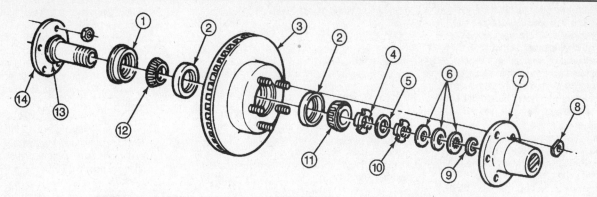

27.8 An exploded view of the manual locking hub and wheel bearings (4WD models)

1 Grease seal	4 Inner locknut	7 Hub	10 Outer locknut	13 Dust seal
2 Wheel bearing cup	5 Washer	8 Retainer washer	11 Outer wheel bearing	14 Wheel spindle
3 Brake disc	6 Axleshaft spacer(s)	9 Snap-ring	12 Inner wheel bearing	

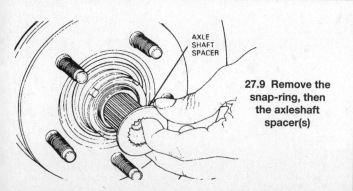

27.9 Remove the snap-ring, then the axleshaft spacer(s)

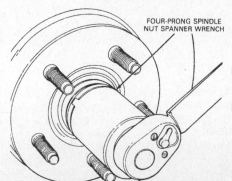

27.10 A four-pronged socket is required to remove the outer locknut; the nut is very tight, so don't use makeshift tools

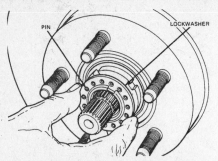

27.16 Align the pin in the locknut with one of the holes in the lockwasher; if necessary, adjust the locknut position slightly to align the pin with a hole

27.21 To remove the internal components from a manual locking hub, insert a small screwdriver behind the retaining ring and work it gently out of its groove - DO NOT remove the screw from the plastic dial

most auto parts stores **(see illustration)**. *Note: This nut is very tight. Don't try to remove it with a makeshift tool.*

11 Remove the inner wheel bearing locknut with the spanner wrench. Be sure the notch in the wrench is positioned over the locknut pin.

12 Perform Steps 9 through 20 of Section 26 to repack the bearings. Be sure to use the type of grease listed in this Chapter's Specifications.

13 Install the inner locknut on the spindle and tighten it to the specified torque (Step 1 in this Chapter's Specifications).

14 Spin the brake disc several turns in each direction to seat the bearings.

15 Loosen the inner locknut 1/4-turn, then retighten it to the specified torque (Step 3 in this Chapter's Specifications).

16 Install the lockwasher and align the locknut pin with one of the holes in the lockwasher **(see illustration)**. If necessary, turn the inner locknut slightly to align the hole and pin.

17 Install the outer locknut and tighten it to the specified torque (Step 4 in this Chapter's Specifications).

18 Lubricate the needle bearing spacer and needle bearing (if equipped) with the same grease used for the wheel bearings. Install them on the spindle.

19 Install the axleshaft spacer.
20 Install the snap-ring on the spindle.
21 Remove the lock ring that secures the inner components in the locking hub **(see illustration)**. **Caution:** *Don't remove the screw from the plastic dial.*
22 Remove the internal assembly, spring and clutch gear. Lubricate the components with the specified grease.
23 Reassemble the hub and install the lock ring.
24 Install the locking hub on the wheel studs and secure it with the retainer washers. Proceed to Step 35.

Automatic locking hubs

Refer to illustrations 27.25a, 27.25b, 27.26, 27.27a, 27.27b, 27.27c, 27.28, 27.30, 27.33a, 27.33b and 27.34

25 Using snap-ring pliers, carefully expand the snap-ring just enough to remove it from the end of the spindle shaft **(see illustrations)**. **Note:** *If you don't have snap-ring pliers, carefully pry the snap-ring off with a screwdriver. Hold a finger against the snap-ring as shown in the illustration so the snap-ring doesn't fly off.*

26 Remove the axleshaft spacer(s) **(see illustration)**.
27 Carefully pull the plastic cam assembly from the wheel bearing adjusting nut **(see illustration)**. **Caution:** *Don't pry the cam off or you may damage it. If it's hard to remove, try turning it as you pull. Pull off the two plastic thrust spacers and remove the locking key with a magnet* **(see illustrations)**. **Note:** *The thrust spacers are thin and flexible, with a tendency to cock sideways and jam on the spindle. Hook your fingernails behind the spacers and pull evenly at two or more points. If necessary, rotate the adjusting nut slightly to relieve pressure on the locking key.*
28 Remove the wheel bearing adjusting nut from the spindle. It may be loose enough to turn with fingers. If not, use a 2-3/8 inch hex socket **(see illustration)**. **Caution:** *Be sure to remove the locking key before you remove the nut or the spindle threads will be damaged.* **Note:** *The socket may be 3/4-inch drive. If so, an adapter can be used so the socket can be turned with a 1/2-inch drive tool.*
29 Perform Steps 9 through 20 of Section 26 to repack the wheel bearings. Be sure to use the correct grease (listed in this Chapter's Specifications).
30 Install the wheel bearing adjusting nut **(see illustration)**. While spinning the brake disc, tighten the nut to the specified torque (Step 1 in this Chapter's Specifications).
31 Loosen the nut 1/4-turn.
32 Retighten the nut to the final torque (Step 3 in this Chapter's Specifications).
33 Line up the center of the spindle keyway slot with the closest slot in the wheel bearing adjusting nut **(see illustration)**. If necessary, tighten the adjusting nut so the next slot aligns with the keyway slot. Be sure the slots line up exactly.

a) *Install the locking key in the keyway slot, under the adjusting nut. Don't force the key in or it will be damaged. If it is difficult to insert, make sure the slots are lined up exactly.*

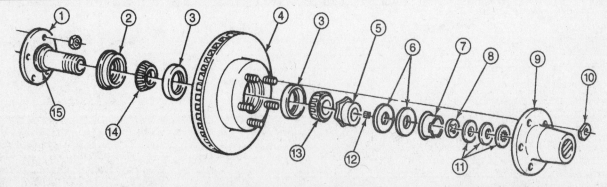

27.25a An exploded view of the automatic locking hub and wheel bearings (4WD models)

1 Wheel spindle	5 Adjusting nut	9 Hub	12 Locking key
2 Grease seal	6 Thrust spacers	10 Retainer washer	13 Outer wheel bearing
3 Wheel bearing cup	7 Retainer	11 Axleshaft spacers (later models)	14 Inner wheel bearing
4 Brake disc	8 Axleshaft spacer		15 Dust seal

Chapter 1 Tune-up and routine maintenance

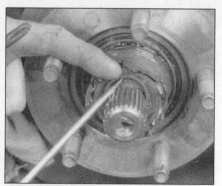

27.25b The snap-ring can be lifted off with a screwdriver if snap-ring pliers aren't available - hold the center of the snap-ring with a finger as shown so it doesn't fly off

27.26 Remove the axleshaft spacer

27.27a Remove the cam from the spindle

27.27b Remove the two plastic thrust spacers

27.27c Remove the locking key (arrow) with a magnet - don't try to unscrew the nut before removing the key

27.28 On automatic locking hubs, loosen the wheel bearing adjusting nut with a 2-3/8 inch hex locknut wrench

b) Install the two thrust spacers (see illustration 27.27b).
c) Line up the key in the fixed cam with the keyway slot in the spindle, then push the cam on over the adjusting nut (see illustration).
d) Install the axleshaft spacer.
e) Install the snap-ring on the end of the spindle.

34 Line up the three legs on the automatic locking hub with the pockets in the cam (see illustration), then install the locking hub and secure it with the retainer washers.

All models

35 Install the wheel and lug nuts. Lower the vehicle and tighten the lug nuts to the torque listed in this Chapter's Specifications.
36 Check the endplay of the wheel on the spindle and measure the amount of torque required to turn the hub. Compare your findings with this Chapter's Specifications. If the measurements are incorrect, readjust the wheel bearings.

27.30 Install the wheel bearing adjusting nut

27.33a The adjusting nut has internal slots - line up one of them with the keyway slot

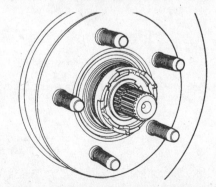

27.33b Position the cam over the wheel bearing adjusting nut - use extreme care to align the key accurately with the slot in the spindle

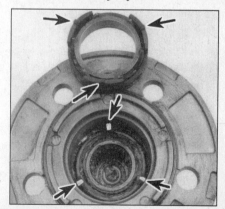

27.34 Align the legs in the automatic locking hub with the pockets on the cam when installing the hub

28 Windshield wiper blade check and replacement (every 12,000 miles or 12 months, whichever comes first)

1 Road film can build up on the wiper

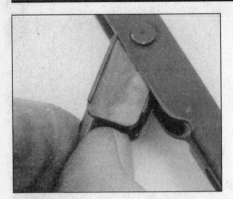

28.5a Squeeze the retaining lever . . .

28.5b . . . push the wiper blade down the arm to disengage the hook and separate the blade from the arm

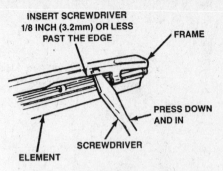

28.7 To remove a wiper element, pry it out of the retaining claw at either end of the blade

blades and affect their efficiency, so they should be washed regularly with a mild detergent solution.

Check

2 The windshield wiper and blade assembly should be inspected periodically. Even if you don't use your wipers, the sun and elements will dry out the rubber portions, causing them to crack and break apart. If inspection reveals hardened or cracked rubber, replace the wiper blades. If inspection reveals nothing unusual, wet the windshield, turn the wipers on, allow them to cycle several times, then shut them off. An uneven wiper pattern across the glass or streaks over clean glass indicate that the blades should be replaced.

3 The operation of the wiper mechanism can loosen the fasteners, so they should be checked and tightened, as necessary, at the same time the wiper blades are checked (see Chapter 12 for further information regarding the wiper mechanism).

Wiper blade replacement

Refer to illustrations 28.5a and 28.5b

4 Park the wiper blades in a convenient position to be worked on. To do this, run the wipers, then turn the ignition key to Off when the wiper blades reach the desired position.

5 Lift the blade slightly from the windshield. Squeeze the retaining lever to release the blade **(see illustration)**, unhook the wiper arm from the blade **(see illustration)** and take the blade off. **Caution:** *Do not press too hard on the spring lock or it will be distorted.*

6 Slide the new blade onto the wiper arm hook until the blade locks. Make sure the spring lock secures the blade to the pin.

Wiper element replacement

Refer to illustrations 28.7 and 28.10

7 Insert a screwdriver blade between the wiper blade and element **(see illustration)**. Twist the screwdriver clockwise while pressing in and down to separate the element from the end retaining claw.

8 Slide the element out of the remaining retaining claws.

9 Starting at either end of the second retaining claw, slide a new element into the second retaining claw (not the one closest to the end of the blade). Slide it through the other retaining claws until it reaches the end of the blade.

10 Bend the element and slide it back into the claw at the end of the blade **(see illustration)**.

29 Cooling system servicing (draining, flushing and refilling) (every 24,000 miles or 24 months, whichever comes first)

Refer to illustration 29.4

Warning: *Do not allow antifreeze to come in contact with your skin or painted surfaces of the vehicle. Rinse off spills immediately with plenty of water. Antifreeze is highly toxic if ingested. Never leave antifreeze lying around in an open container or in puddles on the floor; children and pets are attracted by it's sweet smell and may drink it. Check with local authorities about disposing of used antifreeze. Many communities have collection centers which will see that antifreeze is disposed of safely.*

1 Periodically, the cooling system should be drained, flushed and refilled to replenish the antifreeze mixture and prevent formation of rust and corrosion, which can impair the performance of the cooling system and cause engine damage. When the cooling system is serviced, all hoses and the radiator cap should be checked and replaced if necessary.

Draining

2 Apply the parking brake and block the wheels. If the vehicle has just been driven, wait several hours to allow the engine to cool down before beginning this procedure.

3 Once the engine is completely cool, remove the radiator cap.

4 Move a large container under the radiator drain to catch the coolant **(see illustration)**. Attach a 3/8-inch diameter hose to the drain fitting to direct the coolant into the container, then open the drain fitting (a pair of pliers may be required to turn it).

5 While the coolant is draining, check the condition of the radiator hoses, heater hoses and clamps (see Section 22 if necessary).

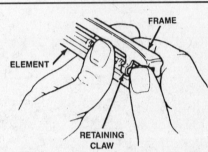

28.10 To install an element, slide it into three of the retaining claws, then bend it back and slide it into the retaining claw at the end of the blade

6 Replace any damaged clamps or hoses (see Chapter 3 for detailed replacement procedures).

Flushing

7 Once the system is completely drained, flush the radiator with fresh water from a garden hose until the water runs clear at the drain. The flushing action of the water will remove sediments from the radiator but will not remove rust and scale from the engine and cooling tube surfaces.

8 These deposits can be removed by the chemical action of a cleaner such as Ford

29.4 The radiator drain is located on the bottom of the radiator; connect a length of rubber hose to the drain and let it hang into the drain pan

Chapter 1 Tune-up and routine maintenance

30.2a Loosen the clamp (arrow) and disconnect the air outlet tube and hose (4.0L engine shown)

30.2b Unplug the electrical connector for the Mass Air Flow sensor (4.0L engine shown)

30.2c Label and disconnect the vacuum lines at the rear of the cover, remove the cover retaining screws and slide the cover sideways out of the retaining tab slots (4.0L engine shown)

30.3 Lift the cover off and take the element out of the housing

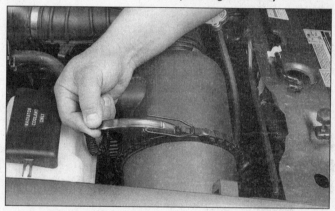

30.9 On 1995 and later models, unsnap the clip . . .

Premium Cooling System Flush, or equivalent. Follow the procedure outlined in the manufacturer's instructions. If the radiator is severely corroded, damaged or leaking, it should be removed (see Chapter 3) and taken to a radiator repair shop.;

9 The heater core should be back-flushed whenever the cooling system is flushed. To do this, disconnect the heater return hose from the thermostat housing or engine. Slide a female garden hose fitting into the heater hose and secure it with a clamp. This will allow you to attach a garden hose securely.

10 Attach the end of a garden hose to the fitting you installed in the heater hose.

11 Disconnect the heater inlet hose and position it to act as a drain.

12 Turn the water on and off several times to create a surging action through the heater core. Then turn the water on full force and allow it to run for approximately five minutes.

13 Turn off the water and disconnect the garden hose from the female fitting. Remove the fitting from the heater return hose, then reconnect the hoses to the engine.

14 Remove the overflow hose from the coolant recovery reservoir. Drain the reservoir and flush it with clean water, then reconnect the hose.

Refilling

15 Close and tighten the radiator drain. Install and tighten the block drain plug(s).

16 Place the heater temperature control in the maximum heat position.

17 Slowly add new coolant (a 50/50 mixture of water and antifreeze) to the radiator until it is full. Add coolant to the reservoir up to the lower mark.

18 Leave the radiator cap off and run the engine in a well-ventilated area until the thermostat opens (coolant will begin flowing through the radiator and the upper radiator hose will become hot).

19 Turn the engine off and let it cool. Add more coolant mixture to bring the coolant level back up to the lip on the radiator filler neck.

20 Squeeze the upper radiator hose to expel air, then add more coolant mixture if necessary. Replace the radiator cap.

21 Start the engine, allow it to reach normal operating temperature and check for leaks.

30 Air filter replacement (every 30,000 miles or 30 months, whichever comes first)

1 If the vehicle is driven in dust, the filter should be checked every 3,000 miles and replaced if it's dirty. In any case, the filter should be replaced at least every 30,000 miles, even if it looks clean.

1993 and 1994 models

Refer to illustrations 30.2a, 30.2b, 30.2c and 30.3

2 Disconnect the PCV hose from the air outlet tube, then disconnect the air outlet tube from the air cleaner cover. Disconnect the electrical connector from the Mass Air Flow (MAF) sensor. Label and disconnect the vacuum lines at the rear of the cover **(see illustrations)**.

3 Remove the cover retaining screws. Disengage the cover tabs from the slots and lift off the cover **(see illustration)**.

4 Lift the element out of the housing.

5 Wipe the inside of the air cleaner housing with a clean cloth.

6 Place the new air filter element in the housing. If the element is marked TOP be sure the marked side faces up.

7 Reinstall the cover and install the screws.

8 Connect the electrical connector, vacuum lines, air outlet tube and PCV hose.

1995 and later models

Refer to illustrations 30.9 and 30.10

9 Unsnap the cover retaining clip **(see illustration)**.

30.10 ... remove the cover and pull the element out

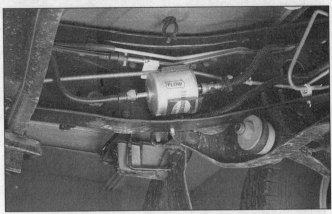

31.1 The inline fuel filter is mounted in the left frame rail; make sure the fuel line fittings are clean before disconnecting them (see Chapter 4 for disconnection procedures)

10 Detach the cover and pull out the element **(see illustration)**.
11 Wipe the inside of the air cleaner housing with a clean cloth.
12 Place the new air filter element in the housing.
13 Reinstall the cover and secure the clip.

31 Fuel filter replacement (every 30,000 miles or 30 months, whichever comes first)

Refer to illustration 31.1

Warning: *Gasoline is extremely flammable, so take extra precautions when you work on any part of the fuel system. Don't smoke or allow open flames or bare light bulbs near the work area, and don't work in a garage where a natural gas-type appliance (such as a water heater or a clothes dryer) with a pilot light is present. Since gasoline is carcinogenic, wear latex gloves when there's a possibility of being exposed to fuel, and, if you spill any fuel on your skin, rinse it off immediately with soap and water. Mop up any spills immediately and do not store fuel-soaked rags where they could ignite. When you perform any kind of work on the fuel system, wear safety glasses and have a Class B type fire extinguisher on hand.*

1 The fuel filter is mounted within the left frame rail **(see illustration)**. Ford states that replacement of the fuel filter at 30,000 miles is recommended, but not required, for California vehicles.
2 Obtain a new fuel filter before starting. **Warning:** *Be sure the new filter is specifically designed for your engine.* Fuel injection system filters are built to withstand high pressure, and as a result, often cost more than filters meant for use in carbureted systems. Filters meant for carbureted systems may burst due to the high pressure. Also, be sure the new filter includes replacement hairpin clips (if used). Ford recommends against reusing the clips. **Warning:** *Before removing the fuel filter, the fuel system pressure must be relieved. See Chapter 4.*
3 Position the front end of the vehicle higher than the rear to prevent fuel siphoning. Remove the gas cap, then reinstall it after relieving the fuel system pressure.
4 Inspect the hose fittings at both ends of the filter to see if they're clean. If more than a light coating of dust is present, clean the fittings before proceeding.
5 Disconnect the push-connect fittings from the filter (see Chapter 4).
6 Note which way the arrow on the filter is pointing - the new filter must be installed the same way. Loosen the clamp screw and detach the filter from the bracket.
7 Install the new filter in the bracket with the arrow pointing in the right direction and tighten the clamp screw securely.
8 Carefully connect each fuel line to the filter (see Chapter 4).
9 Start the engine and check for fuel leaks.

32 Driveshaft and driveaxle yoke lubrication (4WD models) (every 30,000 miles or 30 months, whichever comes first)

At the specified intervals, the slip yokes on the driveshafts and the right front driveaxle should be lubricated (see this Chapter's Specifications for the correct lubricant). This requires removal of the driveshafts and driveaxle (see Chapter 8 for the procedures).

Chapter 2 Part A
Four-cylinder engines

Contents

	Section
Auxiliary shaft (1993 and 1994 models) - removal, inspection and installation	8
Camshaft - removal, inspection and installation	7
CHECK ENGINE light	See Chapter 6
Cylinder compression check	See Chapter 2D
Cylinder head - removal and installation	12
Drivebelt check, adjustment and replacement	See Chapter 1
Engine mounts - removal and installation	17
Engine oil and filter change	See Chapter 1
Engine overhaul - general information	See Chapter 2D
Engine - removal and installation	See Chapter 2D
Exhaust manifold - removal and installation	11
Flywheel/driveplate - removal and installation	15
Front oil seals - replacement	9
General information	1
Intake manifold - removal and installation	10
Oil pan - removal and installation	13
Oil pump - removal and installation	14
Rear main oil seal - replacement	16
Repair operations possible with the engine in the vehicle	2
Spark plug replacement	See Chapter 1
Timing belt - removal and installation	6
Top Dead Center (TDC) for number one piston - locating	3
Valve cover - removal and installation	4
Valve train components - removal and installation	5
Water pump - removal and installation	See Chapter 3

Specifications

General
Cylinder numbers (front-to-rear) 1-2-3-4
Firing order 1-3-4-2

Camshaft
Lobe lift
  1993 and 1994 (intake and exhaust) .... 0.2381 inch
  1995 on (intake and exhaust) .... 0.2163 inch
Bearing journal diameter .... 1.7713 to 1.7720 inch
Bearing clearance .... 0.001 to 0.003 inch
Endplay .... 0.001 to 0.007 inch

Torque specifications
Ft-lbs
(unless otherwise indicated)

Camshaft sprocket bolt .... 50 to 71
Crankshaft pulley/sprocket bolt .... 103 to 133
Auxiliary shaft sprocket bolt .... 28 to 40
Oil pump sprocket bolt .... 28 to 40
Rear camshaft retaining plate bolt .... 72 to 108 in-lbs
Auxiliary shaft retaining plate screws .... 72 to 108 in-lbs
Timing belt outer cover bolts .... 72 to 108 in-lbs
Timing belt tensioner adjustment bolt .... 14 to 21
Timing belt tensioner pivot bolt .... 28 to 40
Valve cover bolts .... 72 to 108 in-lbs
Cylinder head bolts (in sequence - **see illustration 12.33**)
  1993
    Step 1 .... 50 to 60
    Step 2 .... 80 to 90
  1994 on
    Step 1 50
    Step 2 .... Tighten additional 90 to 100-degrees
Flywheel/driveplate bolts .... 56 to 64

Cylinder numbering and coil terminal location

Torque specifications (continued)

Ft-lbs (unless otherwise indicated)

Intake manifold bolts	
1993 and 1994	
Step 1	60 to 84 in-lbs
Step 2	14 to 21
1995 on	
Step 1	84 to 120 in-lbs
Step 2	19 to 28
Exhaust manifold bolts	
1993 and 1994	
Step 1	60 to 84 in-lbs
Step 2	16 to 23
1995 on	
Step 1	15 to 22
Step 2	45 to 59
Oil pan-to-engine block bolts	
1993 through 1998	96 to 120 in-lbs
1999 and 2000	127 to 141 in-lbs
Oil pan-to-transmission bolts	29 to 40
Oil pump bolts	
1993 and 1994	14 to 21
1995 through 1998	84 to 120 in-lbs
1999 and 2000	127 to 141 in-lbs
Oil pump pick-up tube bolts	14 to 21
Engine mount-to-block bolts	45 to 60
Engine mount-to-frame bolts	
Left side	71 to 94
Right side	52 to 67

1 General information

This Part of Chapter 2 is devoted to in-vehicle repair procedures for the OHC four-cylinder engines.

Information concerning engine removal and installation, as well as engine block and cylinder head overhaul, is in Part D of this Chapter.

The following repair procedures are based on the assumption that the engine is installed in the vehicle. If the engine has been removed from the vehicle and mounted on a stand, many of the Steps included in this Part of Chapter 2 will not apply.

The Specifications included in this Part of Chapter 2 apply only to the engine and procedures in this Part. The specifications necessary for rebuilding the block and cylinder head are found in Part D.

2 Repair operations possible with the engine in the vehicle

Many major repair operations can be accomplished without removing the engine from the vehicle.

Clean the engine compartment and the exterior of the engine with some type of pressure washer before any work is done. A clean engine will make the job easier and will help keep dirt out of the internal areas of the engine.

Depending on the components involved, it may be a good idea to remove the hood and engine cover to improve access to the engine as repairs are performed (refer to Chapter 11 if necessary).

If vacuum, exhaust, oil or coolant leaks develop, indicating a need for gasket or seal replacement, the repairs can generally be made with the engine in the vehicle. The intake and exhaust manifold gaskets, oil pan gasket and cylinder head gasket are all accessible with the engine in place.

Engine components such as the intake and exhaust manifolds, the auxiliary shaft, the water pump, the starter motor, the alternator, the distributor and the fuel system components can be removed for repair with the engine in place.

Since the cylinder head can be removed without removing the engine, camshaft and valve component servicing can also be accomplished with the engine in the vehicle.

In extreme cases caused by a lack of necessary equipment, repair or replacement of piston rings, pistons, connecting rods and rod bearings is possible with the engine in the vehicle. However, this practice is not recommended because of the cleaning and preparation work that must be done to the components involved.

3 Top Dead Center (TDC) for number one piston - locating

Refer to illustrations 3.5a and 3.5b

Note: *The four-cylinder engine is not equipped with a distributor. Piston position may be determined by feeling for compression at the number one spark plug hole, then aligning the timing marks as described in Step 5.*

1 Top Dead Center (TDC) is the highest point in the cylinder that each piston reaches as it travels up-and-down when the crankshaft turns. Each piston reaches TDC on the compression stroke and again on the exhaust stroke, but TDC generally refers to piston position on the compression stroke.

2 Positioning the piston at TDC is an essential part of many other repair procedures discussed in this manual.

3 Before beginning this procedure, be sure to place the transmission in Neutral and apply the parking brake or block the rear wheels. Remove the spark plugs (see Chapter 1). Disable the ignition system by disconnecting the front and rear wiring harness connectors from the ignition coil pack, located near the front of the engine.

4 In order to bring any piston to TDC, the crankshaft must be turned using one of the methods outlined below. When looking at the front of the engine, normal crankshaft rotation is clockwise.

a) *The preferred method is to turn the crankshaft with a socket and ratchet attached to the bolt threaded into the front of the crankshaft.*

b) *A remote starter switch, which may save some time, can also be used. Follow the instructions included with the switch. Once the piston is close to TDC, use a socket and ratchet as described in the previous paragraph.*

c) *If an assistant is available to turn the ignition switch to the Start position in short bursts, you can get the piston close to TDC without a remote starter switch. Make sure your assistant is out of the vehicle, away from the ignition switch, then use a socket and ratchet as described in Paragraph a) to complete the procedure.*

5 Remove the access plug(s) from the camshaft drivebelt cover. Rotate the crankshaft until the timing pointer aligns with the "TC" mark on the crankshaft pulley (see

Chapter 2 Part A Four-cylinder engines

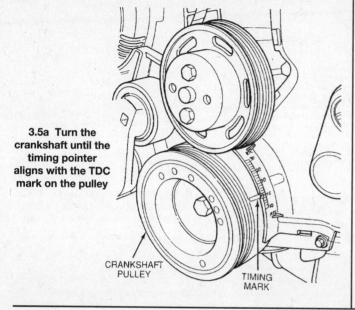

3.5a Turn the crankshaft until the timing pointer aligns with the TDC mark on the pulley

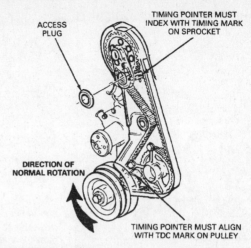

3.5b With the number one piston on the compression stoke, the crankshaft pulley TDC mark will align with the timing pointer and the camshaft timing pointer will align with the index mark on the camshaft sprocket (visible through the access hole)

illustration). Look through the access hole in the camshaft drivebelt cover and confirm that the camshaft timing pointer aligns with the timing mark on the camshaft sprocket **(see illustration)**. If the timing mark is not visible, the piston is at Bottom Dead Center (BDC). Rotate the crankshaft exactly 360-degrees, and look through the access hole again to confirm that the camshaft timing pointer aligns with the timing mark on the camshaft sprocket. TDC occurs when both camshaft and crankshaft pointers are aligned. On 1995 and later models, also confirm that the diamond mark on the oil pump sprocket (visible through the lower access hole in the camshaft drivebelt cover) and the diamond pointer are aligned. TDC occurs when the camshaft, crankshaft and oil pump pointers are aligned.

6 After the number one piston has been positioned at TDC on the compression stroke, TDC for any of the remaining pistons can be located by turning the crankshaft 180-degrees at a time and following the firing order.

4 Valve cover - removal and installation

Refer to illustration 4.6

Removal

1 Disconnect the negative cable from the battery.
2 Remove the accelerator control splash shield.
3 Remove EGR supply tube (see Chapter 6).
4 Identify and tag the various wires and hoses which cross over the valve cover, including the spark plug wires. Do this carefully as they all must be repositioned correctly during installation.

5 Remove the throttle linkage and the throttle body assembly (see Chapter 4).
6 Remove the bolts and separate the cover from the engine **(see illustration)**. It may be necessary to break the gasket seal by tapping the cover with a soft-face hammer. If it's really stuck, use a knife, gasket scraper or chisel to remove it, but be very careful not to damage the gasket sealing surfaces of the cover or head.

Installation

7 Place clean rags in the camshaft gallery to keep foreign material out of the engine.
8 Remove all traces of gasket material from the cover and head. Be careful not to nick or gouge the surfaces. Clean the mating surfaces with lacquer thinner or acetone.
9 Reinstall the cover with a new gasket - no sealant is required. Install the bolts and tighten them to the torque listed in this Chapter's Specifications in a crisscross pattern.
10 Reinstall the upper intake manifold (see Chapter 4) and the remaining components previously removed or disconnected.
11 Since the battery has been disconnected, the engine may not run properly for about 10 miles or so until the powertrain control module relearns its adaptive strategy and adjusts for current engine conditions.

5 Valve train components - removal and installation

Refer to illustrations 5.7, 5.8 and 5.17
Note: *Broken valve springs and defective valve stem seals can be replaced without removing the cylinder head. Two special tools and a compressed air source are normally required to perform this operation, so read through this Section carefully and rent or buy the tools before beginning the job. If com-*

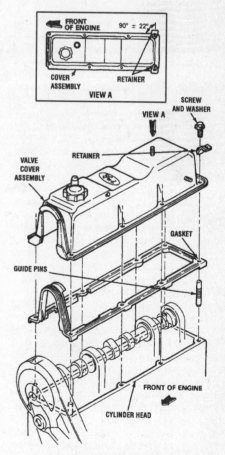

4.6 Typical valve cover and gasket installation details

pressed air isn't available, a length of nylon rope can be used to keep the valves from falling into the cylinder during this procedure.

Chapter 2 Part A Four-cylinder engines

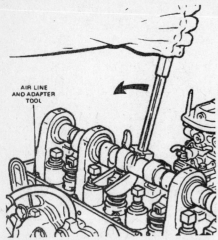

5.7 While compressed air holds the valve closed, a special tool is used to collapse the spring so the cam follower and lash adjuster can be removed

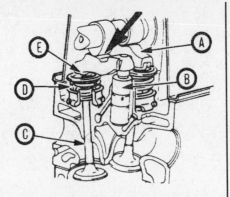

5.8 Valve components - note that in this view, the camshaft lobe is contacting the cam follower (arrow), which compresses the valve spring and opens the valve

A Cam follower
B Valve lash adjuster
C Valve
D Valve spring
E Valve retainer

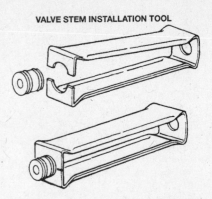

5.17 Use a valve stem seal installation tool, available at most auto parts stores, to seat the valve seal onto the valve guide

Removal

1 Refer to Section 4 and remove the valve cover.
2 Remove the spark plug from the cylinder which has the defective component. If all of the valve stem seals are being replaced, all of the spark plugs should be removed (see Chapter 1).
3 Turn the crankshaft until the piston in the affected cylinder is at top dead center on the compression stroke (see Section 3 for instructions). If you're replacing all of the valve stem seals, begin with cylinder number one and work on the valves for one cylinder at a time. Move from cylinder-to-cylinder following the firing order sequence (1-3-4-2).
4 Thread an adapter into the spark plug hole and connect an air hose from a compressed air source to it. Most auto parts stores can supply the air hose adapter. **Note:** *Many cylinder compression gauges utilize a screw-in fitting that may work with your air hose quick-disconnect fitting.*
5 Apply compressed air to the cylinder.
6 If you don't have access to compressed air, an alternative method can be used. Position the piston at a point just before TDC on the compression stroke, then feed a long piece of nylon rope through the spark plug hole until it fills the combustion chamber. Be sure to leave the end of the rope hanging out of the engine so it can be removed easily. Use a large breaker bar and socket to rotate the crankshaft in the normal direction of rotation until slight resistance is felt.
7 Stuff shop rags into the cylinder head holes above and below the valves to prevent parts and tools from falling into the engine, then use a valve spring compressor to compress the spring **(see illustration)**.
8 With the spring compressed, slide the cam follower over the lash adjuster to remove it **(see illustration)**. Also remove the lash adjuster, if necessary.
9 Still compressing the spring, remove the keepers. Release the valve spring tool and remove the spring retainer, damper assembly and valve spring.
10 Remove and discard the valve stem seal. **Note:** *If air pressure fails to hold the valve in the closed position during this operation, the valve face or seat is probably damaged. If so, the cylinder head will have to be removed for additional repair operations.*
11 Wrap a rubber band or tape around the top of the valve stem so the valve will not fall into the cylinder, then release the air pressure.
12 Inspect the valve stem for damage. Rotate the valve in the guide and check the end for eccentric movement, which would indicate that the valve is bent.
13 Move the valve up-and-down in the guide and make sure it doesn't bind. If the valve stem binds, either the valve is bent or the guide is damaged. In either case, the head will have to be removed for repair.

Installation

14 Reapply air pressure to the cylinder to retain the valve in the closed position, then remove the tape or rubber band from the valve stem. Lubricate the valve stem with engine oil.
15 Place the plastic seal protector (provided with the valve seals) over the end of the valve stem.
16 Start the valve stem seal carefully over the cap and push the seal down until the seal body contacts the top of the guide.
17 Remove the plastic cap and use a seal installation tool to seat the seal onto the valve guide **(see illustration)**.
18 Install the valve spring, damper and retainer, compress the spring and install the keepers.
19 Disconnect the air hose and remove the adapter from the spark plug hole. If a rope was used in place of air pressure, pull it out of the cylinder. **Caution:** *Do not turn the crankshaft backwards to release the rope.* Turning the crankshaft backwards may cause the timing belt to jump time.
20 Using the spring compressor to compress the spring and valve, install the lash adjuster and cam follower. Make sure the lash adjuster has released completely before rotating the camshaft.
21 Reinstall the various components previously removed.
22 Start and run the engine, then check for oil leaks and unusual sounds coming from the valve cover area.

6 Timing belt - removal and installation

Refer to illustrations 6.10, 6.11a, 6.11b and 6.14

Removal

1 Disconnect the negative cable from the battery.
2 Remove the spark plugs and position the number one piston at Top Dead Center (TDC) on the compression stroke (see Section 3). After positioning the number one piston at TDC, DO NOT turn the crankshaft until the timing belt is reinstalled.
3 Drain the cooling system (see Chapter 3).
4 Remove the drivebelts (see Chapter 1).
5 Remove the fan shroud, the fan and the water pump pulley (see Chapter 3).
6 Remove the upper radiator hose and the thermostat housing (see Chapter 3). **Note:** *Due to the lack of space at the front of the engine, access to the components may be difficult. If additional clearance is desired, remove the radiator.*
7 Without disconnecting the hoses, unbolt the air conditioning compressor from the bracket and position the compressor aside. Unbolt the bracket from the engine and position the power steering pump and bracket assembly aside.
8 Remove outer timing belt cover.
9 Remove the crankshaft drivebelt pulley and belt guide.
10 Remove the four bolts and one screw

Chapter 2 Part A Four-cylinder engines

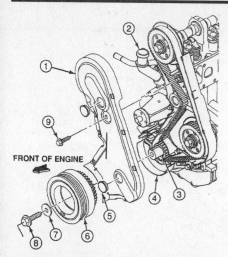

6.10 Timing belt cover installation details

1 Outer timing belt cover
2 Coolant outlet
3 Crankshaft
4 Inner timing belt cover
5 Access plug
6 Crankshaft pulley
7 Washer
8 Bolt
9 Bolt

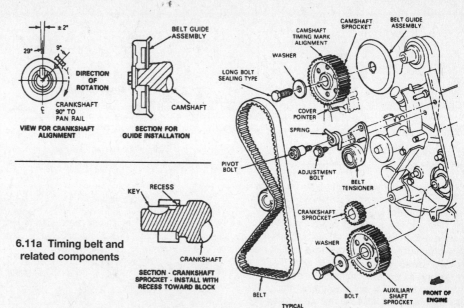

6.11a Timing belt and related components

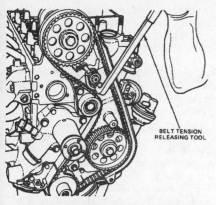

6.11b Release the timing belt tension with the special tool. If the tool is not available, a prybar wedged between the roll pin and tensioner may work

and detach the outer timing belt cover **(see illustration)**.

11 Loosen the belt tensioner adjustment bolt and the tensioner pivot bolt. Position a belt tension releasing tool, available at most auto parts stores, on the tensioner and release the belt tension **(see illustrations)**. Tighten the adjustment bolt just enough to hold the tensioner in the retracted position.

12 Remove the timing belt and inspect it for wear and damage. If it's worn or damaged, replace the belt. If in doubt about the belt condition, install a new one. If the belt breaks during engine operation, extensive damage may occur!

13 Check the sprockets for wear, cracks and burrs on the teeth. If the sprockets are worn or damaged, they can be replaced after removing the mounting bolts. Make sure the new sprockets are installed correctly (see illustration 6.11a). Tighten the mounting bolts to the torque listed in this Chapter's Specifications. Install a new camshaft sprocket bolt or use Teflon sealing tape on the old bolt.

Installation

14 Align the timing mark on the camshaft sprocket with the pointer on the inner timing belt cover. Make sure the crankshaft sprocket is still at TDC (the crankshaft sprocket keyway should be at the 12 O'clock position). On 1993 and 1994 Federal models with a manual transmission, and all 1993 and 1994 models with an automatic transmission, it is not necessary to align the auxiliary shaft sprocket. On 1993 and 1994 California models with manual transmissions, install a Synchronizer Positioner (Ford tool no. T93P-12200-A) onto the auxiliary shaft sprocket and rotate the sprocket until the tool notch engages the notch in the synchronizer bowl. On 1994 and later models, align the timing mark on the oil pump pulley with the pointer on the inner timing belt cover **(see illustration)**.

15 Install the timing belt over the crankshaft sprocket, then over the auxiliary shaft/oil

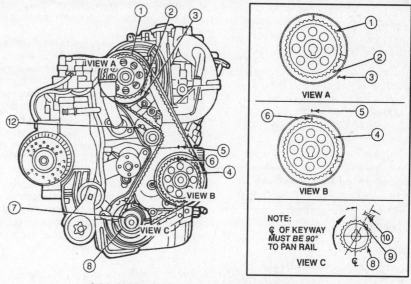

6.14 1995 and later models timing mark details

1 Camshaft sprocket
2 Camshaft sprocket timing mark
3 Pointer on inner timing cover
4 Oil pump sprocket
5 Pointer on inner timing cover
6 Timing mark on oil pump sprocket
7 Crankshaft keyway
8 Crankshaft sprocket
9 Timing mark on crankshaft sprocket
10 Pointer on inner timing belt cover

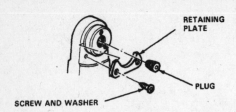

7.9 The camshaft retaining plate is held in place with two screws

pump sprocket. Pull the belt up (pulling slack from between camshaft sprocket and the auxiliary shaft/oil pump sprocket and install it on the camshaft sprocket in a counterclockwise direction so no slack will be between the two sprockets. Align the belt fore and aft on the sprockets.

16 Loosen the tensioner adjustment bolt to allow the tensioner to move against the belt. If the spring does not have enough tension to move the roller against the belt (the belt hangs loose), it may be necessary to insert a pry bar between the tensioner and roll pin and push the roller against the belt.

17 Slowly rotate the crankshaft two complete revolutions in the direction of normal rotation to make sure the belt is seated and to remove any slack. While turning the crankshaft, feel and listen for valve-to-piston contact. Don't force the crankshaft - if binding is felt, recheck all work or seek professional advice!

18 Rotate the tensioner assembly against the belt with approximately 30 to 33 ft-lbs of pressure and tighten the tensioner adjustment and pivot bolts to the torque listed in this Chapter's Specifications. Recheck the alignment of the timing marks.

19 Install the timing belt cover and tighten the mounting bolts to the torque listed in this Chapter's Specifications.

20 The remaining installation steps are the reverse of removal. Be sure to tighten the crankshaft pulley/sprocket bolt to the torque listed in this Chapter's Specifications.

21 Recheck the timing marks for proper alignment, one last time, before installing the belt cover access plugs.

7 Camshaft - removal, inspection and installation

Refer to illustrations 7.9 and 7.12

Removal

1 Disconnect the negative cable at the battery.
2 Remove the valve cover, the timing belt cover and the timing belt (see Sections 4 and 6).
3 Remove the spring clip from the hydraulic valve lash adjuster end of each cam follower, if equipped.
4 Using a valve spring compressor, com-

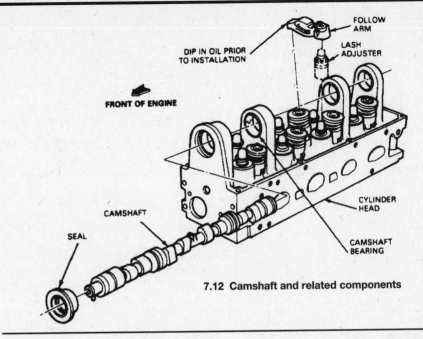

7.12 Camshaft and related components

press the valve springs and remove all the cam followers (see Section 5). Keep the cam followers in order so they can be reinstalled in their original locations.

5 Lift out the hydraulic lash adjusters, keeping each one with its respective cam follower.
6 Remove the camshaft sprocket mounting bolt and washer. A long screwdriver placed through one of the holes in the sprocket will prevent the camshaft from turning.
7 Draw off the sprocket with a puller, then remove the belt guide (if equipped).
8 Remove the sprocket locating pin from the end of the camshaft.
9 Remove the camshaft retaining plate from the rear bearing pedestal (see illustration).
10 Raise the vehicle and support it securely on jackstands.
11 Position a floor jack under the engine. Place a block of wood on the jack pad. Remove the left and right engine mount bolts and nuts. Raise the engine as high as it will go. Place wood blocks between the engine mounts and frame brackets. Lower the engine on the wood blocks and remove the jack. **Warning:** *DO NOT place your hands under the engine where they would be crushed if the jack failed!*
12 Remove the camshaft seal (see Section 9) and withdraw the camshaft (see illustration). Be very careful not to damage the camshaft bearings and journals as it's pulled out.

Inspection

Camshaft and bearings

13 After the camshaft has been removed from the engine, cleaned with solvent and dried, inspect the bearing journals for uneven wear, pits and galling. If the journals are damaged, the bearing inserts in the head are probably damaged as well. Both the camshaft and bearings will have to be replaced with new ones. Measure the inside diameter of each camshaft bearing and record the results (take two measurements, 90 degrees apart, at each bearing).

14 Measure the camshaft bearing journals with a micrometer to determine if they're excessively worn or out-of-round. If they're more than 0.005-inch out-of-round, the camshaft should be replaced with a new one. Subtract the bearing journal diameters from the corresponding bearing inside diameter measurements to obtain the oil clearance. If it's excessive, new bearings must be installed. **Note:** *Camshaft bearing replacement requires special tools and expertise that place it outside the scope of the do-it-yourselfer. Take the head to an automotive machine shop to ensure that the job is done correctly.*

15 Check the camshaft lobes for heat discoloration, score marks, chipped areas, pitting and uneven wear. If the lobes are in good condition the camshaft can be reused. Refer to Chapter 2 Part D for further camshaft inspection procedures.

16 Make sure the camshaft oil passages are clear and clean.

17 To check the thrust plate for wear, install the camshaft in the cylinder head and position the thrust plate at the rear. Using a dial indicator, check the total endplay by prying the camshaft carefully back-and-forth. If the endplay is outside the limit specified in this Chapter's Specifications, replace the thrust plate with a new one.

Cam followers

18 Check the faces, or rollers, of the cam followers (which bear on the camshaft lobes) for signs of pitting, score marks and other forms of wear. They should fit snugly on the lash adjuster.

19 Inspect the face which bears on the

Chapter 2 Part A Four-cylinder engines

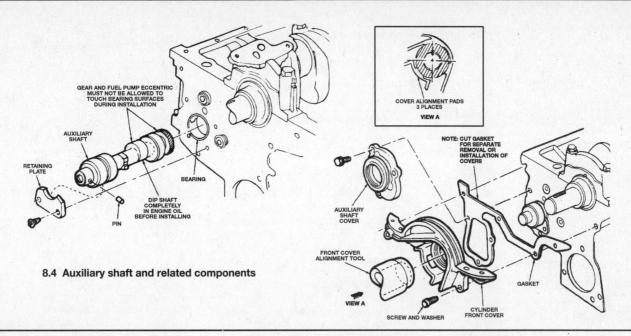

8.4 Auxiliary shaft and related components

valve stem. If it's pitted, the cam follower must be replaced with a new one.
20 If excessive cam follower wear is evident (and possibly excessive camshaft lobe wear), it may be due to a malfunction of the lubrication system. If more than one cam follower is excessively worn, replace the camshaft and all the cam followers. This also applies where excessive camshaft lobe wear is found.
21 During any operation which requires removal of the cam followers, inspect the lash adjusters for damage or wear.

Installation

22 Liberally coat the camshaft journals and bearings with engine assembly lube or moly-base grease, then carefully install the camshaft in the cylinder head.
23 Install the retaining plate and screws.
24 Lubricate the new camshaft oil seal with engine oil and carefully tap it into place at the front of the cylinder head with a large socket and a hammer. Make sure it enters the bore squarely and seats completely.
25 Install the belt guide and pin at the front end of the camshaft, then carefully tap on the sprocket.
26 Install a new sprocket bolt or use Teflon sealing tape on the threads of the old one.
27 Coat the hydraulic lash adjusters with engine assembly lube or moly-base grease, then install them in their original locations.
28 Coat the camshaft lobes and cam followers with engine assembly lube or moly-base grease.
29 Install the timing belt (see Section 6).
30 Compress each valve spring and position each cam follower on its respective valve end and adjuster. Install the retaining spring clips, if equipped.
31 Install the remaining components in the reverse order of removal.

8 Auxiliary shaft (1993 and 1994 models) - removal, inspection and installation

Refer to illustration 8.4

Removal

1 Detach the negative battery cable from the battery. Remove the crankshaft pulley. Remove the bolts/screws and detach the outer timing belt cover **(see illustration 6.10)**.
2 Remove the timing belt (see Section 6). Loosen the auxiliary shaft sprocket bolt. If the shaft turns, immobilize the sprocket by inserting a large screwdriver or 3/8-inch drive extension through one of the sprocket holes.
3 Pull off the sprocket (a puller may be required), and remove the pin from the shaft.
4 Remove the three bolts and detach the auxiliary shaft cover **(see illustration)**.
5 Remove the screws and detach the retaining plate.
6 Withdraw the shaft. If it's tight, reinstall the bolt and washer. Use a pry bar and spacer block to pry out the shaft. Be extremely careful not to damage the bearings as you pull the shaft out of the block.

Inspection

7 Examine the auxiliary shaft bearing for pits and score marks. Replacement must be done by a shop (although it's easy to remove the old bearing, correct installation of the new one requires special tools). The auxiliary shaft may show signs of wear on the bearing journal or the eccentric. Score marks and damage to the bearing journals cannot be removed by grinding. If in doubt, ask a dealer service department to check the auxiliary shaft and give advice on replacement. Examine the gear teeth for wear and damage. If either is evident, a replacement shaft must be obtained.

Installation

8 Dip the auxiliary shaft in engine oil before installing it in the block. Tap it in gently with a soft-face hammer to ensure that it's seated. Install the retaining plate and the auxiliary shaft cover.
9 The remainder of the procedure is the reverse of removal. Make sure that the auxiliary shaft pin is in place before installing the sprocket. Tighten the sprocket mounting bolt to the specified torque.

9 Front oil seals - replacement

Note: *The camshaft, crankshaft, auxiliary shaft, and oil pump oil seals are all replaced the same way, using the same tools, after the appropriate sprocket has been removed.*

1 Disconnect the negative battery cable from the battery.
2 Remove the timing belt cover and timing belt (see Section 6). **Note:** *Due to the lack of space at the front of the engine, access to the components may be difficult. If additional clearance is desired, remove the radiator.*
3 To remove the camshaft sprocket, auxiliary shaft sprocket (1993 and 1994 models), or oil pump sprocket (1995 and later models), refer to Section 7, 8 or 13. Ford manufactures a special puller designed for this purpose. To remove the crankshaft sprocket, use an appropriate gear puller.
4 A special tool, available at most auto parts stores, may be used to remove all the seals. When using the tool, be sure the jaws are gripping the thin edge of the seal very tightly before operating the screw portion of the tool. If the tool isn't available, a hammer and a small chisel may be used to remove the seal(s), if care is exercised.
5 Clean the seal bore and shaft surface prior to installation of the new seal. Apply a

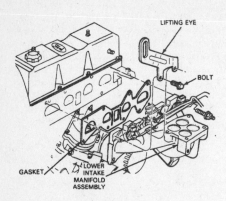

10.8 Typical intake manifold installation details

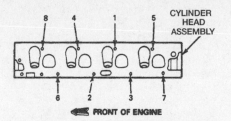

10.13 Intake manifold tightening sequence

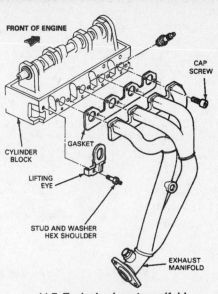

11.7 Typical exhaust manifold installation details

thin layer of grease to the outer edge of the new seal(s).
6   Install the seal(s) with a special tool made for this job, available at most auto parts stores. If the special tool isn't available, you may be able to use a large deep socket and a hammer or a short piece of pipe, the sprocket bolt and a large flat washer.
7   Install the components removed to gain access to the seals. **Caution:** *Always use a new camshaft sprocket bolt or seal the old bolt threads with Teflon sealing tape.*

10 Intake manifold - removal and installation

Refer to illustrations 10.8 and 10.13

Removal

1   Relieve the fuel system pressure (see Chapter 4).
2   Disconnect the negative battery cable from the battery.
3   Drain the cooling system (see Chapter 1).
4   Remove the air cleaner and ducts (see Chapter 4).
5   Label and disconnect the vacuum hoses from the throttle body and emissions devices attached to the intake manifold.
6   Remove the throttle body, fuel injection wiring harness, fuel supply manifold, throttle cable, speed control cable (if equipped) and upper intake manifold (see Chapter 4). Disconnect the EGR tube from the EGR valve.
7   Remove the four bottom mounting bolts from the intake manifold.
8   Remove the four upper mounting bolts from the manifold. Note that the front two bolts also secure the engine lifting eye **(see illustration)**.
9   Detach the manifold from the engine.

Installation

10   Clean the manifold and head mating surfaces and check the manifold for cracks.
Note: *The mating surfaces of the head and manifold must be perfectly clean when the manifold is installed. Gasket removal solvents in aerosol cans are available at most auto parts stores and may be helpful when removing old gasket material that's stuck to the head and manifold. Be careful not to scrape or gouge the sealing surfaces.*
11   Clean and oil the manifold mounting bolts.
12   Position the new intake manifold gasket on the head, using a light coat of gasket sealant to hold it in place.
13   Position the intake manifold, along with the lifting eye, on the head. Install the fasteners and tighten them in sequence, to the torque listed in this Chapter's Specifications in two steps **(see illustration)**.
14   The remainder of installation is the reverse of removal.
15   Reconnect the battery and refill the cooling system.
16   Start the engine and let it run while checking for coolant, fuel and vacuum leaks.

11 Exhaust manifold - removal and installation

Refer to illustrations 11.7 and 11.9

Removal

1   Disconnect the negative battery cable from the battery.
2   Remove the air cleaner and duct assembly if needed to gain access.
3   Remove the EGR tube fitting at the exhaust manifold and loosen at the EGR tube (see Chapter 6).
4   Disconnect the oxygen sensor wiring harness connector.
5   Remove the bolt attaching the heater hoses to the valve cover, if equipped.
6   Raise the vehicle and support it securely on jackstands. Remove the two bolts attaching the exhaust pipe to the manifold (see Chapter 4).
7   Remove the eight exhaust manifold mounting bolts **(see illustration)** and detach the manifold from the engine.
8   Clean the manifold and cylinder head mounting surfaces with a gasket scraper, then wipe them off with a cloth saturated with lacquer thinner or acetone. Clean the bolt threads with a wire brush.

Installation

9   Using a new gasket, install the manifold and the eight bolts attaching it to the cylinder head. Tighten the bolts (in two or three steps) to the torque listed in this Chapter's Specifications in the sequence shown **(see illustration)**.
10   Install the bolts attaching the exhaust pipe to the exhaust manifold.
11   Position the heater hoses on the valve cover.
12   Reconnect the oxygen sensor wire.
13   Install the EGR line at the exhaust manifold.
14   Install the air cleaner and duct assembly.
15   Reconnect the battery cable.

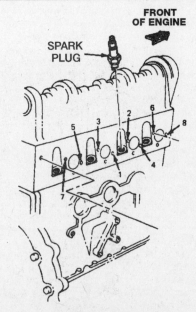

11.9 Exhaust manifold bolt tightening sequence

Chapter 2 Part A Four-cylinder engines

12.31 Be sure to turn the camshaft until the pin is in the 5 o'clock position as shown here before installing the head

12 Cylinder head - removal and installation

Refer to illustrations 12.31 and 12.33

Removal

1 Relieve the fuel system pressure (see Chapter 4). Disconnect the cable from the negative battery terminal and position the number one piston at top dead center (TDC) on the compression stroke (see Section 3).
2 Drain the cooling system (see Chapter 1). Remove the air cleaner assembly and the air duct. Remove the drivebelt (see Chapter 1).
3 Remove the bolt retaining the heater hose to the valve cover, if equipped.
4 Remove the spark plug wires from the spark plugs and remove the ignition coil packs and wires as an assembly. Remove the spark plugs.
5 Label and disconnect any electrical connectors or vacuum hoses attached to components on the cylinder head.
6 Remove the dipstick tube.
7 Remove the upper intake manifold (see Chapter 4).
8 Remove the valve cover (see Section 4).
9 Remove the air conditioning compressor from its mounting bracket and set it aside without disconnecting the hoses. Remove the bracket and set it, and the power steering pump, aside. Remove the alternator bracket mounting bolts and set the alternator and bracket aside.
10 Disconnect the upper radiator hose at both ends and remove it from the engine.
11 Remove the radiator fan/clutch assembly, water pump pulley and timing belt front cover.
12 Confirm that all camshaft and crankshaft marks are still aligned, then loosen the timing belt tensioner (see Section 5).
13 Remove the timing belt from the camshaft and auxiliary shaft or oil pump sprockets (see Section 6).
14 Remove the heat stove from the exhaust manifold.
15 Remove the exhaust manifold mounting bolts (see Section 11). The manifold itself can remain in place.
16 Remove the timing belt tensioner bolts.
17 Remove the timing belt tensioner spring stop from the cylinder head.
18 Disconnect the oil pressure sending unit wiring harness connector.
19 Refer to Section 10 and remove the intake manifold.
20 Using a new head gasket, outline the cylinders and bolt pattern on a piece of cardboard. Be sure to indicate the front of the engine for reference. Punch holes at the bolt locations.
21 Loosen the cylinder head mounting bolts in 1/4-turn increments until they can be removed by hand. Store the bolts in the cardboard holder as they're removed; this will ensure that they're reinstalled in their original locations.
22 Lift the head off the engine. If it's stuck, don't pry between the head and block. Instead, rap the head with a soft-face hammer or a block of wood and a hammer to break the gasket seal.
23 Place the head on a block of wood to prevent damage to the gasket surface. Refer to Part D for cylinder head disassembly, inspection and valve service procedures. Make sure you check the cylinder head for warpage.

Installation

24 If a new cylinder head is being installed, transfer all external parts from the old cylinder head to the new one.
25 The mating surfaces of the cylinder head and block must be perfectly clean when the head is installed.
26 Use a gasket scraper to remove all traces of carbon and old gasket material, then clean the mating surfaces with lacquer thinner or acetone. If there's oil on the mating surfaces when the head is installed, the gasket may not seal correctly and leaks may develop. Use a vacuum cleaner to remove any debris that falls into the cylinders.
27 Check the block and head mating surfaces for nicks, deep scratches and other damage. If damage is slight, it can be removed with a file; if it's excessive, machining may be the only alternative.
28 Use a tap of the correct size to chase the threads in the head bolt holes. Mount each bolt in a vise and run a die down the threads to remove corrosion and restore the threads. Dirt, corrosion, sealant and damaged threads will affect critical head bolt torque readings.
29 Position the new gasket over the dowel pins in the block. Do not use a gasket sealer.
30 Confirm that the crankshaft is at TDC as described in Section 3. If it is not, rotate the crankshaft in the normal direction (see Section 3) until the "TC" mark aligns with the pointer. Pistons 1 and 4 should be at TDC.
31 Before installing the cylinder head, turn the camshaft until the pin is in the five o'clock position - this must be done to avoid damage to the valves when the head is installed **(see illustration)**. Check the camshaft lobes to verify proper camshaft position. The lobes for Cylinders No. 1 and 4 should point almost directly out. The lobes for Cylinders No. 2 and 3 should point up at about a 45-degree angle. This ensures that all 8 valves are seated, and cannot collide with the pistons.
32 Carefully position the head on the block without disturbing the gasket.
33 Install the cylinder head bolts and tighten them in sequence **(see illustration)** to the torque listed in this Chapter's Specifications.

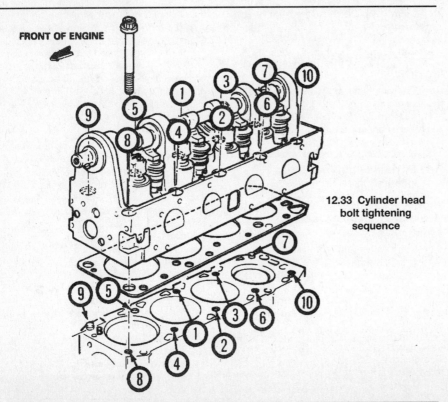

12.33 Cylinder head bolt tightening sequence

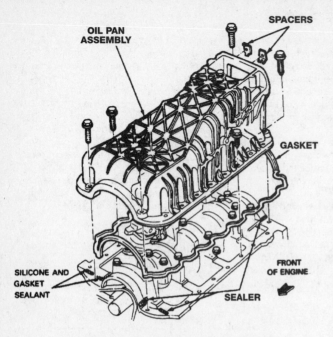

13.7 Oil pan and gasket installation details

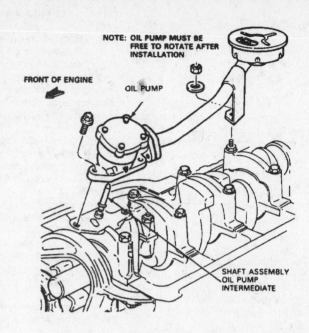

14.1 Oil pump mounting details (1993 and 1994 models)

34　Return the camshaft to its original TDC position.
35　The remainder of installation is the reverse of the removal procedure.
36　Change the engine oil and filter (see Chapter 1).

13 Oil pan - removal and installation

Refer to illustration 13.7
Note: *The engine must be removed from the vehicle to remove the oil pan.*

Removal
1　Drain the oil and remove the oil filter.
2　Remove the engine as described in Part D. Mount the engine in a suitable engine stand, right-side up.
3　Remove the oil pan mounting bolts and lower the pan.

Installation
4　Clean the oil pan and inspect for damage. Remove the spacers, if any, that are attached to the oil pan transmission mounting pad.
5　Clean the oil pan gasket surface on the cylinder block and the oil pan flange. These surfaces must be perfectly clean to prevent oil leaks.
6　Clean the oil pump pick-up tube screen.
7　Apply RTV sealant to six points; at the joints between the front cover and engine block and rear main bearing cap and engine block **(see illustration)**. Position the gasket inside the channel of the oil pan.
8　Place the oil pan onto the engine and install the oil pan mounting bolts and hand tighten them only at this time. They will be tightened to final torque after the engine and transmission are mated to ensure correct alignment.
9　Install the engine in the vehicle as described in Part D.
10　Tighten the two oil pan-to-transmission bolts to the torque listed in this Chapter's Specifications, then loosen them 1/2 turn.
11　Tighten the oil pan-to-engine block bolts to the torque listed in this Chapter's Specifications. Work around the pan, tightening each bolt a little at a time.
12　Retighten the oil pan-to-transmission bolts to the torque listed in this Chapter's Specifications.
13　The remainder of the installation is the reverse of the removal procedure.
14　Be sure to refill the engine with the proper quantity and grade oil (see Chapter 1).

14 Oil pump - removal and installation

1993 and 1994 models

Refer to illustration 14.1
Note: *The engine must be removed to gain access to the oil pump on these models.*
1　Remove the oil pan as described in Section 13. Remove the oil pump and pick-up screen mounting bolts and separate the pump from the engine block **(see illustration)**.
2　If there is any possibility that the pump is faulty or if an engine overhaul is being performed, replace the pump. A faulty pump can ruin an otherwise good engine.
3　Prime the pump before installation. Hold it with the pick-up tube up and pour a few ounces of clean oil into the inlet screen. Turn the pump driveshaft by hand until oil comes out the outlet.
4　Install the pump intermediate shaft in the engine block. Be sure the shaft engages the auxiliary shaft gear. Install the pump assembly and tighten the bolts to the torque listed in this Chapter's Specifications. Install the oil pan (see Section 13).
5　Install the engine and refill the engine with the proper quantity and grade oil (see Chapter 1), then start the engine and be sure the oil pressure comes up. If it doesn't come up within 10 or 15 seconds, shut off the engine immediately and find the cause of the problem. **Caution:** *Continued running of the engine without oil pressure will severely damage the moving parts!*
6　Check for oil leaks.

1995 and later models

Refer to illustration 14.12, 14.13 and 14.17
7　Remove the timing belt as described in Section 3.
8　Loosen the oil pump sprocket bolt. If the shaft turns, immobilize the sprocket by inserting a large screwdriver or 3/8-inch drive extension through one of the sprocket holes.
9　Pull off the sprocket (a puller may be required), and remove the key from the shaft.
10　Disconnect the camshaft position sensor (CMP) wiring connector.
11　Remove the two screws securing the CMP sensor and remove the sensor (see Chapter 6).
12　Remove the four oil pump mounting bolts **(see illustration)**.
13　Loosen the oil pump by prying between the cylinder block and the tab on the oil

Chapter 2 Part A Four-cylinder engines

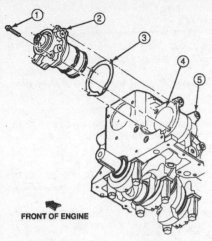

14.12 Oil pump mounting details (1995 and later)

1. Bolt
2. Oil pump
3. Gasket
4. Oil pump cavity
5. Engine block

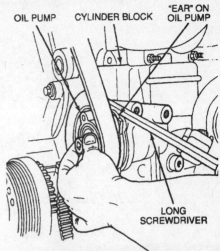

14.13 Pry between the engine block and the "ear" to free the oil pump

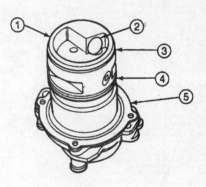

14.17 Oil pump components

1. Oil Pump
2. Inlet passage
3. O-ring
4. O-ring
5. Gasket

the torque listed in this Chapter's Specifications.
22 Check engine oil level on the dipstick. Top up as necessary.
23 Prime the oil pump by connecting it to a drill motor and rotating it until there is some resistance. This may take 10 or more revolutions of the sprocket.
24 Install the CMP sensor and connect the electrical connector.
25 Rotate the oil pump sprocket until the two timing marks align.
26 Verify that the crankshaft, camshaft and oil pump sprockets are properly aligned at TDC (see Section 6).
27 Install the timing belt and outer cover (see Section 6).

15 Flywheel/driveplate - removal and installation

Refer to illustration 15.3

1 Remove the transmission (see Chapter 7, Part A).
2 Remove the pressure plate and the clutch disc from the rear of the flywheel as described in Chapter 8 (manual transmission models).
3 Lock the flywheel/driveplate using a large screwdriver meshed with the starter ring gear and remove the six bolts securing the flywheel to the crankshaft (see illustration). The bolts should be removed a quarter turn at a time in a diagonal and progressive manner.
4 Mark the mating position of the flywheel/driveplate and crankshaft flange and then remove the flywheel.
5 If necessary, remove the engine rear cover plate bolts and ease the rear cover plate over the two dowel pins. Remove the cover plate.
6 If the flywheel ring gear is badly worn or has teeth missing, it should be replaced by an automotive machine shop.

pump (see illustration). **Caution:** *Do not pry on any sealing surfaces.*
14 Remove the oil pump and the associated gasket.
15 Clean the exterior of the oil pump and the mating surface on the cylinder block.
16 Make sure the passage at the back of the oil pump bore in the cylinder block is clean and unobstructed.
17 Install new O-rings on the oil pump body and install a new gasket (see illustration).
18 Pour 8 ounces of engine oil into the inlet passage to prime the oil pump.
19 Install the oil pump in the cylinder block being careful not to spill the primer oil.
20 Install the oil pump mounting bolts and tighten them to the torque listed in this Chapter's Specifications.
21 Insert the key in the oil pump shaft and install the oil pump sprocket. Install the bolt and washer and tighten the sprocket bolt to

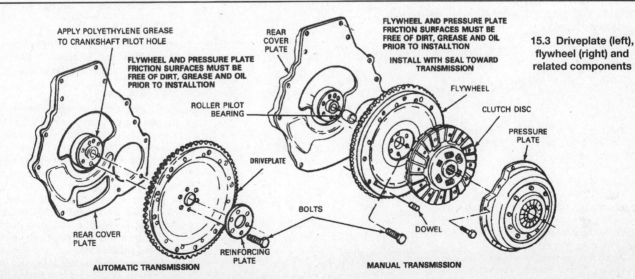

15.3 Driveplate (left), flywheel (right) and related components

2A-12　Chapter 2 Part A　Four-cylinder engines

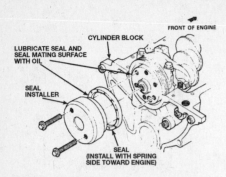

16.5 A special tool is recommended to install the rear main seal

7　Check for cracks and other damage.
8　Installation of the flywheel/driveplate and clutch is the reverse of the removal procedure with the following additions:
a)　Apply oil-resistant sealer sparingly to the bolt threads.
b)　Tighten the mounting bolts to the torque listed in this Chapter's Specifications.
9　For further information on clutch servicing and pilot bearing replacement, see Chapter 8.

16　Rear main oil seal - replacement

Refer to illustration 16.5
1　Remove the transmission and clutch and pressure plate if equipped, to get at the oil seal in the back of the engine (Chapters 7, Part A and 8).
2　Remove the flywheel or driveplate (see Section 15).
3　Use a small punch to make two holes on opposite sides of the seal and install small sheet metal screws in the seal. Pry on the screws with two large screwdrivers until the seal is removed from the engine. It may be necessary to place small blocks of wood against the block to provide a fulcrum point for prying. **Caution:** *Be careful not to scratch or otherwise damage the crankshaft oil seal surface.*
4　Apply a thin film of engine oil to the mating edges of the seal and the seal bore in the block, as well as the seal lips.
5　Position the seal on a seal installer, available at most auto parts stores, or its equivalent, then position the tool and seal at the rear of engine **(see illustration)**. Install the seal with the spring side toward the engine. Alternate tightening of the bolts to seat the seal properly.
6　If the special seal installer is not available, use a section of the appropriate diameter plastic pipe and gently tap the seal into place with a soft-face hammer.
7　The remainder of the installation is the reverse of the removal procedure.

17　Engine mounts - removal and installation

Refer to illustration 17.8
1　Engine mounts seldom require attention, but broken or deteriorated mounts should be replaced immediately or the added strain placed on the driveline components may cause damage or wear.

Check

2　During the check, the engine must be raised slightly to remove the weight from the mounts.
3　Raise the vehicle and support it securely on jackstands, then position a jack under the engine oil pan. Place a large block of wood between the jack head and the oil pan, then carefully raise the engine just enough to take the weight of the mounts. **Warning:** *DO NOT place any part of your body under the engine when it's supported only by a jack!*
4　Check the mounts to see if the rubber is cracked, hardened or separated from the metal plates. Sometimes the rubber will split right down the center.
5　Check for relative movement between the mount plates and the engine or frame (use a large screwdriver or pry bar to attempt to move the mounts). If movement is noted, lower the engine and tighten the mount fasteners.
6　Rubber preservative should be applied to the mounts to slow deterioration.

Replacement

7　Disconnect the negative battery cable from the battery, then raise the vehicle and support it securely on jackstands (if not already done).
8　Remove the fasteners and detach the mount from the frame bracket **(see illustration)**.
9　Raise the engine slightly with a jack or hoist (make sure the fan doesn't hit the radiator or shroud). Remove the mount-to-block nuts/bolts and detach the mount.
10　Installation is the reverse of removal. Use thread locking compound on the mount fasteners and be sure to tighten them securely.

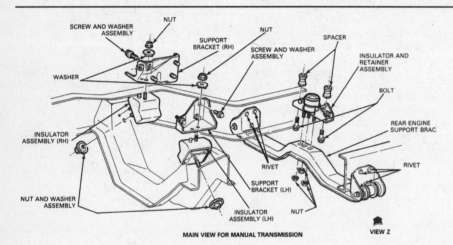

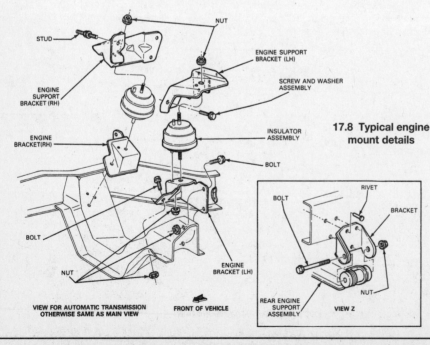

17.8 Typical engine mount details

Chapter 2 Part B
3.0L V6 engine

Contents

	Section		Section
Camshaft - removal, inspection and installation	14	Oil pan - removal and installation	15
CHECK ENGINE light	See Chapter 6	Oil pump and pick-up - removal and installation	16
Crankshaft front oil seal - replacement	10	Rear main oil seal - replacement	17
Cylinder compression check	See Chapter 2D	Repair operations possible with the engine in the vehicle	2
Cylinder heads - removal and installation	9	Rocker arms and pushrods - removal, inspection and installation	5
Drivebelt check, adjustment and replacement	See Chapter 1		
Engine oil and filter change	See Chapter 1	Spark plug replacement	See Chapter 1
Engine overhaul - general information	See Chapter 2D	Timing chain cover - removal and installation	11
Engine - removal and installation	See Chapter 2D	Timing chain and sprockets - check, removal and installation	12
Exhaust manifolds - removal and installation	8	Top Dead Center (TDC) for number one piston - locating	3
Engine mounts - removal and installation	19	Valve covers - removal and installation	4
Flywheel/driveplate - removal and installation	18	Valve lifters - removal, inspection and installation	13
General information	1	Valve springs, retainers and seals - replacement	6
Intake manifold - removal and installation	7		

Specifications

General
Cylinder numbers (front-to-rear)
- Left (driver's) side 4-5-6
- Right side 1-2-3

Firing order 1-4-2-5-3-6

Camshaft
- Lobe lift (intake and exhaust) .... 0.260 inch
- Endplay .... 0.007 inch
- Journal diameter .... 2.0074 to 2.0084 inches
- Bearing inside diameter .... 2.0094 to 2.0104 inches
- Bearing oil clearance .... 0.001 to 0.003 inch

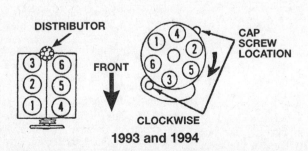

1993 and 1994

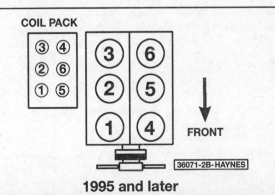

1995 and later

Cylinder numbering, distributor rotation and coil terminal location

Torque specifications

Ft-lbs (unless otherwise indicated)

Camshaft sprocket bolt	40 to 51
Camshaft thrust plate bolts	72 to 96 in-lbs
Cylinder head bolts (in sequence - **see illustration 9.20**)	
1993 through 1998	
Step 1	33 to 41
Step 2	63 to 73
1999 and 2000	
Step 1	59
Step 2	Loosen one full turn
Step 3	34 to 40
Step 4	63 to 73
Flywheel/driveplate bolts	54 to 64
Engine mount-to-block bolts	33 to 44
Engine mount-to-crossmember nuts	52 to 54
Exhaust manifold bolts	
Step 1	96 in-lbs
Step 2	15 to 22
Intake manifold bolts	
Step 1	132 in-lbs
Step 2	19 to 24
Lifter guide plate retainer bolts	84 to 120 in-lbs
Oil pan bolts	84 to 108 in-lbs
Oil pump-to-block bolts	15 to 22
Oil pump pick-up tube-to-oil pump	15 to 22
Rocker arm fulcrum bolts	
Step 1	60 to 132 in-lbs
Step 2	18 to 26
Timing chain cover bolts	
6mm bolts	72 to 96 in-lbs
8mm bolts	15 to 22
Valve cover bolts	84 to 108 in-lbs
Vibration damper bolt	92 to 122

1 General information

This Part of Chapter 2 is devoted to in-vehicle repair procedures. All information concerning engine removal and installation and engine block and cylinder head servicing can also be found in Part D of this Chapter.

The repair procedures included in this Part are based on the assumption that the engine is still installed in the vehicle. Therefore, if this information is being used during a complete engine overhaul - with the engine already out of the vehicle and on a stand - many of the steps included here will not apply.

The specifications included in this Part of Chapter 2 apply only to the engine and procedures found here. For specifications regarding engines other than the 3.0 liter V6, see Part A or C, whichever applies. Part D of Chapter 2 contains the specifications necessary for engine block and cylinder head rebuilding procedures.

2 Repair operations possible with the engine in the vehicle

Many major repair operations can be accomplished without removing the engine from the vehicle.

Clean the engine compartment and the exterior of the engine with some type of pressure washer before any work is done. A clean engine will make the job easier and will help keep dirt out of the internal areas of the engine.

If oil or coolant leaks develop, indicating a need for gasket or seal replacement, the repairs can generally be made with the engine in the vehicle. The cylinder head gaskets, intake and exhaust manifold gaskets, timing cover gaskets and the crankshaft oil seals are accessible with the engine in place. The oil pan is accessible with the engine in place on 1993 models, but the engine must be removed first on 1994 and later models.

Exterior engine components, such as the water pump, the starter motor, the alternator, the distributor and the EFI components, as well as the intake and exhaust manifolds, can be removed for repair with the engine in place.

Since the cylinder heads can be removed without removing the engine, valve component servicing can also be accomplished with the engine in the vehicle.

Replacement of, repairs to or inspection of the timing chain and sprockets and the oil pump are all possible with the engine in place (except on 4WD models).

In extreme cases caused by a lack of necessary equipment, repair or replacement of piston rings, pistons, connecting rods and rod bearings is possible with the engine in the vehicle. However, this practice is not recommended because of the cleaning and preparation work that must be done to the components involved.

3 Top Dead Center (TDC) for number one piston - locating

1 Top Dead Center (TDC) is the highest point in the cylinder that each piston reaches as it travels up-and-down when the crankshaft turns. Each piston reaches TDC on the compression stroke and again on the exhaust stroke, but TDC generally refers to piston position on the compression stroke. The timing marks on the vibration damper installed on the front of the crankshaft are referenced to the number one piston at TDC.

2 Positioning the piston(s) at TDC is an essential part of many procedures such as rocker arm removal, timing chain and sprocket replacement and distributor removal.

3 In order to bring any piston to TDC, the crankshaft must be turned using one of the methods outlined below. When looking at the front of the engine, normal crankshaft rotation is clockwise.

  a) *The preferred method is to turn the crankshaft with a large socket and breaker bar attached to the vibration damper bolt that is threaded into the front of the crankshaft.*

  b) *A remote starter switch, which may save some time, can also be used. Attach the switch leads to the switch and battery terminals on the solenoid. Once the piston is close to TDC, use a socket and breaker bar as described in the previous paragraph.*

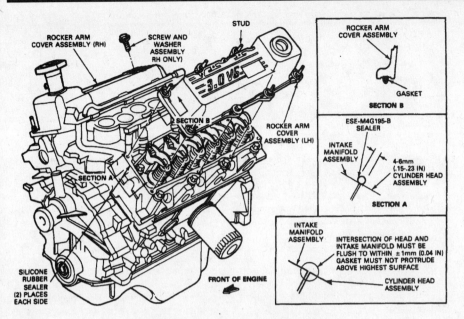

4.11 Rocker arm cover installation details

c) *If an assistant is available to turn the ignition switch to the Start position in short bursts, you can get the piston close to TDC without a remote starter switch. Use a socket and breaker bar as described in Paragraph a) to complete the procedure.*

1993 and 1994 models

4 Before beginning this procedure, be sure to place the transmission in Neutral and apply the parking brake or block the rear wheels. Remove the spark plugs (see Chapter 1). Disconnect and ground the coil wire at the distributor cap to disable the ignition system.

5 Using a felt tip pen, make a mark on the distributor housing directly below the number one spark plug wire terminal on the distributor cap.

6 Remove the distributor cap as described in Chapter 1.

7 Turn the crankshaft (see Step 3 above) until the zero on the vibration damper is aligned with the pointer on the timing cover. The timing pointer and vibration damper are located low on the front of the engine, near the pulley that turns the drivebelt.

8 The rotor should now be pointing directly at the mark on the distributor housing. If it is 180-degrees off, the piston is at TDC on the exhaust stroke.

9 To get the piston to TDC on the compression stroke, turn the crankshaft one complete turn (360-degrees) clockwise. The rotor should now be pointing at the mark. When the rotor is pointing at the number one spark plug wire terminal in the distributor cap (which is indicated by the mark on the housing) and the timing marks are aligned, the number one piston is at TDC on the compression stroke.

10 After the number one piston has been positioned at TDC on the compression stroke, TDC for any of the remaining cylinders can be located by turning the crankshaft 120-degrees at a time and following the firing order (refer to the Specifications).

1995 and later models

Note: *The 3.0L engine used in 1995 and later models is not equipped with a distributor. Piston position must be determined by feeling for compression at the number one spark plug hole, then aligning the ignition timing marks as described in Step 5.*

11 Before beginning this procedure, be sure to place the transmission in Neutral and apply the parking brake or block the rear wheels. Remove the spark plugs (see Chapter 1). Disable the ignition system by disconnecting the wiring harness connector from the ignition coil pack, mounted to the intake manifold.

12 Hold your finger over the number one spark plug hole. Slowly turn the crankshaft (see Step 3 above) until you feel compression at the number one spark plug hole, then continue to turn the crankshaft until the TDC notch is aligned with the pointer (located at the front of the engine).

13 After the number one piston has been positioned at TDC on the compression stroke, TDC for any of the remaining pistons can be located by turning the crankshaft 120-degrees at a time and following the firing order shown in the specifications for this Chapter.

4 Valve covers - removal and installation

Removal

Refer to illustration 4.11

1 Disconnect the negative cable at the battery.

2 Disconnect the spark plug wires on the side(s) you are taking off, leaving them attached to the wire separators. If they are not numbered, tag them so they won't get mixed up on reassembly.

3 Remove the spark plug wire separators from the valve cover studs.

4 If the right valve cover is being removed, disconnect the engine harness connectors, remove the fuel charging wiring standoffs from the inboard valve cover studs and remove the PCV closure hose from the oil fill adapter.

5 If the left cover is being removed, remove the throttle body or the upper intake manifold as required (see Chapter 4), the PCV valve and the fuel charging wiring standoffs from the inboard valve cover studs.

6 Make a sketch noting the location of valve cover bolts and studs so that they can be installed correctly later, then remove them. Use a deep socket to remove the studs.

7 Try bumping the valve cover with a rubber mallet to loosen it. If that doesn't work easily, use a thin-bladed knife to cut the RTV sealant. **Caution:** *Do not cut the integral gasket. Cut the RTV sealant only.* Gently lift the cover off and make sure that the RTV sealant does not pull the integral gasket from the cover. Cut the RTV as necessary to protect the gasket.

8 Carefully roll the integral gasket from its channel in the valve cover. If in good condition, the gasket can be reused indefinitely.

Installation

9 Clean the gasket channel thoroughly to remove oil and dirt. Clean the mating surface on the cylinder head. Make sure to remove all residual RTV sealant. Lightly oil all fastener threads.

10 Install the original gasket or a new one in the valve cover channel and carefully align the stud and bolt holes. Install the bolts and studs in their holes according to the sketch you make earlier. Hold the fastener with a socket and gently push the gasket down around the fastener until it seats fully in the channel. **Note:** *Make sure the gasket lies flat in the valve cover channel, with no bulges, or it will leak.*

11 Apply a bead of RTV sealant at the cylinder head to intake manifold rail step (two places per rail) **(see illustration)**.

12 Install the valve cover straight onto the cylinder head. If it shifts after contacting the sealant, the gasket could be pulled from the channel and leak.

13 Tighten all fasteners to the torque listed in this Chapter's Specifications. Work around the cover in several steps.

14 The remainder of the installation is the reverse of removal.

15 Reconnect the battery, start the engine and check for oil and vacuum leaks.

5 Rocker arms and pushrods - removal, inspection and installation

Refer to illustrations 5.2, 5.3a and 5.3b

Removal

1 Remove the valve covers as described in Section 4.

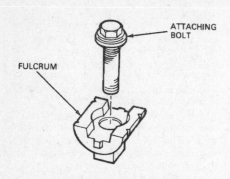

5.2 Rocker arm attaching bolt and fulcrum

5.3a Loosen the bolt (arrow) and pivot the rocker arm to the side to remove the pushrod

5.3b A perforated cardboard box can be used to store the pushrods to ensure installation in their original locations

2 Remove the rocker arm fulcrum bolts **(see illustration)**.
3 Remove the rocker arms. If only the pushrods are being removed, loosen the fulcrum bolts and rotate the rocker arms out of the way of the pushrods **(see illustration)**. Keep the pushrods organized to ensure that they are reinstalled in their original locations **(see illustration)**.

Inspection

4 Check the rocker arms for excessive wear, cracks and other damage, especially where the pushrods and valve stems contact the rocker arm faces.
5 Make sure the hole at the pushrod end of the rocker arm is open.
6 Check the rocker arm fulcrum contact area for wear and galling. If the rocker arms are worn or damaged, replace them with new ones and use new fulcrum seats as well.
7 Inspect the pushrods for cracks and excessive wear at the ends. Roll each pushrod across a flat surface to see if they are bent.

Installation

8 Apply engine oil or assembly lube to the top of the valve stem and the pushrod guide in the cylinder head.
9 Apply engine oil to the rocker arm fulcrum seat and the fulcrum seat socket in the rocker arm.

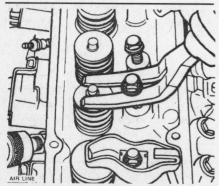

6.8 Apply air pressure to the cylinder, compress the valve spring and remove the keepers

10 Install the pushrods.
11 Install the rocker arms, fulcrums, and fulcrum bolts.
12 Turn the crankshaft until the lifter is all the way down.
13 Tighten the fulcrum bolts to the torque listed in this Chapter's Specifications in two stages.
14 Replace the valve covers and gaskets as described in Section 4.
15 Start the engine and check for roughness and/or noise.

6   Valve springs, retainers and seals - replacement

Refer to illustrations 6.8, 6.14 and 6.15
Note: *Broken valve springs and defective valve stem seals can be replaced without removing the cylinder head. Two special tools and a compressed air source are normally required to perform this operation, so read through this Section carefully and rent or buy the tools before beginning the job. If compressed air is not available, a length of nylon rope can be used to keep the valves from falling into the cylinder during this procedure.*
1 Refer to Section 4 and remove the valve cover from the affected cylinder head. If all of the valve stem seals are being replaced, remove both valve covers.
2 Remove the spark plug from the cylinder which has the defective component. If all of the valve stem seals are being replaced, all of the spark plugs should be removed.
3 Turn the crankshaft until the piston in the affected cylinder is at top dead center on the compression stroke (refer to Section 3 for instructions). If you are replacing all of the valve stem seals, begin with cylinder number one and work on the valves for one cylinder at a time. Move from cylinder-to-cylinder following the firing order sequence.
4 Thread an adapter into the spark plug hole and connect an air hose from a compressed air source to it. Most auto parts stores can supply the air hose adapter. **Note:** *Many cylinder compression gauges utilize a screw-in fitting that may work with your air*

hose quick-disconnect fitting.
5 Remove the bolt, pivot and rocker arm for the valve with the defective part and pull out the pushrod. If all of the valve stem seals are being replaced, all of the rocker arms and pushrods should be removed (refer to Section 5).
6 Apply compressed air to the cylinder. The valves should be held in place by the air pressure.
7 If you do not have access to compressed air, an alternative method can be used. Position the piston at a point approximately 45-degrees before TDC on the compression stroke, then feed a long piece of nylon rope through the spark plug hole until it fills the combustion chamber. Be sure to leave the end of the rope hanging out of the engine so it can be removed easily. Use a large breaker bar and socket to rotate the crankshaft in the normal direction of rotation until slight resistance is felt.
8 Stuff shop rags into the cylinder head holes above and below the valves to prevent parts and tools from falling into the engine, then use a valve spring compressor to compress the spring/damper assembly **(see illustration)**. Remove the keepers with small needle-nose pliers or a magnet. **Note:** *A couple of different types of tools are available for compressing the valve springs with the head in place. One type grips the lower spring coils and presses on the retainer as the knob is turned, while the other type utilizes the rocker arm bolt for leverage. Both types work very well, although the lever type is usually less expensive.*
9 Remove the spring retainer and valve spring assembly, then remove the umbrella-type guide seal **Note:** *If air pressure fails to hold the valve in the closed position during this operation, the valve face or seat is probably damaged. If so, the cylinder head will have to be removed for additional repair operations.*
10 Wrap a rubber band or tape around the top of the valve stem so the valve will not fall into the combustion chamber, then release the air pressure. **Note:** *If a rope was used instead of air pressure, turn the crankshaft slightly in the direction opposite normal rotation.*

Chapter 2 Part B 3.0L V6 engine

6.14 Carefully place the seal in position, then tap it into place with a deep socket and light hammer

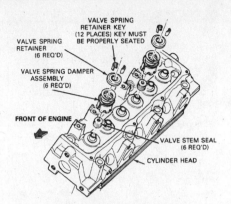

6.15 Valve seals, springs and related components

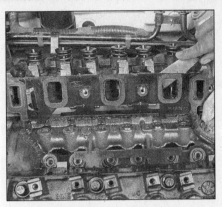

7.11 Use a scraper to remove the intake manifold gaskets

11 Inspect the valve stem for damage. Rotate the valve in the guide and check the end for eccentric movement, which would indicate that the valve is bent.

12 Move the valve up-and-down in the guide and make sure it does not bind. If the valve stem binds, either the valve is bent or the guide is damaged. In either case, the head will have to be removed for repair.

13 Reapply air pressure to the cylinder to retain the valve in the closed position, then remove the tape or rubber band from the valve stem. If a rope was used instead of air pressure, rotate the crankshaft in the normal direction of rotation until slight resistance is felt.

14 Lubricate the valve stem with engine oil and install a new seal onto the guide **(see illustration)**. *Caution: The intake and exhaust seals are different. Make sure you install the correct seal on each valve. On 1995 and later models the intake valve seal has a silver band and the exhaust seal has a red band.*

15 Install the spring/damper assembly in position over the valve **(see illustration)**.

16 Install the valve spring retainer. Compress the valve spring assembly.

17 Position the keepers in the upper groove. Apply a small dab of grease to the inside of each keeper to hold it in place if necessary. Remove the pressure from the spring tool and make sure the keepers are seated.

18 Disconnect the air hose and remove the adapter from the spark plug hole. If a rope was used in place of air pressure, pull it out of the cylinder.

19 Refer to Section 5 and install the rocker arm(s) and pushrod(s).

20 Install the spark plug(s) and hook up the wire(s).

21 Refer to Section 4 and install the valve cover(s).

22 Start the engine, then check for oil leaks and unusual sounds coming from the valve cover area.

7 Intake manifold - removal and installation

Refer to illustrations 7.11, 7.13, 7.14, 7.15, and 7.18

Removal

1 Relieve the fuel pressure (see Chapter 4).

2 Disconnect the negative cable at the battery.

3 Drain the cooling system (see Chapter 1).

4 Disconnect the fuel rails and lines. Cap the lines to prevent entry of dirt.

5 Remove the throttle body (see Chapter 4).

6 Label and disconnect all wiring, vacuum and coolant hoses from the intake manifold.

7 Remove the distributor (1993 and 1994 models - see Chapter 5) or camshaft position sensor (1995 and later models - see Chapter 6).

8 Remove the pushrod from number three cylinder intake valve (see Section 5).

9 Remove the manifold attaching bolts/studs (this requires a Torx driver bit), noting the location of studs for reinstallation.

10 Remove the intake manifold. It may be necessary to pry between the manifold and block, using the water pump lug as a fulcrum. Use care to avoid damaging the machined surfaces. *Caution: Do not pry between the gasket surfaces or leaks will develop.* The fuel rails and injectors may remain in place. Lift the manifold out of the engine compartment.

Installation

11 Clean all traces of old gasket material from the mating surfaces **(see illustration)**. Use a solvent such as lacquer thinner or acetone, if necessary.

12 Lightly oil all threaded fasteners prior to assembly.

13 Apply RTV sealant to the intersections of the cylinder head and block **(see illustration)**.

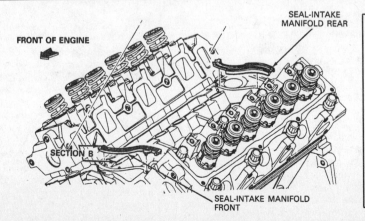

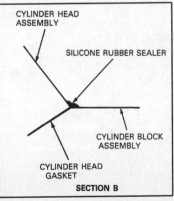

7.13 Apply RTV sealant in the four locations shown before installing the end seals

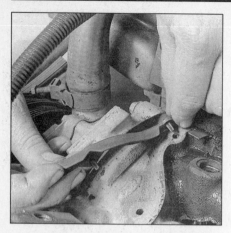

7.14 The end seals have locating pins which must be pressed into place

7.15 Be sure the locking tabs on the gaskets are engaged

14 Install the front and rear manifold end seals **(see illustration)**.
15 Position the intake manifold gaskets in place over the tabs on the cylinder heads **(see illustration)**.
16 Apply RTV sealer on top of the end seals. **Note:** *Do not allow the sealer to dry prior to installation.*
17 Carefully lower the manifold into place. Avoid displacing the seals and gaskets.
18 Install the bolts and studs, tightening them in the sequence shown to the torque listed in this Chapter's Specifications **(see illustration)**.
19 If removed, install the thermostat housing with a new gasket (see Chapter 3).
20 The remainder of the installation is the reverse of the removal procedure. If equipped with a camshaft position sensor, install it as described in Chapter 6, Section 5.
21 Refill the cooling system and connect the battery.
22 Run the engine and check for vacuum and fluid leaks.

8 Exhaust manifolds - removal and installation

Refer to illustrations 8.6 and 8.13

Removal

Warning: *Allow the engine to cool completely before following this procedure.*

1 Disconnect the negative cable at the battery.
2 Raise the vehicle and support it securely on jackstands.
3 Spray the exhaust pipe fasteners and exhaust manifold fasteners with rust penetrant and allow it to soak in for about 10 minutes. If there is any resistance during removal, stop and respray the fasteners to prevent them from breaking off.
4 Unbolt the exhaust pipe(s) from the appropriate manifold(s) as described in Chapter 4.

Right side

5 Remove the right side spark plugs and secure the wires out of the way (see Chapter 1).
6 Unbolt and remove the manifold **(see illustration)**.

Left side

7 Remove the oil dipstick tube bracket nut and either rotate the tube out of the way or remove it completely.
8 Remove the EGR tube (see Chapter 6).
9 Unbolt and remove the manifold.

Installation

10 Clean the mating surfaces of the head and manifold.
11 Clean and lightly oil the threads of all fasteners.
12 Install the manifold and tighten all fasteners snugly. Starting in the center and working out to the ends, tighten the bolts to the torque listed in this Chapter's Specifications.

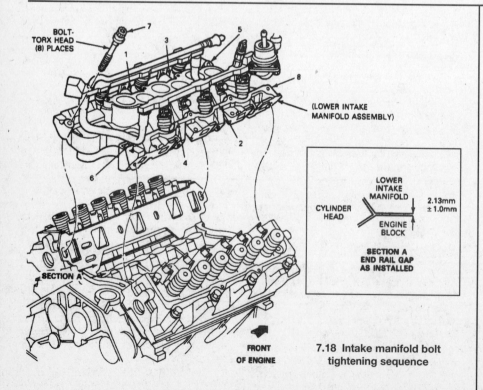

7.18 Intake manifold bolt tightening sequence

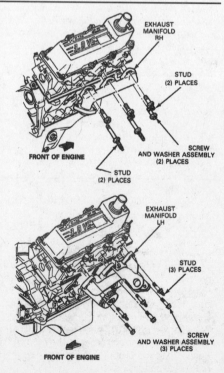

8.6 Exhaust manifold installation details

Chapter 2 Part B 3.0L V6 engine

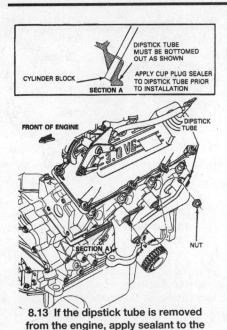

8.13 If the dipstick tube is removed from the engine, apply sealant to the lower section and press it into the block until seated

9.14 Once the bolts are removed, pry the head loose at a point where the gasket surfaces won't be damaged

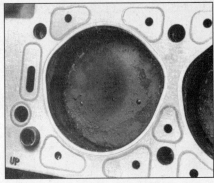

9.19 Position the new gasket over the dowels - make sure the UP (shown) or TOP mark is visible

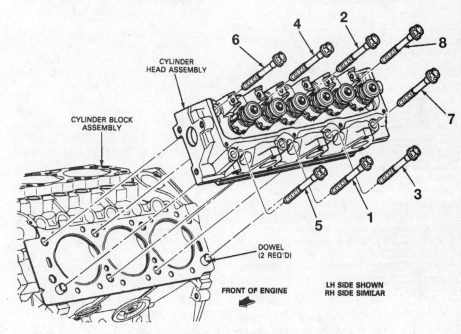

9.20 Cylinder head bolt tightening sequence

13 Reinstall all parts removed for access. If the dipstick was removed, apply RTV sealant to the base of the dipstick tube **(see illustration)**. Be sure to press the dipstick tube all the way into the block until seated.
14 Connect the exhaust pipe to the manifold(s).
15 Lower the vehicle and connect the battery cable.
16 Run the engine and check for exhaust leaks.

9   Cylinder heads - removal and installation

Refer to illustrations 9.14, 9.19 and 9.20

Removal

1 Disconnect the negative cable at the battery.
2 Drain the cooling system (see Chapter 1), including removing the engine block drain plugs.
3 Remove the air cleaner outlet tube and air cleaner assembly (see Chapter 4).
4 Remove the intake manifold (see Section 7).
5 Remove the drivebelts (see Chapter 1).
6 If removing the left cylinder head, remove the air conditioning compressor mounting bolts from the bracket. Without disconnecting the hoses, carefully lift the compressor off and secure it aside.
7 Remove the power steering pump mounting bracket bolts. Leave the pump and hoses connected and secure the pump and bracket assembly aside.
8 Remove the ignition coil bracket and dipstick tube.
9 Remove the alternator and the mounting bracket from the right cylinder head (see Chapter 4).
10 Remove the ground strap and throttle cable bracket.
11 Remove the exhaust manifold(s) (see Section 8).
12 Remove the valve cover(s) (see Section 4).
13 Loosen the rocker arms and remove pushrods (see Section 5).
14 Unbolt and remove the cylinder heads. Pry them up at the point shown **(see illustration)**. **Caution:** *Do not pry the heads at the sealing surfaces.*

Installation

15 Thoroughly clean and inspect the gasket mating surfaces for cracks and warpage. It well help to lay a shop cloth over the lifters and other exposed engine areas to prevent old gasket material and other debris from falling in the engine.
16 Service the head(s) as needed (see Chapter 2 Part D).
17 Be sure all bolt and bolt hole threads are clean, dry and in good condition.
18 Lightly oil all bolt and stud threads prior to installation.
19 Carefully position the head gaskets on the block using the dowels for alignment **(see illustration)**. Notice that the left and right head gaskets are different. **Caution:** *Do not use any kind of gasket sealant material on the head gaskets.*
20 Install the heads and head bolts and tighten them in two steps following the sequence shown **(see illustration)** to the torque listed in this Chapter's Specifications.
21 Installation of the remaining parts is the reverse of removal.
22 Refill cooling system (see Chapter 1).
23 Run the engine and check for leaks and proper operation.

10.4 Remove the vibration damper with the appropriate puller

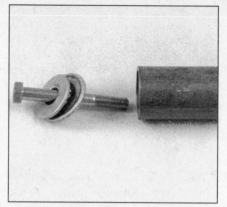

10.8 A piece of pipe, a bolt and several large washers can be used to fabricate a seal installation tool

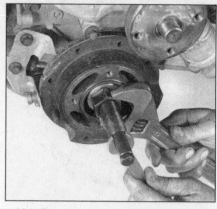

10.9 Press the crankshaft damper into place with an installation tool

10 Crankshaft front oil seal - replacement

Refer to illustrations 10.4, 10.8 and 10.9

1 Disconnect the negative cable at the battery.
2 Remove the radiator (see Chapter 3).
3 Remove the drivebelts (see Chapter 1) and the crankshaft pulley.
4 Remove the vibration damper bolt. Install a puller and remove the vibration damper **(see illustration)**.
5 Carefully pry the old seal from the cover using a seal remover or a screwdriver. Do not nick or scratch the crankshaft or seal bore.
6 Thoroughly clean the seal area.
7 Apply engine oil to the seal lips. Slide the seal into place with the spring side facing the engine.
8 Press the seal into place using a seal installer or a large socket and hammer. A suitable installation tool can be fabricated as shown **(see illustration)**. Be sure the seal seats against the timing chain cover.
9 Reinstall the vibration damper using a damper installation tool **(see illustration)**. Tighten the bolt to the torque listed in this Chapter's Specifications.
10 The remainder of the installation is the reverse of removal.
11 Refill the cooling system and run the engine, checking for leaks.

11 Timing chain cover - removal and installation

Refer to illustrations 11.10 and 11.12

Removal

Note: *The engine must be removed from the vehicle to gain access to the oil pan on 4WD models.*

1 Disconnect the negative cable at the battery.
2 Drain the oil and coolant and remove the drivebelt (see Chapter 1).
3 Remove the fan shroud, fan/clutch assembly, water pump pulley and lower radiator hose (see Chapter 3).
4 Remove the air cleaner intake duct.
5 Remove the air conditioning compressor mounting bolts from the bracket. Without disconnecting the hoses, carefully lift the compressor off and secure it aside. Remove the power steering pump mounting bracket bolts. Leave the pump and hoses connected and secure the pump and bracket assembly aside.
6 Remove the alternator (see Chapter 5) and the alternator bracket.
7 Remove the water pump (see Chapter 3). **Note:** *It's possible to leave the water pump bolted to the timing cover and remove them both together - do not remove bolts 11 through 15* **(see illustration 11.12)**.
8 Remove the crankshaft pulley and damper (see Section 10).
9 Remove the oil pan (see Section 15).
10 Unbolt the timing chain cover and remove it **(see illustration)**.

Installation

11 While the cover is off, replace the front oil seal. Drive the old one out from the rear. Place the cover on a block of wood and install a new seal using a hammer and a suitable driver.

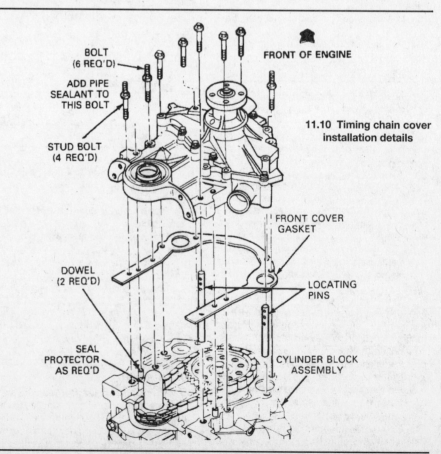

11.10 Timing chain cover installation details

Chapter 2 Part B 3.0L V6 engine

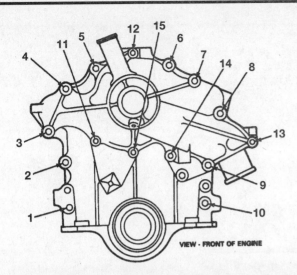

11.12 The timing chain cover and water pump bolts are different diameters and lengths - bolts one through ten secure the timing chain; bolts 11 through 15 secure the water pump

1* M8x1.25x72.25
2* M8x1.25x72.25
3* M8x1.25x70.0 (1993 and 1994); M8x1.25x73.0 (1995 on)
4 M8x1.25x70.0 (1993 and 1994); M8x1.25x73.0 (1995 on)
5 M8x1.25x42.0 (1993 and 1994); M8x1.25x73.0 (1995 on)
6 M8x1.25x99.3 (1993 and 1994); M8x1.25x104.3 (1995 on)
7 M8x1.25x70.0 (1993 and 1994); M8x1.25x73.0 (1995 on)
8 M8x1.25x70.0 (1993 and 1994); M8x1.25x73.0 (1995 on)
9 M8x1.25x70.0 (1993 and 1994); M8x1.25x42.0 (1995 on)
10 M8x1.25x42.0 (1993 and 1994); M8x1.25x52.0 (1995 on)
11 M6x1.0x25.0 (1993 and 1994); M6x1.0x28.5 (1995 on)
12 M6x1.0x25.0 (1993 and 1994); M6x1.0x28.5 (1995 on)
13 M6x1.0x25.0 (1993 and 1994); M6x1.0x28.5 (1995 on)
14 M6x1.0x25.0 (1993 and 1994); M6x1.0x28.5 (1995 on)
15 M6x1.0x25.0 (1993 and 1994); M6x1.0x28.5 (1995 on)

* Apply pipe sealant (Ford part no. D6AZ-19558-A or equivalent) to the threads of these three bolts

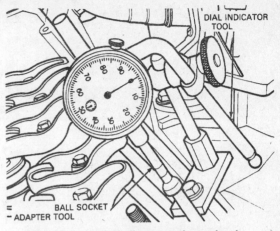

12.3 Mount a dial indicator on the pushrod to detect valve lifter movement

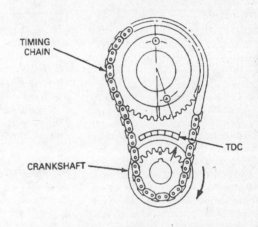

12.4 Turn the crankshaft clockwise to TDC, removing the slack from the timing chain

12 Install the cover and tighten all fasteners to the torque listed in this Chapter's Specifications. All the 6mm diameter bolts are the same length. However, the 8mm are of several different lengths (see illustration). **Note: Use Teflon pipe sealant (Ford D6AZ-19558-A or equivalent) on all cover bolts which go into the water jacket (bolt numbers 1, 2 and 3). Apply a light film of oil to the other fasteners.**
13 The remainder of the installation is the reverse of removal.
14 Refill the cooling system and run the engine, checking for leaks.

12 Timing chain and sprockets - check, removal and installation

Refer to illustrations 12.3, 12.4, 12.11, 12.12 and 12.14

Check

1 Remove the left (driver's side) valve cover (see Section 4).
2 Loosen the number 5 cylinder exhaust valve rocker arm (4th one back) and rotate it aside.
3 Mount a dial indicator on the end of the pushrod (see illustration).
4 Turn the crankshaft clockwise until the number 1 piston is at Top Dead Center (see Section 3). The damper timing mark should point to TDC on the timing degree indicator. This will take up slack on the right side of the chain (see illustration).
5 Zero the dial indicator.
6 Slowly turn the crankshaft counterclockwise until the slightest movement is seen on the dial indicator. Check to see how far the damper TDC timing mark has moved away from the pointer.
7 If the reading on the timing indicator exceeds 6-degrees, replace the timing chain and sprockets.

Removal

8 Turn the crankshaft clockwise until the number one piston is at Top Dead Center (see Section 3).
9 Disconnect the negative cable at the battery.
10 Remove the timing chain cover (see Section 11).
11 Unbolt the camshaft sprocket (see illustration). Be sure the engine does not turn during bolt removal. Carefully remove the sprocket with the chain still on it.

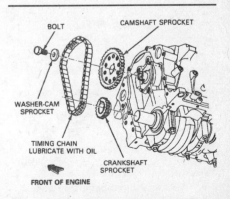

12.11 Timing chain and related components

Chapter 2 Part B 3.0L V6 engine

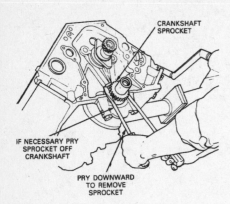

12.12 Use two screwdrivers, as shown, to pry the sprocket off the crankshaft

12 Pry the crankshaft sprocket off **(see illustration)**. Do not lose the small crankshaft key.
13 Check the timing chain and sprockets for visible wear and damage. Replace as a set if necessary.

Installation

14 Slide both sprockets and the chain onto the camshaft and crankshaft with the timing marks aligned **(see illustration)**. Tap the crankshaft sprocket into place with a brass drift and hammer.
15 Install the camshaft bolt and washer and tighten the bolts to the torque listed in this Chapter's Specifications. **Caution:** *The camshaft bolt has a drilled oil passage. Be sure that passage is open before installation. DO NOT substitute any other type bolt.*
16 Apply oil to the timing chain and sprockets.
17 Install the timing chain cover (see Section 11) and the remaining components removed for installation.

13.3 A clearly labeled box with subdividers is a handy way to keep the lifters in order so that they can be reinstalled in the lifter bores they were removed from

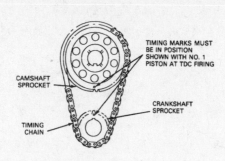

12.14 Align the timing marks on the camshaft and crankshaft sprockets

13 Valve lifters - removal, inspection and installation

Refer to illustrations 13.2, 13.3, 13.4, 13.7a and 13.7b

Removal and installation

1 Remove the intake manifold (see Section 7) and pushrods (see Section 5).
2 Loosen the two bolts securing the roller lifter guide plate retainer and lift the retainer off **(see illustration)**. Remove all the guide plates.
3 Remove the lifters with a magnet and place them in a rack in sequence so they can be replaced in their original location **(see illustration)**.
4 If stuck, carefully use a lifter puller tool to remove them by rotating them back and forth to loosen them from gum or varnish deposits **(see illustration)**. Carburetor cleaner can be used to loosen these deposits. **Caution:** *Do no use pliers to remove the lifers unless you are replacing them. Pliers can easily damage the lifters.*
5 Clean the lifters and apply engine oil to both the lifters and bores before installing. Be

13.4 The lifters in an engine that has accumulated many miles may have to be removed with a special tool

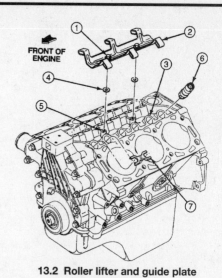

13.2 Roller lifter and guide plate installation details

1 Bolt
2 Guide plate retainer
3 Lifter bore
4 Washer
5 Guide plate retainer threaded hole
6 Lifter
7 Guide plate

sure to install the lifters in their original bores. Align the flats on the lifter body and install the guide plates. Install the retainer and tighten the retainer bolts to the torque listed in this Chapter's Specifications.
6 Install the pushrods and the intake manifold.

Lifter inspection

7 Check that the roller on each lifter turns freely and shows no signs of pitting, scoring or other wear **(see illustration)**. Check the pushrod seat in each lifter for wear. Also make sure each lifter fits in its original bore without excessive looseness **(see illustration)**. Wear is only normally encountered at very high mileages or in cases of neglected

13.7a The roller on roller lifters must turn freely - check for wear and excessive play as well

Chapter 2 Part B 3.0L V6 engine

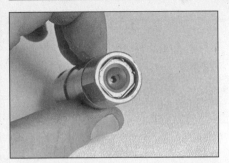

13.7b Check the pushrod seat in the top of each lifter for wear

engine lubrication.
8 Press down on the plunger using a Phillips screwdriver to make sure it operates freely without binding.
9 If the lifters show signs of wear or binding they must be replaced.
10 See Chapter 2 Part D for further information.

14 Camshaft - removal, inspection and installation

Refer to illustration 14.9

Note: Refer to Chapter 2, Part C and check the camshaft lobe lift as described, before removing the camshaft from the engine block.

Removal

1 Remove the intake manifold (see Section 7).
2 Remove the valve covers (see Section 4).
3 Loosen the rocker arm bolts and rotate the rocker arms to the side.
4 If the pushrods are to be reused, remove them from the engine and place them in a holder to keep them in order (see Section 5).
5 Remove the valve lifters from the engine (see Section 13).
6 Remove the timing chain cover (see Section 11).
7 Unbolt the camshaft sprocket. Be sure you do not turn the engine while doing so.
8 Remove the timing chain and sprockets (see Section 12).
9 Remove the bolts securing the camshaft thrust plate to the engine block **(see illustration)**.
10 Slowly withdraw the camshaft from the engine, being careful not to nick, scrape or otherwise damage the bearings with the cam lobes.

Inspection

11 See Chapter 2, Part D for the camshaft and lifter inspection procedures.

Installation

12 Lubricate the camshaft journals liberally with engine oil and apply engine assembly lube to the cam lobes.
13 Slide the camshaft into position, being careful not to scrape or nick the bearings.

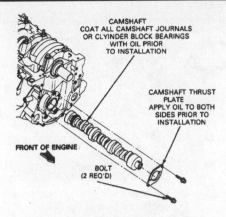

14.9 Camshaft installation details

14 Install the camshaft thrust plate. Tighten the thrust plate retaining bolts to the torque listed in this Chapter's Specifications. Install the timing chain and sprockets.
15 Install the hydraulic valve lifters in their original bores if the original lifters are being used. Make sure they are coated with engine assembly lube.
16 The remainder of the installation procedure is the reverse of removal. On models equipped with a camshaft position sensor, install it as described in Chapter 6, Section 5).

15 Oil pan - removal and installation

2WD models

Refer to illustration 15.12

1 Disconnect the negative cable at the battery.
2 Remove the oil dipstick.
3 Unbolt the fan shroud and lay it over the fan (see Chapter 3).
4 If you're working on a 1993 or 1994 model, remove the distributor (see Chapter 5). **Caution:** *If the distributor is not removed, it may be damaged when you raise the engine.*

5 Detach the engine mounts from the frame.
6 Raise the vehicle and support it securely on jackstands.
7 If equipped, disconnect the low oil level sensor by removing the retainer clip at the sensor and then separating the connector from the sensor.
8 Drain the oil and install a new oil filter (see Chapter 1).
9 Remove the starter motor (see Chapter 5).
10 Remove the inspection cover from the front of the transmission.
11 Remove the passenger side axle beam (see Chapter 10). **Caution:** *Be sure to remove the brake caliper and tie it up out of the way.*
12 Unbolt the oil pan **(see illustration)**. Using a hoist, raise the engine approximately two inches to allow enough clearance to remove the oil pan. If it is stuck on, gently pry it off, taking care not to bend or distort the pan sealing flange. **Caution:** *The oil pan is a tight fit between the transmission spacer plate and the oil pump pick-up tube. Be careful not to damage the pick-up tube when you remove the pan. If necessary, remove the oil pump mounting bolt and lower the oil pump into the oil pan.*
13 Thoroughly clean the gasket surfaces of the block and pan.
14 Install a new gasket on the oil pan using contact adhesive to hold it in place.
15 Apply a 4 to 5 mm (1/4-inch) bead of RTV sealer to the junctions of the rear main bearing cap and engine block and also to the timing chain cover-to-block surface **(see illustration 15.12)**. Instructions are included with the new gasket. Don't allow the sealant to cure before installing the pan.
16 Position the pan on the engine. Install the oil pan bolts and tighten them to the torque listed in this Chapter's Specifications.
17 Reinstall any parts removed and refill with engine oil.
18 Run the engine and check for leaks.

4WD models

Note: The engine must be removed from the

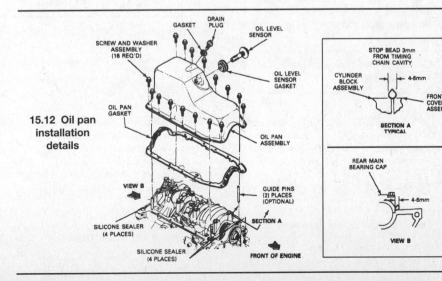

15.12 Oil pan installation details

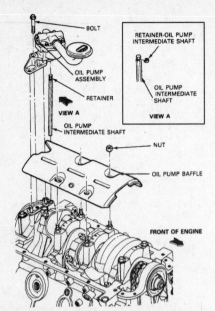

16.2 Oil pump and related components

17.4 Do not scratch the crankshaft when punching a hole in the seal

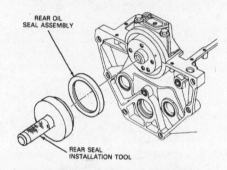

17.10a Rear seal installation

17.6 You may be able to gently pry the seal out, but don't scratch the sealing surfaces

17.10b If the special tool isn't available, tap around the seal, slowly working it into position

vehicle to gain access to the oil pan on 4WD models.
19 Remove the engine as described in Part D of this Chapter. Mount the engine on a suitable engine stand, right-side up.
20 Unbolt the oil pan. If it is stuck on, gently pry it off, taking care not to bend or distort the pan sealing flange.
21 Thoroughly clean the gasket surfaces of the block and pan.
22 Install a new gasket on the oil pan using contact adhesive to hold it in place.
23 Apply a 4 to 5 mm (1/4-inch) bead of RTV sealer to the junctions of the rear main bearing cap and engine block and also to the timing chain cover-to-block. Instructions are included with the new gasket. Don't allow the sealant to cure before installing the pan.
24 Position the pan on the engine. Install the oil pan bolts and tighten them to the torque listed in this Chapter's Specifications.
25 Install the engine.
26 Reinstall any parts removed and refill with engine oil.
27 Run the engine and check for leaks.

16 Oil pump and pick-up - removal and installation

Refer to illustration 16.2
Note: *We recommend oil pump replacement any time the engine is rebuilt and when the pump has high mileage or is otherwise suspect.*
1 Remove the oil pan as described in Section 15.
2 While supporting the oil pump, remove the pump-to-rear main bearing cap bolt **(see illustration)**.
3 Lower the pump and remove it along with the pump driveshaft.
4 If a new oil pump is installed, make sure the pump driveshaft is mated with the shaft inside the pump.
5 Position the pump on the engine and make sure the hex in the upper end of the driveshaft is aligned with the lower end of the distributor or camshaft position sensor shaft. The distributor or camshaft position sensor drives the oil pump, so it is absolutely essential that the components mate properly.
6 Install the mounting bolt and tighten it to the torque listed in this Chapter's Specifications.
7 Install the oil pan and add oil.
8 The remainder of installation is the reverse of removal.

17 Rear main oil seal - replacement

Refer to illustration 17.4, 17.6, 17.10a and 17.10b
1 Remove the transmission (see Chapter 7).
2 On manual transmission equipped vehicles, remove the clutch assembly (see Chapter 8).
3 Remove the flywheel/driveplate.
4 Using a sharp awl, carefully punch one hole into the seal between the lip and the engine block **(see illustration)**.
5 Screw in the threaded end of a special tool available at most auto parts stores, or equivalent. Pull the seal out with the tool.
6 If the special tool is not available, you may be able to carefully pry out the old seal using a screwdriver **(see illustration)**. Take care to prevent nicking or scratching the seal bore or crankshaft.
7 Thoroughly clean and inspect the oil seal area.
8 Lubricate the new seal with engine oil.
9 Start the seal into its recess with the spring side facing the engine.
10 Press the seal into place with a special tool available at most auto parts stores, or equivalent **(see illustration)**. You can also tap the seal in place using a hammer and hardwood drift or section of the appropriate size plastic pipe **(see illustration)**. Work slowly around the seal until it is flush with the rear face of the block.
11 Install the flywheel/driveplate.
12 The remainder of installation is the reverse of removal.
13 Run the engine and check for leaks.

18 Flywheel/driveplate - removal and installation

Refer to Chapter 2 Part C for the flywheel/driveplate removal and installation procedures, but use the torque specifications found in this Chapter's Specifications.

19 Engine mounts - removal and installation

Refer to Chapter 2 Part C for engine mount removal and installation procedures, but use the torque specifications found in this Chapter's Specifications.

Chapter 2 Part C
4.0L V6 engine

Contents

Section		Section	
Camshaft - removal, inspection and installation	14	Oil pump - removal and installation	16
CHECK ENGINE light	See Chapter 6	Oil pan and baffle - removal and installation	15
Compression check	See Chapter 2D	Repair operations possible with the engine in the vehicle	2
Crankshaft pulley and front oil seal - removal and installation	10	Rocker arms and pushrods - removal, inspection and installation	5
Crankshaft oil seals - replacement	17		
Cylinder heads - removal and installation	9	Spark plug replacement	See Chapter 1
Drivebelt check, adjustment and replacement	See Chapter 1	Timing chain and sprockets - inspection, removal and installation	12
Engine mounts - check and replacement	19		
Engine oil and filter change	See Chapter 1	Timing chain cover - removal and installation	11
Engine overhaul - general information	See Chapter 2D	Top Dead Center (TDC) for number one piston - locating	3
Engine - removal and installation	See Chapter 2D	Valve covers - removal and installation	4
Exhaust manifolds - removal and installation	8	Valve lifters - removal, inspection and installation	13
Flywheel/driveplate - removal and installation	18	Valve springs, retainers and seals - replacement	6
General information	1	Valves - servicing	See Chapter 2D
Intake manifold - removal and installation	7	Water pump - removal and installation	See Chapter 3

Specifications

General
Cylinder numbers (front-to-rear)
   Left (driver's side) 4-5-6
   Right side 1-2-3
Firing order 1-4-2-5-3-6

Camshaft
Lobe lift (intake and exhaust)
   1993 through 1998 .... 0.2756 inch
   1999 and 2000 .... 0.2720 inch
Endplay
   Standard .... 0.0008 to 0.004 inch
   Service limit .... 0.009 inch
Journal diameter (standard)
   No. 1 .... 1.951 to 1.952 inch
   No. 2 .... 1.937 to 1.938 inch
   No. 3 .... 1.922 to 1.923 inch
   No. 4 .... 1.907 to 1.908 inch
Bearing inside diameter (standard)
   No. 1 .... 1.954 to 1.955 inch
   No. 2 .... 1.939 to 1.940 inch
   No. 3 .... 1.919 to 1.920 inch
   No. 4 .... 1.924 to 1.925 inch
Journal-to-bearing (oil) clearance
   Standard .... 0.001 to 0.0026 inch
   Service limit .... 0.006 inch

Oil pan-to-transmission spacer thickness
Coded yellow .... 0.010 inch
Coded blue .... 0.020 inch
Coded pink .... 0.030 inch

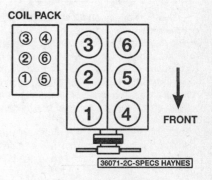

Cylinder numbering and coil terminal location

Torque specifications

	Ft-lbs (unless otherwise indicated)
Camshaft sprocket bolt	44 to 50
Camshaft thrust plate bolts	84 to 120 in-lbs
Crankshaft pulley bolt	
Step 1	30 to 37
Step 2	Tighten an additional 90-degrees
Cylinder head bolts (in sequence - **see illustration 9.18**)	
Step 1	22 to 26
Step 2	52 to 56
Step 3	Tighten an additional 90-degrees
Flywheel/driveplate bolts	
Step 1	108 to 132 in-lbs
Step 2	50 to 55
Engine mount insulator-to-frame nuts	71 to 94
Engine mount insulator-to-bracket nuts	65 to 85
Engine mount bracket-to-block bolts	45 to 60
Exhaust manifold bolts	19
Exhaust pipe-to-manifold nuts	20
Lower intake manifold bolts/nuts	
1993 through 1997	
Step 1	36 in-lbs
Step 2	72 in-lbs
Step 3	132 in-lbs
Step 4	180 in-lbs
1998 through 2000	
Step 1	27 in-lbs
Step 2	89 in-lbs
Step 3	120 in-lbs
Step 4	144 in-lbs
Oil pump drive gear bolt	156 to 180 in-lbs
Oil pump pick-up tube-to-pump bolts	84 to 120 in-lbs
Oil pump-to-block bolts	156 to 180 in-lbs
Oil pan-to-block bolts	60 to 84 in-lbs
Timing chain cover bolts	156 to 180 in-lbs
Timing chain guide bolts	84 to 108 in-lbs
Timing chain tensioner bolts	84 to 108 in-lbs
Rocker arm shaft support bolts	
Step 1	24
Step 2	Tighten an additional 90-degrees
Valve cover bolts	60 to 72 in-lbs

1 General information

This Part of Chapter 2 is devoted to in-vehicle repair procedures for the 4.0L V6 engine. Procedures such as timing chain and sprocket and oil pan removal are also included, however they require removal of the engine from the vehicle. All information concerning engine removal and installation and engine block and cylinder head overhaul can be found in Part D of this Chapter.

The following repair procedures are based on the assumption that the engine is installed in the vehicle. If the engine has been removed from the vehicle and mounted on a stand, many of the steps outlined in this Part of Chapter 2 will not apply.

The Specifications included in this Part of Chapter 2 apply only to the procedures contained in this Part. Part D of Chapter 2 contains the Specifications necessary for cylinder head and engine block rebuilding.

2 Repair operations possible with the engine in the vehicle

Many major repair operations can be accomplished without removing the engine from the vehicle.

Clean the engine compartment and the exterior of the engine with some type of degreaser before any work is done. It will make the job easier and help keep dirt out of the internal areas of the engine.

Depending on the components involved, it may be helpful to remove the hood to improve access to the engine as repairs are performed (refer to Chapter 11 if necessary). Cover the fenders to prevent damage to the paint. Special pads are available, but an old bedspread or blanket will also work.

If vacuum, exhaust, oil or coolant leaks develop, indicating a need for gasket or seal replacement, the repairs can generally be made with the engine in the vehicle. The intake and exhaust manifold gaskets and cylinder head gaskets are all accessible with the engine in place. **Note:** *Removing the oil pan on a 4.0L engine requires removing the engine from the vehicle.*

Exterior engine components, such as the intake and exhaust manifolds, the water pump, the starter motor, the alternator, the distributor and the fuel system components can be removed for repair with the engine in place.

Since the cylinder heads can be removed without pulling the engine, valve component servicing can also be accomplished with the engine in the vehicle. Replacement of the timing chain and sprockets requires removal of the oil pan, so it is not possible with the engine in the vehicle.

3 Top Dead Center (TDC) for number one piston - locating

Refer to illustrations 3.5a and 3.5b
Note: *The 4.0L engine is not equipped with a*

Chapter 2 Part C 4.0L V6 engine

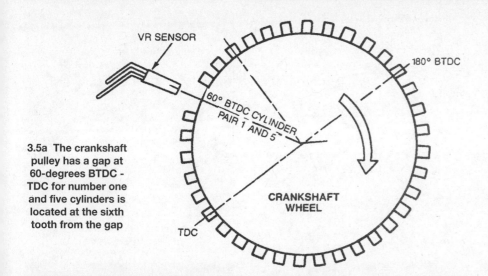

3.5a The crankshaft pulley has a gap at 60-degrees BTDC - TDC for number one and five cylinders is located at the sixth tooth from the gap

distributor. Piston position must be determined by feeling for compression at the number one spark plug hole, then aligning the ignition timing marks as described in Step 5.

1 Top Dead Center (TDC) is the highest point in the cylinder that each piston reaches as it travels up-and-down when the crankshaft turns. Each piston reaches TDC on the compression stroke and again on the exhaust stroke, but TDC generally refers to piston position on the compression stroke.

2 Positioning the piston(s) at TDC is an essential part of many other repair procedures discussed in this manual.

3 Before beginning this procedure, be sure to place the transmission in Neutral and apply the parking brake or block the rear wheels. Remove the spark plugs (see Chapter 1). Disable the ignition system by disconnecting the electrical connector from the ignition coil pack, located above the left valve cover.

4 In order to bring any piston to TDC, the crankshaft must be turned using one of the methods outlined below. When looking at the front of the engine, normal crankshaft rotation is clockwise.

a) The preferred method is to turn the crankshaft with a socket and ratchet attached to the bolt threaded into the front of the crankshaft.
b) A remote starter switch, which may save some time, can also be used. Follow the instructions included with the switch. Once the piston is close to TDC, use a socket and ratchet as described in the previous paragraph.
c) If an assistant is available to turn the ignition switch to the Start position in short bursts, you can get the piston close to TDC without a remote starter switch. Make sure your assistant is out of the vehicle, away from the ignition switch, then use a socket and ratchet as described in Paragraph a) to complete the procedure.

5 The crankshaft pulley has 35 teeth, evenly spaced every 10-degrees around the pulley, and a gap where a 36th tooth would be. The gap is located at 60-degrees Before Top Dead Center (BTDC) **(see illustration)**. Turn the crankshaft (see Paragraph 4) until you feel compression at the number one spark plug hole, then turn it slowly until the sixth tooth from the missing tooth is aligned with the Variable Reluctance (VR) sensor and the TDC notch is aligned with the pointer (located at the front of the engine) **(see illustration)**.

6 After the number one piston has been positioned at TDC on the compression stroke, TDC for any of the remaining pistons can be located by turning the crankshaft and following the firing order.

4 Valve covers - removal and installation

Refer to illustrations 4.7, 4.8, 4.9a, 4.9b, 4.9c, 4.12, 4.16, 4.20 and 4.22

Removal

1 Disconnect the negative cable from the battery.

2 Remove the accelerator control splash shield and air cleaner outlet tube. On 1996 models, disconnect the three vacuum hoses from the intake manifold vacuum outlet fitting.

Right valve cover

3 Remove the alternator and ignition coil pack (refer to Chapter 5).

4 Remove the bolt that secures the air conditioning line above the upper intake manifold (if not already done when removing the coil pack).

5 Detach the spark plug wires from the valve cover clips.

6 Carefully pry the two wiring harnesses loose from the valve cover with a tool such as a door panel clip remover (available at auto parts stores).

7 Disconnect the vacuum hose at the coupling above the valve cover **(see illustration)**.

8 Carefully lift the engine wiring harness clip with your thumb at the point shown to separate it from the valve cover **(see illustration)**. Don't pull on the harness.

3.5b The pointer on the front of the engine lines up with a notch in the crankshaft pulley to indicate TDC

4.7 On the passenger's side, disconnect the vacuum hose at the coupling above the valve cover

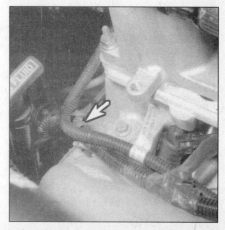

4.8 Lift the wiring harness off the stud with a thumbnail - don't pry it

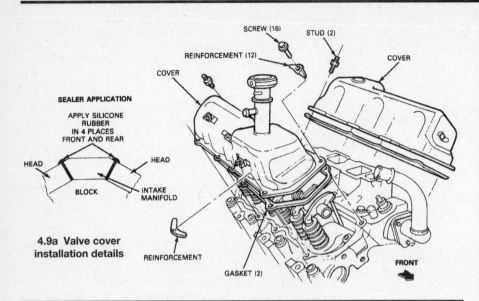

4.9a Valve cover installation details

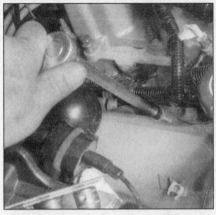

4.9b Remove the valve cover bolts with a ratchet and extension . . .

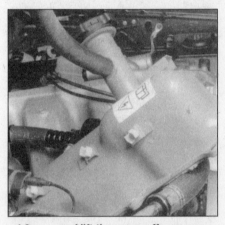

4.9c . . . and lift the cover off - you may have to angle it out, as show here - tap it gently with a soft-face mallet if necessary to break the gasket seal

9 Remove the valve cover bolts and reinforcing plates. Lift the valve cover off. Tap it gently with a soft-face hammer if necessary to break the gasket seal **(see illustrations)**.

Left valve cover

10 If you haven't already done so, remove the bolt from the air conditioning line above the upper intake manifold.
11 Disconnect the electrical connector for the air conditioning compressor clutch. Carefully pry the wiring harness from the back of the compressor with a door panel clip remover or similar tool.
12 Remove the mounting bolts from the air conditioning compressor (see Chapter 3). Lift the air conditioning compressor, then position it out of the way **(see illustration)**. **Warning:** *DO NOT disconnect any refrigerant lines!*
13 Disconnect the brake booster vacuum hose.
14 Label and disconnect the vacuum hoses from the tee on the plenum **(see illustration 10.6 in Chapter 1)**. Remove the upper intake manifold (see Chapter 4).
15 Detach the PCV hose. Remove the PCV hose and valve (see Chapter 1).
16 Carefully pry the wiring harnesses away from the valve cover with a door trim panel remover or similar tool **(see illustration)**. Place the harnesses out of the way.
17 Disconnect the driver's side spark plug wires from the plugs and detach them from the clips on the valve cover. Position the wires out of the way.
18 Carefully lift the engine wiring harness clip with your thumb to separate it from the valve cover. Don't pull on the harness.
19 Remove the bolt that secures the fuel line clip to the front of the engine. Move the fuel line just enough to provide access to the front valve cover bolt. **Warning:** *DO NOT disconnect any fuel lines without first relieving fuel pressure (see Chapter 4).*
20 Remove the valve cover bolts and reinforcing plates **(see illustration 4.9a)**. Lift the valve cover off **(see illustration)**. Tap it gently with a soft-face hammer if necessary to break the gasket seal.

Installation

21 Clean the gasket surfaces on the intake manifold, cylinder head and valve cover. Use a scraper to remove the pieces of old gasket material, then wipe off all residue with lacquer thinner or acetone.
22 Most valve cover gaskets are equipped

4.12 Disconnect the compressor electrical connector from the back of the compressor, then remove the mounting bolts and shift the compressor forward as shown - DO NOT disconnect any refrigerant lines

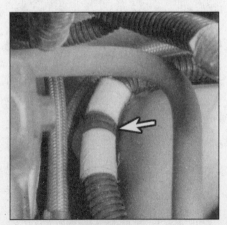

4.16 Wiring harnesses are secured to the valve cover by clips

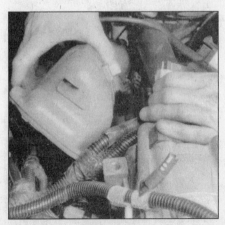

4.20 Move the wiring harnesses out of the way and lift the valve cover off

Chapter 2 Part C 4.0L V6 engine

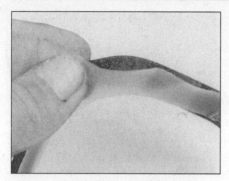

4.22 Just before installing the new gasket, peel the plastic film off the self-sticking sealant on the valve-cover side of the gasket

5.2b Remove the bolts evenly, two turns at a time, to prevent the shafts from being bent by valve spring pressure

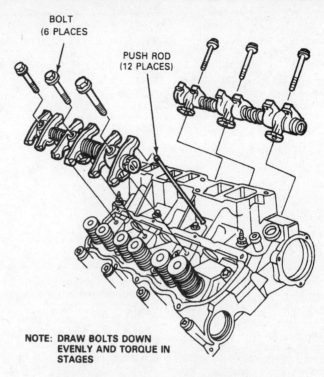

5.2a Rocker arm shaft assembly installation details

with self-sticking sealant on the valve cover side. Pull the plastic film off the gasket **(see illustration)** and stick the gasket to the valve cover.

23 **Note:** *In this step, apply silicone sealant to one side of the engine at a time (if both valve covers were removed), then install the valve cover.* Apply silicone sealant to the seam where the cylinder head joins the intake manifold **(see illustration 4.9a)**. On 1993 through 1995 models, apply a 1/8-inch ball of sealant to the valve cover bolt holes on the outer (exhaust) side of the cylinder head. This is not necessary on 1996 and later models.

24 The remainder of installation is the reverse of the removal Steps. Tighten the valve cover bolts evenly, starting with the center bolts and working out, to the torque listed in this Chapter's Specifications.

5 Rocker arms and pushrods - removal, inspection and installation

Removal

Refer to illustrations 5.2a, 5.2b, 5.3, 5.4a and 5.4b

1 Refer to Section 4 and remove the valve cover(s).

2 Loosen the rocker arm shaft support bolts two turns at a time, starting with the center bolt and working out, until the bolts can be removed by hand **(see illustrations)**.
3 Lift the rocker shaft assembly off the cylinder head **(see illustration)**. The pins will hold the components together. Mark each shaft assembly so it can be returned to it's original location.
4 Lift the pushrods out of the engine **(see illustration)**. Place the pushrods in order in a holder **(see illustration)** so they can be returned to their original positions. Be sure to store them so you can reinstall them with the same end facing up.

Inspection

Refer to illustration 5.5

5 Remove the pins and disassemble the

5.3 Once the bolts are loose, lift the rocker assembly off the engine

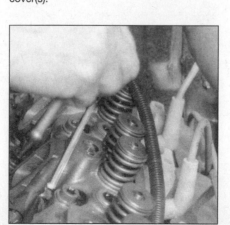

5.4a Pull the pushrods out of the lifters ..

5.4b ... and place them in a labeled holder so they can be returned to their original positions

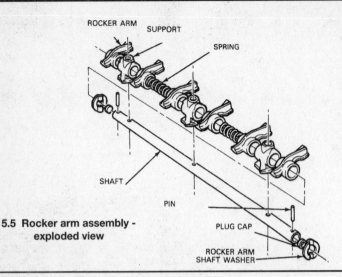

5.5 Rocker arm assembly - exploded view

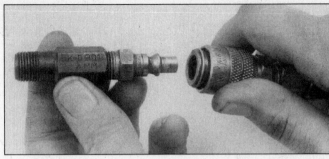

6.4 This is what the air hose adapter that threads into the spark plug hole looks like - they're commonly available from auto parts stores

rocker arm assembly **(see illustration)**. Place the parts in order on a clean workbench. Be sure you don't mix up the parts - they must be reassembled in exactly the same order they were before disassembly.

6 Check each rocker arm for wear, cracks and other damage, especially where the pushrods and valve stems contact the rocker arm faces.

7 Make sure all oil holes are open and not plugged. Plugging can be cleared with a piece of wire.

8 Check each rocker arm bore, and its corresponding position on the rocker shaft, for wear, cracks and galling. If the rocker arms or shaft are damaged, replace them with new ones.

9 Inspect the pushrods for cracks and excessive wear at the ends. Roll each pushrod across a piece of plate glass to see if it's bent (if it wobbles, it's bent).

10 If necessary, remove the plug from each end of the rocker shaft. Drill into one plug and insert a long steel rod through it to knock out the other plug. Knock out the first plug in the same manner.

11 If the rocker arm shaft plugs were removed, tap in new ones with a hammer and suitable drift.

12 Assemble the rocker assembly **(see illustration 5.5)**. Lubricate at friction points (pushrod ends, rocker arm bores and ends) with engine assembly lube.

13 Install new cotter pins in the ends of the rocker shaft. Be sure the rocker shaft oil holes will face down when the shaft is installed. The position of the oil holes is indicated by a notch on the front of each shaft.

Installation

14 Coat each end of each pushrod with assembly lube, then install them in the engine. If you are reinstalling the original pushrods, be sure to return them to their original positions.

15 Coat the rocker arm pads with assembly lube.

16 Install the rocker arm assembly on the engine. The notch on the front end of each shaft should face down.

17 Position the rocker arm ball ends in the pushrods.

18 Tighten the rocker shaft support bolts two turns at a time, working from the center bolt out, to the torque listed in this Chapter's Specifications.

19 The remainder of installation is the reverse of the removal Steps.

6 Valve springs, retainers and seals - replacement

Refer to illustrations 6.4, 6.9a, 6.9b and 6.10

Note: *Broken valve springs and defective valve stem seals can be replaced without removing the cylinder heads. Two special tools and a compressed air source are normally required to perform this operation, so read through this Section carefully and rent or buy the tools before beginning the job. If compressed air isn't available, a length of nylon rope can be used to keep the valves from falling into the cylinder during this procedure.*

1 Refer to Section 4 and remove the valve cover from the affected cylinder head. If all of the valve stem seals are being replaced, remove both valve covers.

2 Remove the spark plug from the cylinder which has the defective component. If all of the valve stem seals are being replaced, all of the spark plugs should be removed.

3 Turn the crankshaft until the piston in the affected cylinder is at top dead center on the compression stroke (refer to Section 3 for instructions). If you're replacing all of the valve stem seals, begin with cylinder number one and work on the valves for one cylinder at a time. Move from cylinder-to-cylinder following the firing order sequence (see this Chapter's Specifications).

4 Thread an adapter into the spark plug hole and connect an air hose from a compressed air source to it. Most auto parts stores can supply the air hose adapter **(see illustration)**. **Note:** *Many cylinder compression gauges utilize a screw-in fitting that may work with your air hose quick-disconnect fitting.*

5 Remove the rocker assembly on the affected side of the engine (see Section 5). If all of the valve stem seals are being replaced, remove both rocker assemblies.

6 Apply compressed air to the cylinder. **Warning:** *The piston may be forced down by compressed air, causing the crankshaft to turn suddenly. If the wrench used when positioning the number one piston at TDC is still attached to the bolt in the crankshaft nose, it could cause damage or injury when the crankshaft moves.*

7 The valves should be held in place by the air pressure.

8 If you don't have access to compressed air, an alternative method can be used. Position the piston at a point approximately 45-degrees before TDC on the compression stroke, then feed a long piece of nylon rope through the spark plug hole until it fills the combustion chamber. Be sure to leave the end of the rope hanging out of the engine so it can be removed easily. Use a large ratchet and socket to rotate the crankshaft in the normal direction of rotation until slight resistance is felt.

9 Stuff shop rags into the cylinder head holes above and below the valves to prevent parts and tools from falling into the engine, then use a valve spring compressor to compress the spring. Remove the keepers with small needle-nose pliers or a magnet **(see illustration)**. **Note:** *A couple of different*

6.9a Compress the valve spring, then remove the keepers

Chapter 2 Part C 4.0L V6 engine

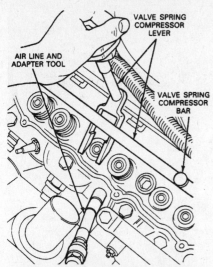

6.9b Here's the lever-type tool and bar used to compress the valve spring - these are special Ford tools, but equivalents may be available from automotive tool companies

types of tools are available for compressing the valve springs with the head in place. One type grips the lower spring coils and presses on the retainer as the knob is turned, while the other type, shown here **(see illustration)**, utilizes a bar installed in place of the rocker arm shaft for leverage.

10 Remove the spring retainer and valve spring, then remove the valve stem seal **(see illustration)**. **Note:** *If air pressure fails to hold the valve in the closed position during this operation, the valve face or seat is probably damaged. If so, the cylinder head will have to be removed for additional repair operations.*

11 Wrap a rubber band or tape around the top of the valve stem so the valve won't fall into the combustion chamber, then release the air pressure. **Note:** *If a rope was used instead of air pressure, turn the crankshaft slightly in the direction opposite normal rotation.*

12 Inspect the valve stem for damage. Rotate the valve in the guide and check the end for eccentric movement, which would indicate that the valve is bent.

13 Move the valve up-and-down in the guide and make sure it doesn't bind. If the valve stem binds, either the valve is bent or the guide is damaged. In either case, the head will have to be removed for repair.

14 Reapply air pressure to the cylinder to retain the valve in the closed position, then remove the tape or rubber band from the valve stem. If a rope was used instead of air pressure, rotate the crankshaft in the normal direction of rotation until slight resistance is felt.

15 Lubricate the valve stem with engine oil and install a new stem seal over the valve guide to the depth shown **(see illustration 6.10)**. Use a special tool, available at most auto parts stores, or a deep socket of the appropriate size and gently tap the seal into place.

16 Install the spring in position over the valve.

17 Install the valve spring retainer. Compress the valve spring and carefully position the keepers in the groove. Apply a small dab of grease to the inside of each lock to hold it in place.

18 Remove the pressure from the spring tool and make sure the keepers are seated.

19 Disconnect the air hose and remove the adapter from the spark plug hole. If a rope was used in place of air pressure, pull it out of the cylinder.

20 Refer to Section 5 and install the rocker arm assembly.

21 Install the spark plugs and hook up the wires.

22 Refer to Section 4 and install the valve covers.

23 Start the engine, then check for oil leaks and unusual sounds coming from the valve cover area.

7 Intake manifold - removal and installation

Refer to illustrations 7.6a, 7.6b, 7.8, 7.9 and 7.11

Removal

1 Relieve the fuel system pressure (see Chapter 4). Disconnect the negative cable from the battery.

2 Remove the ignition coil and mounting bracket (see Chapter 5). Remove the upper intake manifold (see Chapter 4).

3 Remove the valve covers (see Section 4). If it's in the way, remove the camshaft position sensor, if equipped (see Chapter 6, Section 5).

4 Disconnect the electrical connectors from the various sensors threaded into the intake manifold. Disconnect the electrical connectors from the fuel injectors and disconnect the fuel lines to the fuel rail (see Chapter 4).

5 Disconnect the heater hose from the front of the intake manifold. Either remove the thermostat and housing (see Chapter 3) or disconnect the upper radiator hose from the outlet fitting.

6 Remove the intake manifold nuts and bolts and lift the manifold off **(see illustrations)**. If it's stuck, tap it lightly with a soft-

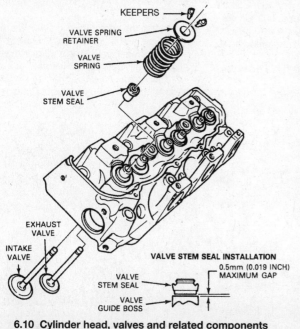

6.10 Cylinder head, valves and related components

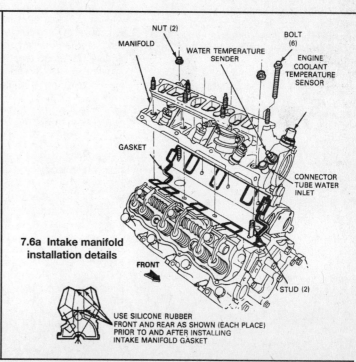

7.6a Intake manifold installation details

7.6b Remove the eight intake manifold bolts and nuts

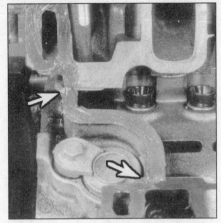

7.8 Before installing the intake manifold gasket, apply a film of RTV sealant around the coolant ports as well as the front sealing surface - be sure to apply a bead to the corners (arrows) at the front and rear of the manifold

7.9 As soon as the sealant is applied, install the gasket and apply a bead of sealant around the coolant ports and in the four corners

face hammer to break the gasket seal. If necessary, pry the manifold off, but pry between a casting protrusion and the engine - don't pry against gasket surfaces.

Installation

7 Clean away all traces of old gasket material. Remove oil and dirt with a cloth and solvent, such as acetone or lacquer thinner.

8 Apply RTV sealant around the coolant ports (the smaller ports at the ends of the heads) and to the front sealing surface. Also apply a bead of sealant to the four corners where the manifold meets the engine **(see the accompanying illustration and illustration 7.6a)**. **Note:** *Install the manifold gasket immediately after applying the sealant. If allowed to set up (approximately 15 minutes), the sealant won't work properly.*

9 Install the manifold gasket **(see illustration)**, then apply RTV sealant onto the gasket, around the coolant ports and to the four corners.

10 Install the manifold over the studs. Install the nuts and bolts and tighten them finger-tight.

11 Tighten the nuts and bolts in the four stages listed in this Chapter's Specifications, following the sequence **(see illustration)**.

12 The remainder of installation is the reverse of the removal steps.

13 Run the engine and check for oil, coolant and vacuum leaks.

8 Exhaust manifolds - removal and installation

Removal

Refer to illustration 8.7

Warning: *Allow the engine to cool completely before beginning this procedure.*

1 Disconnect the cable from the negative battery terminal.

2 Spray the exhaust pipe fasteners and exhaust manifold fasteners with rust penetrant and allow it to soak in for about 10 minutes. If there is any resistance during removal, stop and respray the fasteners to prevent them from breaking off.

3 Remove the spark plug wires from the spark plugs and secure them out of the way (see Chapter 1). Be sure to mark the spark plug wires so they don't get mixed-up.

4 Unbolt the exhaust pipe from the manifold as described in Chapter 4. **Note:** *For access to the exhaust pipe nuts, it may be necessary to raise the vehicle and support it securely on jackstands.*

Left manifold

5 Remove the oil dipstick tube bracket from the engine.

6 If the power steering pump hoses obstruct manifold removal, disconnect them (see Chapter 10). Cap the hoses and fittings to keep out dirt and place the hose ends out of the way.

7 Unbolt and remove the manifold from the cylinder head **(see illustration)**.

Right manifold

8 Detach the heater hose bracket and disconnect the heater hoses (see Chapter 3).

9 Unbolt and remove the manifold from the cylinder head **(see illustration 8.7)**.

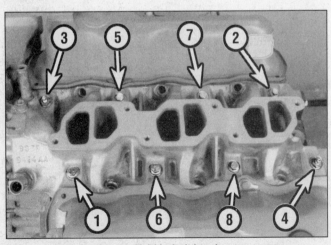

7.11 Intake manifold bolt tightening sequence

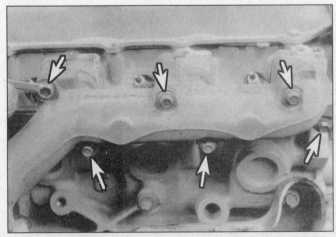

8.7 Remove the exhaust manifold bolts (arrows)

Chapter 2 Part C 4.0L V6 engine

9.12a Remove the head bolts with a T55 Torx bit - don't use an Allen wrench, since it may round out the bolt heads

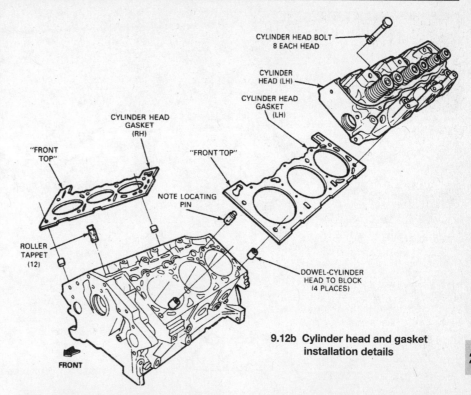

9.12b Cylinder head and gasket installation details

Installation

10 Using a scraper, thoroughly clean the mating surfaces on the cylinder head, manifold and exhaust pipe. Remove residue with a solvent such as acetone or lacquer thinner.

11 **Note:** *These engines were originally assembled without exhaust manifold gaskets and can be reassembled the same way so long as the mating surfaces are perfectly flat and not damaged in any way. Warped or damaged manifolds may require a gasket or machining.* Apply some graphite grease to the cylinder head mating surface and place the manifold on the cylinder head. Tighten the bolts evenly to the torque listed in this Chapter's Specifications.

12 Connect the exhaust pipe to the manifold and tighten the nuts evenly to the torque listed in this Chapter's Specifications.

13 The remainder of installation is the reverse of the removal steps. If the heater hoses were disconnected, fill the cooling system (see Chapter 1).

14 Run the engine and check for exhaust leaks.

9 Cylinder heads - removal and installation

Removal

Refer to illustrations 9.12a, 9.12b and 9.13

Note: *Head bolt removal requires a Torx bit. Obtain the necessary tool before starting. DO NOT try to use an Allen wrench since it may round out the bolt head.*

1 Disconnect the negative cable from the battery.
2 Drain the cooling system (see Chapter 3).
3 Remove the valve covers (see Section 4).
4 Remove the rocker arms and pushrods (see Section 5).
5 Remove the intake manifold (see Section 7).
6 Remove the drivebelt (see Chapter 1).
7 If removing the left cylinder head, detach the air conditioning compressor from the engine and place it out of the way **(see illustration 4.12). Caution:** *DO NOT disconnect any refrigerant lines!* Detach the power steering pump and bracket and set them out of the way (see Chapter 10). Don't disconnect the power steering hoses.
8 If removing the right cylinder head, remove the alternator and bracket (see Chapter 5).
9 If there's a wiring harness attached to the rear of either cylinder head, unwrap or cut the tape to detach the harness from the retainer.
10 Remove the spark plugs (see Chapter 1).
11 Remove the exhaust manifolds (see

9.13 If necessary, pry the head loose; pry against a casting protrusion so the gasket surfaces won't be damaged

9.14 Remove all traces of the old gasket with a scraper, taking care not to gouge the sealing surfaces

Section 8).

12 Remove and discard the cylinder head bolts **(see illustrations)**. The bolts must be replaced with new ones whenever they are removed. Loosen the bolts in several stages.

13 Lift the cylinder head off the engine. If it's difficult to remove, carefully pry it off. Pry against a casting protrusion, not against gasket surfaces **(see illustration)**.

Installation

Refer to illustrations 9.14, 9.17 and 9.18

14 Thoroughly remove all traces of gasket material with a gasket scraper and clean all parts with solvent **(see illustration)**. Use a rag and acetone or lacquer thinner to remove

9.17 Be sure the FRONT TOP mark on the head gasket is positioned correctly

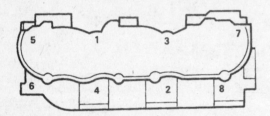

9.18 Cylinder head bolt tightening sequence

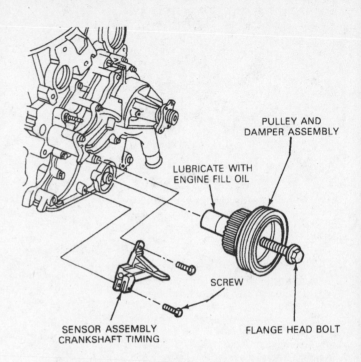

10.2 Crankshaft pulley/damper assembly installation details

any traces of oil from the gasket mating surfaces. See Chapter 2 Part D for cylinder head inspection procedures.

15 Use a tap of the correct size to chase the threads in the head bolt holes.

16 Recheck all head bolt holes and cylinder bores for any traces of coolant, oil or other foreign matter. Remove as needed.

17 Position the new gaskets over the dowel pins on the block. Don't use sealant on the gaskets. Be sure the FRONT TOP marks are positioned correctly **(see illustration)**.

18 Install the heads and new head bolts. Tighten the bolts to the torque listed in this Chapter's Specifications in the sequence shown **(see illustration)**.

19 Refer to Section 7 and apply silicone sealant to the block and cylinder head mating surfaces. Install the intake manifold gasket, then apply silicone sealant to the gasket at same locations.

20 Position the intake manifold on the engine (see Section 7). Install the intake manifold bolts and tighten them to the torque listed in this Chapter's Specifications, in the sequence shown in **illustration 7.11**.

21 The remainder of installation is the reverse of the removal steps.

22 Run the engine and check for oil, coolant and vacuum leaks.

10 Crankshaft pulley and front oil seal - removal and installation

Refer to illustrations 10.2, 10.3a and 10.3b

Crankshaft pulley removal and installation

1 Remove the drivebelt (see Chapter 1).

2 Remove the crankshaft pulley bolt **(see illustration)**.

3 Insert a longer, smaller-diameter bolt into the crankshaft so the puller will have something to bottom against, then remove

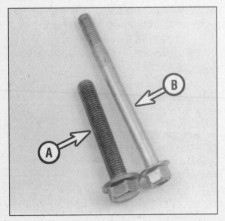

10.3a To remove the crankshaft pulley, remove the pulley bolt (A) and install a longer bolt that's smaller in diameter (B) into the crankshaft . . .

10.3b . . . so the puller screw (arrow) will have something to push against

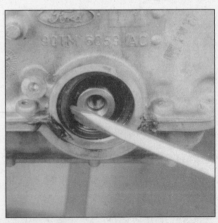

10.5 To remove the front cover seal with the cover on the engine, pry it out with a screwdriver, taking care not to gouge the cover

Chapter 2 Part C 4.0L V6 engine

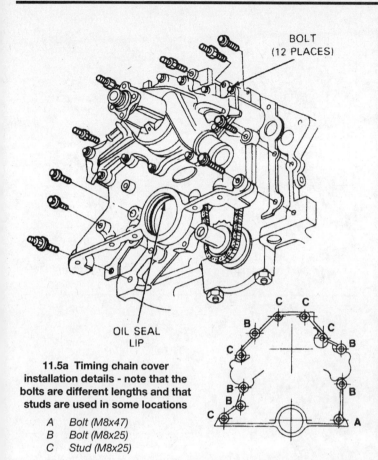

11.5a Timing chain cover installation details - note that the bolts are different lengths and that studs are used in some locations

- A Bolt (M8x47)
- B Bolt (M8x25)
- C Stud (M8x25)

11.5b Label the studs and brackets so they can be returned to their original locations

11.6 Remove the cover from the engine - if it's stuck, recheck to make sure all bolts and studs have been removed

the crankshaft pulley with a puller, designed for this job, available at most auto parts stores. **(see illustrations)**. DO NOT pry the pulley off or use an impact puller!

4 Using clean engine oil, lubricate the surface of the pulley where the front seal rides. Install the crankshaft pulley. Don't hammer the pulley on. Tighten the pulley bolt to the torque listed in this Chapter's Specifications.

Front oil seal replacement

Refer to illustration 10.5

5 Remove the crankshaft pulley (see Steps 1 through 3 above). Carefully pry the oil seal out with a screwdriver **(see illustration)**.
6 Clean the seal bore and check it for nicks or gouges.
7 Coat the lip of the new seal with clean engine oil and drive it into the bore with a socket or large piece of pipe slightly smaller in diameter than the seal. The open side of the seal faces into the engine. Install the crankshaft pulley (see Step 4 above).

11 Timing chain cover - removal and installation

Removal

Refer to illustration 11.5a, 11.5b and 11.6

Note: *This procedure requires the engine be*

removed from the vehicle to properly reseal the timing chain cover-to-oil pan joint.

1 Disconnect the cable from the negative battery terminal. Drain the engine oil and cooling system (see Chapter 1).
2 Remove the engine as described in Part D.
3 Remove the crankshaft pulley (see Section 10). Also remove the crankshaft position sensor (see Chapter 6).
4 Remove the water pump (see Chapter 3).
5 Remove the brackets, studs and cover bolts **(see illustrations)**. **Note:** *The bolts are different lengths, and bolts and studs go in different locations. Label them so they can be installed in the correct holes.*
6 Pull the cover straight off the engine **(see illustration)**. If it's stuck, tap it lightly with a soft-face hammer or pry it carefully to break the gasket seal. Don't use excessive force or you'll crack the cover. If it's difficult to remove, check to make sure you've removed all the bolts.
7 Remove the oil pan (see Section 15).

Installation

8 Thoroughly clean and inspect all the parts. Use a gasket scraper to remove all traces of gasket material. Remove oil film with a solvent such as lacquer thinner or acetone.
9 Apply RTV sealant to the gasket mating surfaces. Install the guide sleeves (if

removed). Install the front cover and start the bolts two or three turns by hand. Note that the bolts are different lengths; be sure to install them in the correct holes. If the labels made during removal are missing or illegible, refer to **illustration 11.5a**.
10 Tighten the cover bolts evenly to the torque listed in this Chapter's Specifications.
11 Install the crankshaft position sensor (see Chapter 6).
12 Install the oil pan (see Section 15).
13 Install the crankshaft pulley (see Step 4 above).
14 Install the engine.
15 The remainder of installation is the reverse of the removal steps.
16 Run the engine and check for oil or coolant leaks.

12 Timing chain and sprockets - inspection, removal and installation

Note: *This procedure requires the engine be removed from the vehicle to properly reseal the timing chain cover-to-oil pan joint.*

Camshaft endplay check

Refer to illustration 12.4

1 Remove the timing chain cover (see

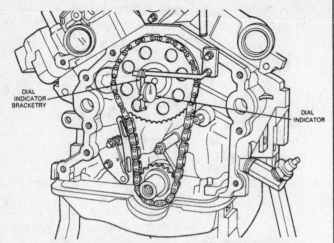

12.4 To check camshaft endplay, position the dial indicator with its tip against the camshaft sprocket bolt

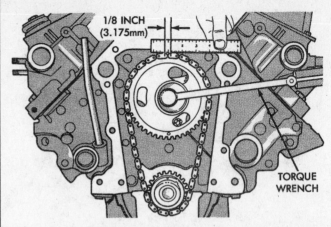

12.8 Position a ruler over the chain to measure timing chain deflection while applying pressure with a torque wrench to the camshaft sprocket bolt

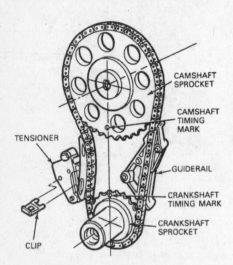

12.19 Align the camshaft and crankshaft timing marks as shown

12.20a Remove the timing chain and sprockets together

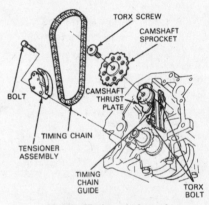

12.20b Timing chain and related components

Section 11).
2 Remove the rocker arm shafts (see Section 5).
3 Push the camshaft to the rear as far as it will go.
4 Install a dial indicator with its pointer on the camshaft sprocket bolt **(see illustration)**. Set the indicator to zero.
5 Pry the camshaft forward with a large screwdriver or prybar between the camshaft sprocket and the block. Note the dial indicator reading and compare with the endplay listed in this Chapter's Specifications. Replace the camshaft thrust plate (see Section 14) if endplay is excessive.

Timing chain and tensioner inspection

Refer to illustration 12.8

6 Remove the timing chain cover (see Section 11).
7 Align the crankshaft and camshaft timing marks **(see illustration 12.19)**
8 Mark a reference point on the top chain link and position a ruler over the top of the chain **(see illustration)**.
9 Attach a torque wrench to the camshaft sprocket bolt and apply 15 ft-lbs of torque in the counterclockwise direction. Don't allow the crankshaft to rotate.
10 Note the point where your mark aligns on the ruler.
11 Apply 15 ft-lbs of torque in the clockwise direction and note the amount of movement.
12 The measurement is the timing chain deflection. If deflection is more than 1/8-inch, replace the timing chain (see below).
13 Check the tensioner for wear and damage. If tensioner face wear is excessive, replace the tensioner. Before reinstalling the tensioner, it must be retracted (see Step 28). Tighten the tensioner bolts to the torque listed in this Chapter's Specifications.

Chain and sprocket

Removal

Refer to illustrations 12.19, 12.20a, 12.20b

14 Disconnect the cable from the negative battery terminal. Drain the cooling system and engine oil (see Chapter 1).
15 Remove the engine as described in Part D. Remove the oil pan (see Section 15).
16 Replace the oil filter (see Chapter 1).
17 Remove the drivebelt (see Chapter 1).
18 Remove the timing chain cover (see Section 11).
19 Turn the crankshaft until the timing marks are aligned **(see illustration)**. Remove the camshaft sprocket Torx bolt and the crankshaft sprocket key. Also remove the timing chain tensioner.
20 Remove the sprockets together with the chain **(see illustrations)**. Do not disturb the crankshaft or camshaft while the timing chain and sprockets are removed.
21 If necessary, remove the Torx bolts that secure the chain guide and remove it from the block.

Installation

Refer to illustration 12.28

22 Be sure the crankshaft and camshaft are positioned so the sprocket timing marks will align correctly after the sprockets are installed **(see illustration 12.19)**.

Chapter 2 Part C 4.0L V6 engine

12.28 To retract the timing chain tensioner so it can be installed, squeeze the tensioner pad and use a sharp-tipped probe to push the ratchet mechanism down, then in to release the latch, allowing the tensioner to retract

13.4 Store the lifters in a marked box so they can be returned to their original positions

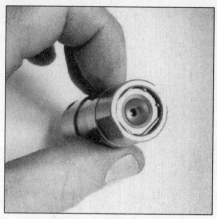

13.8a Check the pushrod seat in the top of each lifter for wear

13.8b The roller on roller lifters must turn freely - check for wear and excessive play as well

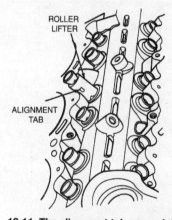

13.11 The alignment tab on each lifter must be positioned in its groove in the lifter bore

23 If the chain guide was removed, install it. Be sure its pin is inserted into the block oil hole, then tighten the guide bolts to the torque listed in this Chapter's Specifications.
24 Place the sprockets in the chain with their timing marks aligned **(see illustration 12.19)**. Install the sprockets and chain together on the crankshaft and camshaft.
25 Install the crankshaft key.
26 Make sure the sprocket timing marks are still aligned **(see illustration 12.19)**. The guide side of the chain must be straight, without any slack, for the marks to align accurately.
27 Install the camshaft sprocket and tighten the bolt to the torque listed in this Chapter's Specifications.
28 Squeeze the tensioner pad with your fingers. At the same time, use a sharp-tipped probe to push the ratchet mechanism down, then in. This will release the latch and let the tensioner retract **(see illustration)**. Retain the tensioner pad in the retracted position by holding it or installing a clip **(see illustration 12.19)** or similar device.
29 Install the timing chain tensioner and allow the tensioner to press against the chain. Tighten the bolts to the torque listed in this Chapter's Specifications.
30 The remainder of installation is the reverse of the removal Steps.
31 Run the engine and check for oil or coolant leaks.

13 Valve lifters - removal, inspection and installation

Refer to illustrations 13.4, 13.8a, 13.8b and 13.11

Removal

1 Remove the intake manifold (see Section 7).
2 Remove the cylinder heads (see Section 9).
3 The lifters protrude from their bores, so if there isn't a lot of varnish buildup, simply pull them out with your fingers. On engines with a lot of sludge and varnish, work the lifters up and down, using carburetor spray cleaner to loosen the deposits. If the lifters are particularly stubborn, special tools designed to grip and remove lifters (Ford tool no. T70P-14151 or equivalent) are manufactured by many tool companies and are widely available.
4 Before removing the lifters, arrange to store them in a clearly labeled box to ensure that they're installed in their original locations **(see illustration)**.
5 Remove the lifters and store them where they won't get dirty.

Inspection

6 Parts for hydraulic valve lifters are not available separately. The work required to remove them from the engine again if cleaning is unsuccessful outweighs any potential savings from repairing them.
7 Clean the lifters with solvent and dry them thoroughly without mixing them up.
8 Check each lifter wall, pushrod seat and roller for scuffing, score marks and uneven wear. Replace any lifter that shows these conditions. If the lifter walls are worn (which isn't very likely), inspect the lifter bores in the engine block as well. If the pushrod seats are worn **(see illustration)**, check the pushrod ends. Make sure each roller turns freely **(see illustration)**.
9 Ford recommends that the lifters and camshaft be replaced as a set if any lifter needs replacement. Never install used lifters unless the original camshaft is used and the lifters can be installed in their original locations!
10 The original lifters, if they're being reinstalled, must be returned to their original locations.

Installation

11 Install the lifters in the bores. Coat them with moly-based grease or engine assembly lube. Note that when the lifters are installed, the alignment tab must fit in the groove in the lifter bore **(see illustration)**.
12 The remainder of installation is the reverse of the removal Steps.

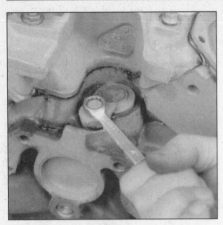

14.18a Remove the oil pump drive assembly bolt

14.18b ... and lift the drive assembly out of the block

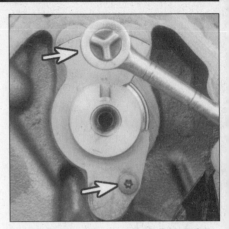

14.27 Remove the camshaft thrust plate bolts with a Torx bit and take the thrust plate off - note which end of the thrust plate is up so it can be installed correctly

14 Camshaft - removal, inspection and installation

Endplay check

1 Refer to Section 12 for this procedure.

Lobe lift check

2 In order to determine the extent of cam lobe wear, the lobe lift should be checked prior to camshaft removal. Refer to Section 4 and remove the valve covers. The rocker arm assembly must also be removed (see Section 5), but leave the pushrods in place.

3 Position the number one piston to TDC on the compression stroke (see Section 3).

4 Beginning with the number one cylinder, mount a dial indicator on the engine and position the plunger in-line with and resting on the first pushrod.

5 Zero the dial indicator, then very slowly turn the crankshaft in the normal direction of rotation until the indicator needle stops and begins to move in the opposite direction. The point at which it stops indicates maximum cam lobe lift.

6 Record this figure for future reference, then reposition the piston at TDC on the compression stroke.

7 Move the dial indicator to the remaining number one cylinder pushrod and repeat the check. Be sure to record the results for each valve.

8 Repeat the check for the remaining valves. Since each piston must be at TDC on the compression stroke for this procedure, work from cylinder-to-cylinder, following the firing order sequence.

9 After the check is complete, compare the results to this Chapter's Specifications. If camshaft lobe lift is less than 0.005-inch for any of the lobes, cam lobe wear has occurred and a new camshaft should be installed.

Removal

Refer to illustrations 14.18a, 14.18b, 14.27 and 14.28

Note: *This procedure requires the engine be removed from the vehicle.*

10 Disconnect the negative cable from the battery.
11 Drain the cooling system and engine oil (see Chapter 1).
12 Remove the drivebelt (see Chapter 1).
13 Remove the engine as described in Part D.
14 Remove the spark plug wires (see Chapter 1).
15 Remove the ignition coil pack and bracket (see Chapter 5).
16 Remove the alternator (see Chapter 5).
17 Position the number one piston to TDC on the compression stroke (see Section 3).
18 On all 1993 models and 1994 Federal emissions models, remove the oil pump drive gear from the block **(see illustrations)**.
19 On 1994 California emissions models and all 1995 and later models, mark the exact orientation of the camshaft position sensor in relation to the block. The camshaft position sensor is installed in the same location as the oil pump drive gear on earlier models. Remove the camshaft position sensor and synchronizer assembly (see Chapter 6, Section 5).
20 Remove the valve covers (see Section 4).
21 Remove the intake manifold (see Section 7). The upper intake manifold can be left attached to the lower manifold.
22 Remove the rocker arms and pushrods (see Section 5).
23 Remove the valve lifters (see Section 13).
24 Remove the oil pan (see Section 15).
25 Remove the timing chain cover (see Section 11).
26 Remove the timing chain and sprockets (see Section 12).
27 Remove the camshaft thrust plate bolts with a Torx bit **(see illustration)** and remove the thrust plate from the block.
28 Carefully pull the camshaft out of the block, rotating it as you pull **(see illustration)**. Don't let the camshaft lobes or journals nick the camshaft bearings.

Inspection

29 After the camshaft has been removed

14.28 As you're withdrawing the camshaft, support it near the block with both hands - be careful not to let the camshaft nick the bearings

from the engine, cleaned with solvent and dried, inspect the bearing journals for uneven wear, pits and galling. If the journals are damaged, the bearing inserts in the engine are probably damaged as well. Both the camshaft and bearings will have to be replaced with new ones. Using a telescoping gauge, measure the inside diameter of each camshaft bearing and record the results (take two measurements, 90-degrees apart, at each bearing).

30 Measure the camshaft bearing journals with a micrometer to determine if they're excessively worn or out-of-round. If they're out-of-round, the camshaft should be replaced with a new one. Subtract the bearing journal diameters from the corresponding bearing inside diameter measurements to obtain the oil clearance. If it's excessive, new bearings must be installed. **Note:** *Camshaft bearing replacement requires special tools and expertise that place it outside the scope of the home mechanic. Take the block to an automotive machine shop to ensure that the job is done correctly.*

Chapter 2 Part C 4.0L V6 engine

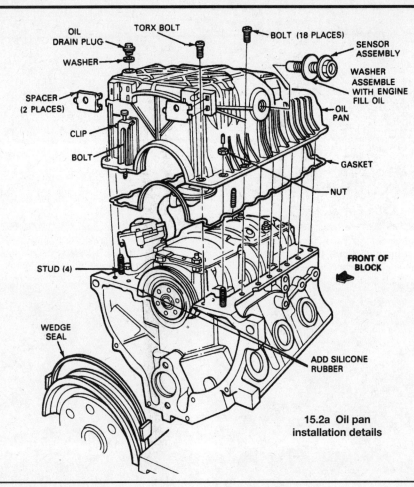

15.2a Oil pan installation details

15.2b Use a Torx driver to remove the two bolts (arrows) adjacent to the crankshaft rear oil seal

15.3 The oil baffle is secured to the main bearing caps

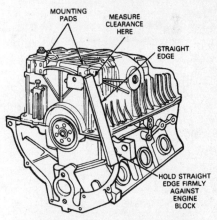

15.5 To select pan-to-transmission spacers, place a straightedge on the transmission mounting surface and each of the pan-to-transmission mounting pads in turn and measure the gap between the straightedge and the mounting pad

31 Check the camshaft lobes for heat discoloration, score marks, chopped areas, pitting, flaking and uneven wear. If the lobes are in good condition the camshaft can be reused.

32 Make sure the camshaft oil passages are clear and clean.

Installation

33 Lubricate the camshaft bearing journals and cam lobes with engine assembly lube.

34 Slide the camshaft into the engine. Support the cam near the block and be careful not to scrape or nick the bearings.

35 Coat both sides of the thrust plate with engine assembly lube, then install it so it covers the main oil gallery (see illustration 14.27). Tighten the bolts to the torque listed in this Chapter's Specifications.

36 The remainder of installation is the reverse of removal. On models with a camshaft position sensor, make sure to install it as described in Chapter 6, Section 5.

15 Oil pan and baffle - removal and installation

Removal

Refer to illustrations 15.2a, 15.2b and 15.3
Note: *This procedure requires the engine be removed from the vehicle.*

1 Remove the engine from the vehicle (see Part D of this Chapter).

2 Remove the oil pan fasteners and remove the pan from the engine **(see illustration)**. Use a Torx bit to remove the bolt on each side of the crankshaft rear oil seal **(see illustration)**.

3 If necessary, remove the baffle fasteners and remove the baffle **(see illustration)**

Installation

Refer to illustrations 15.5, 15.9a, 15.9b, and 15.10
Caution: *Since two of the oil pan bolts are attached to the transmission, spacers are used between the oil pan and transmission. Failure to select the spacers correctly can cause oil pan damage or oil leaks when the pan is installed.*

4 Using a gasket scraper, thoroughly clean all old gasket material from the oil pan and its mounting surface. Remove residue and oil film with a solvent such as acetone or lacquer thinner.

5 Temporarily place the pan on the engine and secure it with four nuts. Place a straightedge across the transmission mounting surface on the cylinder block and one of the mounting pads for the oil pan-to-transmission bolts **(see illustration)**.

6 Measure the gap between the mounting pad and straightedge with a feeler gauge.

7 Move the straightedge to the other mounting pad and measure its gap. Remove the nuts and detach the pan from the engine.

8 Select spacers for the mounting pads to compensate for the gap. Spacers are available in three thickness (see this Chapter's Specifications).

Chapter 2 Part C 4.0L V6 engine

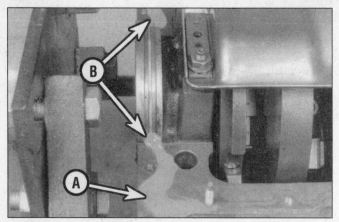

15.9a At the rear of the engine, apply thin beads of sealant to the outside edges of the gasket surface (A) and also to the crankshaft rear oil seal corners (B)

15.9b Apply beads of sealant to the corners of the front seal cover (arrows) and at the seams where the timing chain cover meets the engine block

15.10 The long Torx screw next to the crankshaft rear oil seal passes through a boss on the pan, then through the pan flange and into the block

16.3 Remove the oil pump mounting bolts and separate the pump from the engine

16.4a As you remove the intermediate driveshaft, note the different shapes of its ends and the location of the retainer ring on the shaft

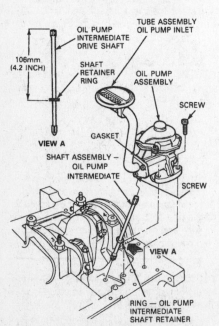

16.4b Oil pump and intermediate drive shaft installation details

9 At the rear of the engine, apply a thin bead of RTV sealant to the outside edge of the gasket surface and another bead to the area around the crankshaft rear oil seal **(see illustration)**. At the front of the engine, apply a bead of sealant to each of the engine block-to-timing-chain cover seams and to the corners of the front seal cover **(see illustration)**.

10 Install a new pan gasket on the engine, followed by the oil pan. Install the pan fasteners and tighten them evenly to the torque listed in this Chapter's Specifications. The long screw next to the crankshaft rear oil seal passes through a boss on the pan, then through the pan flange and into the block **(see illustration)**.

11 Install the spacers on the mounting pads before bolting the engine to the transmission.

16 Oil pump - removal and installation

Removal

Refer to illustrations 16.3, 16.4a, and 16.4b

Note: *This procedure requires the engine be removed from the vehicle.*

1 Remove the engine from the vehicle (see Part D of this Chapter).
2 Remove the oil pan (see Section 15).
3 Remove the oil pump mounting bolts and separate the pump from the engine **(see illustration)**.
4 Withdraw the oil pump intermediate drive shaft **(see illustrations)**. If necessary, unbolt the inlet tube and screen from the pump.

Installation

5 Fill one of the pump ports with clean engine oil and rotate the pump by hand to prime it.
6 If removed, install the oil pump pick-up tube on the pump, using a new gasket. Make sure the screen is clean and free of debris.
7 Insert the intermediate drive shaft into the engine, pointed end first. The pressed-on retaining ring on the shaft should be positioned as shown in **illustration 16.4b**.
8 Install the oil pump on the block. Install the oil pump bolts and tighten them to the torque listed in this Chapter's Specifications.
9 The remainder of installation is the

Chapter 2 Part C 4.0L V6 engine

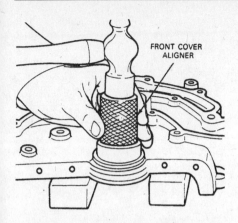

17.5 Drive the seal in squarely and evenly - a large socket or piece of pipe can be used if a seal driver isn't available

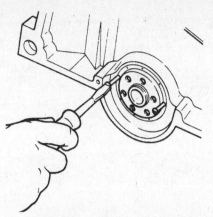

17.11 Thread two sheet metal screws into the crankshaft rear oil seal to pry it out

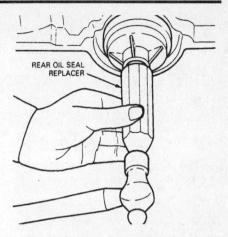

17.15 Install the crankshaft rear oil seal with the open end facing into the engine - use a socket or large piece of pipe if a seal driver isn't available

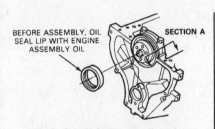

17.16 Crankshaft rear oil seal installation details - the seal must be square with the crankshaft centerline and within the dimensions shown

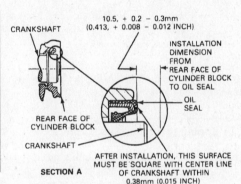

reverse of removal.
10 Run the engine and make sure oil pressure comes up to normal quickly. If it doesn't, stop the engine and find out the cause. Severe engine damage can result from running an engine with insufficient oil pressure!

17 Crankshaft oil seals replacement

Front seal - timing chain cover in place

1 Refer to Section 10 for this procedure.

Front seal - timing chain cover removed

Refer to illustration 17.5

2 Support the timing chain cover on wooden blocks, then drive out the seal with a punch and hammer.
3 Check the seal bore in the timing chain cover for nicks or burrs.
4 Coat the seal lip and outer circumference with engine oil.
5 Position the new seal with its lip facing IN. Drive in the new seal with a seal driver **(see illustration)**. If a seal driver isn't available, use a large socket or piece of pipe the same diameter as the seal.

Rear seal

Refer to illustrations 17.11, 17.15 and 17.16

6 Remove the transmission (see Chapter 7).
7 Remove the clutch (if equipped) (see Chapter 8).
8 Remove the flywheel or driveplate (see Section 17).
9 Remove the engine rear plate.
10 With a sharp awl or similar tool, punch two holes in the seal on opposite sides just above the point where the main bearing cap meets the cylinder block.
11 Thread a sheet metal screw into each hole **(see illustration)**. Pry against the sheet metal screws with two large screwdrivers or small prybars to remove the seal. **Note:** *If you need a fulcrum to pry against, use two small blocks of wood.* **Caution:** *Don't scratch or gouge the crankshaft seal surface.*
12 Clean the oil seal bore in the block and main bearing cap. Check the seal bore and crankshaft sealing surface for nicks or burrs.
13 Apply a thin coat of clean engine oil to the outer diameter of the seal and to the contact surfaces of the seal and crankshaft.
14 Position the seal in the bore with its open end facing IN.
15 Drive the seal in with a seal installer **(see illustration)** until it is securely seated. Use a socket or piece of pipe the same diameter as the seal if a seal driver isn't available.
16 After installation, the seal must be square with the crankshaft centerline and driven into the bore within the dimensions shown **(see illustration)**.

18 Flywheel/driveplate - removal and installation

Refer to illustrations 18.3 and 18.4

1 Raise the vehicle and support it securely on jackstands, then refer to Chapter 7 and remove the transmission. If it's leaking, now would be a very good time to replace the front pump seal/O-ring (automatic transmission only).
2 Remove the pressure plate and clutch disc (see Chapter 8) (manual transmission equipped vehicles). Now is a good time to check/replace the clutch components and pilot bearing.
3 Make alignment marks on the flywheel/driveplate and crankshaft to ensure correct alignment during reinstallation **(see illustration)**.

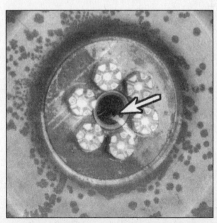

18.3 Make alignment marks on the flywheel/driveplate and crankshaft

18.4 Remove the mounting bolts and take the flywheel/driveplate off the crankshaft - the manual transmission flywheel is heavy, so be sure to support it during removal

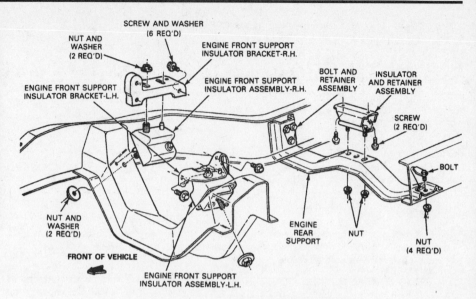

19.7 Engine and transmission mount details

4   Remove the bolts that secure the flywheel/driveplate to the crankshaft **(see illustration)**. If the crankshaft turns, wedge a screwdriver through the starter opening to jam the flywheel.

5   Remove the flywheel/driveplate from the crankshaft. Since the flywheel is fairly heavy, be sure to support it while removing the last bolt.

6   Clean the flywheel to remove grease and oil. Inspect the surface for cracks, rivet grooves, burned areas and score marks. Light scoring can be removed with emery cloth. Check for cracked and broken ring gear teeth. Lay the flywheel on a flat surface and use a straightedge to check for warpage.

7   Clean and inspect the mating surfaces of the flywheel/driveplate and the crankshaft. If the crankshaft rear seal is leaking, replace it before reinstalling the flywheel/driveplate.

8   Position the flywheel/driveplate against the crankshaft. Be sure to align the marks made during removal. Note that some engines have an alignment dowel or staggered bolt holes to ensure correct installation. Before installing the bolts, apply thread locking compound to the threads.

9   Wedge a screwdriver through the starter motor opening to keep the flywheel/driveplate from turning as you tighten the bolts in a criss-cross pattern, to the torque listed in this Chapter's Specifications.

10   The remainder of installation is the reverse of the removal procedure.

19   Engine mounts - check and replacement

Refer to illustration 19.7

1   Engine mounts seldom require attention, but broken or deteriorated mounts should be replaced immediately or the added strain placed on the driveline components may cause damage or wear.

Check

2   During the check, the engine must be raised slightly to remove the weight from the mounts.

3   Raise the vehicle and support it securely on jackstands, then position a jack under the engine oil pan. Place a large block of wood between the jack head and the oil pan, then carefully raise the engine just enough to take the weight off the mounts. **Warning:** *DO NOT place any part of your body under the engine when it's supported only by a jack!*

4   Check the mounts to see if the rubber is cracked, hardened or separated from the metal plates. Sometimes the rubber will split right down the center.

5   Check for relative movement between the mount plates and the engine or frame (use a large screwdriver or prybar to attempt to move the mounts). If movement is noted, lower the engine and tighten the mount fasteners.

Replacement

6   Disconnect the negative battery cable from the battery, then raise the vehicle and support it securely on jackstands (if not already done).

7   Remove the fasteners and detach the mount from the frame bracket **(see illustration)**.

8   Raise the engine slightly with a jack or hoist (make sure the fan doesn't hit the radiator or shroud). Remove the mount-to-block bolts and detach the mount.

9   Installation is the reverse of removal. Use thread locking compound on the mount bolts and be sure to tighten them securely.

10   Rubber preservative should be applied to the mounts to slow deterioration.

Chapter 2 Part D
General engine overhaul procedures

Contents

	Section		Section
Camshaft, lifters and bearings - inspection and replacement	20	Engine overhaul - reassembly sequence	21
CHECK ENGINE light	See Chapter 6	Engine rebuilding alternatives	4
Crankshaft - inspection	18	Engine - removal and installation	6
Crankshaft - installation and main bearing oil clearance check	23	Engine removal - methods and precautions	5
Crankshaft - removal	13	General information	1
Cylinder compression check	3	Initial start-up and break-in after overhaul	2
Cylinder head - cleaning and inspection	9	Main and connecting rod bearings - inspection	19
Cylinder head - disassembly	8	Piston rings - installation	22
Cylinder head - reassembly	11	Pistons/connecting rods - inspection	17
Cylinder honing	16	Pistons/connecting rods - installation and rod bearing oil clearance check	25
Engine block - cleaning	14	Pistons/connecting rods - removal	12
Engine block - inspection	15	Rear main oil seal - installation	24
Engine overhaul - disassembly sequence	7	Valves - servicing	10
Engine overhaul - general information	2		

Specifications

Four-cylinder engine

General

Oil pressure	40 to 60 psi @ 2000 rpm
Compression pressure	Lowest reading cylinder must be within 15 psi of highest reading cylinder (100 psi minimum)

Engine block

Cylinder bore
Diameter	3.7795 to 3.7810 inch
Out-of-round limit	0.0015 inch
Taper limit	0.005 inch

Cylinder head and valves

Cylinder head warpage limit	0.006 inch (overall)
Valve face angle	44-degrees
Valve seat angle	45-degrees
Valve margin width	1/32-inch

Valve stem diameter
- 1993 and 1994
  - Intake: 0.3416 to 0.3423 inch
  - Exhaust: 0.3411 to 0.3418 inch
- 1995 on
  - Intake: 0.2746 to 0.2754 inch
  - Exhaust: 0.2736 to 0.2744 inch

Valve stem-to-guide clearance
- 2.3L engine
  - Intake: 0.0010 to 0.0027 inch
  - Exhaust: 0.0015 to 0.0032 inch
- 2.5L engine
  - Intake: 0.0008 to 0.0027 inch
  - Exhaust: 0.0018 to 0.0037 inch

Valve spring free length
- 1993 and 1994: 1.877 inches
- 1995 on: 2.020 inches

Valve spring installed height
- 1993 and 1994: 1.49 to 1.55 inches
- 1995 on: 1.54 to 1.58 inches

Crankshaft and connecting rods

Crankshaft endplay	0.004 to 0.008 inch
Connecting rod side clearance	0.0035 to 0.0115 inch

Maximum PSI	Minimum PSI	Maximum PSI	Minimum PSI	Maximum PSI	Minimum PSI	Maximum PSI	Minimum PSI
134	101	164	123	194	145	224	168
136	102	166	124	196	147	226	169
138	104	168	126	198	148	228	171
140	105	170	127	200	150	230	172
142	107	172	129	202	151	232	174
144	108	174	131	204	153	234	175
146	110	176	132	206	154	236	177
148	111	178	133	208	156	238	178
150	113	180	135	210	157	240	180
152	114	182	136	212	158	242	181
154	115	184	138	214	160	244	183
156	117	186	140	216	162	246	184
158	118	188	141	218	163	248	186
160	120	190	142	220	165	250	187
162	121	192	144	222	166		

Cylinder compression variation chart - locate your maximum compression reading on the chart and look to the right to find the minimum acceptable compression (then compare it to your lowest reading)

Crankshaft and connecting rods (continued)

Main bearing journal
  Diameter .... 2.2059 to 2.2051 inches
  Out-of-round limit .... 0.0006 inch maximum
  Taper limit .... 0.0006 inch
Main bearing oil clearance
  Nominal .... 0.0008 to 0.0015 inch
  Allowable .... 0.0008 to 0.0026 inch
Connecting rod bearing journal
  Diameter .... 2.0464 to 2.0472 inch
  Out-of-round limit .... 0.0006 inch
  Taper limit
    2.3L .... 0.0006 inch
    2.5L .... 0.0005 inch
Connecting rod bearing oil clearance
  Nominal .... 0.0008 to 0.0015 inch
  Allowable .... 0.0008 to 0.0026 inch

Pistons and rings

2.3L engine
Piston diameter
  1993 and 1994 .... 3.7768 to 3.7779 inches
  1995 on .... 3.7883 to 3.7793 inches
Piston-to-bore clearance
  1993 and 1994 .... 0.0024 to 0.0034 inch
  1995 on .... 0.010 to 0.020 inch
Piston ring side clearance
  Compression rings .... 0.0016 to 0.0033 inch
  Oil ring .... Snug fit
Piston ring end gap
  1993 and 1994
    Compression ring (top) .... 0.010 to 0.020 inch
    Compression ring (bottom) .... 0.010 to 0.020 inch
    Oil ring .... 0.015 to 0.049 inch
  1995 on
    Compression ring (top) .... 0.008 to 0.016 inch
    Compression ring (bottom) .... 0.013 to 0.019 inch
    Oil ring .... 0.010 to 0.030 inch

2.5L engine
Piston diameter
  Code red .... 3.7780 to 3.7785 inch
  Code blue .... 3.7785 to 3.7790 inch
  Code Yellow .... 3.7790 to 3.7795 inch
Piston-to-bore clearance .... 0.001 to 0.002 inch
Piston selection in relation to bore size
  Code red .... 3.7795 to 3.7800 inch
  Code blue .... 3.7800 to 3.7805 inch
  Code yellow .... 3.7805 to 3.7810 inch

Chapter 2 Part D General engine overhaul procedures

Piston ring side clearance
  Compression rings .... 0.0014 to 0.0030 inch
  Oil ring .... Snug fit
Piston ring end gap
  Upper compression ring... 0.008 to 0.018 inch
  Lower compression ring .... 0.013 to 0.023 inch
  Oil ring .... 0.010 to 0.035 inch

Camshaft
Lobe lift (intake and exhaust)
  1993 and 1994 .... 0.2381 inch
  1995 on .... 0.2163 inch
Bearing journal diameter... 1.7713 to 1.7720 inches
Bearing oil clearance
  Standard .... 0.001 to 0.003 inch
  Service limit .... 0.006 inch
Endplay
  Standard .... 0.001 to 0.007 inch
  Service limit .... 0.009 inch

Torque specifications*   Ft-lbs
Main bearing cap bolts
  Step 1 .... 51 to 59
  Step 2 .... 76 to 84
Connecting rod nuts
  Step 1 .... 26 to 30
  Step 2 .... 31 to 36

*Note: *Refer to Part A for additional torque specifications.*

3.0L V6 engine

General
Oil pressure... 40 to 60 psi @ 2500 rpm
Compression pressure .... Lowest reading cylinder must be within 15 psi of highest reading cylinder (100 psi minimum)

Engine block
Cylinder bore
  Diameter .... 3.504 inches
  Out-of-round limit... 0.002 inch
  Taper limit.. 0.002 inch

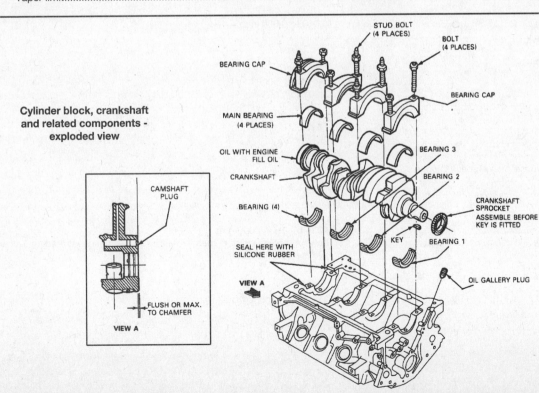

Cylinder block, crankshaft and related components - exploded view

Cylinder head and valves
Cylinder head warpage limit .... 0.006 inch (overall)
Valve face angle 44-degrees
Valve seat angle 45-degrees
Valve margin width 1/32-inch
Valve stem-to-guide clearance
    Intake .... 0.0010 to 0.0027 inch
    Exhaust .... 0.0015 to 0.0032 inch
Valve spring installed height
    1993 and 1994 .... 1.58 inches
    1995 on .... 1.736 to 1.650 inches
Valve spring free length .... 1.84 inches

Crankshaft and connecting rods
Crankshaft endplay .... 0.004 to 0.008 inch
Connecting rod side clearance .... 0.006 to 0.014 inch
Main bearing journal
    Diameter .... 2.5190 to 2.5198 inches
    Out-of-round limit .... 0.0003 inch
    Taper limit .... 0.0006 inch
Main bearing oil clearance
    Nominal .... 0.001 to 0.0014 inch
    Allowable .... 0.0005 to 0.0023 inch
Connecting rod journal
    Diameter .... 2.1253 to 2.1261 inch
    Out-of-round limit .... 0.0003 inch
    Taper limit .... 0.0003 inch
Connecting rod bearing oil clearance
    Nominal .... 0.001 to 0.0014 inch
    Allowable .... 0.00086 to 0.0027 inch

Pistons and rings
Piston diameter
    Coded red .... 3.5024 to 3.5031 inches
    Coded blue .... 3.5035 to 3.5041 inches
    Coded yellow .... 3.5045 to 3.5051 inches
Piston-to-bore clearance .... 0.0012 to 0.0022 inch
Piston ring side clearance
    Compression rings .... 0.0016 to 0.0037 inch
    Oil ring .... Snug fit
Piston ring end gap
    Compression rings .... 0.010 to 0.020 inch
    Oil ring .... 0.010 to 0.049 inch

Camshaft
Lobe lift .... 0.260 inch
Bearing journal diameter .... 2.0074 to 2.0084 inches
Bearing oil clearance .... 0.001 to 0.003 inch
Endplay .... 0.007 inch

Torque specifications*
Ft-lbs
Connecting rod nuts 26
Main bearing cap bolts 59

Note: Refer to Part B for additional torque specifications.

4.0L V6 engine
General
Oil pressure .... 40 to 60 psi @ 2000 rpm
Cylinder compression pressure .... Lowest reading cylinder must be within 15 psi of highest reading cylinder (100 psi minimum)

Engine block
Cylinder bore
    Diameter .... 3.9527 to 3.9543 inches
    Out-of-round limit .... 0.005 inch
    Taper limit .... 0.005 inch

Cylinder head and valves
Cylinder head warpage limit .... 0.006 inch (overall)
Valve face angle 44-degrees
Valve seat angle 45-degrees

Chapter 2 Part D General engine overhaul procedures

Valve margin width	1/32 inch
Valve stem diameter	
Intake	0.3159 to 0.3167 inch
Exhaust	0.3149 to 0.3156 inch
Valve stem-to-guide clearance	
Intake	0.0008 to 0.0025 inch
Exhaust	0.0018 to 0.0035 inch
Valve spring free length	1.91 inch
Valve spring installed height	1-37/64 to 1-39/64 inch

Crankshaft and connecting rods

Crankshaft endplay	0.002 to 0.0125 inch
Connecting rod side clearance	
Standard	0.0002 to 0.0025 inch
Service limit	0.014 inch
Main bearing journal	
Diameter	2.2433 to 2.2441 inches
Out-of-round limit	0.0006 inch
Taper limit	0.0006 inch
Main bearing oil clearance	
Nominal	0.0008 to 0.0015 inch
Allowable	0.0005 to 0.0019 inch
Connecting rod journal	
Diameter (standard)	2.1252 to 2.1260 inches
Out-of-round limit	0.0006 inch
Taper limit	0.0006 inch
Connecting rod bearing oil clearance	
Nominal	0.0003 to 0.0022 inch
Allowable	0.0005 to 0.0024 inch

Pistons and rings

Piston diameter	3.9524 to 3.9531 inches
Piston-to-bore clearance	0.0008 to 0.0019 inch
Piston ring side clearance	
Compression rings	0.0020 to 0.0033 inch
Oil ring	Snug fit
Piston ring end gap	
Compression rings	0.015 to 0.023 inch
Oil ring rails	0.015 to 0.055 inch

Camshaft

Lobe lift	0.272 inch
Bearing journal diameter	
No. 1	1.951 to 1.952 inches
No. 2	1.937 to 1.938 inches
No. 3	1.922 to 1.923 inches
No. 4	1.907 to 1.908 inches
Bearing oil clearance	0.001 to 0.003 inch
Endplay	0.002 to 0.006

Torque specifications*

	Ft-lbs
Main bearing cap bolts	
Step 1	25
Step 2	72
Connecting rod cap nuts	
1993 through 1997	18 to 24
1998 through 2000	
Step 1	15
Step 2	Tighten an additional 90-degrees

*Note: Refer to Part C for additional torque specifications.

1 General information

Warning: *On models so equipped, whenever working in the vicinity of the front grille/bumper, steering wheel, steering column or other components of the airbag system, the system should be disarmed. To do this, perform the following steps:*

a) Turn the ignition switch to Off.

b) Detach the cable from the negative battery terminal, then detach the positive cable. Wait two minutes for the electronic module backup power supply to be depleted.

To enable the system

a) Turn the ignition switch to the Off position.

b) Connect the positive battery cable first, then connect the negative cable.

Included in this portion of Chapter 2 are the general overhaul procedures for the cylinder heads and internal engine components.

The information ranges from advice concerning preparation for an overhaul and the purchase of replacement parts to detailed, step-by-step procedures covering removal and installation of internal engine components and the inspection of parts.

The following Sections have been writ-

2D-6 Chapter 2 Part D General engine overhaul procedures

ten based on the assumption that the engine has been removed from the vehicle. For information concerning in-vehicle engine repair, as well as removal and installation of the external components necessary for the overhaul, see Parts A through C of this Chapter and Section 7 of this Part.

The Specifications included in this Part are only those necessary for the inspection and overhaul procedures which follow. Refer to Parts A through C for additional Specifications.

2 Engine overhaul - general information

Refer to illustrations 2.4a, 2.4b, and 2.4c

It's not always easy to determine when, or if, an engine should be completely overhauled, as a number of factors must be considered.

High mileage is not necessarily an indication that an overhaul is needed, while low mileage doesn't preclude the need for an overhaul. Frequency of servicing is probably the most important consideration. An engine that's had regular and frequent oil and filter changes, as well as other required maintenance, will most likely give many thousands of miles of reliable service. Conversely, a neglected engine may require an overhaul very early in its life.

Excessive oil consumption is an indication that piston rings, valve seals and/or valve guides are in need of attention. Make sure that oil leaks aren't responsible before deciding that the rings and/or guides are bad. Perform a cylinder compression check to determine the extent of the work required (see Section 3).

Check the oil pressure with a gauge installed in place of the oil pressure sending unit **(see illustrations)** and compare it to this Chapter's Specifications. If it's extremely low, the bearings and/or oil pump are probably worn out.

Loss of power, rough running, knocking or metallic engine noises, excessive valve train noise and high fuel consumption rates may also point to the need for an overhaul, especially if they're all present at the same time. If a complete tune-up doesn't remedy the situation, major mechanical work is the only solution.

An engine overhaul involves restoring the internal parts to the specifications of a new engine. During an overhaul, the piston rings are replaced and the cylinder walls are reconditioned (rebored and/or honed). If a rebore is done by an automotive machine shop, new oversize pistons will also be installed. The main bearings, connecting rod bearings and camshaft bearings are generally replaced with new ones and, if necessary, the crankshaft may be reground to restore the journals. Generally, the valves are serviced as well, since they're usually in less-than-perfect condition at this point.

While the engine is being overhauled, other components, such as the starter and alternator, can be rebuilt as well. The end result should be a like-new engine that will give many trouble-free miles. **Note:** *Critical cooling system components such as the hoses, drivebelts, thermostat and water pump MUST be replaced with new parts when an engine is overhauled. The radiator should be checked carefully to ensure that it isn't clogged or leaking (see Chapter 3). Also, we don't recommend overhauling the oil pump - always install a new one when an engine is rebuilt.*

Before beginning the engine overhaul, read through the entire procedure to familiarize yourself with the scope and requirements of the job. Overhauling an engine isn't difficult if you follow all of the instructions carefully, have the necessary tools and equipment and pay close attention to all specifications; however, it is time consuming. Plan on the vehicle being tied up for a minimum of two weeks, especially if parts must be taken to an automotive machine shop for repair or reconditioning. Check on availability of parts and make sure that any necessary special tools and equipment are obtained in advance. Most work can be done with typical hand tools, although a number of precision measuring tools are required for inspecting parts to determine if they must be replaced. Often an automotive machine shop will handle the inspection of parts and offer advice concerning reconditioning and replacement. **Note:** *Always wait until the engine has been completely disassembled and all components, especially the engine block, have been inspected before deciding what service and repair operations must be performed by an automotive machine shop.* Since the block's condition will be the major factor to consider when determining whether to overhaul the original engine or buy a rebuilt one, never purchase parts or have machine work done on other components until the block has been thoroughly inspected. As a general rule, time is the primary cost of an overhaul, so it doesn't pay to install worn or substandard parts.

As a final note, to ensure maximum life and minimum trouble from a rebuilt engine, everything must be assembled with care in a spotlessly clean environment.

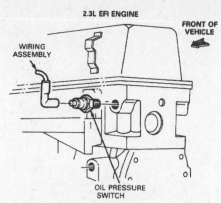

2.4a On four-cylinder engines, the oil pressure sender is located at the left rear of the block

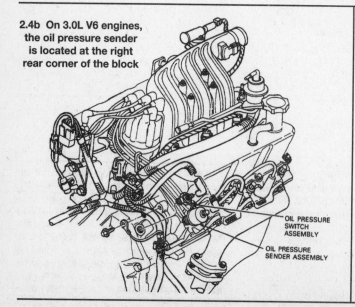

2.4b On 3.0L V6 engines, the oil pressure sender is located at the right rear corner of the block

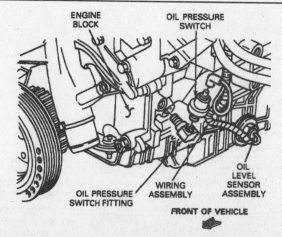

2.4c On 4.0L V6 engines, the oil pressure sender is located on the driver's side of the block next to the water pump inlet - do not confuse the oil pressure sender with the oil level sensor located in the side of the oil pan

Chapter 2 Part D General engine overhaul procedures

3 Cylinder compression check

Refer to illustration 3.8

1 A compression check will tell you what mechanical condition the upper end (pistons, rings, valves, head gaskets) of your engine is in. Specifically, it can tell you if the compression is down due to leakage caused by worn piston rings, defective valves and seats or a blown head gasket. **Note:** *The engine must be at normal operating temperature and the battery must be fully charged for this check.*

2 Begin by cleaning the area around the spark plugs before you remove them (compressed air should be used, if available, otherwise a small brush or even a bicycle tire pump will work). The idea is to prevent dirt from getting into the cylinders as the compression check is being done.

3 On V6 engines, remove all of the spark plugs from the engine (see Chapter 1).

4 On 2.3L four-cylinder engines with twin plug ignition, remove all four spark plugs from one side only. If you remove all eight plugs, compression will escape out the open spark plug hole.

5 Block the throttle wide open.

6 Disable the ignition system. On distributor ignitions, detach the coil secondary wire. On distributorless ignitions, detach the electrical connector from the coil pack (see Chapter 5).

7 Disable the fuel pump circuit (see Chapter 4, Section 2).

8 Install the compression gauge in the number one spark plug hole **(see illustration)**.

9 Crank the engine over at least seven compression strokes and watch the gauge. The compression should build up quickly in a healthy engine. Low compression on the first stroke, followed by gradually increasing pressure on successive strokes, indicates worn piston rings. A low compression reading on the first stroke, which doesn't build up during successive strokes, indicates leaking valves or a blown head gasket (a cracked head could also be the cause). Deposits on the undersides of the valve heads can also cause low compression. Record the highest gauge reading obtained.

10 Repeat the procedure for the remaining cylinders and compare the results to this Chapter's Specifications.

11 Add some engine oil (about three squirts from a plunger-type oil can) to each cylinder, through the spark plug hole, and repeat the test.

12 If the compression increases after the oil is added, the piston rings are definitely worn. If the compression doesn't increase significantly, the leakage is occurring at the valves or head gasket. Leakage past the valves may be caused by burned valve seats and/or faces or warped, cracked or bent valves.

13 If two adjacent cylinders have equally low compression, there's a strong possibility that the head gasket between them is blown. The appearance of coolant in the combustion chambers or the crankcase would verify this condition.

14 If one cylinder is 20-percent lower than the others, and the engine has a slightly rough idle, a worn exhaust lobe on the camshaft could be the cause.

15 If the compression is unusually high, the combustion chambers are probably coated with carbon deposits. If that's the case, the cylinder head(s) should be removed and decarbonized.

16 If compression is way down or varies greatly between cylinders, it would be a good idea to have a leak-down test performed by an automotive repair shop. This test will pinpoint exactly where the leakage is occurring and how severe it is.

3.8 A gauge with a threaded fitting for the spark plug hole is preferred over the type that requires hand pressure to maintain the seal during the compression check

4 Engine rebuilding alternatives

The do-it-yourselfer is faced with a number of options when performing an engine overhaul. The decision to replace the engine block, piston/connecting rod assemblies and crankshaft depends on a number of factors, with the number one consideration being the condition of the block. Other considerations are cost, access to machine shop facilities, parts availability, time required to complete the project and the extent of prior mechanical experience on the part of the do-it-yourselfer.

Some of the rebuilding alternatives include:

Individual parts - If the inspection procedures reveal that the engine block and most engine components are in reusable condition, purchasing individual parts may be the most economical alternative. The block, crankshaft and piston/connecting rod assemblies should all be inspected carefully. Even if the block shows little wear, the cylinder bores should be surface honed.

Short block - A short block consists of an engine block with a crankshaft, camshaft, timing chain and sprockets and piston/connecting rod assemblies already installed. All new bearings are incorporated and all clearances will be correct. The existing valve train components, cylinder head(s) and external parts can be bolted to the short block with little or no machine shop work necessary.

Long block - A long block consists of a short block plus an oil pump, oil pan, cylinder heads, valve covers, valve train components and timing chain cover. All components are installed with new bearings, seals and gaskets incorporated throughout. The installation of manifolds and external parts is all that's necessary.

Give careful thought to which alternative is best for you and discuss the situation with local automotive machine shops, auto parts dealers and experienced rebuilders before ordering or purchasing replacement parts.

5 Engine removal - methods and precautions

If you've decided that an engine must be removed for overhaul or major repair work, several preliminary steps should be taken.

Locating a suitable place to work is extremely important. Adequate work space, along with storage space for the vehicle, will be needed. If a shop or garage isn't available, at the very least a flat, level, clean work surface made of concrete or asphalt is required.

Cleaning the engine compartment and engine before beginning the removal procedure will help keep tools clean and organized.

An engine hoist or A-frame will also be necessary. Make sure the equipment is rated in excess of the combined weight of the engine and accessories. Safety is of primary importance, considering the potential hazards involved in lifting the engine out of the vehicle.

If the engine is being removed by a novice, a helper should be available. Advice and aid from someone more experienced would also be helpful. There are many instances when one person cannot simultaneously perform all of the operations required when lifting the engine out of the vehicle.

Plan the operation ahead of time. Arrange for or obtain all of the tools and equipment you'll need prior to beginning the job. Some of the equipment necessary to perform engine removal and installation safely and with relative ease are (in addition to an engine hoist) a heavy duty floor jack, complete sets of wrenches and sockets as described in the front of this manual, wooden blocks and plenty of rags and cleaning solvent for mopping up spilled oil, coolant and gasoline. If the hoist must be rented, make sure that you arrange for it in advance and perform all of the operations possible without it beforehand. This will save you money and time.

Plan for the vehicle to be out of use for quite a while. A machine shop will be required to perform some of the work which the do-it-yourselfer can't accomplish without special equipment. These shops often have a

Chapter 2 Part D General engine overhaul procedures

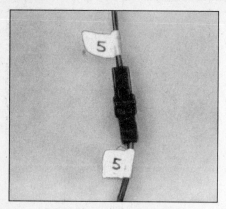

6.5a Label each wire before unplugging the connector

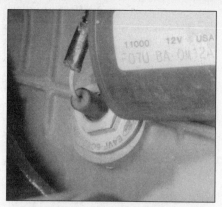

6.5b The oil level sensor is mounted in the oil pan, forward of the starter (4.0L V6 engine)

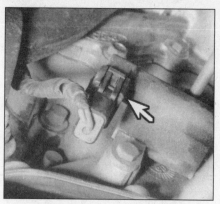

6.5c The electrical connector for the crankshaft position sensor is located near the crankshaft pulley (arrow)

busy schedule, so it would be a good idea to consult them before removing the engine in order to accurately estimate the amount of time required to rebuild or repair components that may need work.

Always be extremely careful when removing and installing the engine. Serious injury can result from careless actions. Plan ahead, take your time and a job of this nature, although major, can be accomplished successfully.

6 Engine - removal and installation

Warning: *If vehicle is equipped with airbags, refer to Chapter 12 to disarm the airbag system prior to performing any work described below.*

Removal

Refer to illustrations 6.5a, 6.5b, 6.5c, 6.15 and 6.24

Warning 1: *Some models are equipped with airbags. Always disconnect the negative battery cable, then the positive battery cable and wait 2 minutes before working in the vicinity of the impact sensors, steering column or instrument panel to avoid the possibility of accidental deployment of the airbag, which could cause personal injury (see Chapter 12).*

Warning 2: *The air conditioning system is under high pressure. Do not loosen any hose fittings or remove any components until after the system has been discharged. Air conditioning refrigerant should be properly discharged into an EPA-approved recovery/recycling unit at a dealer service department or an automotive air conditioning repair facility. Always wear eye protection when disconnecting air conditioning system fittings.*

Warning 3: *Gasoline is extremely flammable, so take extra precautions when you work on any part of the fuel system. Don't smoke or allow open flames or bare light bulbs near the work area, and don't work in a garage where a natural gas-type appliance (such as a water heater or a clothes dryer) with a pilot light is present. Since gasoline is carcinogenic, wear latex gloves when there's a possibility of contact with fuel, and, if you spill any*

fuel on your skin, rinse it off immediately with soap and water. Mop up any spills immediately and do not store fuel-soaked rags where they could ignite. The fuel system is under constant pressure, so, if any fuel lines are to be disconnected, the fuel pressure in the system must be relieved first (see Chapter 4 for more information). When you perform any kind of work on the fuel system, wear safety glasses and have a Class B type fire extinguisher on hand.

1 Refer to Chapter 4 and relieve the fuel system pressure, then disconnect the negative cable from the battery.

2 Cover the fenders and cowl and remove the hood (see Chapter 11). Special pads are available to protect the fenders, but an old bedspread or blanket will also work.

3 Remove the air cleaner assembly.

4 Drain the cooling system (see Chapter 1).

5 Label the vacuum lines, emissions system hoses, wiring connectors, ground straps and fuel lines to ensure correct reinstallation, then detach them. Pieces of masking tape with numbers or letters written on them work well **(see illustration)**. If there's any possibility of confusion, make a sketch of the engine compartment and clearly label the lines, hoses and wires. Don't forget the following:

a) *Oil level sensor connector, if equipped (in the oil pan on the driver's side of the vehicle)* **(see illustration)**.
b) *Wiring harness at the rear of the driver's side cylinder head (V6 engines).*
c) *Hose bracket at the lower front corner of the engine on the passenger's side (V6 engines).*
d) *Crankshaft position sensor connector near the crankshaft pulley, if equipped* **(see illustration)**.

6 Label and detach all coolant hoses from the engine.

7 Remove the drivebelt (see Chapter 1).

8 Remove the cooling fan, shroud and radiator (see Chapter 3).

9 Disconnect the fuel lines running from the engine to the chassis (see Chapter 4). Plug or cap all open fittings/lines.

10 Disconnect the throttle linkage (and TV linkage/speed control cable, if equipped) from the engine (see Chapter 4).

11 Unbolt the power steering pump (see Chapter 10). Leave the lines/hoses attached and make sure the pump is kept in an upright position in the engine compartment (use wire or rope to restrain it out of the way).

12 On air-conditioned models, unbolt the compressor (see Chapter 3) and set it aside. DO NOT disconnect the hoses!

13 Drain the engine oil (Chapter 1) and remove the filter.

14 Remove the starter motor (see Chapter 5).

15 Remove the alternator (see Chapter 5). Remove the alternator bracket **(see illustration)**.

16 Disconnect the exhaust system from the engine (see Chapter 4).

17 If you're working on a vehicle with an automatic transmission, refer to Chapter 7, Part B and remove the torque converter-to-driveplate fasteners.

18 Support the transmission with a jack. Position a block of wood between the jack head and transmission to prevent damage to the transmission. Special transmission jacks with safety chains are available - use one if possible.

19 Attach an engine sling or a length of chain to the lifting brackets on the engine.

20 Roll the hoist into position and connect the sling to it. Take up the slack in the sling or chain, but don't lift the engine. **Warning:** *DO NOT place any part of your body under the engine when it's supported only by a hoist or other lifting device.*

21 Remove the transmission-to-engine block bolts. Be sure to remove the two bolts that secure the engine oil pan to the transmission.

22 Remove the engine mount-to-frame bolts.

23 Recheck to be sure nothing is still connecting the engine to the transmission or vehicle. Disconnect anything still remaining.

24 Raise the engine slightly. It should move easily if it's completely disconnected. If it doesn't, it's stuck - DO NOT try to pry it loose or force it loose with the hoist, since it could shift suddenly and cause injury or damage.

Chapter 2 Part D General engine overhaul procedures

Find out why it won't move and fix the problem before continuing. Carefully work the engine forward to separate it from the transmission. If you're working on a vehicle with an automatic transmission, be sure the torque converter stays in the transmission (clamp a pair of vise-grips to the housing to keep the converter from sliding out). If you're working on a vehicle with a manual transmission, the input shaft must be completely disengaged from the clutch. Slowly raise the engine out of the engine compartment (see illustration). Check carefully to make sure nothing is hanging up.

25 Remove the clutch (if equipped - see Chapter 8) and flywheel/driveplate and mount the engine on an engine stand.

Installation

26 Check the engine and transmission mounts. If they're worn or damaged, replace them.

27 If you're working on a manual transmission equipped vehicle, install the clutch and pressure plate (see Chapter 8). Now is a good time to install a new clutch.

28 Carefully lower the engine into the engine compartment - make sure the engine mounts line up.

29 If you're working on an automatic transmission equipped vehicle, remove the vise grips and guide the torque converter into the crankshaft following the procedure outlined in Chapter 7, Part B.

30 If you're working on a manual transmission equipped vehicle, apply a dab of high-temperature grease to the input shaft and guide it into the crankshaft pilot bearing until the bellhousing is flush with the engine block.

31 Install the transmission-to-engine bolts and tighten them securely. **Caution:** *DO NOT use the bolts to force the transmission and engine together!*

32 Reinstall the remaining components in the reverse order of removal.

33 Add coolant, oil, power steering and transmission fluid as needed.

34 Run the engine and check for leaks and proper operation of all accessories, then install the hood and test drive the vehicle.

7 Engine overhaul - disassembly sequence

1 It's much easier to disassemble and work on the engine if it's mounted on a portable engine stand. A stand can often be rented quite cheaply from an equipment rental yard. Before the engine is mounted on a stand, the flywheel/driveplate should be removed from the engine.

2 If a stand isn't available, it's possible to disassemble the engine with it blocked up on the floor. Be extra careful not to tip or drop the engine when working without a stand.

3 If you're going to obtain a rebuilt engine, all external components must come off first, to be transferred to the replacement engine,

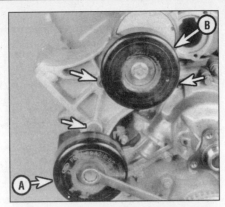

6.15 To remove the alternator bracket, remove the lower idler pulley bolt and pulley (A), but don't remove the drivebelt adjustment pulley (B) - then remove the three bracket bolts (arrows); one bolt is hidden behind the adjustment pulley (V6 engine shown)

just as they will if you're doing a complete engine overhaul yourself. These include:

Alternator and brackets (if not removed during engine removal)
Emissions control components
Coil pack, spark plug wires and spark plugs
Thermostat and housing cover
Water pump
EFI components
Intake/exhaust manifolds
Oil filter
Engine mounts
Engine rear plate
Clutch and flywheel/driveplate

Note: *When removing the external components from the engine, pay close attention to details that may be helpful or important during installation. Note the installed position of gaskets, seals, spacers, pins, brackets, washers, bolts and other small items.*

4 If you're obtaining a short block, which consists of the engine block, crankshaft, pistons and connecting rods all assembled, then the cylinder head, oil pan and oil pump will have to be removed as well. See *Engine rebuilding alternatives* for additional information regarding the different possibilities to be considered.

5 If you're planning a complete overhaul, the engine must be disassembled and the internal components removed in the following order:

Four-cylinder engine

Clutch and flywheel or driveplate
Valve cover
Intake and exhaust manifolds
Timing belt, sprockets and covers
Cylinder head
Camshaft and related components
Oil pan
Front cover
Oil pump
Piston/connecting rod assemblies
Crankshaft and main bearings

6.24 Carefully raise the engine from the engine compartment

V6 engines

Clutch and flywheel or driveplate
Valve covers
Rocker assemblies and pushrods
Intake and exhaust manifolds
Cylinder heads
Oil pan
Oil pump
Timing chain cover and cover plate
Timing chain and sprockets
Valve lifters
Camshaft
Piston/connecting rod assemblies
Crankshaft and main bearings

6 Before beginning the disassembly and overhaul procedures, make sure the following items are available. Also, refer to *Engine overhaul - reassembly sequence* for a list of tools and materials needed for engine reassembly.

Common hand tools
Small cardboard boxes or plastic bags for storing parts
Gasket scraper
Ridge reamer
Vibration damper puller
Micrometers
Telescoping gauges
Dial indicator set
Valve spring compressor
Cylinder surfacing hone
Piston ring groove cleaning tool
Electric drill motor
Tap and die set
Wire brushes
Oil gallery brushes
Cleaning solvent

8 Cylinder head - disassembly

Refer to illustrations 8.2, 8.3 and 8.4
Note: *New and rebuilt cylinder heads are commonly available for most engines at dealerships and auto parts stores. Due to the fact that some specialized tools are necessary for the disassembly and inspection procedures, and replacement parts may not be readily available, it may be more practical and economical for the home mechanic to purchase replacement*

8.2 A small plastic bag, with an appropriate label, can be used to store the valve train components so they can be kept together and reinstalled in the original location

8.3 Use a valve spring compressor to compress the spring, then remove the keepers from the valve stem

8.4 If the valve won't pull through the guide, deburr the edge of the stem end and the area around the top of the keeper groove with a file or whetstone

heads rather than taking the time to disassemble, inspect and recondition the originals.

1   Cylinder head disassembly involves removal of the intake and exhaust valves and related components. On V6 engines, remove the rocker arm shaft assemblies (see Part A through C of this Chapter). On four-cylinder engines, remove the camshaft, followers and lash adjusters. Label the parts or store them separately so they can be reinstalled in their original locations.

2   Before the valves are removed, arrange to label and store them, along with their related components, so they can be kept separate and reinstalled in the same valve guides they are removed from **(see illustration)**.

3   Compress the springs on the first valve with a spring compressor and remove the keepers **(see illustration)**. Carefully release the valve spring compressor and remove the retainer and spring.

4   Pull the valve out of the head, then remove the oil seal from the guide. If the valve binds in the guide won't pull through, push it back into the head and deburr the area around the keeper groove with a fine file or whetstone **(see illustration)**.

5   Repeat the procedure for the remaining valves. Remember to keep all the parts for each valve together so they can be reinstalled in the same locations.

6   Once the valves and related components have been removed and stored in an organized manner, the head should be thoroughly cleaned and inspected. If a complete engine overhaul is being done, finish the engine disassembly procedures before beginning the cylinder head cleaning and inspection process.

---

9   Cylinder head - cleaning and inspection

Refer to illustrations 9.12, 9.14, 9.15, 9.16, 9.17 and 9.18

1   Thorough cleaning of the cylinder head(s) and related valve train components, followed by a detailed inspection, will enable you to decide how much valve service work must be done during the engine overhaul. **Note:** *If the engine was severely overheated, the cylinder head is probably warped (see Step 12).*

Cleaning

2   Scrape all traces of old gasket material and sealing compound off the head gasket, intake manifold and exhaust manifold sealing surfaces. Be very careful not to gouge the cylinder head. Special gasket removal solvents that soften gaskets and make removal much easier are available at auto parts stores.

3   Remove all built up scale from the coolant passages.

4   Run a stiff wire brush through the various holes to remove deposits that may have formed in them.

5   Run an appropriate size tap into each of the threaded holes to remove corrosion and thread sealant that may be present. If compressed air is available, use it to clear the holes of debris produced by this operation. **Warning:** *Wear eye protection when using compressed air!*

6   Clean the rocker arm pivot stud threads with a wire brush.

7   Clean the cylinder head with solvent and dry it thoroughly. Compressed air will speed the drying process and ensure that all holes and recessed areas are clean. **Note:** *Decarbonizing chemicals are available and may prove very useful when cleaning cylinder heads and valve train components. They are very caustic and should be used with caution. Be sure to follow the instructions on the container.*

8   Clean the rocker arms, fulcrums or shafts, bolts and pushrods with solvent and dry them thoroughly (don't mix them up during the cleaning process). Compressed air will speed the drying process and can be used to clean out the oil passages.

9   Clean all the valve springs, spring seats, keepers and retainers with solvent and dry them thoroughly. Do the components from one valve at a time to avoid mixing up the parts.

10   Scrape off any heavy deposits that may have formed on the valves, then use a motorized wire brush to remove deposits from the valve heads and stems. Again, make sure the valves don't get mixed up.

Inspection

Note: *Be sure to perform all of the following inspection procedures before concluding that machine shop work is required. Make a list of the items that need attention.*

Cylinder head

11   Inspect the head very carefully for cracks, evidence of coolant leakage and other damage. If cracks are found, check with an automotive machine shop concerning repair. If repair isn't possible, a new cylinder head should be obtained.

12   Using a straightedge and feeler gauge, check the head gasket mating surface for warpage **(see illustration)**. If the warpage exceeds the limit listed in this Chapter's Specifications, it can be resurfaced at an automotive machine shop. **Note:** *If the heads are resurfaced, the intake manifold flanges will also require machining.*

13   Examine the valve seats in each of the combustion chambers. If they're pitted, cracked or burned, the head will require valve service that's beyond the scope of the home mechanic.

14   Check the valve stem-to-guide clearance by measuring the lateral movement of the valve stem with a dial indicator attached securely to the head **(see illustration)**. The valve must be in the guide and approximately 1/16-inch off the seat. The total valve stem movement indicated by the gauge needle must be divided by two to obtain the actual clearance. After this is done, if there's still some doubt regarding the condition of the valve guides they should be checked by an automotive machine shop (the cost should be minimal).

Valves

15   Carefully inspect each valve face for uneven wear, deformation, cracks, pits and burned areas **(see illustration)**. Check the

Chapter 2 Part D General engine overhaul procedures   2D-11

9.12 Check the cylinder head gasket surface for warpage by trying to slip a feeler gauge under the straightedge (see this Chapter's Specifications for the maximum warpage)

9.14 A dial indicator can be used to determine the valve stem-to-guide clearance (move the valve stem as indicated by the arrows)

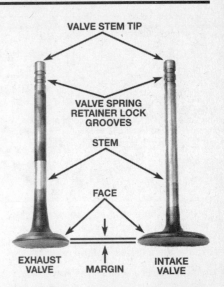

9.15 Check for valve wear at the points shown here

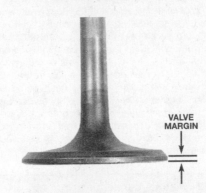

9.16 The margin width on each valve must be as specified - if no margin exists, the valve cannot be reused

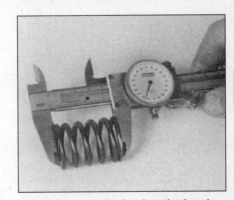

9.17 Measure the free length of each valve spring with a dial or vernier caliper

9.18 Check each valve spring for squareness

valve stem for scuffing and galling and the neck for cracks. Rotate the valve and check for any obvious indication that it's bent. Look for pits and excessive wear on the end of the stem. The presence of any of these conditions indicates the need for valve service by an automotive machine shop.

16 Measure the margin width on each valve (see illustration). Any valve with a margin narrower than 1/32-inch will have to be replaced with a new one.

Valve components

17 Check each valve spring for wear (on the ends) and pits. Measure the free length and compare it to the Specifications (see illustration). Any springs that are shorter than specified have sagged and should not be reused. The tension of all springs should be checked with a special fixture before deciding that they're suitable for use in a rebuilt engine (take the springs to an automotive machine shop for this check).

18 Stand each spring on a flat surface and check it for squareness (see illustration). If any of the springs are distorted or sagged, replace all of them with new parts.

19 Check the spring retainers and keepers for obvious wear and cracks. Any questionable parts should be replaced with new ones, as extensive damage will occur if they fail during engine operation.

Cam followers (four-cylinder engine)

20 Clean all the parts thoroughly. Make sure that all oil passages are open.

21 Inspect the pad at the valve end of the follower and the camshaft end pad for indications of scuffing or abnormal wear. If the pad is grooved, replace the follower. Do not attempt to true the surface by grinding.

Rocker arm components (V6 engines)

22 Check the rocker arm faces (the areas that contact the pushrod ends and valve stems) for pits, wear, galling, score marks and rough spots. Check the rocker arm shafts as well. Look for cracks in each rocker arm and shaft.

23 Inspect the pushrod ends for scuffing and excessive wear. Roll each pushrod on a flat surface, like a piece of plate glass, to determine if it's bent.

24 Any damaged or excessively worn parts must be replaced with new ones.

25 If the inspection process indicates that the valve components are in generally poor condition and worn beyond the limits specified, which is usually the case in an engine that's being overhauled, reassemble the valves in the cylinder head and refer to Section 10 for valve servicing recommendations.

10 Valves - servicing

1 Because of the complex nature of the job and the special tools and equipment needed, servicing of the valves, the valve seats and the valve guides, commonly known as a valve job, should be done by a professional.

2 The home mechanic can remove and disassemble the heads, do the initial cleaning and inspection, then reassemble and deliver them to a dealer service department or an automotive machine shop for the actual service work. Doing the inspection will enable you to see what condition the head and valvetrain components are in and will ensure that you know what work and new parts are required when dealing with an automotive machine shop.

3 The dealer service department, or auto-

11.7 Be sure to check the valve spring installed height for each valve (the distance from the top of the seat/shims to the underside of the retainer

12.1 A ridge reamer is required to remove the ridge from the top of each cylinder - do this before removing the pistons!

12.3 Connecting rod side clearance is measured with feeler gauges inserted between the rod cap and crankshaft

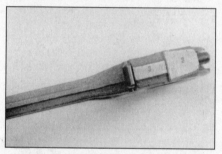

12.4 Check the connecting rod and cap for identification marks - if they aren't visible, mark the rods and caps with a hammer and punch to indicate which cylinder they came from (be sure to mark the rod and cap on the same side so they can be reassembled correctly)

motive machine shop, will remove the valves and springs, recondition or replace the valves and valve seats, recondition the valve guides, check and replace the valve springs, spring retainers and keepers (as necessary), replace the valve seals with new ones, reassemble the valve components and make sure the installed spring height is correct. The cylinder head gasket surface will also be resurfaced if it's warped.

4   After the valve job has been performed by a professional, the head will be in like-new condition. When the head is returned, be sure to clean it again before installation on the engine to remove any metal particles and abrasive grit that may still be present from the valve service or head resurfacing operations. Use compressed air, if available, to blow out all the oil holes and passages

11 Cylinder head - reassembly

Refer to illustration 11.7

1   Regardless of whether or not the head was sent to an automotive repair shop for valve servicing, make sure it's clean before beginning reassembly.
2   If the head was sent out for valve servicing, the valves and related components will already be in place. Go to Step 7.
3   Lubricate and install each valve, then install new seals on each of the valve stems, as described in Parts A through C. Don't twist or cock the seals during installation or they won't seat properly on the valve stems.
4   Drop the spring seat or shim(s) over the valve guide and set the valve spring and retainer in place.
5   Compress the springs with a valve spring compressor and carefully install the keepers in the upper groove, then slowly release the compressor and make sure the keepers seat properly. Apply a small dab of grease to each keeper to hold it in place if necessary.
6   Repeat the procedure for the remaining valves. Be sure to return the components to their original locations - don't mix them up!
7   Check the installed valve spring height with a ruler graduated in 1/32-inch increments or a dial caliper. If the head was sent out for service work, the installed height should be correct (but don't automatically assume that it is). The measurement is taken from the top of each spring seat or shim(s) to the bottom of the retainer **(see illustration)**. If the height is greater than the figure listed in this Chapter's Specifications, shims can be added under the springs to correct it. **Caution:** *Don't, under any circumstances, shim the springs to the point where the installed height is less than specified.*

12 Pistons/connecting rods - removal

Refer to illustrations 12.1, 12.3, 12.4 and 12.6
Note: *Prior to removing the piston/connecting rod assemblies, remove the cylinder head(s), the oil pan and the oil pump by referring to the appropriate Sections in Chapter 2.*
1   Use your fingernail to feel if a ridge has formed at the upper limit of ring travel (about 1/4-inch down from the top of each cylinder). If carbon deposits or cylinder wear have produced ridges, they must be completely removed with a special tool **(see illustration)**. Follow the manufacturer's instructions provided with the tool. Failure to remove the ridges before attempting to remove the piston/connecting rod assemblies may result in piston breakage.
2   After the cylinder ridges have been removed, turn the engine upside-down so the crankshaft is facing up.
3   Before the connecting rods are removed, check the endplay with feeler gauges. Slide them between the first connecting rod and the crankshaft throw until the play is removed **(see illustration)**. The endplay is equal to the thickness of the feeler gauge(s). If the endplay exceeds the service limit, new connecting rods will be required. If new rods (or a new crankshaft) are installed, the endplay may fall under the specified minimum (if it does, the rods will have to be machined to restore it - consult an automotive machine shop for advice if necessary). Repeat the procedure for the remaining connecting rods.
4   Check the connecting rods and caps for identification marks **(see illustration)**. If they aren't plainly marked, use a small center-punch to make the appropriate number of indentations on each rod and cap (1, 2, 3, etc.).
5   Loosen each of the connecting rod cap nuts 1/2-turn at a time until they can be removed by hand. Remove the number one connecting rod cap and bearing insert. Don't drop the bearing insert out of the cap.
6   Slip a short length of plastic or rubber hose over each connecting rod cap bolt to protect the crankshaft journal and cylinder wall as the piston is removed **(see illustration)**.
7   Remove the bearing insert and push the connecting rod/piston assembly out through the top of the engine. Use a wooden hammer handle to push on the upper bearing surface in the connecting rod. If resistance is felt, double-check to make sure that all of the ridge was removed from the cylinder.
8   Repeat the procedure for the remaining cylinders.
9   After removal, reassemble the connecting rod caps and bearing inserts in their respective connecting rods and install the cap nuts finger tight. Leaving the old bearing inserts in place until reassembly will help pre-

Chapter 2 Part D General engine overhaul procedures

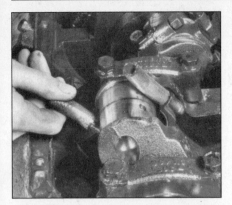

12.6 Place short pieces of hose over the connecting rod studs to protect the crankshaft and cylinder walls

13.1 Checking crankshaft endplay with a dial indicator

13.3 Checking crankshaft endplay with a feeler gauge

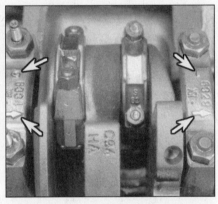

13.4 The main bearing caps are numbered from the front to the rear of the engine and have cast-in arrow that point to the front of the engine when the caps are installed

14.1a A hammer and large punch can be used to knock the core plugs sideways in their bores

14.1b Pull the core plugs from the block with pliers

vent the connecting rod bearing surfaces from being accidentally nicked or gouged.
10 Don't separate the pistons from the connecting rods (see Section 17 for additional information).

13 Crankshaft - removal

Refer to illustrations 13.1, 13.3 and 13.4
Note: *The crankshaft can be removed only after the engine has been removed from the vehicle. It's assumed that the flywheel or driveplate, vibration damper, timing chain, oil pan, oil pump and piston/connecting rod assemblies have already been removed.*

1 Before the crankshaft is removed, check the endplay. Mount a dial indicator with the stem in line with the crankshaft and just touching one of the crank throws **(see illustration)**.
2 Push the crankshaft all the way to the rear and zero the dial indicator. Next, pry the crankshaft to the front as far as possible and check the reading on the dial indicator. The distance that it moves is the endplay. If it's greater than the limit listed in this Chapter's Specifications, check the crankshaft thrust surfaces for wear. If no wear is evident, new main bearings should correct the endplay.
3 If a dial indicator isn't available, feeler gauges can be used. Gently pry or push the crankshaft all the way to the front of the engine. Slip feeler gauges between the crankshaft and the front face of the thrust main bearing (number 3 bearing) to determine the clearance **(see illustration)**.
4 Check the main bearing caps to see if they're marked to indicate their locations. They should be numbered consecutively from the front of the engine to the rear **(see illustration)**. If they aren't, mark them with number stamping dies or a center-punch. Main bearing caps generally have a cast-in arrow, which points to the front of the engine. Loosen the main bearing cap bolts 1/4-turn at a time each, until they can be removed by hand. Note if any stud bolts are used and make sure they're returned to their original locations when the crankshaft is reinstalled.
5 Gently tap the caps with a soft-face hammer, then separate them from the engine block. If necessary, use the bolts as levers to remove the caps. Try not to drop the bearing inserts if they come out with the caps.
6 Carefully lift the crankshaft out of the engine. It may be a good idea to have an assistant available, since the crankshaft is quite heavy. With the bearing inserts in place in the engine block and main bearing caps, return the caps to their respective locations on the engine block and tighten the bolts finger tight.

14 Engine block - cleaning

Refer to illustrations 14.1, 14.8 and 14.10
Caution: *The core plugs (also known as freeze plugs or soft plugs) may be difficult or impossible to retrieve if they're driven into the block coolant passages.*

1 Remove all the core plugs from the engine block. To do this, knock one side of the plug into the block with a hammer and punch, then grasp it with large pliers and pull it out **(see illustrations)**.
2 Using a gasket scraper, remove all traces of gasket material from the engine block. Be very careful not to nick or gouge the gasket sealing surfaces.
3 Remove the main bearing caps and separate the bearing inserts from the caps and the engine block. Tag the bearings, indicating which cylinder they were removed from and whether they were in the cap or the block, then set them aside.
4 Remove all of the threaded oil gallery plugs from the block. The plugs are usually very tight - they may have to be drilled out and the holes retapped. Use new plugs when the engine is reassembled.
5 If the engine is extremely dirty it should be taken to an automotive machine shop to be steam cleaned or hot tanked.
6 After the block is returned, clean all oil

14.8 All bolt holes in the block - particularly the main bearing cap and head bolt holes - should be cleaned and restored with a tap (be sure to remove debris from the holes after this is done)

14.10 A large socket on an extension can be used to drive the new core plugs into the bores

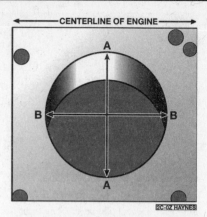

15.4a Measure the diameter of each cylinder just under the wear ridge (A), at the center (B) and at the bottom

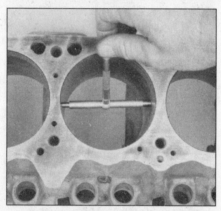

15.4b The ability to "feel" when the telescoping gauge is at the correct point will be developed over time, so work slowly and repeat the check until you're satisfied the bore measurement is accurate

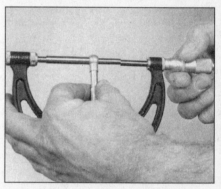

15.4c The gauge is then measured with a micrometer to determine the bore size

holes and oil galleries one more time. Brushes specifically designed for this purpose are available at most auto parts stores. Flush the passages with warm water until the water runs clear, dry the block thoroughly and wipe all machined surfaces with a light, rust preventive oil. If you have access to compressed air, use it to speed the drying process and to blow out all the oil holes and galleries. **Warning:** *Wear eye protection when using compressed air!*

7   If the block isn't extremely dirty or sludged up, you can do an adequate cleaning job with hot soapy water and a stiff brush. Take plenty of time and do a thorough job. Regardless of the cleaning method used, be sure to clean all oil holes and galleries very thoroughly, dry the block completely and coat all machined surfaces with light oil.

8   The threaded holes in the block must be clean to ensure accurate torque readings during reassembly. Run the proper size tap into each of the holes to remove rust, corrosion, thread sealant or sludge and restore damaged threads **(see illustration)**. If possible, use compressed air to clear the holes of debris produced by this operation. Now is a good time to clean the threads on the head bolts and the main bearing cap bolts as well.

9   Reinstall the main bearing caps and tighten the bolts finger tight.

10   After coating the sealing surfaces of the new core plugs with Permatex no. 2 sealant, install them in the engine block **(see illustration)**. Make sure they're driven in straight and seated properly or leakage could result. Special tools are available for this purpose, but a large socket, with an outside diameter that will just slip into the core plug, a 1/2-inch drive extension and a hammer will work just as well.

11   Apply non-hardening sealant (such as Permatex no. 2 or Teflon pipe sealant) to the new oil gallery plugs and thread them into the holes in the block. Make sure they're tightened securely.

12   If the engine isn't going to be reassembled right away, cover it with a large plastic trash bag to keep it clean.

15 Engine block - inspection

Refer to illustrations 15.4a, 15.4b and 15.4c

1   Before the block is inspected, it should be cleaned as described in Section 14.

2   Visually check the block for cracks, rust and corrosion. Look for stripped threads in the threaded holes. It's also a good idea to have the block checked for hidden cracks by an automotive machine shop that has the special equipment to do this type of work. If defects are found, have the block repaired, if possible, or replaced.

3   Check the cylinder bores for scuffing and scoring.

4   Measure the diameter of each cylinder at the top (just under the ridge area), center and bottom of the cylinder bore, parallel to the crankshaft axis **(see illustrations)**.

5   Next, measure each cylinder's diameter at the same three locations across the crankshaft axis. Compare the results to this Chapter's Specifications.

6   If the required precision measuring tools aren't available, the piston-to-cylinder clearances can be obtained, though not quite as accurately, using feeler gauge stock. Feeler gauge stock comes in 12-inch lengths and various thickness and is generally available at auto parts stores.

7   To check the clearance, select a feeler gauge and slip it into the cylinder along with the matching piston. The piston must be positioned exactly as it normally would be. The feeler gauge must be between the piston and cylinder on one of the thrust faces (90-degrees to the piston pin bore).

8   The piston should slip through the cylinder (with the feeler gauge in place) with moderate pressure.

9   If it falls through or slides through easily, the clearance is excessive and a new piston will be required. If the piston binds at the lower end of the cylinder and is loose toward the top, the cylinder is tapered. If tight spots are encountered as the piston/feeler gauge is rotated in the cylinder, the cylinder is out-of-round.

10   Repeat the procedure for the remaining pistons and cylinders.

11   If the cylinder walls are badly scuffed or scored, or if they're out-of-round or tapered beyond the limits given in this Chapter's Specifications, have the engine block rebored and honed at an automotive machine shop. If a rebore is done, oversize pistons and rings will be required.

12   If the cylinders are in reasonably good

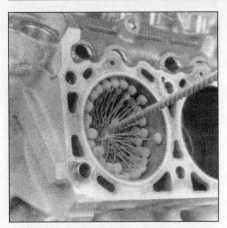

16.3a A "bottle brush" hone will produce better results if you've never honed cylinders before

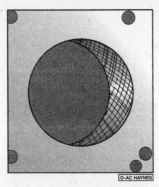

16.3b The cylinder hone should leave a smooth, crosshatch pattern with the lines intersecting at approximately a 60-degree angle

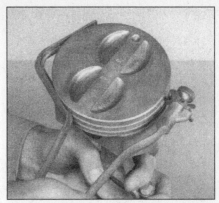

17.4a The piston ring grooves can be cleaned with a special tool, as shown here . . .

17.4b . . . or a section of a broken ring

condition and not worn to the outside of the limits, and if the piston-to-cylinder clearances can be maintained properly, then they don't have to be rebored. Honing is all that's necessary (see Section 16).

16 Cylinder honing

Refer to illustrations 16.3a and 16.3b

1   Prior to engine reassembly, the cylinder bores must be honed so the new piston rings will seat correctly and provide the best possible combustion chamber seal. **Note:** *If you don't have the tools or don't want to tackle the honing operation, most automotive machine shops will do it for a reasonable fee.*
2   Before honing the cylinders, install the main bearing caps and tighten the bolts to the torque listed in this Chapter's Specifications.
3   Two types of cylinder hones are commonly available - the flex hone or "bottle brush" type and the more traditional surfacing hone with spring-loaded stones. Both will do the job, but for the less experienced mechanic the "bottle brush" hone will probably be easier to use. You'll also need some kerosene or honing oil, rags and an electric drill motor. Proceed as follows:

a) Mount the hone in the drill motor, compress the stones and slip it into the first cylinder **(see illustration)**. Be sure to wear safety goggles or a face shield!
b) Lubricate the cylinder with plenty of honing oil, turn on the drill and move the hone up-and-down in the cylinder at a pace that will produce a fine crosshatch pattern on the cylinder walls. Ideally, the crosshatch lines should intersect at approximately a 60-degree angle **(see illustration)**. Be sure to use plenty of lubricant and don't take off any more material than is absolutely necessary to produce the desired finish. **Note:** *Piston ring manufacturers may specify a smaller crosshatch angle than the traditional 60-degrees - read and follow any instructions included with the new rings.*
c) Don't withdraw the hone from the cylinder while it's running. Instead, shut off the drill and continue moving the hone up-and-down in the cylinder until it comes to a complete stop, then compress the stones and withdraw the hone. If you're using a "bottle brush" type hone, stop the drill motor, then turn the chuck in the normal direction of rotation while withdrawing the hone from the cylinder.
d) Wipe the oil out of the cylinder and repeat the procedure for the remaining cylinders.

4   After the honing job is complete, chamfer the top edges of the cylinder bores with a small file so the rings won't catch when the pistons are installed. Be very careful not to nick the cylinder walls with the end of the file.
5   The entire engine block must be washed again very thoroughly with warm, soapy water to remove all traces of the abrasive grit produced during the honing operation. **Note:** *The bores can be considered clean when a lint-free white cloth - dampened with clean engine oil - used to wipe them out doesn't pick-up any more honing residue, which will show up as gray areas on the cloth. Be sure to run a brush through all oil holes and galleries and flush them with running water.*
6   After rinsing, dry the block and apply a coat of light rust preventive oil to all machined surfaces. Wrap the block in a plastic trash bag to keep it clean and set it aside until reassembly.

17 Pistons/connecting rods - inspection

Refer to illustrations 17.4a, 17.4b, 17.10 and 17.11

1   Before the inspection process can be carried out, the piston/connecting rod assemblies must be cleaned and the original piston rings removed from the pistons. **Note:** *Always use new piston rings when the engine is reassembled.*
2   Using a piston ring installation tool, carefully remove the rings from the pistons. Be careful not to nick or gouge the pistons in the process.
3   Scrape all traces of carbon from the top of the piston. A hand-held wire brush or a piece of fine emery cloth can be used once the majority of the deposits have been scraped away. Do not, under any circumstances, use a wire brush mounted in a drill motor to remove deposits from the pistons. The piston material is soft and may be eroded away by the wire brush.
4   Use a piston ring groove cleaning tool to remove carbon deposits from the ring grooves. If a tool isn't available, a piece broken off the old ring will do the job. Be very careful to remove only the carbon deposits - don't remove any metal and do not nick or scratch the sides of the ring grooves **(see illustrations)**.
5   Once the deposits have been removed, clean the piston/rod assemblies with solvent and dry them with compressed air (if available). Make sure the oil return holes in the back sides of the ring grooves are clear.
6   If the pistons and cylinder walls aren't damaged or worn excessively and if the engine block is not rebored, new pistons won't be necessary. Normal piston wear appears as even vertical wear on the piston thrust surfaces and slight looseness of the top ring in its groove. New piston rings, however, should always be used when an engine is rebuilt.

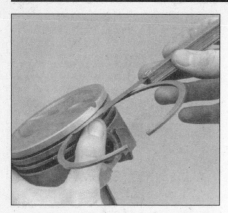

17.10 Check the ring side clearance with a feeler gauge at several points

17.11 Measure the piston diameter at a 90-degree angle to the piston pin and in line with it

18.1 The oil holes should be chamfered so sharp edges don't gouge or scratch the new bearings

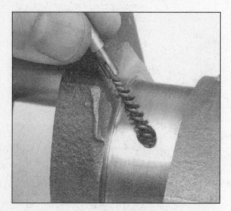

18.2 Use a wire or stiff plastic bristle brush to clean the oil passages in the crankshaft

18.4 Rubbing a penny lengthwise on each journal will reveal its condition - if copper rubs off and is embedded in the crankshaft, the journals should be reground

7   Carefully inspect each piston for cracks around the skirt, at the pin bosses and at the ring lands.

8   Look for scoring and scuffing on the thrust faces of the skirt, holes in the piston crown and burned areas at the edge of the crown. If the skirt is scored or scuffed, the engine may have been suffering from overheating and/or abnormal combustion, which caused excessively high operating temperatures. The cooling and lubrication systems should be checked thoroughly. A hole in the piston crown is an indication that abnormal combustion (preignition) was occurring. Burned areas at the edge of the piston crown are usually evidence of spark knock (detonation). If any of the above problems exist, the causes must be corrected or the damage will occur again. The causes may include intake air leaks, incorrect fuel/air mixture, incorrect ignition timing and EGR system malfunctions.

9   Corrosion of the piston, in the form of small pits, indicates that coolant is leaking into the combustion chamber and/or the crankcase. Again, the cause must be corrected or the problem may persist in the rebuilt engine.

10   Measure the piston ring side clearance by laying a new piston ring in each ring groove and slipping a feeler gauge in beside it **(see illustration)**. Check the clearance at three or four locations around each groove. Be sure to use the correct ring for each groove - they are different. If the side clearance is greater than the figure listed in this Chapter's Specifications, new pistons will have to be used.

11   Check the piston-to-bore clearance by measuring the bore (see Section 15) and the piston diameter. Make sure the pistons and bores are correctly matched. Measure the piston across the skirt, at a 90-degree angle to and in line with the piston pin **(see illustration)**. Subtract the piston diameter from the bore diameter to obtain the clearance. If it's greater than specified, the block will have to be rebored and new pistons and rings installed.

12   Check the piston-to-rod clearance by twisting the piston and rod in opposite directions. Any noticeable play indicates excessive wear, which must be corrected. The piston/connecting rod assemblies should be taken to an automotive machine shop to have the pistons and rods resized and new pins installed.

13   If the pistons must be removed from the connecting rods for any reason, they should be taken to an automotive machine shop. While they are there have the connecting rods checked for bend and twist, since automotive machine shops have special equipment for this purpose. **Note:** *Unless new pistons and/or connecting rods must be installed, do not disassemble the pistons and connecting rods.*

14   Check the connecting rods for cracks and other damage. Temporarily remove the rod caps, lift out the old bearing inserts, wipe the rod and cap bearing surfaces clean and inspect them for nicks, gouges and scratches. After checking the rods, replace the old bearings, slip the caps into place and tighten the nuts finger tight. **Note:** *If the engine is being rebuilt because of a connecting rod knock, be sure to install new rods.*

18 Crankshaft - inspection

Refer to illustrations 18.1, 18.2, 18.4, 18.6 and 18.8

1   Remove all burrs from the crankshaft oil holes with a stone, file or scraper **(see illustration)**.

2   Clean the crankshaft with solvent and dry it with compressed air (if available). Be sure to clean the oil holes with a stiff brush **(see illustration)** and flush them with solvent.

3   Check the main and connecting rod bearing journals for uneven wear, scoring, pits and cracks.

4   Rub a penny across each journal several times **(see illustration)**. If a journal picks up copper from the penny, it's too rough and must be reground.

5   Check the rest of the crankshaft for cracks and other damage. It should be magnafluxed to reveal hidden cracks - an automotive machine shop will handle the procedure.

6   Using a micrometer, measure the diameter of the main and connecting rod journals and compare the results to this Chapter's Specifications **(see illustration)**. By measuring the diameter at a number of points around each journal's circumference, you'll be able to determine whether or not the journal is out-of-round. Take the measurement at each end of the journal, near the crank throws, to determine if the journal is tapered.

7   If the crankshaft journals are damaged, tapered, out-of-round or worn beyond the

limits given in the Specifications, have the crankshaft re-ground by an automotive machine shop. Be sure to use the correct size bearing inserts if the crankshaft is reconditioned.

8   Check the oil seal journals at each end of the crankshaft for wear and damage. If the seal has worn a groove in the journal, or if it's nicked or scratched **(see illustration)**, the new seal may leak when the engine is reassembled. In some cases, an automotive machine shop may be able to repair the journal by pressing on a thin sleeve. If repair isn't feasible, a new or different crankshaft should be installed.

9   Refer to Section 19 and examine the main and rod bearing inserts.

19 Main and connecting rod bearings - inspection

Refer to illustration 19.1

1   Even though the main and connecting rod bearings should be replaced with new ones during the engine overhaul, the old bearings should be retained for close examination, as they may reveal valuable information about the condition of the engine **(see illustration)**.

2   Bearing failure occurs because of lack of lubrication, the presence of dirt or other foreign particles, overloading the engine and corrosion. Regardless of the cause of bearing failure, it must be corrected before the engine is reassembled to prevent it from happening again.

3   When examining the bearings, remove them from the engine block, the main bearing caps, the connecting rods and the rod caps and lay them out on a clean surface in the same general position as their location in the engine. This will enable you to match any bearing problems with the corresponding crankshaft journal.

4   Dirt and other foreign particles get into the engine in a variety of ways. It may be left in the engine during assembly, or it may pass through filters or the PCV system. It may get into the oil, and from there into the bearings. Metal chips from machining operations and normal engine wear are often present. Abrasives are sometimes left in engine components after reconditioning, especially when parts are not thoroughly cleaned using the proper cleaning methods. Whatever the source, these foreign objects often end up embedded in the soft bearing material and are easily recognized. Large particles will not embed in the bearing and will score or gouge the bearing and journal. The best prevention for this cause of bearing failure is to clean all parts thoroughly and keep everything spotlessly clean during engine assembly. Frequent and regular engine oil and filter changes are also recommended.

5   Lack of lubrication (or lubrication breakdown) has a number of interrelated causes. Excessive heat (which thins the oil), overloading (which squeezes the oil from the bearing

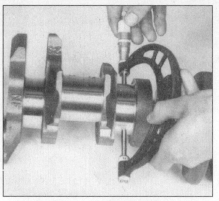

18.6 Measure the diameter of each crankshaft journal at several points to detect taper and out-of-round conditions

18.8 If the seals have worn grooves in the crankshaft journals, or if the seal contact surfaces are nicked or scratched, the new seals will leak

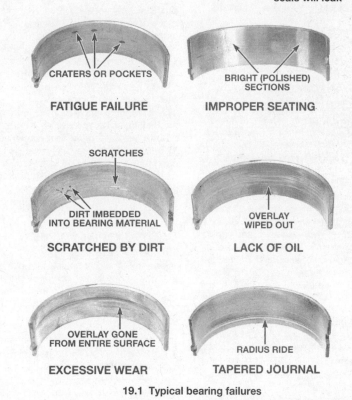

19.1 Typical bearing failures

face) and oil leakage or throw off (from excessive bearing clearances, worn oil pump or high engine speeds) all contribute to lubrication breakdown. Blocked oil passages, which usually are the result of misaligned oil holes in a bearing shell, will also oil starve a bearing and destroy it. When lack of lubrication is the cause of bearing failure, the bearing material is wiped or extruded from the steel backing of the bearing. Temperatures may increase to the point where the steel backing turns blue from overheating.

6   Driving habits can have a definite effect on bearing life. Full throttle, low speed operation (lugging the engine) puts very high loads on bearings, which tends to squeeze out the oil film. These loads cause the bearings to flex, which produces fine cracks in the bearing face (fatigue failure). Eventually the bearing material will loosen in pieces and tear away from the steel backing. Short trip driving leads to corrosion of bearings because insufficient engine heat is produced to drive off the condensed water and corrosive gases. These products collect in the engine oil, forming acid and sludge. As the oil is carried to the engine bearings, the acid attacks and corrodes the bearing material.

7   Incorrect bearing installation during engine assembly will lead to bearing failure as well. Tight fitting bearings leave insufficient bearing oil clearance and will result in oil starvation. Dirt or foreign particles trapped behind a bearing insert result in high spots on the bearing which lead to failure.

2D-18   Chapter 2 Part D   General engine overhaul procedures

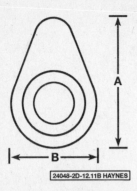

20.1 To verify camshaft lobe lift, measure the major (A) and minor (B) diameters of each lobe with a micrometer or vernier caliper - subtract each minor diameter from the major diameter to arrive at lobe lift

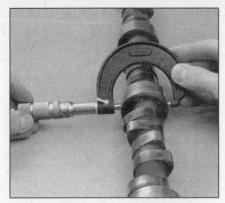

20.5 Use a micrometer to check the camshaft journals for wear, taper and out-of-round conditions

20.9 The roller on roller lifters must turn freely - check for wear and excessive play as well

20 Camshaft, lifters and bearings - inspection and replacement

Camshaft inspection

Refer to illustration 20.1 and 20.5

1 Measure the major (A) and minor (B) diameters of each cam lobe **(see illustration)**. The difference between the readings is the lobe lift. Compare your measurement to the lobe lift in this Chapter's Specifications.
2 If the lobe lift is 0.005-inch less than specified, replace the camshaft and all the lifters or followers.
3 After the camshaft is removed from the engine, cleaned with solvent and dried, inspect the bearing journals for uneven wear, pitting and evidence of seizure. If the journals are damaged, the bearing inserts are probably damaged as well. Both the camshaft and bearings will have to be replaced with new ones.
4 Inspect the bearing journals for uneven wear, pitting or evidence of seizure. If the journals are damaged, the bearing inserts are probably damaged as well. Both the camshaft and bearings will have to be replaced with new ones.
5 Measure the camshaft bearing journals with a micrometer **(see illustration)** to determine if they are excessively worn or out-of-round. If they are more than 0.001-inch out-of-round, the camshaft should be replaced with a new one.
6 Check the camshaft lobes for heat discoloration, score marks, chipped areas, pitting and uneven wear.
7 If the lobes are in good condition and if the lobe lift measurements are as specified, the camshaft can be reused.

Lifter inspection

Refer to illustration 20.9

8 Clean the lifters with solvent and dry them thoroughly without mixing them up.
9 Check that the roller on each lifter turns freely and shows no signs of pitting, scoring or other wear **(see illustration)**. Check the pushrod seat in each lifter for wear. Also make sure each lifter fits in its original bore without excessive looseness. Wear is only normally encountered at very high mileage or in cases of neglected engine lubrication.
10 Press down on the plunger using a Phillips screwdriver to make sure it operates freely without binding.
11 If the lifters show signs of wear or binding they must be replaced as a complete set and the camshaft must be replaced as well.
12 If new lifters are being installed, a new camshaft must also be installed. If a new camshaft is installed, then use new lifters as well. Never install used lifters unless the original camshaft is used and the lifters can be installed in their original locations.

Bearing replacement

13 Camshaft bearing replacement requires special tools and expertise that place it outside the scope of the do-it-yourselfer. Take the block (V6) or head (four-cylinder) to an automotive machine shop to ensure that the job is done correctly.

21 Engine overhaul - reassembly sequence

1 Before beginning engine reassembly, make sure you have all the necessary new parts, gaskets and seals as well as the following items on hand:
   Common hand tools
   A 1/2-inch drive torque wrench
   Piston ring installation tool
   Piston ring compressor
   Vibration damper installation tool
   Short lengths of rubber or plastic hose to fit over connecting rod bolts
   Plastigage
   Feeler gauges
   A fine-tooth file
   New engine oil
   Engine assembly lube or moly-base grease
   Gasket sealant
   Thread locking compound
2 In order to save time and avoid problems, engine reassembly must be done in the following general order:

Four-cylinder engine

   New camshaft bearings (must be done by an automotive
      machine shop)
   Piston rings
   Crankshaft and main bearings
   Piston/connecting rod assemblies
   Oil pump
   Front engine cover
   Oil pan
   Camshaft and related components
   Cylinder heads
   Timing belt, sprockets and covers
   Intake and exhaust manifolds
   Valve cover
   Clutch and flywheel or driveplate

V6 engines

   New camshaft bearings (must be done by an automotive
      machine shop)
   Piston rings
   Crankshaft and main bearings
   Piston/connecting rod assemblies
   Camshaft
   Valve lifters
   Timing chain and sprockets
   Timing chain cover
   Oil pump
   Oil pan
   Cylinder heads
   Intake and exhaust manifolds
   Rocker arms and pushrods
   Valve covers
   Clutch and flywheel or driveplate

22 Piston rings - installation

Refer to illustrations 22.3, 22.4, 22.5, 22.9a, 22.9b and 22.12

1 Before installing the new piston rings, the ring end gaps must be checked. It's assumed that the piston ring side clearance has been checked and verified correct (see Section 17).
2 Lay out the piston/connecting rod

Chapter 2 Part D General engine overhaul procedures

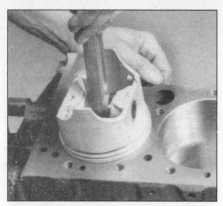

22.3 When checking piston ring end gap, the ring must be square in the cylinder bore (this is done by pushing the ring down with the top of a piston as shown)

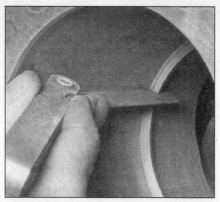

22.4 With the ring square in the cylinder, measure the end gap with a feeler gauge

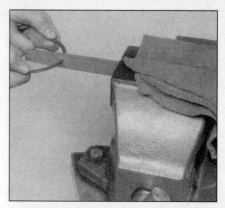

22.5 If the end gap is too small, clamp a file in a vise and file the ring ends (from the outside in only) to enlarge the gap slightly

22.9a Installing the spacer/expander in the oil control ring groove

22.9b DO NOT use a piston ring installation tool when installing the oil ring side rails

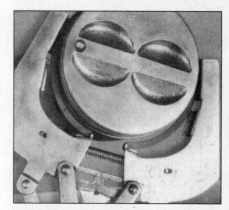

22.12 Installing the compression rings with a ring expander - the mark (arrow) must face up

assemblies and the new ring sets so the ring sets will be matched with the same piston and cylinder during the end gap measurement and engine assembly.

3   Insert the top (number one) ring into the first cylinder and square it up with the cylinder walls by pushing it in with the top of the piston **(see illustration)**. The ring should be near the bottom of the cylinder, at the lower limit of ring travel.

4   To measure the end gap, slip feeler gauges between the ends of the ring until a gauge equal to the gap width is found **(see illustration)**. The feeler gauge should slide between the ring ends with a slight amount of drag. Compare the measurement to this Chapter's Specifications. If the gap is larger or smaller than specified, double-check to make sure you have the correct rings before proceeding.

5   If the gap is too small, it must be enlarged or the ring ends may come in contact with each other during engine operation, which can cause serious damage to the engine. The end gap can be increased by filing the ring ends very carefully with a fine file. Mount the file in a vise equipped with soft jaws, slip the ring over the file with the ends contacting the file face and slowly move the ring to remove material from the ends. When performing this operation, file only from the outside in **(see illustration)**.

6   Excess end gap isn't critical unless it's greater than 0.040-inch. Again, double-check to make sure you have the correct rings for your engine.

7   Repeat the procedure for each ring that will be installed in the first cylinder and for each ring in the remaining cylinders. Remember to keep rings, pistons and cylinders matched up.

8   Once the ring end gaps have been checked/corrected, the rings can be installed on the pistons.

9   The oil control ring (lowest one on the piston) is usually installed first. It's composed of three separate components. Slip the spacer/expander into the groove **(see illustration)**. If an anti-rotation tang is used, make sure it's inserted into the drilled hole in the ring groove. Next, install the lower side rail. Don't use a piston ring installation tool on the oil ring side rails, as they may be damaged. Instead, place one end of the side rail into the groove between the spacer/expander and the ring land, hold it firmly in place and slide a finger around the piston while pushing the rail into the groove **(see illustration)**. Next, install the upper side rail in the same manner.

10 After the three oil ring components have been installed, check to make sure that both the upper and lower side rails can be turned smoothly in the ring groove.

11 The number two (middle) ring is installed next. It's usually stamped with a mark which must face up, toward the top of the piston. **Note:** *Always follow the instructions printed on the ring package or box - different manufacturers may require different approaches.* Do not mix up the top and middle rings, as they have different cross-sections.

12 Use a piston ring installation tool and make sure the identification mark is facing the top of the piston, then slip the ring into the middle groove on the piston **(see illustration)**. Don't expand the ring any more than necessary to slide it over the piston.

13 Install the number one (top) ring in the same manner. Make sure the mark is facing up. Be careful not to confuse the number one and number two rings.

14 Repeat the procedure for the remaining pistons and rings.

23 Crankshaft - installation and main bearing oil clearance check

Refer to illustrations 23.5, 23.11, 23.15, and 23.22

1   Crankshaft installation is the first step in engine reassembly. It's assumed at this point that the engine block and crankshaft have

23.5 The upper main bearing in position (flanged thrust bearing shown) - the oil hole in the bearing must align with the oil hole(s) in the block

23.11 Lay the Plastigage strips (arrow) on the main bearing journals, parallel to the crankshaft centerline

23.15 Compare the width of the crushed Plastigage to the scale on the envelope to determine the main bearing oil clearance (always take the measurement at the widest point of the Plastigage); be sure to use the correct scale - standard and metric ones are included

been cleaned, inspected and repaired or reconditioned.
2 Position the engine with the bottom facing up.
3 Remove the main bearing cap bolts and lift out the caps. Lay them out in the proper order to ensure correct installation.
4 If they're still in place, remove the original bearing inserts from the block and the main bearing caps. Wipe the bearing surfaces of the block and caps with a clean, lint-free cloth. They must be kept spotlessly clean.

Main bearing oil clearance check

5 Clean the back sides of the new main bearing inserts and lay one in each main bearing saddle in the block. If one of the bearing inserts from each set has a large groove in it, make sure the grooved insert is installed in the block. Lay the other bearing from each set in the corresponding main bearing cap. Make sure the tab on the bearing insert fits into the recess in the block or cap. **Caution:** *The oil holes in the block must line up with the oil holes in the bearing insert* **(see illustration).** Do not hammer the bearing into place and don't nick or gouge the bearing faces. No lubrication should be used at this time.
6 The flanged thrust bearing must be installed in the third cap and saddle (counting from the front of the engine).
7 Clean the faces of the bearings in the block and the crankshaft main bearing journals with a clean, lint-free cloth.
8 Check or clean the oil holes in the crankshaft, as any dirt here can go only one way - straight through the new bearings.
9 Once you're certain the crankshaft is clean, carefully lay it in position in the main bearings.
10 Before the crankshaft can be permanently installed, the main bearing oil clearance must be checked.
11 Cut several pieces of the appropriate size Plastigage (they must be slightly shorter than the width of the main bearings) and place one piece on each crankshaft main bearing journal, parallel with the journal axis

(see illustration).
12 Clean the faces of the bearings in the caps and install the caps in their respective positions (don't mix them up) with the arrows pointing toward the front of the engine. Don't disturb the Plastigage.
13 Starting with the center main and working out toward the ends, tighten the main bearing cap bolts, in three steps, to the torque listed in this Chapter's Specifications. Don't rotate the crankshaft at any time during this operation.
14 Remove the bolts and carefully lift off the main bearing caps. Keep them in order. Don't disturb the Plastigage or rotate the crankshaft. If any of the main bearing caps are difficult to remove, tap them gently from side-to-side with a soft-face hammer to loosen them.
15 Compare the width of the crushed Plastigage on each journal to the scale printed on the Plastigage envelope to obtain the main bearing oil clearance **(see illustration).** Check the Specifications to make sure it's correct.
16 If the clearance is not as specified, the bearing inserts may be the wrong size - oversize or undersize - (which means different ones will be required). Before deciding that different inserts are needed, make sure that no dirt or oil was between the bearing inserts and the caps or block when the clearance was measured. If the Plastigage was wider at one end than the other, the journal may be tapered (refer to Section 18).
17 Carefully scrape all traces of the Plastigage material off the main bearing journals and/or the bearing faces. Use your fingernail or the edge of a credit card - don't nick or scratch the bearing faces.

Final crankshaft installation

18 Carefully lift the crankshaft out of the engine.
19 Clean the bearing faces in the block, then apply a thin, uniform layer of moly-base grease or engine assembly lube to each of the bearing surfaces. Be sure to coat the thrust faces as well as the journal face of the thrust bearing.
20 Make sure the crankshaft journals are clean, then lay the crankshaft back in place in the block.
21 Clean the faces of the bearings in the caps, then apply lubricant to them.
22 Install the caps in their respective positions with the arrows pointing toward the front of the engine. Apply sealant into the corners and on the mating surfaces where the rear main cap meets the cylinder block **(see illustration).**
23 Install the bolts.
24 Tighten all except the thrust bearing cap bolts to the specified torque (work from the center out and approach the final torque in three steps).
25 Tighten the thrust bearing cap bolts to 10-to-12 ft-lbs.
26 Tap the ends of the crankshaft forward and backward with a lead or brass hammer to line up the main bearing and crankshaft thrust surfaces.
27 Retighten all main bearing cap bolts to the specified torque, starting with the center main and working out toward the ends. Also, install the rear main bearing oil seal over the end of the crankshaft. Be sure it is installed to

23.22 Before installing the rear main cap, apply sealant to the cap mating surfaces and in the corners where the cap meets the block

the correct depth (see Chapter 2 Part A, B or C).
28 On manual transmission-equipped models, install a new pilot bearing in the end of the crankshaft (see Chapter 8).
29 Rotate the crankshaft a number of times by hand to check for any obvious binding.
30 The final step is to check the crankshaft endplay with a feeler gauge or a dial indicator as described in Section 13. The endplay should be correct if the crankshaft thrust faces aren't worn or damaged and new bearings have been installed.

24 Rear main oil seal - installation

See Chapter 2A, 2B or 2C.

25 Pistons/connecting rods - installation and rod bearing oil clearance check

Refer to illustrations 25.5, 25.9a, 25.9b, 25.9c, 25.11, 25.13 and 25.17

1 Before installing the piston/connecting rod assemblies, the cylinder walls must be perfectly clean, the top edge of each cylinder must be chamfered, and the crankshaft must be in place.
2 Remove the cap from the end of the number one connecting rod (refer to the marks made during removal). Remove the original bearing inserts and wipe the bearing surfaces of the connecting rod and cap with a clean, lint-free cloth. They must be kept spotlessly clean.

Connecting rod bearing oil clearance check

3 Clean the back side of the new upper bearing insert, then lay it in place in the connecting rod. Make sure the tab on the bearing fits into the recess in the rod. Don't hammer the bearing insert into place and be very careful not to nick or gouge the bearing face.

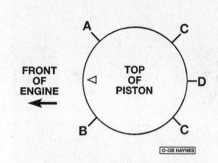

25.5 Ring end gap positions- Align the oil ring spacer gap at D, the oil ring side rails at C (one inch either side of the pin centerline), and the compression rings at A and B, one inch either side of the pin centerline

Don't lubricate the bearing at this time.
4 Clean the back side of the other bearing insert and install it in the rod cap. Again, make sure the tab on the bearing fits into the recess in the cap, and don't apply any lubricant. It's critically important that the mating surfaces of the bearing and connecting rod are perfectly clean and oil free when they're assembled.
5 Position the piston ring gaps at intervals around the piston **(see illustration)**.
6 Slip a section of plastic or rubber hose over each connecting rod cap bolt.
7 Lubricate the piston and rings with clean engine oil and attach a piston ring compressor to the piston. Leave the skirt protruding about 1/4-inch to guide the piston into the cylinder. The rings must be compressed until they're flush with the piston.
8 Rotate the crankshaft until the number one connecting rod journal is at BDC (bottom dead center) and apply a coat of engine oil to the cylinder walls.
9 With the arrow on top of the piston **(see illustration)** facing the front of the engine,

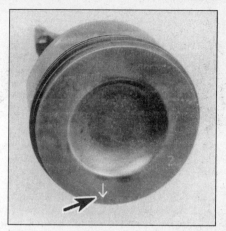

25.9a The arrow in the top of each piston must face the FRONT of the engine as the pistons are installed

gently insert the piston/connecting rod assembly into the number one cylinder bore and rest the bottom edge of the ring compressor on the engine block. Be sure the oil squirt hole in each connecting rod faces the proper direction when the rod is installed **(see illustrations)**. *Caution: If the oil squirt hole faces the wrong way when the arrow on the piston points toward the front, the rod and piston have been assembled incorrectly. Correct the problem before assembling further.*
10 Tap the top edge of the ring compressor to make sure it's contacting the block around its entire circumference.
11 Gently tap on the top of the piston with the end of a wooden hammer handle **(see illustration)** while guiding the end of the connecting rod into place on the crankshaft journal. The piston rings may try to pop out of the ring compressor just before entering the cylinder bore, so keep some down pressure on the ring compressor. Work slowly, and if any resistance is felt as the piston enters the cylinder, stop immediately. Find out what's hanging up and fix it before proceeding. Do not, for

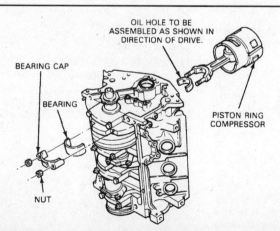

25.9b Note the direction of the oil squirt hole in the connecting rod - be sure the corresponding hole in the bearing insert aligns with the connecting rod hole

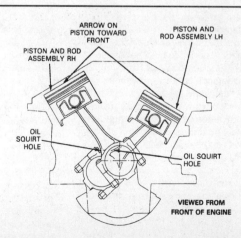

25.9c The oil squirt holes in the connecting rods face in the directions shown

2D-22 Chapter 2 Part D General engine overhaul procedures

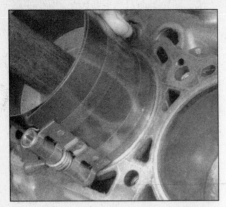

25.11 Drive the piston gently into the cylinder bore with the end of a wooden or plastic hammer handle

25.13 Lay the Plastigage strips (arrow) on the main bearing journals, parallel to the crankshaft centerline

25.17 Measuring the width of the crushed Plastigage to determine the rod bearing oil clearance (be sure to use the correct scale - standard and metric ones are included)

any reason, force the piston into the cylinder - you might break a ring and/or the piston.

12 Once the piston/connecting rod assembly is installed, the connecting rod bearing oil clearance must be checked before the rod cap is permanently bolted in place.

13 Cut a piece of the appropriate size Plastigage slightly shorter than the width of the connecting rod bearing and lay it in place on the number one connecting rod journal, parallel with the journal axis **(see illustration)**.

14 Clean the connecting rod cap bearing face, remove the protective hoses from the connecting rod bolts and install the rod cap. Make sure the mating mark on the cap is on the same side as the mark on the connecting rod.

15 Install the nuts and tighten them to the torque listed in this Chapter's Specifications, working up to it in three steps. **Note:** *Use a thin-wall socket to avoid erroneous torque readings that can result if the socket is wedged between the rod cap and nut. If the socket tends to wedge itself between the nut and the cap, lift up on it slightly until it no longer contacts the cap. Do not rotate the crankshaft at any time during this operation.*

16 Remove the nuts and detach the rod cap, being very careful not to disturb the Plastigage.

17 Compare the width of the crushed Plastigage to the scale printed on the Plastigage envelope to obtain the oil clearance **(see illustration)**. Compare it to the Specifications to make sure the clearance is correct.

18 If the clearance is not as specified, the bearing inserts may be the wrong size (which means different ones will be required). Before deciding that different inserts are needed, make sure that no dirt or oil was between the bearing inserts and the connecting rod or cap when the clearance was measured. Also, recheck the journal diameter. If the Plastigage was wider at one end than the other, the journal may be tapered (refer to Section 18).

Final connecting rod installation

19 Carefully scrape all traces of the Plastigage material off the rod journal and/or bearing face. Be very careful not to scratch the bearing - use your fingernail or the edge of a credit card.

20 Make sure the bearing faces are perfectly clean, then apply a uniform layer of clean moly-base grease or engine assembly lube to both of them. You'll have to push the piston into the cylinder to expose the face of the bearing insert in the connecting rod - be sure to slip the protective hoses over the rod bolts first.

21 Slide the connecting rod back into place on the journal, remove the protective hoses from the rod cap bolts, install the rod cap and tighten the nuts to the specified torque. Again, work up to the torque in three steps.

22 Repeat the entire procedure for the remaining pistons/connecting rods.

23 The important points to remember are:

a) *Keep the back sides of the bearing inserts and the insides of the connecting rods and caps perfectly clean when assembling them.*
b) *Make sure you have the correct piston/rod assembly for each cylinder.*
c) *The notch or mark on the piston must face the front of the engine.*
d) *Lubricate the cylinder walls with clean oil.*
e) *Lubricate the bearing faces when installing the rod caps after the oil clearance has been checked.*

24 After all the piston/connecting rod assemblies have been properly installed, rotate the crankshaft a number of times by hand to check for any obvious binding.

25 As a final step, the connecting rod endplay must be checked. Refer to Section 12 for this procedure.

26 Compare the measured endplay to the Specifications to make sure it's correct. If it was correct before disassembly and the original crankshaft and rods were reinstalled, it should still be right. If new rods or a new crankshaft were installed, the endplay may be inadequate. If so, the rods will have to be removed and taken to an automotive machine shop for resizing.

26 Initial start-up and break-in after overhaul

Warning: *Have a fire extinguisher handy when starting the engine for the first time.*

1 Once the engine has been installed in the vehicle, double-check the engine oil and coolant levels.

2 With the spark plugs out of the engine and the ignition system disabled (see Section 3), crank the engine until oil pressure registers on the gauge or the light goes out.

3 Install the spark plugs, hook up the plug wires and restore the ignition system functions (see Section 3).

4 Start the engine. It may take a few moments for the fuel system to build up pressure, but the engine should start without a great deal of effort. **Note:** *If backfiring occurs through the carburetor or throttle body, recheck the valve timing and ignition timing.*

5 After the engine starts, it should be allowed to warm up to normal operating temperature. While the engine is warming up, make a thorough check for fuel, oil and coolant leaks.

6 Shut the engine off and recheck the engine oil and coolant levels.

7 Drive the vehicle to an area with minimum traffic, accelerate from 30 to 50 mph, then allow the vehicle to slow to 30 mph with the throttle closed. Repeat the procedure 10 or 12 times. This will load the piston rings and cause them to seat properly against the cylinder walls. Check again for oil and coolant leaks.

8 Drive the vehicle gently for the first 500 miles (no sustained high speeds) and keep a constant check on the oil level. It is not unusual for an engine to use oil during the break-in period.

9 At approximately 500 to 600 miles, change the oil and filter.

10 For the next few hundred miles, drive the vehicle normally. Do not pamper it or abuse it.

11 After 2000 miles, change the oil and filter again and consider the engine broken in.

Chapter 3
Cooling, heating and air conditioning systems

Contents

	Section		Section
Air conditioning accumulator - removal and installation	17	Cooling system servicing	See Chapter 1
Air conditioning compressor - removal and installation	15	General information	1
Air conditioning condenser - removal and installation	16	Heater and air conditioning control assembly - removal and installation	10
Air conditioning evaporator - replacement	18		
Air conditioning system - check and maintenance	14	Heater blower motor and circuit - check and replacement	12
Antifreeze - general information	2	Heater control cables - check and adjustment	11
Coolant level check	See Chapter 1	Heater core - replacement	13
Coolant reservoir - removal and installation	6	Heater - general information	9
Coolant temperature sending unit - check and replacement	7	Radiator - removal and installation	4
Cooling fan and viscous clutch - inspection, removal and installation	5	Thermostat - check and replacement	3
		Water pump - check and replacement	8
Cooling system check	See Chapter 1		

Specifications

General

Thermostat
- Type ..... Wax pellet
- Opening temperature ..... 192-degrees to 199-degrees F
- Fully open ..... 226-degrees F

Coolant temperature switch closing temperature (dash light on) ..... 249-degrees F
Air conditioning system refrigerant capacity ..... 28 ounces

Torque specifications

Ft-lbs (unless otherwise indicated)

Cooling fan-to-clutch bolts ..... 50 to 70 in-lbs
Fan clutch-to-water pump nut or bolts
- Four-cylinder engine ..... 13 to 17
- V6 engines ..... 30 to 100
Thermostat housing bolts ..... 120 in-lbs
Water pump bolts
- Four-cylinder engine ..... 15 to 22
- 3.0L V6 engine ..... 84 in-lbs
- 4.0L V6 engine ..... 72 to 108 in-lbs
Accumulator pressure switch
- Metal base ..... 60 to 120 in-lbs
- Plastic base ..... Hand tighten only

1 General information

Warning: On models so equipped, whenever working in the vicinity of the front grille/bumper, steering wheel, steering column or other components of the airbag system, the system should be disarmed. To do this, perform the following steps:

a) Turn the ignition switch to Off.
b) Detach the cable from the negative battery terminal, then detach the positive cable. Wait two minutes for the electronic module backup power supply to be depleted.

To enable the system

a) Turn the ignition switch to the Off position.
b) Connect the positive battery cable first, then connect the negative cable.

Engine cooling system

All vehicles covered by this manual employ a pressurized engine cooling system with thermostatically controlled coolant circulation. An impeller type water pump mounted on the front of the block pumps coolant through the engine. The coolant flows around each cylinder and toward the rear of the engine. Cast-in coolant passages direct coolant around the intake and exhaust ports, near the spark plug areas and in close proximity to the exhaust valve guides.

A wax pellet type thermostat is located in a housing near the front of the engine. During warm up, the closed thermostat prevents coolant from circulating through the radiator. As the engine nears normal operating temperature, the thermostat opens and allows hot coolant to travel through the radiator, where it's cooled before returning to the engine.

The cooling system is sealed by a pressure type radiator cap, which raises the boiling point of the coolant and increases the cooling efficiency of the radiator. If the system pressure exceeds the cap pressure relief value, the excess pressure in the system forces the spring-loaded valve inside the cap off its seat and allows the coolant to escape through the overflow tube into a coolant reservoir. When the system cools the excess coolant is automatically drawn from the reservoir back into the radiator.

The coolant reservoir serves as both the point at which fresh coolant is added to the cooling system to maintain the proper fluid level and as a holding tank for overheated coolant.

This type of cooling system is known as a closed design because coolant that escapes past the pressure cap is saved and reused.

Heating system

The heating system consists of a blower fan and heater core located in the heater box, the hoses connecting the heater core to the engine cooling system and the heater/air conditioning control head on the dashboard. Hot engine coolant is circulated through the heater core. When the heater mode is activated, a flap door opens to expose the heater box to the passenger compartment. A fan switch on the control head activates the blower motor, which forces air through the core, heating the air.

Air conditioning system

The air conditioning system consists of a condenser mounted in front of the radiator, an evaporator mounted adjacent to the heater core, a compressor mounted on the engine, a filter-drier (accumulator) which contains a high pressure relief valve and the plumbing connecting all of the above components.

A blower fan forces the warmer air of the passenger compartment through the evaporator core (sort of a radiator-in-reverse), transferring the heat from the air to the refrigerant. The liquid refrigerant boils off into low pressure vapor, taking the heat with it when it leaves the evaporator.

2 Antifreeze/coolant - general information

Warning: *Do not allow antifreeze to come in contact with your skin or painted surfaces of the vehicle. Rinse off spills immediately with plenty of water. Antifreeze is highly toxic if ingested. Never leave antifreeze lying around in an open container or in puddles on the floor; children and pets are attracted by it's sweet smell and may drink it. Check with local authorities about disposing of used antifreeze. Many communities have collection centers which will see that antifreeze is disposed of safely. Never dump used anti-freeze on the ground or pour it into drains.*

Note: *Non-Toxic antifreeze is now manufactured and available at local auto parts stores,*

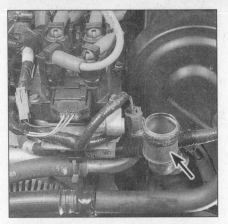

3.8a Location of the thermostat housing on four-cylinder engines (arrow) (upper radiator hose removed for clarity)

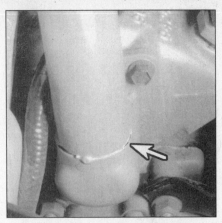

3.8b Location of the thermostat housing on 4.0L V6 engines

but even these types should be disposed of properly.

The cooling system should be filled with a water/ethylene glycol based antifreeze solution which will prevent freezing down to at least -20-degrees F (even lower in cold climates). It also provides protection against corrosion and increases the coolant boiling point.

The cooling system should be drained, flushed and refilled at least every other year (see Chapter 1). The use of antifreeze solutions for periods of longer than two years is likely to cause damage and encourage the formation of rust and scale in the system.

Before adding antifreeze to the system, check all hose connections. Antifreeze can leak through very minute openings.

The exact mixture of antifreeze to water which you should use depends on the relative weather conditions. The mixture should contain at least 50-percent antifreeze, but should never contain more than 70-percent antifreeze. Consult the mixture ratio chart on the antifreeze container before adding coolant.

3 Thermostat - check and replacement

Warning: *Do not remove the radiator cap, drain the coolant or replace the thermostat until the engine has cooled completely.*

Check

1 Before assuming the thermostat is to blame for a cooling system problem, check the coolant level, drivebelt tension (see Chapter 1) and temperature gauge (or light) operation.
2 If the engine seems to be taking a long time to warm up (based on heater output or temperature gauge operation), the thermostat is probably stuck open. Replace the thermostat with a new one.
3 If the engine runs hot, use your hand to check the temperature of the upper radiator hose. If the hose isn't hot, but the engine is, the thermostat is probably stuck closed, preventing the coolant inside the engine from escaping to the radiator. Replace the thermostat. **Caution:** *Don't drive the vehicle without a thermostat. The computer may stay in open loop and emissions and fuel economy will suffer.*
4 If the upper radiator hose is hot, it means that the coolant is flowing and the thermostat is open. Consult the Troubleshooting section at the front of this manual for cooling system diagnosis.

Replacement

Refer to illustrations 3.8a, 3.8b, 3.11a, 3.11b and 3.11c

5 Disconnect the negative battery cable from the battery.
6 Remove the air cleaner air duct from the throttle body and air cleaner,
7 Drain the cooling system (see Chapter 1). If the coolant is relatively new or in good condition (see Chapter 1), save it and reuse it.
8 Follow the upper radiator hose to the engine to locate the thermostat housing **(see illustrations)**. The housing is located as follows:

   a) *Four-cylinder engines: on the front of the engine.*
   b) *V6 engine: near the front of the intake manifold.*

9 Loosen the hose clamp, then detach the hose from the fitting. If it's stuck, grasp it near the end with a pair of adjustable pliers and twist it to break the seal, then pull it off. If the hose is old or deteriorated, cut it off and install a new one. **Caution:** *Do not cut through to the hose fitting. A deep score may cause a leak.*
10 If the outer surface of the large fitting that mates with the hose is deteriorated (corroded, pitted, etc.), it may be damaged further by hose removal. If it is, the thermostat housing cover will have to be replaced.
11 Remove the bolts and detach the housing cover. If the cover is stuck, tap it with a soft-face hammer to jar it loose. Be prepared for some coolant to spill out as the gasket

Chapter 3 Cooling, heating and air conditioning systems

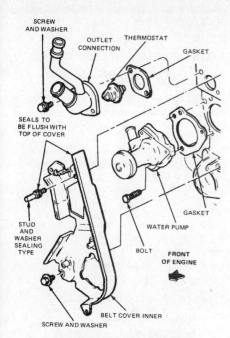

3.11a Thermostat, housing and water pump - exploded view (four-cylinder engines - typical)

with the bridge section toward the outlet housing cover. On 4.0L engines, make sure the air release valve is in the up position **(see illustration 3.11c)**.
16 Install the housing cover and bolts. Tighten the bolts to the torque listed in this Chapter's Specifications.
17 Reattach the hose to the fitting and tighten the hose clamp securely.
18 Refill the cooling system (see Chapter 1).
19 Install the air cleaner air duct
20 Start the engine and allow it to reach normal operating temperature. Keep the system topped up as necessary. As air is purged from the cooling system, more coolant may be needed. Check for leaks and proper thermostat operation (as described in Steps 2 through 4).

4 Radiator - removal and installation

Refer to illustrations 4.3a, 4.3b, 4.5, 4.6 and 4.8
Warning: *If vehicle is equipped with airbags, refer to Chapter 12 to disarm the airbag system prior to performing any work described below.*
Warning: *Wait until the engine is completely cool before beginning this procedure.*
1 Disconnect the negative battery cable from the battery.
2 Drain the cooling system (see Chapter 1). If the coolant is clean, you may save it and reuse it.
3 Loosen the hose clamps and slide them back on the hoses, then detach the radiator hoses from the fittings **(see illustrations)**. If they're stuck, grasp each hose near the end with a pair of adjustable pliers and twist it to break the seal, then pull it off - be careful not to distort the radiator fittings! If the hoses are

seal is broken **(see illustrations)**.
12 Note how it's installed (which end is facing toward the engine), then remove the thermostat. It may be necessary to rotate the thermostat to free it.
13 Stuff a rag into the engine opening, then remove all traces of corrosion from the housing and cover with a gasket scraper. Remove the rag from the opening and clean the mating surfaces with lacquer thinner or acetone.
14 On 4.0L engines, make sure the sealing ring is positioned correctly on the thermostat **(see illustration 3.11c)**.
15 Install the new thermostat in the housing

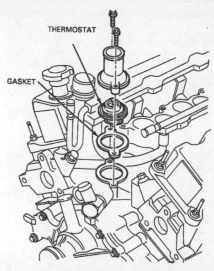

3.11b Thermostat, housing and gasket - exploded view (3.0L V6 engines)

old or deteriorated, cut them off and install new ones. **Caution:** *Do not cut through to the hose fitting. A deep score may cause a leak.*
4 Disconnect the reservoir hose from the radiator filler neck.

4.3a Use a socket or screwdriver to loosen the hose clamp (arrow)

4.3b On spring-type hose clamps, squeeze the clamp with pliers and slide it back along the hose

3.11c Thermostat, housing and gasket - exploded view (4.0L V6 engines)

4.5 Remove the shroud mounting bolts and place the shroud over the fan

4.6 On automatic transmission models, hold the cooler line fittings with a backup wrench and unscrew the fittings

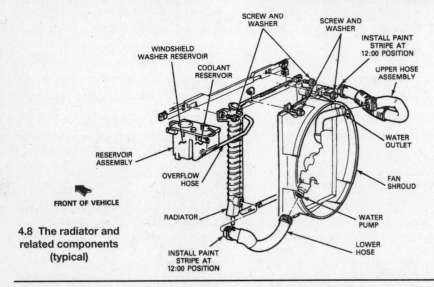

4.8 The radiator and related components (typical)

5 Remove the screws that attach the fan shroud to the radiator and slide the shroud toward the engine **(see illustration)**. Let the shroud rest on the fan.

6 If the vehicle is equipped with an automatic transmission, hold the cooler fittings with a backup wrench and disconnect the lines from the radiator **(see illustration)**. Use a drip pan to catch spilled fluid.

7 Plug the lines and fittings to avoid spillage and contamination. Automatic transmission internal parts require very clean conditions.

8 Remove the radiator mounting bolts **(see illustration)**.

9 Carefully lift the radiator up and out of its lower mounting pads or grommets. Remove the radiator and be careful not to spill coolant on the vehicle or scratch the paint.

10 With the radiator removed, it can be inspected for leaks and damage. If it needs repair, have a radiator shop or dealer service department perform the work as special techniques are required.

11 Bugs and dirt can be removed from the radiator with compressed air and a soft brush. Don't bend the cooling fins while doing this. If you are careful, you can straighten out the fins to permit better flow, but in most cases this is not necessary.

12 Check the radiator mounts for deterioration and make sure there's nothing in them when the radiator is installed.

13 Installation is the reverse of the removal procedure with the following additions:

a) After installation, fill the cooling system with the proper mixture of antifreeze and water (see Chapter 1).
b) Start the engine and check for leaks. Allow the engine to reach normal operating temperature, indicated by the upper radiator hose becoming hot. Recheck the coolant level and add more if required.
c) If you're working on an automatic transmission equipped vehicle, check and add transmission fluid as needed (see Chapter 1).

5 Cooling fan and viscous clutch - inspection, removal and installation

Warning: *To avoid possible injury or damage, DO NOT operate the engine with a damaged fan. Plastic fans crack and can fly apart when the engine is revved. Do not attempt to repair fan blades - replace a damaged fan with a new one.*

Viscous clutch inspection

1 Disconnect the negative cable from the battery.
2 Rock the fan fore and aft by hand to check for excessive bearing play.
3 With the engine cold, turn the fan by hand. The fan should turn freely.
4 Visually inspect for substantial fluid leakage from the clutch assembly. If problems are noted, replace the clutch assembly.
5 Reconnect the battery cable and warm up the engine. With the engine completely warmed up, turn off the ignition switch and disconnect the negative battery cable from the battery. Turn the fan by hand. Some drag should be evident. If the fan turns easily, the viscous clutch is not working. Replace the fan clutch.

Removal and installation

Four-cylinder engine

Refer to illustration 5.9.

6 Disconnect the negative cable from the battery.
7 Disconnect the reservoir hose from the radiator filler neck.
8 Remove the screws securing the shroud to the radiator and slide the shroud toward the engine (see Section 4).
9 Remove the four screws and washers attaching the fan/clutch assembly to the water pump drive belt hub **(see illustration)**.
10 Lift the fan/clutch assembly out of the engine compartment.
11 Inspect the fan blades for damage and defects. Replace it if necessary by removing the screws securing the fan to the clutch assembly. **Caution:** *If the fan clutch is stored, position it with the radiator side facing down. Otherwise, fluid may leak out.*
12 Installation is the reverse of removal. Be sure to tighten the fan and clutch mounting bolts evenly and securely.

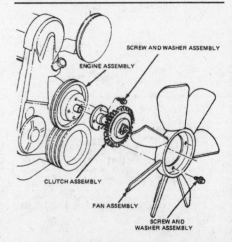

5.9 Fan and clutch assembly (four-cylinder engine) exploded view

Chapter 3 Cooling, heating and air conditioning systems

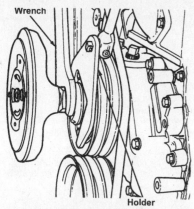

5.16a To loosen the large nut that holds the fan clutch to the water pump on V6 models, use the tools shown . . .

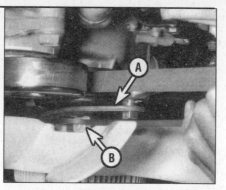

5.16b . . . hold the pulley from turning (A) and turn the nut counterclockwise to loosen it (B)

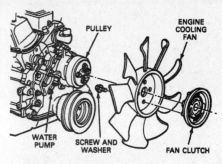

5.18 Exploded view of the fan and clutch assembly (typical V6 model)

V6 engines

Refer to illustrations 5.16a, 5.16b and 5.18.

13 Disconnect the negative cable from the battery.
14 Disconnect the reservoir hose from the radiator filler neck.
15 Remove the screws securing the shroud to the radiator and slide the shroud toward the engine (see Section 4).
16 Remove the large nut securing the clutch hub assembly to the water pump drivebelt hub. **Caution:** *This nut has left-hand threads and must be turned clockwise for removal.* Two special tools are necessary, a fan clutch pulley holder and a fan clutch nut wrench, available at most auto parts stores that carry tools **(see illustrations)**. If the special tools are not available, use a strap wrench to hold the assembly from rotating and use a large adjustable wrench to unscrew the large nut.
17 Lift the fan/clutch assembly out of the engine compartment.
18 Inspect the fan blades for damage and defects. Replace it if necessary by removing the bolts securing the fan to the clutch assembly **(see illustration)**. If the fan clutch is stored, position it with the radiator side facing down.
19 Installation is the reverse of the removal steps. Be sure to tighten the fan and clutch nut evenly and to the torque listed in this Chapter's Specifications.

6 Coolant reservoir - removal and installation

Refer to illustration 6.4

1 Disconnect the radiator overflow hose from the base of the coolant reservoir (see Chapter 1).
2 Connect a hose to the fitting on the coolant reservoir and drain the coolant from the reservoir.
3 Disconnect the electrical connector for the windshield washer pump.
4 Remove the reservoir attachment to the inner fender panel **(see illustration)**.
5 Lift the reservoir out of the engine compartment.
6 Installation is the reverse of removal. Refill the reservoir with coolant (see Chapter 1).

7 Coolant temperature sending unit - check and replacement

General Information

Refer to illustration 7.3.

1 There are two engine coolant temperature senders. One drives the temperature gauge or warning light on the instrument panel; this is generally referred to as the *sending unit*. The other is used by the computer to determine engine coolant temperature and adjust fuel/air mixture accordingly; this is generally referred to as a *sensor*. The two are easily distinguished, since the sending unit for the temperature gauge has only one wire going to it, while the unit that provides information to the computer has two or three wires to it.
2 On four-cylinder engines, the sending unit for the gauge or light is threaded into the driver's side of the engine block, at the rear, just below the cylinder head. **Caution:** *Do not mistake the temperature sending unit for the oil pressure sending unit, which is just above the coolant temperature sending unit, threaded into the cylinder head.*
3 On V6 engines, the sending unit for the gauge is located on the front lower part of the intake manifold, near the thermostat housing **(see illustration)**.

Check

4 To check the temperature sending unit, disconnect the electrical connector. Measure the resistance across the temperature sending unit electrical connector terminal and engine ground. The resistance should be between approximately 75 ohms (full cold) and 10 ohms (full hot).

Replacement

5 Allow the cooling system to completely cool down, then open the radiator filler cap to relieve any pressure in the cooling system. Install the cap. This will minimize coolant loss during this procedure.
6 Disconnect the electrical connector from the sending unit.
7 Wrap the threads of the new sending unit with Teflon tape.
8 Remove the old sending unit from the engine. Place your finger over the hole in the engine to minimize coolant loss.
9 Immediately install the new sending unit and tighten it securely.
10 Connect the sending unit electrical connector.
11 Check coolant level and top up if neces-

6.4 The coolant reservoir is secured to the inner fender panel by screws

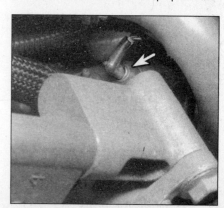

7.3 A typical temperature sending unit (arrow)

Chapter 3 Cooling, heating and air conditioning systems

8.4 A typical water pump weep hole (arrow) - on some models the hole is on the bottom of the pump

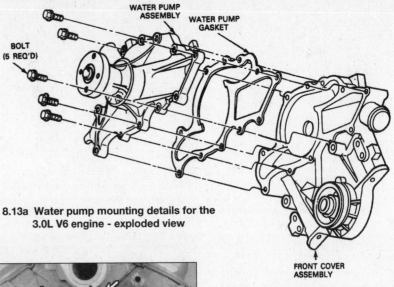

8.13a Water pump mounting details for the 3.0L V6 engine - exploded view

8.13b Locations of the water pump bolts on the 4.0L V6 engine (arrows)

sary (see Chapter 1).
12 Start the engine and check for leaks.

8 Water pump - check and replacement

Check

Refer to illustration 8.4

1 The water pump is driven by the accessory drivebelt on all models.
2 A failure in the water pump can cause serious engine damage due to overheating.
3 There are two ways to check the operation of the water pump while it's installed on the engine. If the pump is defective, it should be replaced with a new or rebuilt unit.
4 Water pumps are equipped with weep or vent holes. If a failure occurs in the pump seal, coolant will leak from the hole **(see illustration)**. In most cases you'll need a flashlight to find the hole on the water pump to check for leaks.
5 If the water pump shaft bearings fail there may be a howling sound at the front of the engine while it's running. Shaft wear can be felt if the water pump pulley is rocked up and down. Don't mistake drivebelt slippage, which causes a squealing sound, for water pump bearing failure.

Replacement

Refer to illustrations 8.13a and 8.13b
Warning 1: *Wait until the engine is completely cool before beginning this procedure.*
Warning 2: *The air conditioning system is under high pressure. Do not loosen any hose fittings or remove any components until after the system has been discharged by a dealer service department or service station. Always wear eye protection when disconnecting air conditioning system fittings.*

6 Disconnect the negative battery cable from the battery.
7 Drain the cooling system (see Chapter 1). If the coolant is clean, you may save it and reuse it.
8 Remove the cooling fan and shroud (see Sections 4 and 5).
9 Remove the drivebelt (see Chapter 1) and the pulley at the end of the water pump shaft.
10 Loosen the clamps and detach the hoses from the water pump. If they're stuck, grasp each hose near the end with a pair of adjustable pliers and twist it to break the seal, then pull it off. If the hoses are deteriorated, cut them off and install new ones.
11 On four-cylinder engines, remove the outer timing belt cover (see Chapter 2A).
12 On V6 engines, if necessary, remove the alternator and its mounting bracket from the water pump (see Chapter 5). Without disconnecting the air-conditioning lines, remove the air-conditioning compressor mounting bracket with the compressor and power steering pump still attached **(see illustration 15.6)**. Set the assembly aside without disconnecting any lines.
13 Remove the bolts and detach the water pump from the engine **(see the accompanying illustrations and illustration 3.11a)**. Note the locations of the various lengths and different types of bolts as they're removed to ensure correct installation.
14 Clean the bolt threads and the threaded holes in the engine to remove corrosion and sealant.
15 Compare the new pump to the old one to make sure they're identical.
16 Remove all traces of old gasket material from the engine gasket surface with a gasket scraper.
17 Clean the engine and new water pump gasket mating surfaces with lacquer thinner or acetone.
18 Apply a thin coat of RTV sealant to the engine side of the new gasket.
19 Apply a thin layer of RTV sealant to the gasket mating surface of the new pump, then carefully mate the gasket and the pump. Slip a couple of bolts through the pump mounting holes to hold the gasket in place.
20 Carefully attach the pump and gasket to the engine and thread the bolts into the holes finger tight.
21 Install the remaining bolts (if they also hold the alternator bracket in place, be sure to reposition the bracket at this time). Tighten them evenly to the torque listed in this Chapter's Specifications in 1/4-turn increments. Don't overtighten them or the pump may be distorted. **Note:** *On 3.0L engines, some of the water pump bolts also secure the timing chain cover to the block. Refer to Chapter 2B for additional information.*
22 Reinstall all parts removed for access to the pump.
23 Refill the cooling system and check the drivebelt tension (see Chapter 1). Run the engine and check for leaks.

9 Heater - general information

The heater circulates engine coolant through a small radiator (heater core) in the passenger compartment. Air is drawn in through an opening in the cowl, then blown (by the blower motor) through the heater core to pick up heat. The heated air is blended with varying amounts of unheated air to regulate the temperature. The heated air is then blown into the passenger compartment. Various doors in the heater control the flow of air to the floor and through the instrument panel louvers and defroster outlets.

Chapter 3 Cooling, heating and air conditioning systems

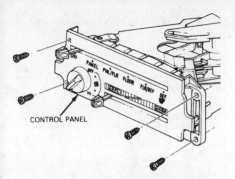

10.3 Heater control assembly mounting screws (1993 and 1994 models shown, later models similar)

10 Heater and air conditioning control assembly - removal and installation

Refer to illustration 10.3

Removal

1 Disconnect the negative cable from the battery. On 1994 and earlier models, remove the ashtray and the retainer bracket. On 1995 and later models, remove the radio (see Chapter 12).
2 Remove the instrument cluster trim panel. Carefully pull the panel straight back approximately one inch, then up.
3 Remove the four screws securing the control panel to the instrument panel **(see illustration)**.
4 Pull the control assembly through the opening in the instrument panel far enough to allow removal of the electrical connections. Carefully spread the clips on the electrical connectors and disconnect the connectors. On 1995 and later models, the assembly can now be removed from the vehicle.
5 Use a small screwdriver and remove the vacuum harness from the vacuum switch on the side of the control assembly.
6 At the back of the control assembly, use a screwdriver or needle-nose pliers and disconnect the temperature and function cable from the white control bracket, if equipped.
7 At the bottom of the control assembly, disconnect the temperature control cable from the control, if equipped.
8 Pull enough cable through the instrument opening until the function cable can be held vertical to the control assembly, then remove the cable end from the function lever.
9 Remove the control assembly from the instrument panel.

Installation

10 Pull the control cables through the instrument panel opening approximately eight inches.
11 Carefully bend and attach the function cable.
12 Attach the black temperature control cable. Be sure that the end of the cable is seated securely with the T-top pin on the control assembly.
13 Install the electrical wire harness connectors.
14 Connect the dual terminal on the vacuum hose to the vacuum switch on the control assembly.
15 Position the control assembly into the instrument panel and install the four mounting screws.
16 Complete the installation by installing the various trim panels and instrument panel items previously removed. Check for proper operation and adjust the cables if necessary (see Section 11).
17 Installation is the reverse of removal.

11 Heater control cables - check and adjustment

Check

1 Move the control lever or turn the control knob all the way from left to right.
2 If the control lever or control knob stops before the end and bounces back, the cables are out of adjustment.

Adjustment

Refer to illustrations 11.4a and 11.4b

3 Squeeze the tabs on either side of the glove box door to disengage it, then let the door hang down to provide access to the control cables.
4 Working through the glove box opening, remove the cable jacket from its metal attaching clip on top of the heater assembly **(see illustrations)**. The cable ends should remain attached to the door cams and/or crank arms at this time.
5 To adjust the temperature control cable, set the temperature lever or knob to the Cool position and hold it there.
6 Push gently on the black cable jacket to seat the blend door (push until you feel resistance).
7 Reinstall the cable to the clip by pushing the jacket into the clip from the top until it snaps in.
8 To adjust the function control cable, set the function control lever or knob to Defrost and hold it there.
9 Pull on the white cam jacket until the cam travel stops.
10 Reinstall the cable to the clip by pushing

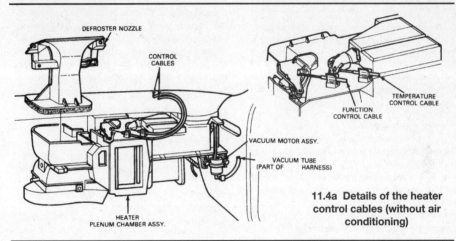

11.4a Details of the heater control cables (without air conditioning)

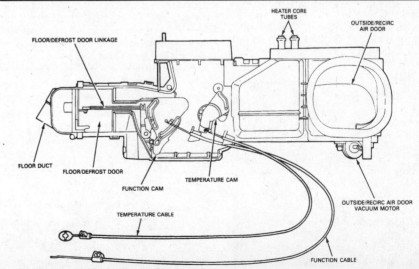

11.4b Details of the heater control cables (1993 and 1994 models with air conditioning)

12.3 The electrical connector is on the right side of the blower motor (arrow)

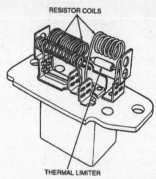

12.8 The blower motor resistor is located on the heater plenum chamber, next to the blower motor - the resistor coils are in the blower airstream inside the chamber

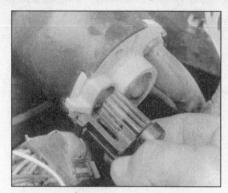

12.15 Push down on the electrical connector locking tab and disconnect the connector

the jacket into the clip from the top until it snaps in.
11 Run the system on High and actuate the levers, checking for proper operation. Readjust if necessary.
12 Install the glove box door.

12 Heater blower motor and circuit - check and replacement

Warning: *If vehicle is equipped with airbags, refer to Chapter 12 to disarm the airbag system prior to performing any work described below.*

Circuit checks

Basic circuit check

Refer to illustration 12.3

1 Check the fuse and all connections in the circuit for looseness and corrosion. Make sure the battery is fully charged.
2 Backprobe the blower motor connector with two large straight-pins (leave the connector plugged in). The blower motor is located in the engine compartment on the passenger side of the firewall **(see illustration)**.
3 Connect a voltmeter across the two pins on the connector.
4 With the transmission in Park, the parking brake securely applied, turn the ignition switch to the On position (engine not running).

5 Have an assistant turn the blower switch through each position and note the voltage readings. Changes in voltage indicates that the blower switch should be able to vary the motor speed at different positions.
6 If there is voltage at the connector, but the motor does not operate, you can test the motor separately. With the connector disconnected, connect one terminal on the motor to a good ground. Connect the other terminal to the battery positive terminal through a fused jumper. The motor should operate at its highest speed. If it does, check the condition of the connectors and connecting wires. If the motor still does not operate, the motor is probably faulty.

Blower motor resistor assembly check

Refer to illustration 12.8.

7 The blower resistor assembly is located on the heater case to the right of the blower motor. There are several resistance elements mounted on the resistor board to provide four blower speeds.
8 Remove the blower resistor from its mounting location and visually check for damage. Check the resistor block for continuity between all terminals. If the thermal limiter circuit has been opened as a result of excessive heat, it should be replaced with the identical part **(see illustration)**.

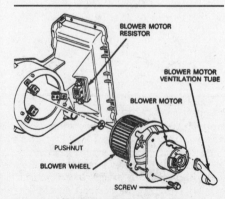

12.17 An exploded view of the blower motor assembly

Blower switch check

9 Turn the ignition switch to the Off position.
10 Disconnect the blower resistor electrical connector.
11 Using an ohmmeter, check for continuity to ground at each wire terminal with the blower switch in its corresponding position (see the wiring diagrams at the end of Chapter 12). **Caution:** *Make sure the ignition switch is OFF before probing the Hi speed circuit or damage to your meter may result.*
12 If there's an open (no continuity) in any one of the circuits, either the switch is defective or the wiring harness is open. Remove the control panel (see Section 10) and check the switch for continuity. If there's no continuity in all of the switch positions, check for continuity in the ground circuit.
13 If the switch is defective, replace the control panel assembly.

Blower motor replacement

Refer to illustrations 12.15, 12.16 and 12.17.
Warning: *If vehicle is equipped with airbags, refer to Chapter 12 to disarm the airbag system prior to performing any work described below.*

14 Disconnect the negative battery cable.
15 Disconnect the blower motor connector **(see illustration)**.
16 Remove the blower motor mounting screws and remove the motor **(see illustration)**.
17 Remove the blower motor wheel hub pushnut and the clamp from the shaft. Pull the blower motor wheel from the shaft **(see illustration)**.
18 Installation is the reverse of these steps.

13 Heater core - replacement

Removal

All models

Refer to illustration 13.5
Warning: *If vehicle is equipped with airbags, refer to Chapter 12 to disarm the airbag system prior to performing any work described below.*

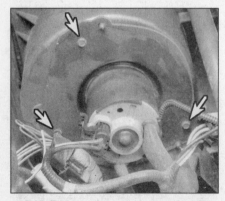

12.16 Remove the screws securing the heater blower motor (arrows)

Chapter 3 Cooling, heating and air conditioning systems

13.5 Disconnect the heater hoses from the fittings at the firewall (arrow)

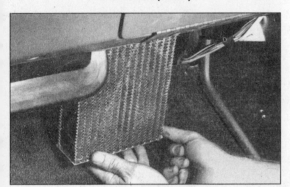

13.9 Pull the heater core to the rear and down to remove it from the plenum

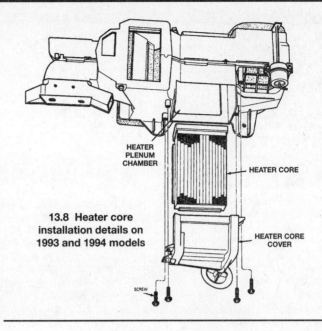

13.8 Heater core installation details on 1993 and 1994 models

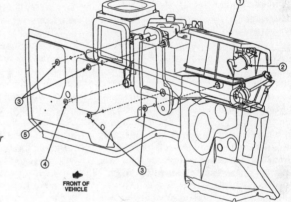

13.11 An exploded view of the 1995 and later heater assembly

1 Heater plenum chamber
2 Vacuum control motor
3 Nuts
4 Stud
5 Firewall

1 Allow the cooling system to completely cool down.
2 Using a thick cloth for protection, turn the radiator filler cap to the first stop.
3 Step back and let the pressure release.
4 Once the pressure has been released, completely drain the cooling system (see Chapter 1).
5 Working within the engine compartment, loosen the clamps on the heater hoses at the engine compartment side of the firewall **(see illustration)**. Twist the hoses and carefully separate them from the heater core tubes.
6 Plug or cap the heater core tubes to prevent coolant from spilling into the passenger compartment when the heater core is removed.
7 Place a plastic sheet on the floor of the vehicle to prevent stains if the coolant spills.

1993 and 1994 models

Refer to illustrations 13.8 and 13.9

8 Working in the passenger compartment, remove the four screws securing the heater core access cover to the plenum assembly **(see illustration)**.
9 Carefully pull the heater core to the rear and down and remove it from the plenum assembly **(see illustration)**.

1995 and later models

Refer to illustration 13.11

10 Remove the instrument panel (see Chapter 12).
11 Remove the heater plenum chamber **(see illustration)**.

12 Separate the heater core from the heater plenum chamber.

Installation

13 Installation for all models is the reverse of the removal procedure, with the following additions:
  a) Fill the cooling system (see Chapter 1).
  b) Run the engine, check for leaks and test the heater.

14 Air conditioning system - check and maintenance

Refer to illustration 14.5.
Warning: *The air conditioning system is under high pressure. Do not loosen any hose fittings or remove any components until after the system has been discharged by a dealer service department or service station. Always wear eye protection when disconnecting air conditioning system fittings.*
Note: *1993 through 1994 air conditioning systems use R-12 refrigerant. In 1995 and later models, the air conditioning system was changed to use the new, "environmentally friendly" R-134a refrigerant. Each system uses similar components and locations, but components are NOT interchangeable. All discharging of refrigerant, for parts replacement or maintenance, should be done by an approved air conditioning facility with the proper refrigerant recovery equipment.*

1 The following maintenance checks should be performed on a regular basis to ensure that the air conditioner continues to operate at peak efficiency.
  a) Check the compressor drivebelt. If it's worn or deteriorated, replace it (see Chapter 1).
  b) Check the system hoses. Look for cracks, bubbles, hard spots and deterioration. Inspect the hoses and all fittings for oil bubbles and seepage. If there's any evidence of wear, damage or leaks, replace the hose(s).
  c) Inspect the condenser fins for leaves, bugs and other debris. Use a "fin comb" or compressed air to clean the condenser.

14.5 The WOT relay (arrow) disengages the air conditioning clutch to reduce the engine load during wide open throttle conditions

d) Make sure the system has the correct refrigerant charge.

2 It's a good idea to operate the system for about 10 minutes at least once a month, particularly during the winter. Long term non-use can cause hardening, and subsequent failure, of the seals.

3 Leaks in the air conditioning system are best spotted when the system is brought up to temperature and pressure, by running the engine with the air conditioning ON for five minutes. Shut the engine off and inspect the air conditioning hoses and connections. Traces of oil usually indicate refrigerant leaks.

4 Because of the complexity of the air conditioning system and the special equipment necessary to service it, in-depth troubleshooting and repairs are not included in this manual. However, simple checks and component replacement procedures are provided in this Chapter.

5 If the air conditioning system doesn't operate at all, check the fuse panel and the Wide Open Throttle (WOT) relay, located in the relay box in the engine compartment **(see illustration)**.

6 The most common cause of poor cooling is simply a low system refrigerant charge. If a noticeable drop in cool air output occurs, the following quick check will help you determine if the refrigerant level is low. For more

14.15 Connect the refrigerant recharging hose to the low-pressure charging port

14.12 A basic charging kit for R-134a systems is available at most auto parts stores - it must say R-134a (not R-12) and so should the 12-ounce can of refrigerant

complete information on the air conditioning system, refer to the Haynes Automotive Heating and Air conditioning Manual.

Checking the refrigerant charge

7 Warm the engine up to normal operating temperature.

8 Place the air conditioning temperature selector at the coldest setting and put the blower at the highest setting. Open the doors (to make sure the air conditioning system doesn't cycle off as soon as it cools the passenger compartment).

9 With the compressor engaged - the clutch will make an audible click and the center of the clutch will rotate - feel the evaporator inlet pipe between the fixed orifice tube and the evaporator with one hand while placing your other hand on the metal portion of the hose between the evaporator and the accumulator (see Sections 17 and 18).

10 The pipe leading from the fixed orifice to the evaporator should be cold, and the outlet hose should be slightly colder (3 to 10-degrees F). If the outlet is considerably warmer than the inlet, the system charge is low. The earliest warning that a system is low on refrigerant is the air coming out of the ducts inside the vehicle. If the air isn't as cold as it used to be, the system probably needs a charge. Further inspection or testing of the system requires special tools and techniques and is beyond the scope of this manual.

11 If the inlet pipe has frost accumulation or feels cooler than the accumulator surface, the refrigerant charge is low.

Adding refrigerant

Refer to illustrations 14.12, and 14.15
Note: This procedure applies only to 1995 and later models, which use R-134a refrigerant. R-12 refrigerant used in 1993 and 1994 models is no longer available to the do-it-yourselfer.

12 Buy an automotive charging kit at an auto parts store. A charging kit includes a 12-ounce can of refrigerant, a tap valve and a short section of hose that can be attached between the tap valve and the system low

side service valve **(see illustration)**. Because one can of refrigerant may not be sufficient to bring the system charge up to the proper level, it's a good idea to buy an additional can. Make sure that one of the cans contains red refrigerant dye. If the system is leaking, the red dye will leak out with the refrigerant and help you pinpoint the location of the leak. **Warning:** Never add more than two cans of refrigerant to the system.

13 Hook up the charging kit by following the manufacturer's instructions. **Warning:** DO NOT hook the charging kit hose to the system high side! The fittings on the charging kit are designed to fit only on the low side of the system.

14 Back off the valve handle on the charging kit and screw the kit onto the refrigerant can, making sure first that the O-ring or rubber seal inside the threaded portion of the kit is in place. **Warning:** Wear protective eye wear when dealing with pressurized refrigerant cans.

15 Remove the dust cap from the low-side charging port (just above the coolant recovery tank) and attach the quick-connect fitting on the kit hose **(see illustration)**.

16 Warm up the engine and turn on the air conditioner. Keep the charging kit hose away from the fan and other moving parts. The charging process requires the compressor to be running. Open the vehicle doors to keep the clutch from cycling off. This will keep the compressor engaged.

17 Turn the valve handle on the kit until the stem pierces the can, then back the handle out to release the refrigerant. You should be able to hear the rush of gas. Add refrigerant to the low side of the system until both the accumulator surface and the evaporator inlet pipe feel about the same temperature. Allow stabilization time between each addition.

18 If you have an accurate thermometer, you can place it in the center air conditioning duct inside the vehicle and keep track of the outlet air temperature. A charged system that is working properly should cool to 40-degrees F. If the ambient (outside) air temperature is very high, say 110-degrees F, the duct air temperature may be as high as 60-degrees F, but generally the air conditioning is 30 to 50-degrees F cooler than the ambient air.

19 When the can is empty, turn the valve handle to the closed position and release the connection from the low-side port. Replace the dust cap.

20 Remove the charging kit from the can and store the kit for future use with the piercing valve in the UP position, to prevent inadvertently piercing the can on the next use.

15 Air conditioning compressor - removal and installation

Refer to illustrations 15.3, 15.5 and 15.6.
Warning: The air conditioning system is under high pressure. DO NOT disassemble any part of the system (hoses, compressor,

Chapter 3 Cooling, heating and air conditioning systems

15.3 Disconnect the electrical connector from the front end of the compressor (arrow)

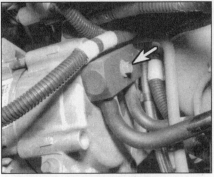

15.5 The refrigerant lines are secured to the rear of the compressor by one bolt (arrow)

line fittings, etc.) until after the system has been depressurized by a dealer service department or service station.
Note: *The accumulator (see Section 17) should be replaced whenever the compressor is replaced.*
1 Have the air conditioning system discharged (see **Warning** above).
2 Disconnect the negative cable from the battery.
3 Disconnect the electrical connector from the compressor **(see illustration)**.
4 Remove the drivebelt (see Chapter 1).
5 Remove the bolt that secures the refrigerant lines to the rear of the compressor. Plug the open fittings to prevent entry of dirt and moisture **(see illustration)**.
6 Unbolt the compressor from the mounting brackets and lift it out of the vehicle **(see illustration)**.
7 If a new compressor is being installed, follow the directions with the compressor regarding the draining of excess oil prior to installation.
8 The clutch may have to be transferred from the original to the new compressor.
9 Installation is the reverse of removal with the following additions:

a) *Replace all O-rings with new ones specifically made for air conditioning system use. Lubricate them with special PAG refrigerant oil designed for use in R-134a systems. Use mineral oil on R-12 systems.* **Caution:** *Do not use the wrong lubricant or you may contaminate the system!*
b) *Have the system evacuated, recharged and leak tested by the shop that discharged it.*

16 Air conditioning condenser - removal and installation

Refer to illustrations 16.6, 16.10a, 16.10b and 16.10c.
Warning 1: *If vehicle is equipped with airbags, refer to Chapter 12 to disarm the airbag system prior to performing any work described below.*
Warning 2 *The air conditioning system is under high pressure. DO NOT disassemble any part of the system (hoses, compressor, line fittings, etc.) until after the system has been depressurized by a dealer service department or service station.*
1 Have the air conditioning system discharged (see **Warning** above).
2 Remove the battery (see Chapter 5).
3 Drain the cooling system (see Chapter 1).
4 Remove the radiator (see Section 4).
5 Disconnect the refrigerant lines from the condenser. This requires a spring lock coupling tool similar to that used for fuel injection lines (see Chapter 4B).
6 Working from below, remove the mounting nuts from the condenser studs **(see illustration)**.
7 Lift the condenser out of the vehicle and plug the lines to keep dirt and moisture out.
8 If the original condenser will be reinstalled, store it with the line fittings on top to prevent oil from draining out.
9 If a new condenser is being installed, pour one ounce of refrigerant oil into it prior to installation. **Caution:** *Use PAG oil for R-134a systems and mineral refrigerant oil for R-12 systems.*
10 Reinstall the components in the reverse order of removal. Be sure the rubber pads are in place under the condenser. Be sure the seals are in place around the condenser **(see illustrations)**.
11 Have the system evacuated, recharged and leak tested by the shop that discharged it.

17 Air conditioning accumulator - removal and installation

Warning: *The air conditioning system is under high pressure. DO NOT disassemble any part of the system (hoses, compressor, line fittings, etc.) until after the system has been depressurized by a dealer service department or service station.*

Removal

Refer to illustration 17.4
1 Have the air conditioning system discharged (see **Warning** above).
2 Disconnect the negative cable from the battery.

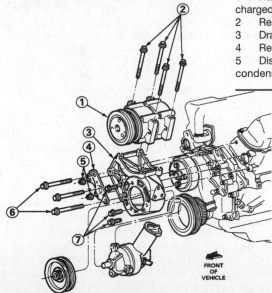

15.6 Air conditioning compressor and related components

1 Compressor
2 Mounting bolts
3 Air conditioning compressor and power steering pump bracket
4 Power steering pump support brace
5 Support brace-to-timing chain cover nuts
6 Power steering pump bolts
7 Power steering pump bolts

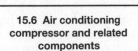

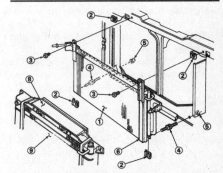

16.6 Air conditioning condenser and related components

1 Condenser
2 U-nut
3 Bolt
4 Stud and washer
5 Nut and washer
6 Seal (some automatic transmission models use two seals)
7 Condenser bottom seal (automatic transmission models only)
8 Condenser top seal
9 Radiator

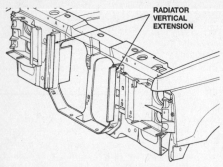

16.10a The condenser has two radiator extensions . . .

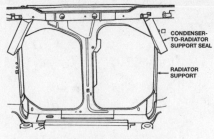

16.10b . . . two seals between the condenser and the radiator support. . .

17.4 The accumulator is mounted on the evaporator case in the engine compartment - disconnect this quick-disconnect fitting (arrow)

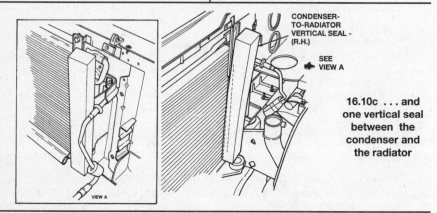

16.10c . . . and one vertical seal between the condenser and the radiator

3 Remove the accumulator as described in Section 17.
4 Disconnect the refrigerant line from the evaporator to the condenser using a spring lock coupling tool. Cap the lines to prevent entry of dirt and moisture.
5 Remove the screws holding the evaporator housing, and pull off the housing **(see illustration).**
6 Remove the evaporator core.
7 Installation is the reverse of removal with the following additions:
  a) Replace all O-rings with new ones specifically made for air conditioning system use. Lubricate them with special PAG refrigerant oil designed for use in R-134a systems. Use mineral oil on R-12 systems. **Caution:** *Do not use the wrong lubricant or you may contaminate the system.*
  b) Have the system evacuated, recharged and leak tested by the shop that discharged it.

3 Unplug the electrical connector from the pressure switch near the top of the accumulator. Unscrew the pressure switch.
4 Disconnect the refrigerant line from the accumulator **(see illustration).** This requires a spring lock coupling tool of the type used for fuel injection system lines (see Chapter 4B).
5 Use a backup wrench to hold the fitting that connects the accumulator to the evaporator core, then disconnect the fitting. Plug the open fittings to prevent entry of dirt and moisture.
6 Remove the mounting bracket screw and the screw that holds the evaporator tube to the accumulator bracket, then lift the accumulator out.
7 If a new accumulator is being installed, remove the Schrader valve and pour the oil out into a measuring cup, noting the amount. Add fresh refrigerant oil to the new accumulator equal to the amount removed from the old unit, plus one ounce. **Caution:** *Use PAG oil for R-134a systems and mineral refrigerant oil for R-12 systems.*

Installation

8 Loosely position the bracket on the new accumulator.
9 Connect the accumulator to the evaporator core, using a new O-ring lubricated with clean refrigerant oil. At the same time, align the bracket with the slot between the evaporator case flanges.
10 Using a backup wrench, tighten the fitting securely.
11 Install the screw that holds the bracket between the case flanges.
12 Tighten the accumulator bracket and install the clip that holds the evaporator inlet tube to the bracket.
13 Place a new O-ring, lubricated with clean refrigerant oil of the correct type, on the pressure switch nipple on the accumulator.
14 Install the pressure switch. If it has a metal base, tighten it to the torque listed in this Chapter's Specifications. If it has a plastic base, tighten it by hand only. Connect the pressure switch electrical connector.
15 Reconnect the negative cable to the battery.
16 Take the vehicle to the shop that discharged the air conditioning system. Have the system recharged and tested for leaks.

18 Air conditioning evaporator - replacement

Refer to illustration 18.5
Warning: *The air conditioning system is under high pressure. DO NOT disassemble any part of the system (hoses, compressor, line fittings, etc.) until after the system has been depressurized by a dealer service department or service station.*
Note: *The accumulator (see Section 17) should be replaced whenever the evaporator is replaced.*

1 Have the air conditioning system discharged (see **Warning** above).
2 Disconnect the negative cable from the battery.

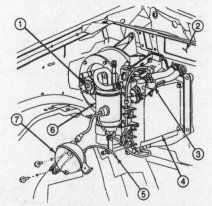

18.5 Remove the screws securing the evaporator housing

1 Blower motor
2 Firewall
3 Heater control valve
4 Evaporator housing
5 Vacuum hose to engine
6 Accumulator
7 Vacuum reservoir

Chapter 4
Fuel and exhaust systems

Contents

	Section		Section
Accelerator cable - removal and installation	10	Fuel lines and fittings - general information	3
Air cleaner housing - removal and installation	9	Fuel pressure relief procedure	2
CHECK ENGINE light	See Chapter 6	Fuel pump - removal and installation	7
Electronic fuel injection system - component check and replacement	12	Fuel pump/fuel pressure - check	4
Electronic fuel injection system - general checks	11	Fuel tank - cleaning and repair	6
Exhaust system service - general information	13	Fuel tank - removal and installation	5
Fuel level sending unit - check and replacement	8	General information	1

Specifications

Fuel pressure
1997 and earlier
  Key on, engine off .... 35 to 45 psi
  At idle
    With vacuum hose connected .... 25 to 35 psi
    With vacuum hose disconnected .... 35 to 45 psi
1998 and later (key on, engine off or engine running)
  All except 3.0L Flex Fuel .... 56 to 72 psi
  3.0L Flex Fuel .... 47 to 63 psi

Fuel injector resistance
All models .... 12 to 16 ohms

Torque specifications   **Ft-lbs** (unless otherwise indicated)
Upper intake manifold nuts/bolts
  Four-cylinder engines
    Step 1 .... 61 to 88 in-lbs
    Step 2 .... 19 to 28
  3.0L V6 engine 19
  4.0L V6 engine .... 15 to 18
Throttle body bolts
  Four-cylinder engines .... 15 to 22
  3.0L V6 engine 19
  4.0L V6 engine .... 76 to 106 in-lbs
Fuel pressure regulator mounting bolts/screws
  Four-cylinder engines .... 27 to 44 in-lbs
  3.0L V6 engine .... 27 to 35 in-lbs
  4.0L V6 engine .... 72 to 96 in-lbs
Fuel rail mounting bolts/studs
  Four-cylinder engines .... 15 to 22
  3.0L V6 engine .... 72 to 96 in-lbs
  4.0L V6 engine .... 84 to 120 in-lbs

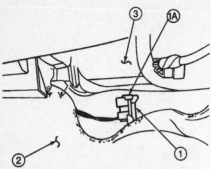

2.1 The inertia switch is located on the passenger side of the transmission hump, on the firewall and under the carpet

1. Inertia switch
1A. Reset button
2. Transmission hump
3. Firewall

1 General information

Fuel system

The fuel system consists of the fuel tank, the fuel pump, an air cleaner assembly, a fuel injection system and the various Teflon hoses, steel lines and fittings connecting the fuel delivery components together.

Vehicles covered by this manual are equipped with the Ford Multiport Fuel Injection (MFI) system. This system uses one fuel injector per cylinder.

The fuel pump is a single high pressure electric pump located within the fuel tank.

Exhaust system

All vehicles are equipped with exhaust manifold(s), a catalytic converter, an exhaust pipe and a muffler. Any component of the exhaust system can be replaced. See Chapter 6 for further details regarding the catalytic converter.

2 Fuel pressure relief procedure

Refer to illustration 2.1

Warning: *The fuel supply lines will remain pressurized for a long period of time after the engine is shut down. The pressure within the fuel system must be relieved before servicing any of the fuel delivery components. Gasoline is extremely flammable, so take extra precautions when you work on any part of the fuel system. Don't smoke or allow open flames or bare light bulbs near the work area, and don't work in a garage where a natural gas-type appliance (such as a water heater or clothes dryer) with a pilot light is present. If you spill any fuel on your skin, rinse it off immediately with soap and water. When you perform any kind of work on the fuel system, wear safety glasses and have a Class B type fire extinguisher on hand.*

1 The fuel pump switch - sometimes called the "inertia switch" - shuts off fuel to

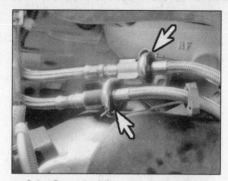

3.1a Some fuel lines use spring lock couplings (these are the fuel inlet and return lines in the engine compartment, viewed through the wheel well)

the engine in the event of a collision. This switch provides an easy way to relieve fuel pressure before servicing any fuel injection component. The switch is located in the passenger compartment on the passenger side of the transmission hump **(see illustration)**.

2 Unplug the inertia switch electrical connector.

3 Crank the engine with the starter for 15 to 20 seconds.

4 The fuel system pressure is now relieved. When you're finished working on the fuel system, reconnect the electrical connector back onto the switch and push the reset button on the top of the switch (just in case it may have "popped").

5 Even though the pressure has been relieved, it's a good idea to wrap a rag around fuel fittings when they are disconnected.

3 Fuel lines and fittings - general information

Warning: *Gasoline is extremely flammable, so take extra precautions when you work on any part of the fuel system. Don't smoke or allow open flames or bare light bulbs near the work area, and don't work in a garage where a natural gas-type appliance (such as a water heater or clothes dryer) with a pilot light is present. Since gasoline is carcinogenic, wear latex gloves when there's a possibility of being exposed to fuel, and, if you spill any fuel on your skin, rinse it off immediately with soap and water. Mop up any spills immediately and do not store fuel-soaked rags where they could ignite. The fuel system is under constant pressure, so, if any fuel lines are to be disconnected, the fuel pressure in the system must be relieved first (see Section 2). When you perform any kind of work on the fuel system, wear safety glasses and have a Class B type fire extinguisher on hand.*

Spring lock couplings - disassembly and reassembly

Refer to illustrations 3.1a, 3.1b, 3.2 and 3.4a through 3.4e

1 The fuel supply and return lines used on

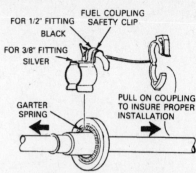

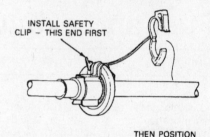

3.1b The lines are secured with tethered clips

these engines have spring lock couplings at some connections **(see illustration)**. The male end of the spring lock coupling, which is girded by two O-rings, is inserted into a female flared end fitting. The coupling is secured by a garter spring which prevents disengagement by gripping the flared end of the female fitting. A clip and tether assembly provides additional security **(see illustration)**.

2 To disconnect the spring lock coupling supply fitting, you'll need to obtain a special spring lock coupling tool or its equivalent **(see illustration)**, available at most auto parts stores, designed for:

 a) 3/8-inch return fitting
 b) 1/2-inch supply fitting
 c) 5/8-inch return fitting

3 Before disconnecting the line, unclip the safety clip **(see illustration 3.1b)**.

4 To disconnect and reconnect the line, **refer to the accompanying illustrations** and follow the information in the captions.

Stainless steel fitting - disassembly and reassembly

Refer to illustrations 3.9 and 3.12

5 All fuel lines outside the engine compartment are either Teflon hose and stainless steel (fuel supply and return lines) or carbon

Chapter 4 Fuel and exhaust systems

3.2 These special tools are required to connect and disconnect engine compartment fuel lines

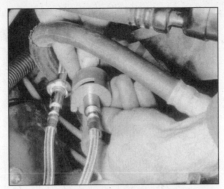

3.4a Open the tool and place it over the line . . .

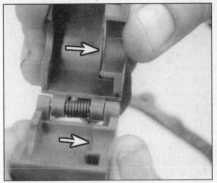

3.4b . . . then slide the tool against the fuel line fitting so the lip inside the tool (arrows) . . .

3.4c . . . pushes against the garter spring (arrow)

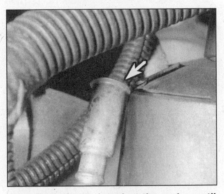

3.4d Push the tool against the spring until the spring slides off the flared female end of the line (arrow)

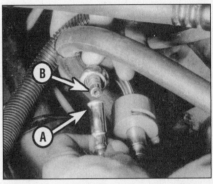

3.4e To reconnect the line, place the flared female end (A) over the O-ring on the male end and push the ends together until the flared end slips under the garter spring (B) - tug on the ends to make sure they're locked, then install the safety clip

steel (vapor line). These lines are fastened to the frame rails with integral clips and connect with stainless steel fittings. This type of fitting consists of a female tube end with internal retaining "fingers" that grasp a formed male tube end. A safety clip holds the two ends of the fitting together.

6   To disconnect the stainless steel fitting, you'll need to obtain a quick-disconnect tool kit, available at most auto parts stores that carry tools.

7   Release the safety clip from the fitting and install the proper quick-disconnect tool on the fitting body.

8   Align the quick-disconnect tool on the fitting and push the tool into the fitting body to release the retaining fingers holding the male tube end. Once the fingers are released, pull the male tube end from the female fitting end and remove the tool.

9   On 4.0L engines, study the **accompanying illustration** carefully before detaching the fuel return line fitting from the pressure regulator.

10 After disassembly, inspect and clean the tube end sealing surfaces. Also inspect the inside of the fitting for any internal damage. If damage is noted, replace the fuel tube.

11 Before reconnecting the fitting, wipe the male tube end with a clean cloth. Check the inside of the fitting to make sure that it's free of dirt and/or obstructions.

12 To reconnect the fitting, align it with the male tube end and push the tube into the fitting. When the fitting is engaged, a definite click will be heard. Pull on the fitting to ensure that it's fully engaged (see illustration).

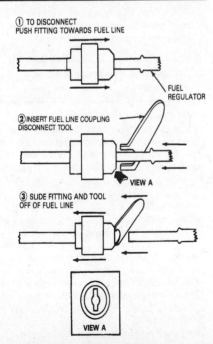

3.9 Disconnecting the pressure regulator fitting

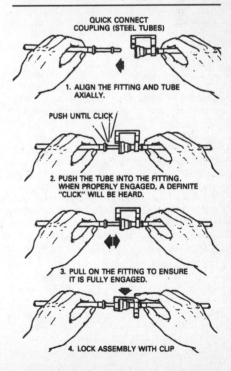

3.12 Connecting stainless steel fittings

Chapter 4 Fuel and exhaust systems

Fuel lines - Teflon-stainless steel and carbon steel

Warning: *Use only the recommended tubing materials. Other types of tubing could fail in service.*

13 Teflon-stainless steel fuel lines must not be repaired using hose and hose clamps. Should the fittings or steel tubing ends become damaged, approved service parts must be used to replace the lines. Splicing such lines with seamless steel tubing or rubber hose is not recommended.

14 If a length of Teflon hose is damaged, it should be replaced only with an approved service part.

4 Fuel pump/fuel pressure - check

Warning: *Gasoline is extremely flammable, so take extra precautions when you work on any part of the fuel system. Don't smoke or allow open flames or bare light bulbs near the work area, and don't work in a garage where a natural gas-type appliance (such as a water heater or clothes dryer) with a pilot light is present. Since gasoline is carcinogenic, wear latex gloves when there's a possibility of being exposed to fuel, and, if you spill any fuel on your skin, rinse it off immediately with soap and water. Mop up any spills immediately and do not store fuel-soaked rags where they could ignite. The fuel system is under constant pressure, so, if any fuel lines are to be disconnected, the fuel pressure in the system must be relieved first (see Section 2). When you perform any kind of work on the fuel system, wear safety glasses and have a Class B type fire extinguisher on hand.*

Note: *Always make sure there is fuel in the tank before assuming the fuel pump is defective.*

General check

1 An electric fuel pump malfunction will usually result in a loss of fuel flow and/or pressure that is often reflected by a corresponding drop in performance (or a no-run condition).

2 Verify the pump actually runs. Remove the fuel filler cap and have an assistant turn the ignition switch to On - you should hear a brief whirring noise as the pump comes on and pressurizes the system. Have the assistant start the engine (if possible). This time you should hear a constant whirring sound from the pump (but it's more difficult to hear with the engine running).

3 If the pump doesn't run (makes no sound), check the fuel pump fuse and relay (see Chapter 12). If the fuse and relay are okay, check the continuity of the inertia switch with an ohmmeter (see Section 2 for the location of the switch). If the inertia switch is good, proceed to the next Step. If the fuse is blown, replace it and see if the pump works (if it blows again, trace the fuel pump circuit for a short to ground). If the fuse doesn't blow but the fuel pump still doesn't work, proceed to the next Step.

4.7 Connect the fuel pressure gauge to the Schrader valve on the fuel rail (arrow) (four-cylinder engine shown)

4 With the ignition off, unplug the electrical connector from the fuel pump. Using a jumper wire, ground the negative terminal of the fuel pump. Using a fused jumper wire, apply 12-volts to the fuel pump feed terminal.

5 If the fuel pump doesn't run, replace it (see Section 7). If it does run, the problem lies somewhere in the electrical circuit to the fuel pump (refer to the wiring diagrams at the back of this manual).

Pressure check

Refer to illustration 4.7

Note: *On 1998 and later models, the fuel pressure regulator is located in the fuel tank along with the fuel pump. The system is equipped with a fuel pressure damper on the fuel rail. The damper is similar in appearance to a fuel pressure regulator, but the damper's only purpose is to reduce fuel pulsations in the fuel rail.*

6 Relieve the fuel pressure (see Section 2).
7 Connect a fuel pressure gauge to the Schrader valve on the fuel rail **(see illustration)**.
8 Turn the ignition switch to the On position. The fuel pump should run for about two seconds, then the pressure should hold steady. Note the reading on the gauge and compare it to the pressure listed in this Chapter's Specifications. On 1997 and earlier models, continue the check at Step 9. On 1998 and later models, proceed to Step 11.
9 Start the engine (if possible) and allow it to idle. The pressure should be lower by 3 to 10 psi (see this Chapter's specifications). Disconnect the vacuum hose from the fuel pressure regulator and compare your reading with the pressure listed in this Chapter's Specifications. If all the pressure readings are within specifications, the system is operating properly.
10 If the pressure did not drop by 3 to 10 psi after starting the engine, apply 10-inches Hg of vacuum to the pressure regulator using a hand-held vacuum pump. If the pressure drops, repair the vacuum source to the regulator. If the pressure doesn't drop, replace the regulator (see Section 12).
11 If the pressure is higher than specified,

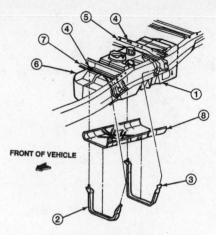

5.6 Fuel tank and related components

1	Fuel tank	4	U-lock nut
2	Fuel tank support strap	5	Crossmember
		6	Heat shield
3	Fuel tank support strap	7	Crossmember
		8	Heat shield

check for a faulty regulator or a pinched or clogged fuel return hose or pipe.

12 If the pressure is lower than specified:
 a) Replace the fuel filter (it may be clogged
 b) Look for a pinched or clogged fuel hose or line between the fuel tank and the fuel rail
 c) On 1997 and earlier models, check the pressure regulator for a malfunction. With the engine running, pinch the fuel return line (if possible). If the pressure rises, replace the fuel pressure regulator.
 d) Look for leaks in the fuel feed line.
 e) Check for leaking injectors.

13 After the testing is done, relieve the fuel system pressure (see Section 2) and remove the fuel pressure gauge.

5 Fuel tank - removal and installation

Refer to illustrations 5.6 and 5.9

Warning: *Gasoline is extremely flammable, so take extra precautions when you work on any part of the fuel system. Don't smoke or allow open flames or bare light bulbs near the work area, and don't work in a garage where a natural gas-type appliance (such as a water heater or clothes dryer) with a pilot light is present. Since gasoline is carcinogenic, wear latex gloves when there's a possibility of being exposed to fuel, and, if you spill any fuel on your skin, rinse it off immediately with soap and water. Mop up any spills immediately and do not store fuel-soaked rags where they could ignite. The fuel system is under constant pressure, so, if any fuel lines are to be disconnected, the fuel pressure in the system must be relieved first (see Section 2). When you perform any kind of work on the fuel system, wear safety glasses and have a Class B type fire extinguisher on hand.*

Chapter 4 Fuel and exhaust systems

Note: The following procedure is much easier to perform if the fuel tank is empty. Some tanks have a drain plug for this purpose. If the tank does not have a drain plug, siphon the gasoline into an approved gasoline container.

1   Remove the fuel tank filler cap to relieve fuel tank pressure.
2   Relieve the fuel system pressure (see Section 2).
3   Detach the negative cable from the battery.
4   If the tank still has fuel in it, siphon it into an approved gasoline container. **Warning:** *Don't start the siphoning action by mouth! Use a siphoning kit, available at most auto parts stores.*
5   Raise the vehicle and place it securely on jackstands.
6   Remove the heat shield, skid plate and the front retaining strap **(see illustration)**.
7   Support the fuel tank with a floor jack or jackstands. Position a piece of wood between the jack head and the fuel tank to protect the tank.
8   Remove the bolt from the rear retaining strap and pivot the strap down until it is hanging out of the way.
9   Loosen the screw clamps at the fill pipe and the vent pipe. Disconnect the pipes at the tank **(see illustration)**.
10   Lower the tank enough to disconnect the vapor valve and fuel line connectors from the sender unit and vapor valve. **Note:** *The fuel feed and return lines and the vapor return line are three different diameters, so reattachment is simplified. If you have any doubts, however, clearly label the three lines and the fittings. Be sure to plug the hoses to prevent leakage and contamination of the fuel system.*
11   Disconnect the fuel pump and fuel gauge sending unit connectors at the rear of the fuel tank.
12   Remove the tank from the vehicle. **Warning:** *Store the tank in a safe place, away from sparks and open flames (read the Warning at the beginning of this Section).*
13   Installation is the reverse of removal. The manufacturer recommends that new tank retaining strap bolts be used.

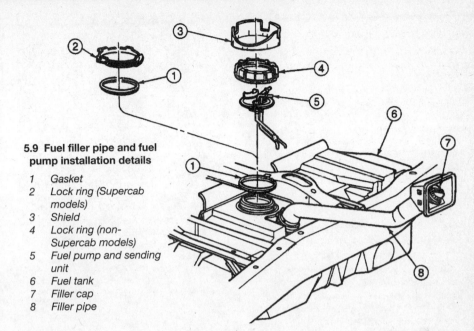

5.9 Fuel filler pipe and fuel pump installation details

1   Gasket
2   Lock ring (Supercab models)
3   Shield
4   Lock ring (non-Supercab models)
5   Fuel pump and sending unit
6   Fuel tank
7   Filler cap
8   Filler pipe

6 Fuel tank - cleaning and repair

1   The polyethylene fuel tank cannot be repaired. No reliable repair procedures are available to correct leaks or damage. Fuel tank replacement is the only approved service. **Warning:** *Even after cleaning and flushing of the fuel tank, explosive fumes can remain and ignite.*
2   If the fuel tank is removed from the vehicle, it should not be placed in an area where sparks or open flames could ignite the fumes coming out of the tank. Be especially careful inside garages where a natural gas-type appliance is located, because the pilot light could cause an explosion.
3   Whenever the fuel tank is steam-cleaned or otherwise serviced, the vapor valve assembly should be replaced. All grommets and seals must be replaced to prevent possible leakage.

7 Fuel pump - removal and installation

Refer to illustration 7.4

Warning: *Gasoline is extremely flammable, so take extra precautions when you work on any part of the fuel system. Don't smoke or allow open flames or bare light bulbs near the work area, and don't work in a garage where a natural gas-type appliance (such as a water heater or clothes dryer) with a pilot light is present. Since gasoline is carcinogenic, wear latex gloves when there's a possibility of being exposed to fuel, and, if you spill any fuel on your skin, rinse it off immediately with soap and water. Mop up any spills immediately and do not store fuel-soaked rags where they could ignite. The fuel system is under constant pressure, so, if any fuel lines are to be disconnected, the fuel pressure in the system must be relieved first (see Section 2). When you perform any kind of work on the fuel system, wear safety glasses and have a Class B type fire extinguisher on hand.*

1   Remove the fuel tank (refer to Section 5).
2   Remove any dirt that has accumulated where the fuel sender/pump is attached to the fuel tank. Don't allow dirt to enter the fuel tank.
3   Turn the lock ring counterclockwise with lock ring wrench (Ford tool no. T90T-9275-A or equivalent tool) until it's loose. Remove the lock ring.
4   Carefully pull the fuel pump/sending unit assembly from the tank **(see illustration)**. Don't bend the arm for the sending unit float.
5   Remove the old lock ring seal and discard it.
6   Clean the fuel pump mounting flange, the tank mounting surface and seal ring groove.
7   Installation is the reverse of the removal procedure with the following additions:
 a) Apply a thin coat of heavy grease to the new lock ring seal to hold it in place during assembly. Install the seal in the lock ring groove.
 b) Install the fuel sender/pump unit into the fuel tank so that the tabs of the unit are positioned into the slots in the fuel tank. Make sure to keep the seal in place during installation until the lock ring has been tightened properly.
 c) Hold the fuel sender/pump unit in place, install and tighten the locking ring clockwise with the lock ring wrench, or equivalent, until the stop is against the retainer ring tab.

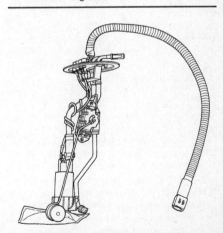

7.4 The fuel pump is combined with the fuel gauge sending unit and is located inside the fuel tank

4-6 Chapter 4 Fuel and exhaust systems

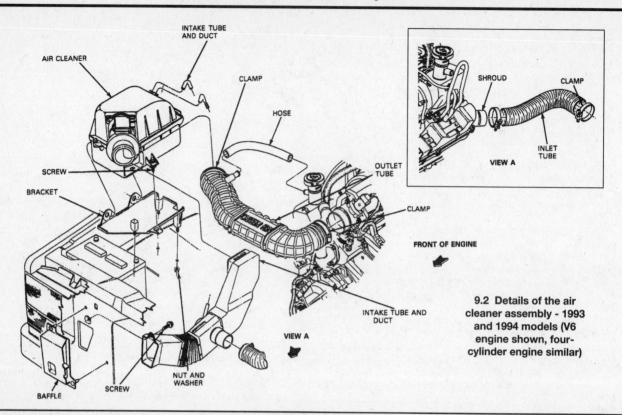

9.2 Details of the air cleaner assembly - 1993 and 1994 models (V6 engine shown, four-cylinder engine similar)

8 Fuel level sending unit - check and replacement

Warning: *Gasoline is extremely flammable, so take extra precautions when you work on any part of the fuel system. Don't smoke or allow open flames or bare light bulbs near the work area, and don't work in a garage where a natural gas-type appliance (such as a water heater or clothes dryer) with a pilot light is present. Since gasoline is carcinogenic, wear latex gloves when there's a possibility of being exposed to fuel, and, if you spill any fuel on your skin, rinse it off immediately with soap and water. Mop up any spills immediately and do not store fuel-soaked rags where they could ignite. The fuel system is under constant pressure, so, if any fuel lines are to be disconnected, the fuel pressure in the system must be relieved first (see Section 2). When you perform any kind of work on the fuel system, wear safety glasses and have a Class B type fire extinguisher on hand.*

Check

Note: *The fuel sending unit consists of a variable resistor controlled by the movement of an attached float in the fuel tank. When the fuel level is low, resistance in the sending unit is low. When the fuel level is high, resistance is high. The position of the float determines the resistance, and the resulting current flow in the circuit is proportional to the fuel level.*

1 Raise the vehicle and support it securely on jackstands.
2 If the fuel level gauge on the instrument panel does not appear to be working properly, first check the operation of the gauge. If the gauge works according to the tests which follow, and the circuit is OK, the problem must lie in the fuel level sending unit.
3 Disconnect the electrical connector for the fuel pump and sending unit. Turn on the ignition key. The needle on the gauge should deflect past the Full mark on the gauge.
4 Using a test light or voltmeter, probe the terminals on the electrical connector to find the one with battery voltage present. Using a jumper wire, ground the terminal that **does not** have battery voltage. The needle on the gauge should swing past the Empty mark. If the gauge works as described it is functioning properly.
5 A more accurate check of the sending unit is possible by removing it from the fuel tank and checking its resistance while manually operating the float arm.

Replacement

6 The fuel sender is an integral part of the fuel pump assembly. Remove and replace the fuel pump and sending unit as described in Section 7.

9 Air cleaner housing - removal and installation

1993 and 1994 models

Refer to illustration 9.2

1 Disconnect the negative cable from the battery.
2 Loosen the hose clamps on the air cleaner outlet tube assembly **(see illustration)**. Remove the tube from the throttle body and the air cleaner housing. Cover the throttle body inlet to prevent the entry of foreign matter.
3 Disconnect the vacuum lines at the temperature sensitive vacuum motor in the air cleaner cover.
4 Disconnect the electrical connector at the Mass Air Flow (MAF) sensor.
5 Remove the screws securing the air cleaner assembly upper half. Label and disconnect the two small hoses from the upper case half and remove the upper half and the air cleaner filter.
6 To remove the remainder of the assembly, perform the following:
   a) Squeeze the protrusions on the air cleaner hot air tube and disconnect the tube from the air cleaner housing.
   b) Remove the bolt securing the air cleaner intake tube assembly to the air cleaner intake deflector next to the radiator panel. Remove the intake tube from the lower half of the air cleaner housing.
   c) Remove the screw securing the lower half of the air cleaner housing to the mounting bracket. Remove the lower half of the air cleaner housing.
7 Installation is the reverse of the removal procedure. Tighten all hose clamps and fittings securely.

1995 and later models

Refer to illustrations 9.9, 9.10 and 9.11

8 Disconnect the negative cable from the battery.
9 Disconnect the crankcase ventilation hose on V6 models **(see illustration)**.

Chapter 4 Fuel and exhaust systems

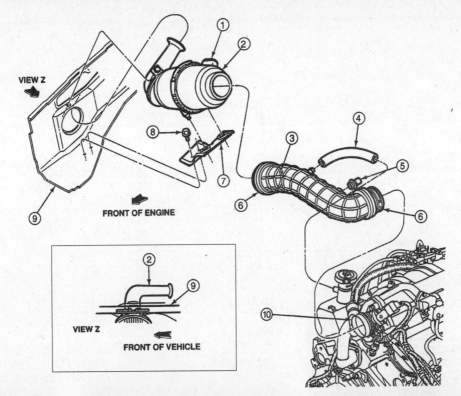

9.9 Details of the air cleaner assembly (1995 and later models)

1. Mass Airflow Sensor
2. Air cleaner housing
3. Air intake duct
4. Crankcase ventilation hose
5. Intake Air Temperature sensor
6. Hose clamp
7. Air cleaner mounting bracket
8. Bolt
9. Inner fender panel
10. Throttle body

9.10 Detach the electrical connector from the Mass Air Flow sensor (arrow) . . .

9.11 . . . and the Intake Air Temperature sensor (arrow)

10.1 Remove the shield over the accelerator cable linkage

10 Disconnect the electrical connector at the Mass Air Flow (MAF) sensor **(see illustration)**.
11 Disconnect the electrical connector at the Intake Air Temperature (IAT) sensor **(see illustration)**.
12 Loosen the air cleaner housing clamps at the engine air cleaner and the throttle body and remove the housing.
13 Installation is the reverse of the removal procedure. Tighten all clamps and fittings securely.

10 Accelerator cable - removal and installation

Refer to illustrations 10.1, 10.2 and 10.3

1 Remove the accelerator linkage shield **(see illustration)**.
2 Disconnect the accelerator cable from the throttle lever ball stud at the throttle body **(see illustration)**.
3 Detach the accelerator cable from its bracket on the intake manifold **(see illustration)**.
4 From inside the vehicle, remove the cable end from the top of the accelerator pedal.
5 Release the retaining clips at the dash panel and push the cable through the dash panel into the engine compartment.
6 Installation is the reverse of the removal procedure, except for the following points:
   a) Make sure that no wiring or hoses contact any moving portion of the accelerator controls.
   b) Before starting the engine, make sure that the throttle operates smoothly without any sticking or binding from the accelerator pedal.

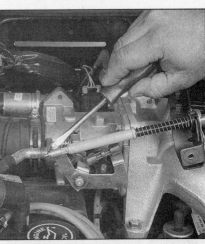

10.2 Pry the end of the accelerator cable from the throttle lever

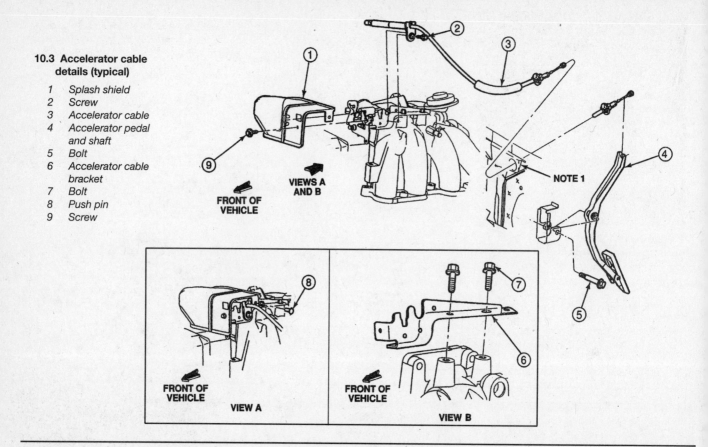

10.3 Accelerator cable details (typical)

1. Splash shield
2. Screw
3. Accelerator cable
4. Accelerator pedal and shaft
5. Bolt
6. Accelerator cable bracket
7. Bolt
8. Push pin
9. Screw

c) With the engine running, make sure that the throttle operates normally and the engine returns to idle when you remove your foot from the accelerator pedal.

11 Electronic fuel injection system - general checks

Warning: *Gasoline is extremely flammable, so take extra precautions when you work on any part of the fuel system. Don't smoke or allow open flames or bare light bulbs near the work area, and don't work in a garage where a natural gas-type appliance (such as a water heater or clothes dryer) with a pilot light is present. Since gasoline is carcinogenic, wear latex gloves when there's a possibility of being exposed to fuel, and, if you spill any fuel on your skin, rinse it off immediately with soap and water. Mop up any spills immediately and do not store fuel-soaked rags where they could ignite. The fuel system is under constant pressure, so, if any fuel lines are to be disconnected, the fuel pressure in the system must be relieved first (see Section 2). When you perform any kind of work on the fuel system, wear safety glasses and have a Class B type fire extinguisher on hand.* The following procedure assumes that the fuel pump works and fuel pressure is adequate (see Section 4).

Preliminary checks

1 Check all electrical connectors that are related to the system. Loose electrical connectors and poor grounds can cause many problems that resemble more serious malfunctions.
2 Check to see that the battery is fully charged, as the control unit and sensors depend on an accurate supply voltage in order to properly meter the fuel.
3 Check the air filter element. A dirty or partially blocked filter will severely impede performance and economy (see Chapter 1).
4 If you find a blown fuse, replace it and see if it blows again. If it does, search for a grounded wire in the harness to the fuel pump (see Chapter 12).

System checks

5 Check the condition of the vacuum hoses connected to the intake manifold.
6 Remove the air intake duct from the throttle body (see Section 9) and check for dirt, carbon or other residue build-up in the throttle body, particularly around the throttle plate. If it is dirty, clean it with carburetor cleaner and a toothbrush.
7 With the engine running, place an automotive stethoscope against each injector, and listen for a clicking sound indicating that it is working. If you don't have a stethoscope, you can use a long screwdriver - place the tip of the screwdriver against the injector body and listen through the handle.
8 If you came across an injector that doesn't make any sound, install a "noid" light in line with the wiring harness of each suspected injector, start the engine and check to see if the noid light flashes. If it does, the injector is receiving proper voltage. If it doesn't flash, further diagnosis is necessary by the dealer or other experienced repair shop.
9 With the engine OFF and the fuel injector electrical connectors disconnected, measure the resistance of each injector. Compare the resistance readings with the values listed in this Chapter's-Specifications.

12 Electronic fuel injection system - component check and replacement

Upper intake manifold and throttle body

Four-cylinder engines

Refer to illustrations 12.5 and 12.9

1 Remove the air cleaner housing (see Section 9).
2 Remove the splash shield from the accelerator linkage **(see illustration 10.1)**. Remove the accelerator cable bracket and disconnect the accelerator cable from the throttle body (see Section 10).
3 Disconnect the electrical connections to

Chapter 4 Fuel and exhaust systems

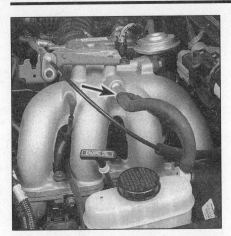

12.5 Disconnect the vacuum hoses from the intake manifold

the throttle position sensor, idle air control valve and EGR vacuum regulator control.
4 Disconnect the PCV hose from the intake manifold.
5 Label and disconnect the vacuum hoses from the manifold vacuum tee (see illustration).
6 Disconnect the EGR transducer hoses from the EGR valve, and remove EGR valve (see Chapter 6).
7 Remove the bolts and studs that secure the upper intake manifold.
8 Thoroughly clean all old gasket material from the mating surfaces of the upper intake manifold and the lower intake manifold.
9 Install a new gasket on the lower intake manifold, then install the upper manifold and tighten the fasteners evenly to the torque listed in this Chapter's Specifications (see illustration).
10 The remainder of installation is the reverse of the removal steps.

3.0L V6 engine

Refer to illustration 12.19

11 Label and disconnect the electrical connectors and vacuum lines between the upper intake manifold and the engine and body.
12 Remove the splash shield from the accelerator linkage. Remove the accelerator cable bracket and disconnect the accelerator cable from the throttle body (see Section 10).
13 Remove the air inlet tube that connects the air cleaner to the throttle body.
14 Label and disconnect the vacuum lines from the upper intake manifold.
15 Detach the PCV valve from the right valve cover and disconnect the hose from the throttle body.
16 Detach the spark plug wires from the ignition coil and move them out of the way.
17 Remove the ignition coil with bracket, the EGR vacuum regulator control and the radio ignition interference capacitor.
18 Disconnect the hoses from the EGR valve, then remove the EGR valve.
19 Remove bolts and studs that secure the upper intake manifold (see illustration), and remove the manifold.

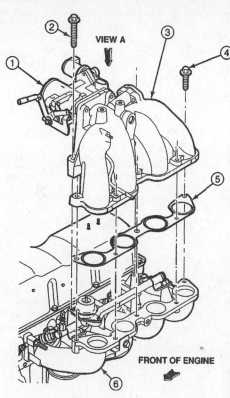

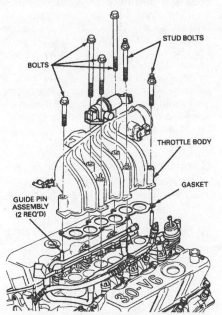

12.19 On the 3.0L engine, remove the bolts and studs securing the upper intake manifold to the lower intake manifold

20 Thoroughly clean all old gasket material from the mating surfaces of the upper intake manifold and the lower intake manifold.
21 Install a new gasket on the lower intake manifold, then install the upper manifold and tighten the fasteners evenly to the torque listed in this Chapter's Specifications.
22 The remainder of installation is the reverse of the removal steps.

12.9 On the four-cylinder engine, remove the bolts and studs securing the upper intake manifold to the lower intake manifold

1 Throttle body
2 Bolt
3 Upper intake manifold
4 Bolt
5 Gasket
6 Lower intake manifold

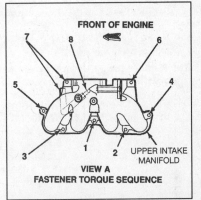

12.30 Unbolt the air conditioning hose bracket from the upper intake manifold and move it aside DO NOT disconnect any refrigerant lines

4.0L V6 engine

Refer to illustrations 12.30, 12.31a, 12.31b and 12.31c

23 Label and disconnect the electrical connectors and vacuum lines between the upper intake manifold, engine and body.
24 Remove the splash shield from the accelerator linkage. Remove the accelerator cable bracket and disconnect the accelerator cable from the throttle body (see Section 10).
25 Remove the air inlet tube that connects the air cleaner to the throttle body.
26 Label and disconnect the vacuum lines from the upper intake manifold.
27 Detach the PCV valve from the right valve cover (see Chapter 6).
28 Detach the spark plug wires from the back of the upper intake manifold.
29 Disconnect the canister purge line at the throttle body fitting.
30 Remove the bolt that secures the air

Chapter 4 Fuel and exhaust systems

12.31a Remove the upper intake manifold nuts with a socket and extension

conditioning refrigerant line at the upper rear part of the manifold (see illustration). DO NOT disconnect any refrigerant lines!

31 Remove the six nuts that secure the upper intake manifold (see illustrations). Note the cap over the front stud. Lift the upper manifold and throttle body together off the fuel rail and lower intake manifold.

32 Thoroughly clean all old gasket material from the mating surfaces of upper intake manifold and the fuel rail.

33 Install a new gasket on the fuel rail, then install the upper manifold and tighten the fasteners evenly to the torque listed in this Chapter's Specifications.

34 The remainder of installation is the reverse of the removal steps.

Throttle body

Refer to illustration 12.39

Note: *This procedure applies to all models except 1997 and earlier 3.0L V6 models. On 1997 and earlier 3.0L V6 models, the throttle body is an integral part of the upper intake manifold.*

35 Remove the splash shield from the accelerator linkage and disconnect the accelerator cable (see Section 10).

36 Disconnect the electrical connector from the throttle position sensor.

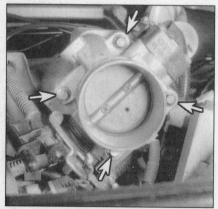

12.39 Disconnect the electrical connector and remove the throttle body mounting bolts

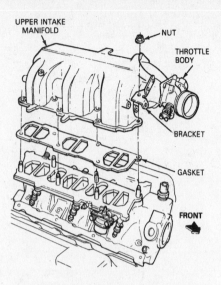

12.31b Details of the upper intake manifold and gasket - 4.0L V6 engine

37 Unclamp and disconnect the air inlet duct at the throttle body.

38 Disconnect the canister purge hose from underneath the throttle body.

39 Remove the four throttle body mounting bolts (see illustration). Remove the throttle body and gasket from the upper intake manifold.

40 Carefully clean all old gasket material from the mating surface of throttle body and upper intake manifold. **Caution:** *Don't let gasket material fall into the manifold.*

41 Place a new gasket on the manifold, then install the throttle body and tighten the screws to the torque listed in this Chapter's Specifications. The remainder of installation is the reverse of the removal steps.

Camshaft position sensor

42 Refer to Chapter 6 for procedures to check and replace the camshaft position sensor.

Throttle Position Sensor (TPS)

43 Refer to Chapter 6 for procedures to check and replace the Throttle Position Sensor (TPS).

Idle Air Control Solenoid (IAC)

44 Refer to Chapter 6 for procedures to check and replace the Intake Air Control solenoid (IAC).

Intake Air Temperature sensor

45 Refer to Chapter 6 for procedures to check and replace the Intake Air Temperature sensor (IAT).

Mass Air Flow (MAF) sensor

46 Refer to Chapter 6 for procedures to check and replace the Mass Air Flow sensor (MAF).

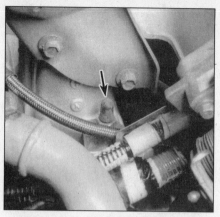

12.31c Note the cap over the front stud (arrow) - remove it before unscrewing the nut

Fuel rail and injectors

Refer to illustrations 12.53a, 12.53b, 12.53c, 12.53d and 12.54

Removal

47 If the engine is very dirty, it may be a good idea to steam clean the engine first to prevent dirt from contaminating exposed fittings and fuel metering orifices.

48 Relieve the fuel system pressure (see Section 2).

49 Disconnect the negative cable from the battery.

50 Remove the air inlet tube the runs between the air cleaner and throttle body.

51 Remove the upper intake manifold as described above.

52 Disconnect the fuel supply and return lines (see Section 3).

53 Remove the bolts that secure the fuel rail (see illustrations). Carefully lift the fuel rail and injectors out with a rocking, pulling motion.

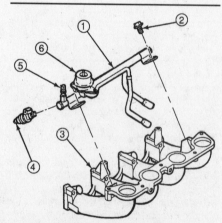

12.53a Fuel rail mounting details - 2.3L four-cylinder engine

1 Fuel rail
2 Bolt
3 Lower intake manifold
4 Fuel injector
5 Fuel pressure test port
6 Fuel pressure regulator

Chapter 4 Fuel and exhaust systems

4-11

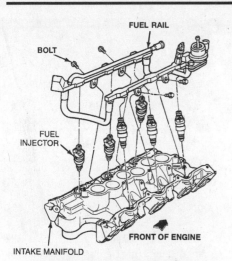

12.53b Fuel rail mounting details - 3.0L V6 engine

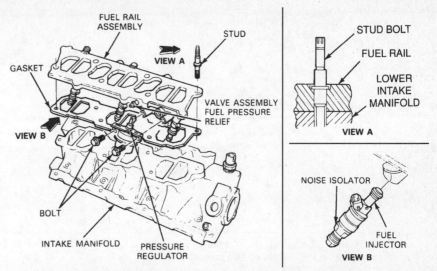

12.53d Details of the fuel rail and injectors (4.0L V6 engine)

12.53c Remove the bolts that secure the fuel rail to the lower intake manifold (4.0L V6 engine)

54 The injectors may come out with the fuel rail or remain attached to the lower intake manifold. Detach the injectors as follows:

a) Remove the injector retaining clip(s), if equipped.
b) Carefully pull the injector out with a rocking motion (see illustration).

Installation

Refer to illustration 12.55

55 Check the injector O-rings for damage or deterioration and replace as needed **(see illustration)**.
56 Inspect the plastic "hat" that covers the end of each injector (if it's not on the injector, it may be in the intake manifold). Replace the hat if damaged or deteriorated.
57 Lubricate the injector O-rings with a light film of engine oil. Each injector uses two O-rings. **Caution:** *Don't lubricate the injector O-rings with silicone grease. It will clog them.*
58 Install the injectors in the fuel rail with a light rocking motion.
59 The remainder of installation is the reverse of the removal steps.
60 Turn the ignition key On and Off several times (without starting the engine) to pressurize the fuel system. Check all fuel system connections for leaks.

Fuel pressure regulator (1997 and earlier models)

Check

Note 1: *This procedure assumes that the fuel filter is in good condition. If the filter is dirty, fuel pressure readings may not be accurate.*
Note 2: *On 1998 and later models, the fuel pressure regulator is located in the fuel tank along with the fuel pump. The system is equipped with a fuel pressure damper on the fuel rail. The damper is similar in appearance to a fuel pressure regulator, but the damper's only purpose is to reduce fuel pulsations in the fuel rail. The fuel pressure regulator is not serviced separately and if defective, must be replaced along with the fuel pump assembly.*
61 Check fuel pressure as described in Section 4. If fuel pressure is not within specifications, replace the fuel pressure regulator.

Replacement

Refer to illustrations 12.68, 12.69 and 12.70

62 Disconnect the negative cable from the battery.
63 Remove the splash shield from the throttle linkage.
64 Remove the air inlet tube that connects the air cleaner to the throttle body.
65 Relieve fuel system pressure (see Section 2).
66 Disconnect the vacuum line from the

12.54 Detach the fuel injectors with a rocking, pulling motion and take them out

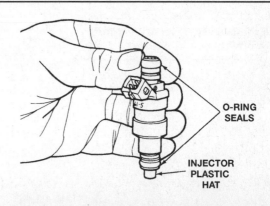

12.55 Each injector uses two O-rings - the injector pintle is protected by a plastic "hat"

12.68 On 2.3L and 3.0L engines, the fuel pressure regulator is held to the fuel rail by two screws

12.69 On 4.0L engines, the fuel pressure regulator is secured by two bolts (arrows)

pressure regulator.

67 Disconnect the fuel return line from the pressure regulator (see Section 3).

68 On 2.3L and 3.0L engines, remove the regulator mounting screws and the regulator **(see illustration)**.

69 On 4.0L engines, remove the regulator mounting bolts and the regulator **(see illustration)**.

70 Remove the O-ring(s) and washer **(see illustration)**.

71 Check the O-ring(s) for cracks or deterioration and replace as needed. Replace the washer whenever it's removed.

72 Check the fuel return line for kinks or worn spots and replace as needed.

73 Lubricate the O-ring(s) with a light coat of engine oil.

74 The remainder of installation is the reverse of the removal steps, with the following additions:

a) Tighten the pressure regulator fasteners to the torque listed in this Chapter's Specifications.

b) To connect the fuel line, push the ends together until the shoulder on the male end touches the fitting on the female end. Install the retainer and snap its two halves together.

13 Exhaust system service - general information

Warning: *Inspection and repair of exhaust system components should be done only after enough time has elapsed after driving the vehicle to allow the system components to cool completely. Also, when working under the vehicle, make sure it is securely supported on jackstands.*

1 The exhaust system consists of the exhaust manifolds, the catalytic converter, the mufflers, the tailpipe and all connecting pipes, brackets, hangers and clamps. The exhaust system is attached to the body with mounting brackets and rubber hangers. If any of these components are improperly installed, excessive noise and vibration will be transmitted to the body.

2 Conduct regular inspections of the exhaust system to keep it safe and quiet. Look for any damaged or bent parts, open seams, holes, loose connections, excessive corrosion or other defects which could allow exhaust fumes to enter the vehicle. Deteriorated exhaust system components should not be repaired; they should be replaced with new parts.

3 If the exhaust system components are extremely corroded or rusted together, welding equipment will probably be required to remove them. The convenient way to accomplish this is to have a muffler repair shop remove the corroded sections with a cutting torch. If, however, you want to save money by doing it yourself (and you don't have a welding outfit with a cutting torch), simply cut off the old components with a hacksaw. If you have compressed air, special pneumatic cutting chisels can also be used. If you do decide to tackle the job at home, be sure to wear safety goggles to protect your eyes from metal chips and work gloves to protect your hands.

4 Here are some simple guidelines to follow when repairing the exhaust system:

a) Work from the back to the front when removing exhaust system components.

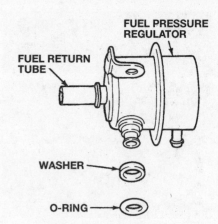

12.70 Remove the O-ring(s) and washer (4.0L shown)

b) Apply penetrating oil to the exhaust system component fasteners to make them easier to remove.

c) Use new gaskets, hangers and clamps when installing exhaust systems components.

d) Apply anti-seize compound to the threads of all exhaust system fasteners during reassembly.

e) Be sure to allow sufficient clearance between newly installed parts and all points on the underbody to avoid overheating the floor pan and possibly damaging the interior carpet and insulation. Pay particularly close attention to the catalytic converter and heat shield.

Chapter 5
Engine electrical systems

Contents

	Section		Section
Alternator - removal and installation	16	General information	1
Battery - emergency jump starting	3	Ignition coil - check, removal and installation	7
Battery - removal and installation	2	Ignition module - removal and installation	12
Battery cables - check and replacement	4	Ignition system - check	6
Battery check and maintenance	See Chapter 1	Ignition system - general information	5
Camshaft position sensor - removal and installation	9	Ignition timing procedure	13
Charging system - check	15	Spark plug replacement	See Chapter 1
Charging system - general information and precautions	14	Spark plug wire replacement	See Chapter 1
CHECK ENGINE light on	See Chapter 6	Starter motor - removal and installation	20
Crankshaft position sensor - removal and installation	8	Starter motor and circuit - in-vehicle check	19
Distributor - removal and installation	10	Starter relay - removal and installation	21
Distributor stator assembly - check and replacement	11	Starting system - general information	18
Drivebelt check, adjustment and replacement	See Chapter 1	Voltage regulator/alternator brushes - replacement	17

Specifications

Ignition system
EDIS coil secondary resistance .... 13 to 15 K-ohms

Battery voltage
Engine off 12-volts
Engine running .... 14 to 15-volts

Alternator brush length
New 1/2-inch
Minimum 1/4-inch

Torque specifications
Alternator mounting bolts .... 30 to 40 ft-lbs
Distributor hold-down bolt .... 17 to 25 ft-lbs
Starter mounting bolts .... 15 to 20 ft-lbs

1 General information

The engine electrical systems include all ignition, charging and starting components. Because of their engine-related functions, these components are considered separately from chassis electrical devices like the lights, instruments, etc.

Be very careful when working on the engine electrical components. They are easily damaged if checked, connected or handled improperly. The alternator is driven by an engine drivebelt which could cause serious injury if your hands, hair or clothes become entangled in it with the engine running. Both the starter and alternator are connected directly to the battery and could arc or even cause a fire if mishandled, overloaded or shorted out.

Never leave the ignition switch on for long periods of time with the engine off. Don't disconnect the battery cables while the engine is running. Correct polarity must be maintained when connecting battery cables from another source, such as another vehicle, during jump starting. Always disconnect the negative cable first and hook it up last or the battery may be shorted by the tool being used to loosen the cable clamps.

Additional safety related information on the engine electrical systems can be found in *Safety first* near the front of this manual. It should be referred to before beginning any operation included in this Chapter.

Chapter 5 Engine electrical systems

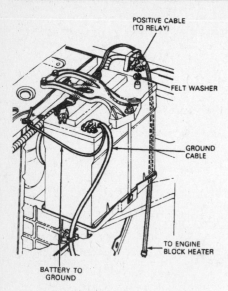

2.1 Disconnect the negative cable first, then the positive cable. Remove the battery hold-down clamps and remove the battery

2 Battery - removal and installation

Refer to illustrations 2.1 and 2.4

1 Disconnect both cables from the battery terminals **(see illustration)**. **Caution:** *Always disconnect the negative cable first and hook it up last or sparks may occur that can cause the battery to explode.*
2 Remove the battery heat shield (if equipped).
3 Locate the battery hold-down clamp at the base of the battery. Remove the bolt and the hold-down clamp.
4 Lift out the battery. Special straps or clamps that attach to the battery are available - lifting and moving the battery is much easier if you use one **(see illustration)**.
5 Installation is the reverse of removal.

6.1 To use a calibrated ignition tester (available at most auto parts stores), simply disconnect a spark plug wire, attach the wire to the tester, clip the tester to a convenient ground and operate the starter. If there is enough power to fire the plug, sparks will be visible in the tester

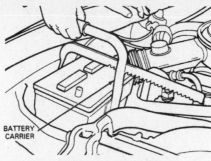

2.4 A battery carrier simplifies battery removal

Make sure the battery cables and battery posts are free of corrosion. Clean them if necessary (see Section 4).

3 Battery - emergency jump starting

Refer to the *Booster battery (jump) starting* procedure at the front of this manual.

4 Battery cables - check and replacement

1 Periodically inspect the entire length of each battery cable for damage, cracked or burned insulation and corrosion. Poor battery cable connections can cause starting problems and decreased engine performance.
2 Check the cable-to-terminal connections at the ends of the cables for cracks, loose wire strands and corrosion. The presence of white, fluffy deposits under the insulation at the cable terminal connection is a sign that the cable is corroded and should be replaced. Check the terminals for distortion, missing mounting bolts and corrosion.
3 When replacing the cables, always disconnect the negative cable first and hook it up last or the battery may be shorted by the tool used to loosen the cable clamps. Even if only the positive cable is being replaced, be sure to disconnect the negative cable from the battery first.
4 Disconnect and remove the cable. Make sure the replacement cable is the same length and wire size (diameter).
5 Clean the threads of the starter relay or ground connection with a wire brush to remove rust and corrosion. Apply a light coat of petroleum jelly to the threads to prevent future corrosion.
6 Attach the cable to the starter relay or ground connection and tighten the mounting nut/bolt securely.
7 Before connecting the new cable to the battery, make sure that it reaches the battery post without having to be stretched. Clean the battery posts and cable ends thoroughly and apply a light coat of petroleum jelly to prevent corrosion (see Chapter 1).
8 Connect the positive cable first, followed by the negative cable.

5 Ignition system - general information

All four-cylinder engines, 4.0L engines, and 1995 and later 3.0L engines are equipped with electronic distributorless ignition systems that have no moving parts. This system includes a crankshaft position sensor (CKP) which provides base timing information and cylinder identification to the Powertrain Control Module (PCM). The PCM determines when to fire the spark plugs based on the CKP information, engine speed, load, temperature and other sensor data. The Ignition Control Module, controlled by the PCM, fires spark plugs for two cylinders at the same time. One cylinder is fired on its compression stroke while the other is fired on its exhaust stroke.

There are some variations within this basic design. 1993 and 1994 models with a 2.3L engine use a dual vane actuator mounted on the crankshaft pulley to provide information to the crankshaft position sensor (CKP). However, 1995 and later four-cylinder engines, 3.0L engines and all 4.0L engines use a 36 tooth pulley on the crankshaft. A missing tooth on this pulley provides information about basic timing and cylinder identification. Another important difference is that four-cylinder engines have twin spark plugs, i.e. two spark plugs per cylinder, and it has two ignition coil packs.

1993 and 1994 3.0L models are equipped with a traditional distributor solid state ignition. The distributor is driven by the camshaft, and does not have either centrifugal or vacuum advance. As the distributor rotates, a Hall Effect vane switch signals the PCM and Ignition Control Module to turn the ignition coil on and off. The PCM determines the desired spark advance electronically. High voltage from the ignition coil is distributed to the spark plugs by a conventional rotor and distributor cap.

6 Ignition system - check

Warning: *Because of the high voltage generated by the ignition system, extreme care should be taken whenever an operation is performed involving ignition components. This not only includes the igniter (electronic ignition), coil, distributor and spark plug wires, but related components such as spark plug connectors, tachometer and other test equipment.*

Distributor ignition system (1993 and 1994 3.0L V6 engines)

Refer to illustrations 6.1 and 6.6

1 If the engine will not start even though it turns over, check for spark at the spark plug by installing a calibrated ignition system tester to one of the spark plug wires **(see illustration)**. **Note:** *The tool is available at most auto parts stores. Be sure to get a tester designed for breakerless type (electronic ignition).*

Chapter 5 Engine electrical systems

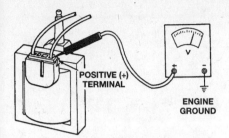

6.6 Check for battery voltage to the coil with the ignition key on

2 Connect the clip on the tester to a ground such as a metal bracket or valve cover bolt, crank the engine and watch the end of the tester for bright blue, well defined sparks.
3 If sparks occur, sufficient voltage is reaching the plugs to fire the engine. However the plugs themselves may be fouled, so remove and check them as described in Chapter 1 or replace them with new ones.
4 If no sparks occur, detach the coil secondary wire from the distributor cap, connect the tester to the coil wire and repeat the test.
5 If sparks occur, the distributor cap, rotor, spark plug wires or spark plugs may be defective. Remove the distributor cap and check the cap, rotor and spark plug wires as described in Chapter 1. Replace defective parts as necessary. If moisture is present in the distributor cap, use contact cleaner or something similar to dry out the cap and rotor, then reinstall the cap and repeat the spark test at the spark plug wire.
6 If no sparks occur at the coil wire, check the primary wire connections at the coil to make sure they are clean and tight. Check for voltage to the ignition coil **(see illustration)**.
7 The coil-to-cap wire may be bad, check the resistance with an ohmmeter and compare it to the Specifications. Make any necessary repairs and repeat the test.
8 If there's still no spark, the ignition coil, module or other internal components may be defective.

Electronic distributorless ignition system

Refer to illustration 6.12

9 If the engine turns over but won't start, disconnect the spark plug lead from any spark plug and attach it to a calibrated ignition tester (available at most auto parts stores). Make sure the tester is designed for Ford ignition systems if a universal tester isn't available.
10 Connect the clip on the tester to a bolt or metal bracket on the engine **(see illustration 6.1)**, crank the engine and watch the end of the tester to see if bright blue, well-defined sparks occur.
11 If sparks occur, sufficient voltage is reaching the spark plug to fire it (repeat the check at the remaining plug wires to verify that all the ignition coils and wires are functioning). However, the plugs themselves may

6.12 Primary wiring to the ignition coil on an electronic distributorless ignition system

be fouled, so remove and check them as described in Chapter 1 or install new ones.
12 If no spark occurs or the spark is intermittent, check for battery voltage to the ignition coil **(see illustration)**. Check for a bad spark plug wire by swapping wires. Check the coils and EDIS ignition module. If the checks fail to identify the problem, have the system diagnosed at a dealer service department or other properly equipped repair facility.

Alternative method (all ignition systems)

Note: *If you're unable to obtain a calibrated ignition tester, the following method will allow you to determine if the ignition system has spark, but it won't tell you if there's enough voltage produced to actually initiate combustion in the cylinders.*

13 Remove the wire from one of the spark plugs. Using an insulated tool, hold the wire about 1/4-inch from a good ground and have an assistant crank the engine.
14 If bright blue, well-defined sparks occur, sufficient voltage is reaching the plug to fire it. However, the plug(s) may be fouled, so remove and check them as described in Chapter 1 or install new ones.
15 If there's no spark, check the remaining wires in the same manner. A few sparks followed by no spark is the same condition as no spark at all.
16 If no sparks occur, remove the distributor cap and check the cap and rotor as described in Chapter 1. If moisture is present, dry out the cap and rotor, then reinstall the cap and rotor and repeat the spark test.
17 If there's still no spark, disconnect the secondary coil wire from the distributor cap, hold it about 1/4-inch from a good engine ground with an insulated tool and crank the engine again.
18 If no sparks occur, check the primary (small) wire connections at the coil to make sure they're clean and tight.
19 If sparks now occur, the distributor cap and rotor (3.0L engines only), plug wire(s) and spark plug(s) may be defective.
20 If there's still no spark, the coil-to-cap wire may be bad (check the resistance with an ohmmeter and compare it to the Specifications). If a known good wire doesn't make any difference in the test results, the ignition coil, module or other internal components may be defective. Refer further testing to a Ford dealer or qualified electrical specialist.

7 Ignition coil - check, removal and installation

Distributor ignition system (1993 and 1994 3.0L V6 engines)

Check

Primary and secondary coil resistance

Refer to illustrations 7.1 and 7.2

Caution: *If the coil terminals touch a ground source, the coil and/or stator (pick-up coil) could be damaged.*

1 With the ignition off, disconnect the wires from the coil. Connect an ohmmeter across the coil primary (small wire) terminals **(see illustration)**. Ford does not specify a resistance, but it should be less than 5.0 ohms. If the resistance is very high or open circuit, the coil may be defective. Take it to your dealer or other competent repair shop for testing before replacing it. They can bench test it for you, often for nothing.
2 Connect an ohmmeter between the negative primary terminal and the secondary terminal **(see illustration)** (the one that the distributor cap wire connects to). Ford does not

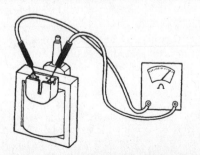

7.1 Checking primary coil resistance on a distributor system (1993 and 1994 3.0L V6 engines)

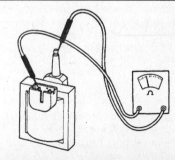

7.2 Checking secondary coil resistance on a distributor ignition system (1993 and 1994 3.0L V6 engines)

7.8 To check the primary resistance of a distributorless coil, connect the probes to the center terminal and outer terminal of the coil. Check all coil packs this way

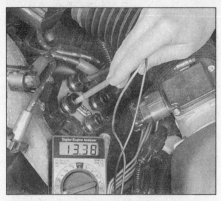

7.9 Check the coil secondary resistance by probing the paired cylinders

7.11 Disconnect the electrical connector from the coil pack

specify a resistance, but it should be relatively high, around 10 to 15K. If it is significantly different or open circuit, the coil may be defective. Take it to your dealer or other competent repair shop for testing before replacing it. They can bench test it for you, often for nothing.

Replacement

3   Detach the cable from the negative terminal of the battery.
4   Detach the connector for the primary coil.
5   Unplug the coil secondary lead.
6   Remove both bracket bolts and detach the coil.
7   Installation is the reverse of removal.

Electronic distributorless ignition system

Check

Refer to illustrations 7.8 and 7.9

8   With the ignition off, disconnect the electrical connector from the coil. Connect an ohmmeter across the coil primary (center terminal) terminal and the outer terminal **(see illustration)**. Ford does not specify a resistance, but it should be less than 5.0 ohms. If the resistance is very high or open circuit, the coil may be defective. Take it to your dealer or other competent repair shop for testing before replacing it. They can bench test it for you, often for nothing.
9   Connect an ohmmeter between the secondary terminals of the coil **(see illustration)**. The resistance should be as listed in this Chapter's Specifications. Remember that cylinders are paired, i.e. two will fire at the same time. On four-cylinder engines 1/4 and 2/3 are paired. On V6 engines, 1/5, 2/6 and 3/4 are paired.

Replacement

Refer to illustrations 7.11 and 7.12

10   Disconnect the negative cable from the battery.
11   Disconnect the ignition coil electrical connector(s) from each individual coil pack **(see illustration)**.

7.12 Squeeze the locking tabs to release the spark plug wires

12   Disconnect the ignition wires by squeezing the locking tabs and twisting while pulling **(see illustration)**. DO NOT just pull on the wires to disconnect them.
13   Remove the bolts securing the ignition coil to the mounting bracket on the engine.
14   Installation is the reverse of the removal procedure with the following additions:

a) *Prior to installing the spark plug lead into the ignition coil, coat the entire interior of the rubber boot with Silicone Dielectric Compound.*
b) *Insert each spark plug wire into the proper terminal of the ignition coil. Push the wire into the terminal and make sure the boots are fully seated and both locking tabs are engaged properly.*

8   Crankshaft position sensor - removal and installation

Refer to Chapter 6, Section 5 for information on the crankshaft position sensor.

9   Camshaft position sensor - removal and installation

Refer to Chapter 6, Section 5 for information on the camshaft position sensor.

10.3a Before removing the distributor, remove the distributor cap and mark the direction in which the rotor is pointing by scribing or painting an alignment mark on the edge of the distributor base, immediately below the rotor tip

10   Distributor - removal and installation

Note: *This procedure applies only to 1993 and 1994 3.0L V6 engines.*

Removal

Refer to illustrations 10.3a and 10.3b

1   Unplug the primary lead from the coil.
2   Unplug the electrical connector for the module. Follow the wires as they exit the distributor to find the connector.
3   Make a mark on the edge of the distributor base directly below the rotor tip and in line with it **(see illustration)**. Mark the distributor base and the engine block to ensure that the distributor will be installed correctly **(see illustration)**.
4   Remove the distributor hold-down bolt and clamp, then pull the distributor straight up to remove it. Be careful not to disturb the intermediate driveshaft. **Note:** *If the crankshaft is turned while the distributor is removed, or it a new distributor is required, the alignment marks will be useless.*

10.3b Paint or scribe an alignment mark on the distributor base and the block to ensure proper reinstallation, then remove the distributor hold down bolt/clamp and lift out the distributor assembly

11.7 With the distributor housing locked securely in a vise lined with shop rags or wood blocks to protect the housing, drive out the roll pin

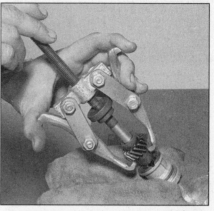

11.8 With the distributor shaft pointing up, use a small puller to separate the drive gear from the shaft

11.9 As soon as you remove the distributor shaft, note how the washer is installed before removing it because it is easily lost

11.12 If the O-ring at the base of the distributor is worn or damaged, replace it with a new one

Installation

5 Insert the distributor into the engine in exactly the same relationship to the block that it was in when removed.;

6 To mesh the helical gears on the camshaft and the distributor, it may be necessary to turn the rotor slightly. If the distributor doesn't seat completely, the hex shaped recess in the lower end of the distributor shaft is not mating properly with the oil pump shaft. Recheck the alignment marks between the distributor base and the block to verify that the distributor is in the same position it was in before removal. Also check the rotor to see if it's aligned with the mark you made on the edge of the distributor base. **Note:** *If the crankshaft has been moved while the distributor is out, locate Top Dead Center (TDC) for the number one piston (see Chapter 2, Part D) and position the distributor and rotor accordingly.*

7 Place the hold-down clamp in position and loosely install the bolt.

8 Install the distributor cap and tighten the cap screws securely.

9 Plug in the module electrical connector.

10 Reattach the spark plug wires to the plugs (if removed).

11 Connect the cable to the negative terminal of the battery.

12 Check and, if necessary, adjust the ignition timing (see Chapter 1) and tighten the distributor hold-down bolt securely.

11 Distributor stator assembly - check and replacement

Note: *This procedure applies only to 1993 and 1994 3.0L V6 engines.*

Check

1 The stator switch cannot be tested. If the ignition module is good, but the system won't function normally, replace the stator assembly.

Replacement

Refer to illustrations 11.7, 11.8, 11.9, 11.12, 11.17a and 11.17b

2 Remove the distributor cap and position it out of the way with the wires attached.

3 Remove the distributor (see Section 10).

4 Remove the rotor.

5 Clamp the gear in a vise using wood blocks to protect the teeth. Remove the two screws holding the rotary vane (armature) and remove the rotary vane. **Caution:** *Do not clamp the rotary vane to remove the screws. Remove the distributor from the vise.*

6 Mark orientation of the gear on the shaft with a felt tip pen.

7 Have an assistant hold the distributor steady in a vise, and drive the roll pin out of the shaft with a pin punch **(see illustration)**.

8 Remove the gear with a small puller **(see illustration)**.

9 Remove the thrust washer **(see illustration)**.

10 Deburr the shaft and polish it lightly with emery paper or crocus cloth so that the shaft will slide out freely from the distributor base, then remove the shaft.

11 Remove two screws securing the stator assembly, then remove the assembly.

12 Inspect the base O-ring for damage. Replace it if necessary **(see illustration)**.

13 Check the distributor base for excessive wear, cracks and other damage. Make sure the bushing shows no evidence of wear, galling or burning. Replace the entire distributor if there is damage.

14 Place the stator assembly in position over the shaft bushing and press is down onto the distributor base until it is completely seated on the posts. Place the stator connector in position. The tab should fit in the notch on the base and the eyelets should align with the screws holes. Install the two screws securing the stator assembly.

15 Apply a light coat of clean motor oil to the shaft and insert the shaft through the base bushing.

16 Mount the distributor in a vise with the lower end up. Line the jaws with rags to protect the distributor base. Place a block of wood under the distributor shaft to keep it from falling out while installing the gear.

17 Align the marks that you made on the gear and the shaft as accurately as possible. Using a deep socket and a brass or plastic mallet, carefully tap the drive gear back onto

the distributor shaft (see illustration). Make sure the roll pin hole in the drive gear an the corresponding hole in the shaft are lined up.
Caution: *If the holes do not align perfectly when the gear is fully installed, you must not try to align them with a punch. You must remove the gear and try again* (see illustration). When the holes align, insert the roll pin. Make sure neither end protrudes from the gear.
18 Install the rotary vane (armature) and secure it with two screws.
19 Turn the shaft, and look for binding. The shaft should turn freely. If the rotary vane contacts the stator at any point, replace the entire distributor.
20 Install the distributor.
21 Connect the wiring harness to the distributor.
22 Install the rotor and distributor cap.
23 Set the initial ignition timing (see Section 13)

12 Ignition module - check, removal and installation

Warning: *The ignition module is a delicate and relatively expensive electronic component. Failure to follow the step-by-step procedures could result in damage to the module and/or other electronic devices, including the EEC-IV microprocessor itself. Additionally, all devices under computer control are protected by a Federally mandated extended warranty. Check with your dealer concerning this warranty before attempting to diagnose and replace the module yourself.*
Caution: *Use an LED test light when testing the ignition module or any other electronic circuit. Do not use an incandescent test light.*

Distributor ignition module (1993 and 1994 3.0L V6 engine only)

Ignition coil primary circuit
1 Unplug the ignition coil electrical connectors and inspect them for dirt, corrosion and damage.
2 Attach an LED test light to the battery positive (+) terminal and to the coil driver output on the coil electrical connector and crank the engine.
3 The test light should flash with each output signal from the ignition control module through the coil primary circuit.

Ignition module checks
Refer to illustration 12.5
4 Reconnect the coil electrical connector.
5 Check for power to the ignition module. Turn the ignition switch ON (engine not running). Using a voltmeter or LED test light, probe terminal number 3 for battery voltage from the ignition switch (see illustration).
6 Check the SPOUT signal. Backprobe either terminal on the SPOUT connector with

11.17a After securing the distributor assembly upside down in a vise, "eyeball" the roll pin holes in the drive gear and shaft, then tap the drive gear onto the shaft with a deep socket and mallet

11.17b If the drive gear and shaft roll pin hole are misaligned, the roll pin cannot be driven through the drive gear and shaft holes. This drive gear must now be pulled off the shaft and realigned

the LED test light, and crank the engine. The light should blink if the SPOUT signal is present. If not check the wiring, and if necessary replace the ignition control module.
7 Check the PIP signal. Probe the PIP signal at the distributor connector (GY/O wire) with the LED test light and crank the engine. The light should blink if the PIP signal is present. If not, check the wiring and if necessary replace the distributor stator.

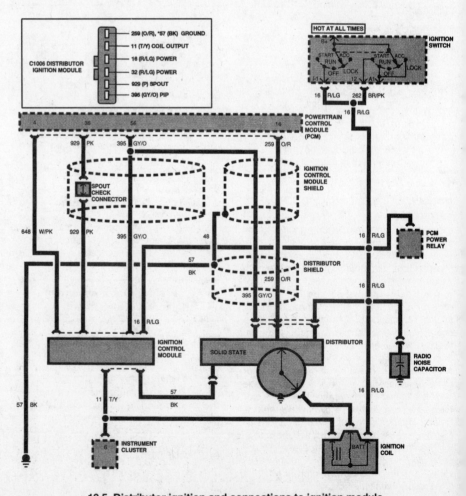

12.5 Distributor ignition and connections to ignition module (1993 and 1994 3.0L V6 engine)

Chapter 5 Engine electrical systems

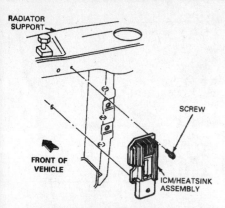

12.9 Mounting details for the ignition module (3.0L distributor models)

Replacement
Refer to illustration 12.9

8 Disconnect the negative cable from the battery.
9 Remove the two screws securing the ignition control module to the radiator support **(see illustration)**.
10 Disconnect the electrical connector.
11 Remove two screws securing the ignition control module to the heatsink.
12 Installation is the reverse of the removal procedure. Before attaching the ignition control module to the heatsink, evenly coat the baseplate of the module with Silicone Dielectric Compound WA-10 or equivalent.

EDIS ignition module (1993 and 1994 2.3L four-cylinder and 4.0L V6 engines)
Refer to illustrations 12.14a and 12.14b

Ignition coil primary circuit
13 Unplug the ignition coil wiring harness connectors and inspect them for dirt, corrosion and damage.
14 Attach an LED test light to the battery positive (+) terminal and to the coil driver outputs on the coil electrical connector and crank the engine.
   a) On four-cylinder engines, check pins 8, 9, 11 and 12 **(see illustration)**.
   b) On V6 engines, check pins 10, 11 and 12 **(see illustration)**.
15 The test light should flash with each output signal from the EDIS ignition module through the coil primary circuit.

Ignition module checks
16 Check for power to the ignition module. Using a voltmeter, check for battery voltage through the power relay **(see illustrations 12.14a and 12.14b)**. With the ignition ON (engine not running), there should be battery voltage.
   a) On four-cylinder engines, check pin 1.
   b) On V6 engines, check pin 8.
17 Check for power to the ignition module. Turn the ignition switch ON (engine not running). Using a voltmeter or LED test light, probe terminal number 1 on four-cylinder engines and terminal number 8 on V6 engines from the ignition switch.
18 Check the SPOUT signal. Backprobe either terminal on the SPOUT connector with the LED test light, and crank the engine. The light should blink if the SPOUT signal is present. If not check the wiring, and if necessary replace the ignition control module.
19 Check the PIP signal. Probe the PIP signal at the PIP wire (GY/O wire) with the LED test light and crank the engine. The light should blink if the PIP signal is present. If not, check the wiring and if necessary replace the crankshaft position sensor.

Replacement
20 Disconnect the negative cable from the battery.
21 Disconnect the electrical connector from the EDIS ignition module.
22 Remove the screws securing the EDIS module to the fenderwell.
23 Installation is the reverse of the removal procedure.

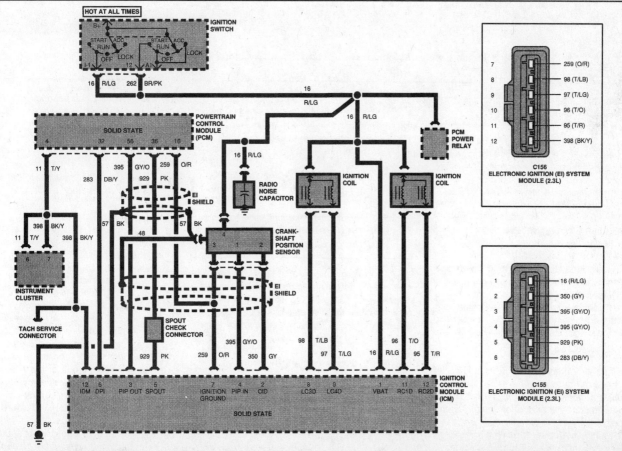

12.14a Schematic of the distributorless ignition system for the 1993 and 1994 2.3L four-cylinder engine

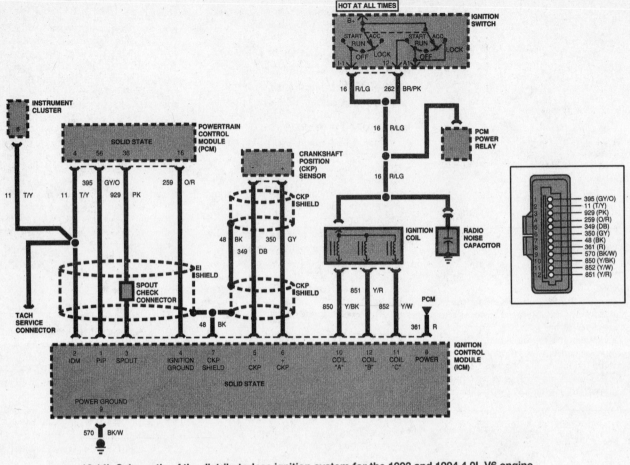

12.14b Schematic of the distributorless ignition system for the 1993 and 1994 4.0L V6 engine

13 Ignition timing procedure

Note: *This applies only to 1993 and 1994 3.0L engines with distributor ignition. Ignition timing adjustment is not possible on other vehicles covered by this manual.*

1 Put the transmission in Neutral or Park.
2 Make sure the air conditioner and heater are OFF.
3 Connect an inductive timing light according to the manufacturer's instructions. Connect a tachometer.
4 Loosen the distributor hold-down bolt enough to allow the distributor to be turned.
5 Disconnect the single wire in-line SPOUT connector or remove the shorting bar from the double wire SPOUT connector. The SPOUT connector is located in-line with the Pink wire going to the Ignition Control Module on the distributor.
6 Start the engine and allow it to warm up to normal operating temperature. **Note:** *Do not use a remote starter. Always use the ignition key to start the engine or ignition timing will not be correct.*
7 Run the engine at the timing rpm specified in this Chapter's specifications.
8 Check the timing with the timing light. If necessary, adjust the timing by rotating the distributor slightly.
9 Reconnect the in-line SPOUT or remove the shorting bar.
10 Rev the engine while watching the timing with the timing light and verify that the timing is advancing.
11 Stop the engine and tighten the distributor hold-down bolt. Remove the timing light.

14 Charging system - general information and precautions

The charging system includes the alternator with integral voltage regulator, an ammeter or charge indicator light, the battery, a fusible link and the wiring between all the components. The charging system supplies electrical power for the ignition system, the lights, the radio, etc. The alternator is driven by a drivebelt at the front of the engine.

The purpose of the integral solid-state voltage regulator is to limit the alternator's voltage to a preset value. This prevents power surges, circuit overloads, etc., during peak voltage output.

The charging system doesn't ordinarily require periodic maintenance. However, the drivebelt, battery, wires and connections should be inspected at the intervals outlined in Chapter 1.

Be very careful when making electrical circuit connections to the vehicle and note the following:

a) When reconnecting wires to the alternator from the battery, be sure to note the polarity.
b) Before using arc welding equipment to repair any part of the vehicle, disconnect the wires from the alternator and the battery terminals.
c) Never start the engine with a battery charger connected.
d) Always disconnect both battery leads before using a battery charger.
e) Never disconnect the battery when the engine is running.
f) The alternator is turned by the engine drivebelt which could cause serious injury if your hands or clothes become entangled in it with the engine running.
g) Because the alternator is connected directly to the battery, it could arc or cause a fire if overloaded or shorted out.
h) Wrap a plastic bag over the alternator and secure it with rubber bands before steam cleaning the engine.

Chapter 5 Engine electrical systems

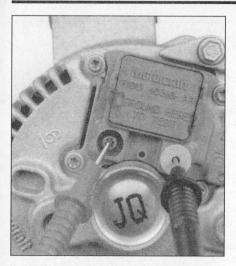

15.8 Connect an ohmmeter between the A and F terminal screws

15.9 Connect the voltmeter negative lead to the alternator housing and the positive lead to the regulator A terminal screw

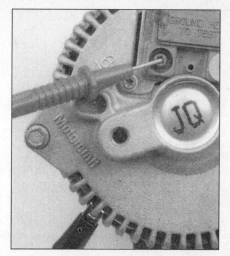

15.11 Connect the voltmeter negative lead to the alternator housing and the positive lead to the F terminal screw

15 Charging system - check

Note: *If the alternator requires service it must be replaced as an assembly. Both new and rebuilt units are readily available. Replacement parts are not available individually.*

1   If a malfunction occurs in the charging circuit, don't automatically assume that the alternator is causing the problem. First check the following items:

   a) *The battery cables where they connect to the battery. Make sure the connections are clean and tight (see Chapter 1).*
   b) *Check the external alternator wiring harness and the connectors at the alternator. They must be in good condition, clean, free of corrosion and tight.*
   c) *Check the drivebelt condition and tension (see Chapter 1).*
   d) *Make sure the alternator mounting and adjustment bolts are tight.*
   e) *Check the fusible link located between the starter relay and the alternator. If it's burned out, determine the cause, repair the circuit and replace the link (see Chapter 12).*
   f) *Run the engine and check the alternator for abnormal noise.*

2   Turn the headlights on for 10 to 15 seconds to remove any surface charge from the battery. Wait three to five minutes for the battery voltage to stabilize before continuing with Step 3.

3   Connect a voltmeter between the battery terminals and check the battery voltage with the engine off. It should be approximately 12-volts.

4   Run the engine at a fast idle (approximately 1500 rpm) and check the battery voltage again after the voltage stops rising. This may take a few minutes. It should now be higher than battery voltage, but not more than 3-volts higher.

5   If the voltage reading is less than the specified charging voltage, perform the Under-voltage test in this Section. If it is higher, perform the Over-voltage test.

6   Turn on the high beam headlights and turn the heater or air conditioner blower to its highest setting. Run the engine at 2,000 rpm and check the voltage reading. It should now be at least 0.5-volt higher than battery voltage. If not, perform the Under-voltage test.

7   If the voltage readings are correct in the preceding Steps, the charging system is working properly. Use a 12-volt test light and the wiring diagrams (see Chapter 12) to check for a battery drain.

Under-voltage test

Refer to illustrations 15.8, 15.9, 15.11, 15.13 and 15.14

8   Unplug the electrical connector from the voltage regulator. Connect an ohmmeter between the A and F terminal screws **(see illustration)**. The ohmmeter should indicate at least 2.4 ohms.

   a) *If the ohmmeter reading is too low, the regulator is defective.*
   b) *If the ohmmeter reading is within specifications, reconnect the electrical connector to the regulator and perform Step 9.*

9   Connect the voltmeter negative lead to the alternator rear housing and the positive lead to the regulator A terminal screw **(see illustration)**. The voltmeter should indicate battery voltage. If not, check the A circuit for breaks or bad connections.

10   Repeat the load test (see Step 6).

11   If the voltmeter indicates battery voltage in Step 9, place the ignition key in the Off position. Connect the voltmeter negative lead to the alternator frame and the positive lead to the F terminal screw **(see illustration)**.

   a) *If the voltmeter indicates no voltage, replace the alternator.*
   b) *If the voltmeter indicates battery voltage, proceed to Step 12.*

12   Turn the ignition key to the Run position, but don't start the engine. Touch the voltmeter negative lead to the rear of the alternator and the positive lead to the F terminal screw on the regulator.

   a) *If the voltmeter indicates more than two volts, perform the I circuit test in this Section.*
   b) *If the voltmeter indicates two volts or less, proceed to Step 13.*

13   Perform the load test (see Step 6) with the voltmeter positive terminal connected to the alternator output stud **(see illustration)**.

   a) *If voltage increases to more than 0.5-volt above battery voltage, check the wiring from alternator to starter relay for breaks or bad connections.*
   b) *If the voltage does not increase to more than 0.5 volt above battery voltage, perform Step 14.*

15.13 Perform the load test with the voltmeter positive lead connected to the alternator output stud

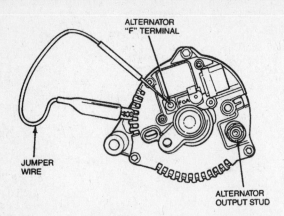

15.14 Using a jumper wire, ground the F terminal to the alternator frame, connect the positive lead of the voltmeter to the alternator output stud and perform the load test described in Step 6

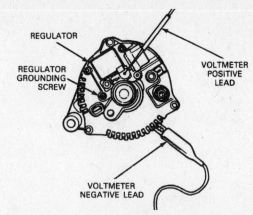

15.16 Connect the voltmeter negative lead to the alternator housing, then connect the positive lead to the A terminal screw and the regulator grounding screw in turn

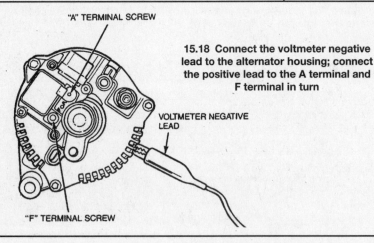

15.18 Connect the voltmeter negative lead to the alternator housing; connect the positive lead to the A terminal and F terminal in turn

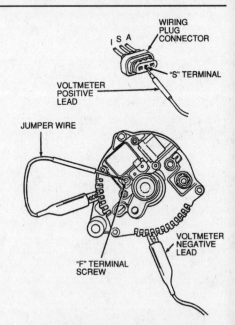

15.22 Connect the voltmeter negative lead to the alternator housing; connect the positive lead to the connector S terminal and A terminal in turn

14   Connect a jumper wire between the alternator rear housing and the regulator F terminal screw **(see illustration)**. Repeat the load test (see Step 6) with the voltmeter positive lead connected to one of the B+ jumper wire terminals.

a) If voltage increases by more than 0.5-volt, the regulator is defective and the alternator must be replaced.
b) If voltage doesn't increase by more than 0.5-volt, replace the alternator.

Over-voltage test

Refer to illustrations 15.16 and 15.18

15   Turn the key to the On position but leave the engine off.
16   Connect the voltmeter negative lead to the alternator rear housing and connect the positive lead to the alternator output connection at the starter relay, then to the regulator A terminal **(see illustration)**. If the voltage readings differ by more than 0.5-volt, check the A circuit for breaks or bad connections.
17   If Step 16 doesn't correct the problem, check for loose regulator grounding screws and tighten as needed.
18   If the voltage reading is still too high, place the ignition key in the Off position. Connect the voltmeter negative lead to the alternator frame. Connect the voltmeter positive lead to the A terminal screw, then to the F terminal screw **(see illustration)**.

a) If the voltage readings at the two screw heads are different, replace the alternator.
b) If the voltage readings at the two screw heads are the same, the regulator is defective and the alternator must be replaced.

S and/or I circuit test

Refer to illustration 15.22

19   Disconnect the alternator electrical connector.
20   Connect one jumper wire from the alternator A terminal to its corresponding terminal in the electrical connector. Connect another jumper wire from the regulator F screw to the alternator housing.
21   Start the engine and let it idle.
22   Connect the voltmeter negative lead to the alternator housing. Connect the positive lead to the connector S terminal and A terminal in turn **(see illustration)**. Voltage at the A terminal should be approximately double the reading at the S terminal.

a) If the readings are as specified, the regulator is defective and the alternator must be replaced.

b) If there is no voltage, check the wiring for breaks or bad connections. If the wiring is good, replace the alternator.

16   Alternator - removal and installation

Refer to illustration 16.6

1   Disconnect the negative cable from the battery.
2   Remove the air cleaner assembly (see Chapter 4)
3   Unplug the electrical connectors from the alternator.
4   Remove the electrical connector bracket (if equipped).

Chapter 5 Engine electrical systems

16.6 Location of the alternator mounting bolts (arrows)

17.3 To remove the voltage regulator/brush assembly, remove the four screws

17.4 Lift the assembly from the alternator

5 Remove the drivebelt (see Chapter 1).
6 Remove the three bolts holding the alternator to its mounting bracket **(see illustration)**.
7 Installation is the reverse of removal.

17 Voltage regulator/alternator brushes - replacement

Refer to illustrations 17.3, 17.4, 17.5 and 17.9

1 Remove the alternator (see Section 16).
2 Set the alternator on a clean workbench.
3 Remove the four voltage regulator mounting screws **(see illustration)**.
4 Detach the voltage regulator/brush holder assembly **(see illustration)**.
5 Detach the rubber plugs and remove the brush lead retaining screws and nuts to separate the brush leads from the holder **(see illustration)**. Note that the screws have Torx heads and require a special screwdriver.
6 After noting the relationship of the brushes to the brush holder assembly, remove both brushes. Don't lose the springs.
7 Install the brushes into the regulator. Make sure the springs are properly compressed and the brushes are properly inserted into the recesses in the brush holder.
8 Install the brush lead retaining screws and nuts.
9 Insert a short section of wire, like a paper clip, through the hole in the voltage regulator **(see illustration)** to hold the brushes in the retracted position during regulator installation.
10 Carefully install the regulator. Make sure the brushes don't hang up on the rotor.
11 Install the voltage regulator screws and tighten them securely.
12 Remove the wire or paper clip.
13 Install the alternator (see Section 16).

18 Starting system - general information

The system is composed of the starter motor, starter relay, battery, ignition switch, safety switch and connecting wires.

17.5 To remove the brushes from the voltage regulator/brush holder assembly, remove the rubber plugs from the two brush lead screws and remove both screws

Turning the ignition key to the Start position actuates the starter relay through the starter control circuit. The starter relay then connects the battery to the starter. The battery supplies the electrical energy to the starter motor, which does the actual work of cranking the engine.

Vehicles equipped with an automatic transmission have a Neutral start switch in the starter control circuit, which prevents operation of the starter unless the shift lever is in Neutral or Park. The circuit on vehicles with a manual transmission prevents operation of the starter motor unless the clutch pedal is depressed.

Never operate the starter motor for more than 15 seconds at a time without pausing to allow it to cool for at least two minutes. Excessive cranking can cause overheating, which can seriously damage the starter.

19 Starter motor and circuit - in-vehicle check

Note: *Before diagnosing starter problems,*

17.9 Before installing the voltage regulator/brush holder assembly, insert a paper clip as shown to hold the brushes in place during installation. After installation, remove the paper clip

make sure the battery is fully charged.
1 If the starter motor doesn't turn at all when the switch is operated, make sure the shift lever is in Neutral or Park (automatic transmission) or the clutch pedal is depressed (manual transmission).
2 Make sure the battery is fully charged and that all cables at the battery and starter relay terminals are not corroded and that they are secure.
3 If the starter motor spins but the engine doesn't turn over, then the drive assembly in the starter motor is slipping and the starter motor must be replaced (see Section 20).
4 If, when the switch is actuated, the starter motor doesn't operate at all but the starter relay operates (clicks), then the problem lies with either the battery, the starter relay contacts, the starter solenoid or the starter motor electrical connections.
5 If the starter relay doesn't click when the ignition switch is actuated, a discharged battery is the most likely cause. If the battery is fully charged, the starter relay circuit may be open or the relay itself could be defective. Check the starter relay circuit or replace the

20.3 The starter wires are connected to the terminals at the top of the starter

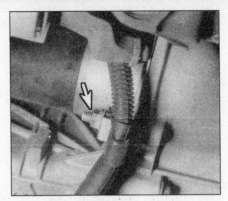

20.4 Remove the lower mounting nut

20.5 Remove the upper mounting bolt and remove the starter

21.1 The starter relay is located between the battery and the power distribution box in the engine compartment

relay (see Section 21).
6 To check the starter relay circuit, remove the push-on electrical connector from the relay (red with blue stripe). Make sure that the connection is clean and secure and the relay bracket is grounded. If the connections are good, check the operation of the relay with a jumper wire. To do this, place the transmission in Park (automatic) or Neutral (manual). Remove the push-on connector from the relay. Connect a jumper wire between the battery positive terminal and the relay main terminal (BAT side) on the starter relay. If the starter motor now operates, the starter relay is okay. The problem is in the ignition switch, Neutral start switch, starter/clutch switch, or in the starting circuit wiring (look for open or loose connections). Reconnect the push-on electrical connector.

7 If the starter motor still doesn't operate, check the starter relay contacts. To do this, momentarily jumper across the two large terminals with red wires at the starter relay. **Caution:** *Use a jumper with the same size wire or larger. If the starter operates, the starter relay is good, and the trouble must be with the starter solenoid; replace the starter. If the starter still does not operate, replace the starter relay* (see Section 21).
8 If the starter motor cranks the engine at an abnormally slow speed, first make sure the battery is fully charged and all terminal connections are clean and tight. Also check the connections at the starter relay and battery ground. Eyelet terminals should not be easily rotated by hand. Also check for a short to ground. If the engine is partially seized, or has the wrong (too heavy) viscosity oil in it, it will crank slowly.

20 Starter motor - removal and installation

Refer to illustrations 20.3, 20.4 and 20.5

1 Disconnect the negative cable from the battery.
2 Raise the vehicle and support it securely on jackstands.
3 Disconnect the starter cable and push-on connector from the starter solenoid terminals **(see illustration)**. **Caution:** *Disconnect the connector at the solenoid S terminal by grasping it and pulling straight off to prevent damage to the connector and terminal.*
4 Remove the starter lower mounting nut **(see illustration)**.
5 Remove the upper bolt. Pull the starter forward, lift the front end of it up to clear the bellhousing and remove the starter from the engine **(see illustration)**.
6 Installation is the reverse of the removal procedure. Tighten the mounting bolts to the torque listed in this Chapter's Specifications.

21 Starter relay - removal and installation

Refer to illustration 21.1

1 The starter relay is located in the engine compartment, mounted between the battery and the power distribution box **(see illustration)**.
2 Disconnect the negative cable from the battery.
3 Label the wires and the terminals on the relay to prevent mix-up during installation.
4 Disconnect the small wire from the starter relay.
5 Disconnect the positive battery cable and the feed cable from the starter relay.
6 Disconnect the starter feed cable from the other terminal on the starter relay.
7 Remove the mounting bolts and remove the relay.
8 Installation is the reverse of the removal procedure with the following additions:
   a) *Prior to installing the relay, clean the mounting surface with a wire brush or sandpaper to provide a good ground.*
   b) *Attach the relay to the inner fender panel and tighten the bolts securely.* **Caution:** *Do not overtighten the bolts - they are self-threading and can easily strip out the holes in the sheet metal.*
   c) *Be sure to connect the cables to the correct terminals on the relay. Refer to the labels made in Step 3.*

Chapter 6
Emissions control systems

Contents

Section		Section
Catalytic converter	10	Exhaust Gas Recirculation (EGR) system 6
CHECK ENGINE light .... See Section 2		General information 1
Diagnostic systems - general information 2		Information sensors 5
Electronic Engine Control (EEC-IV, EEC-V)		Inlet air temperature control system (1993-1994) 9
system and trouble codes 3		Positive Crankcase Ventilation (PCV) system 8
Evaporative Emissions Control System (EECS) 7		Powertrain Control Module (PCM) 4

1 General information

Refer to illustration 1.6

To prevent pollution of the atmosphere from incompletely-burned and evaporating gases, and to maintain good driveability and fuel economy, a number of emission control systems are incorporated. They include the:

   Electronic Engine Control system (EEC-IV or EEC-V)
   Evaporative Emission Control (EECS) system
   Positive Crankcase Ventilation (PCV) system
   Exhaust Gas Recirculation (EGR) system
   Catalytic converter

All of these systems are linked, directly or indirectly, to the emission control system.

The Sections in this Chapter include general descriptions, checking procedures within the scope of the home mechanic and component replacement procedures (when possible) for each of the systems listed above.

Before assuming that an emissions control system is malfunctioning, check the fuel and ignition systems carefully. The diagnosis of some emission-control devices requires specialized tools, equipment and training. If checking and servicing become too difficult or if a procedure is beyond your ability, consult a dealer service department. Remember, the most frequent cause of emissions problems is simply a loose or broken vacuum hose or wire, so always check the hose and wiring connections first.

This doesn't mean, however, that emission control systems are particularly difficult to maintain and repair. You can quickly and easily perform many checks and do most of the regular maintenance at home with common tune-up and hand tools. **Note:** *Because of a Federally mandated extended warranty which covers the emission control system components, check with your dealer about warranty coverage before working on any emissions-related systems. Once the warranty has expired, you may wish to perform some of the component checks and/or replacement procedures in this Chapter to save money.*

Pay close attention to any special precautions outlined in this Chapter. It should be noted that the illustrations of the various systems may not exactly match the system installed on the vehicle you're working on because of changes made by the manufacturer during production or from year-to-year.

A Vehicle Emissions Control Information (VECI) label is located in the engine compartment **(see illustration)**. This label contains important emissions specifications and adjustment information, as well as a vacuum hose schematic with emissions components identified. When servicing the engine or emissions systems, the VECI label in your particular vehicle should always be checked for up-to-date information.

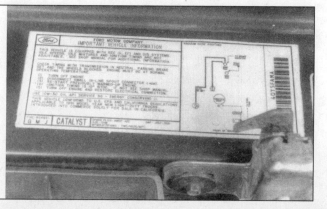

1.6 The Vehicle Emissions Control Information (VECI) label is located in the engine compartment on the radiator support. It contains important emissions specifications and adjustment information specific to your vehicle

Chapter 6 Emissions control systems

2 Diagnostic systems - general information

One of the first automotive applications of computers was self-diagnosis of system and component failures. While early on-board diagnostic computers simply lit a "CHECK ENGINE" light on the dash, present systems must monitor complex interactive emission control systems, and provide enough data to the technician to successfully isolate a malfunction.

The computer's role in self-diagnosing emission control problems has become so important that such computers are now required by Federal law. The requirements of the "first generation" system, later nicknamed OBD I, are incorporated in 1993 and 1994 models with EEC-IV emission control systems. OBD II, a "second generation" system is incorporated in 1995 and later models with EEC-V emission control systems.

The purpose of On Board Diagnostics (OBD) is to ensure that emission related components and systems are functioning properly to reduce emission levels of several pollutants emitted by auto and truck engines. The first step is to detect that a malfunction has occurred which may cause increased emissions. The next step is for the system to notify the driver so that the vehicle can be serviced. The final step is to store enough information about the malfunction so that it can be identified and repaired.

Many of the self-diagnostic tests performed by early OBD systems are retained in OBD II systems, however, they are more sensitive in detecting malfunctions. OBD II systems perform many additional tests not required under the earlier OBD. These include monitoring for engine misfires and detecting deterioration of the catalytic converter.

3 Electronic Engine Control (EEC-IV, EEC-V) system and trouble codes

Refer to illustrations 3.1, 3.20a and 3.20b

General description

1   1993 and 1994 models use the Electronic Engine Control (EEC-IV) system **(see illustration)**. 1995 and later models use the EEC-V system. Both systems consist of an onboard computer, known as the Powertrain Control Module (PCM), and information sensors, which monitor various functions of the engine and send data to the PCM. Based on the data and the information programmed into the computer's memory, the PCM generates output signals to control various engine functions via control relays, solenoids and other output actuators.

2   The PCM, located under the instrument panel, is the "brain" of the EEC system. It receives data from a number of sensors and other electronic components (switches, relays, etc.). Based on the information it receives, the PCM generates output signals to control various relays, solenoids and other actuators. The PCM is specifically calibrated to optimize the emissions, fuel economy and driveability of the vehicle.

3   Because of a Federally mandated extended warranty which covers the EEC system components and because any owner-induced damage to the PCM, the sensors and/or the control devices may void the warranty, it isn't a good idea to attempt diagnosis or replacement of the PCM at home while the vehicle is under warranty. Take the vehicle to a dealer service department if the PCM or a system component malfunctions.

Information sensors

Information sensors provide information to the PCM that would affect engine operation and emissions.

4   When the air-conditioning compressor clutch operates, a signal is sent to the PCM indicating that the compressor has added a load to the engine. The PCM responds by increasing engine idle speed accordingly to compensate.

5   The Intake Air Temperature sensor (IAT), threaded into a runner of the intake manifold (see Section 5), provides the PCM with fuel/air mixture temperature information. The PCM uses this information to control fuel flow, ignition timing and EGR system operation.

6   The Engine Coolant Temperature (ECT) sensor, which is threaded into a coolant passage in the thermostat housing (four-cylinder) or intake manifold (V6), monitors engine coolant temperature. The ECT sends the PCM a constantly varying voltage signal which influences PCM control of the fuel mixture, ignition timing and EGR operation.

7   The Heated Exhaust Gas Oxygen sensors (HEGO), which are threaded into the exhaust manifolds, provide a constant signal to the PCM indicating oxygen content of the exhaust gases as an indication of emissions level. The PCM converts this exhaust gas oxygen content signal to the fuel/air ratio, compares it to the ideal ratio for current engine operating conditions and alters the signal to the injectors accordingly.

8   The Throttle Position Sensor (TPS), which is mounted on the side of the throttle body (see Section 5) and connected directly to the throttle shaft, senses throttle movement and position, then transmits an electrical signal to the PCM that is proportional to throttle position.

9   The Mass Air Flow (MAF) sensor, which is mounted in the air cleaner intake passage, measures the mass of the air entering the engine (see Section 5). Because air mass varies with air temperature (cold air is denser than warm air), measuring air mass provides the PCM with a very accurate way of determining the correct amount of fuel to obtain the ideal fuel/air mixture.

Output devices

Output devices are used by the PCM to change engine operating conditions to minimize emissions.

10   The EEC power relay, which is activated by the ignition switch, supplies battery voltage to the EEC-IV system components when the switch is the Start or Run position.

11   The canister purge solenoid (CANP) switches manifold vacuum to operate the canister purge valve when a signal is received from the PCM. Vacuum opens the purge valve when the solenoid is energized allowing fuel vapor to flow from the canister to the intake manifold.

12   The solenoid-operated fuel injectors are located above the intake ports (see Chapter 4). The PCM controls the length of time the injector is open. The "open" time of the injector determines the amount of fuel delivered. For information regarding injector replacement, refer to Chapter 4.

13   The fuel pump relay is activated by the PCM with the ignition switch in the On position. When the ignition switch is turned to the On position, the relay is activated to supply initial line pressure to the system. For information regarding fuel pump check and replacement, refer to Chapter 4.

14   The PCM uses a signal from the Profile Ignition Pick-Up (PIP) to determine crankshaft position. Ignition timing is determined by the PCM, which then signals the module to fire the coil.

Obtaining codes

15   The diagnostic codes for the EEC systems are arranged in such a way that a series of tests must be completed in order to extract ALL the codes from the system. If one portion of the test is performed without the others, there may be a chance the trouble code that will pinpoint a problem in your particular vehicle will remain stored in the PCM without detection. The tests start first with a Key On, Engine Off (KOEO) test followed by a computed timing test then finally an Engine Running (ER) test.

16   Codes on EEC-IV systems (1993 and 1994 models) can be read with an analog voltmeter, by observing the "Check Engine" light on the dash or by using a scan tool if one is available. Codes on the EEC-V system (1995 and later models) can only be read using a scan tool.

Here is a brief overview of the code-extracting procedures of the EEC-IV system followed by the actual test:

Note: *Before attempting to repair or replace a component, always check vacuum lines and electrical connectors first, as a majority of problems are simple connections and leaking vacuum lines.*

Quick Test - Key On Engine Off (KOEO)

17   The following codes are generated with the key on, engine off "quick test":

Chapter 6 Emissions control systems

Control System Component Locator

3.1 Locations of the EEC information sensors and output actuators (1995 shown, others similar)

Self test codes - These codes are accessed on the test connector by using a jumper wire and an analog voltmeter or the factory diagnostic tool called the Star tester. These codes are also called *Hard Codes*.

Separator pulse codes - After the initial Hard Codes, the system will flash a code 111 and then will flash a series of Soft Codes.

Continuous Memory Codes - These codes indicate a fault that may or may not be present at the time of testing. These codes usually indicate an intermittent failure. Continuous Memory codes are stored in the system and they will flash after the normal Hard Codes. These codes are three digit codes. These codes can indicate chronic or intermittent problems. Also called Soft Codes.

Fast codes - These codes are transmitted 100 times faster than normal codes and can only be read by a Star Tester from Ford or an equivalent SCAN tool.

Engine running codes (KOER) or (ER)

18 **Running tests** - These tests make it possible for the PCM to pick-up a diagnostic trouble code that cannot be set while the engine is in KOEO. These problems usually occur during driving conditions. Some codes are detected by cold or warm running conditions, some are detected at low rpm or high rpm and some are detected at closed throttle

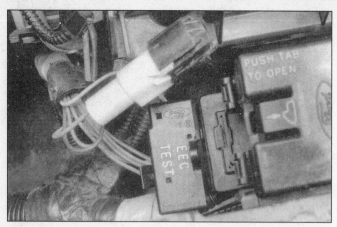

3.20a The EEC diagnostic connector is located behind the fuse box on the passenger side inner fender wall (1993 and 1994 models)

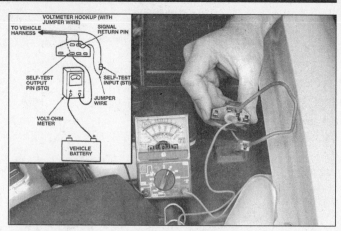

3.20b To read any stored trouble codes, connect a voltmeter and jumper to the Diagnostic Test connector as shown

or WOT.

ID Pulse codes - These codes indicate the type of engine (4 or 6 cylinder) or the correct module and Self Test mode access.

Computed engine timing test - This engine running test determines base timing for the engine and starts the process of allowing the engine to store running codes.

Wiggle test - This engine running test checks the wiring system to the sensors and output actuators as the engine performs.

Cylinder balance test - This engine running test determines injector balance as well as cylinder compression balance. **Note:** *This test should be performed by a dealer service department.*

Beginning the test

19 Apply the parking brake. Position the shift lever in PARK and block the drive wheels. Turn off all electrical loads - air conditioning, radio, heater fan blower etc. Make sure the engine is warmed to operating temperatures (if possible).

20 **Perform the KOEO tests:**

a) *Turn the ignition key off for at least 10 seconds*

b) *Locate the Diagnostic Test connector inside the engine compartment* **(see illustration)** *Connect the positive voltmeter lead onto the positive battery terminal and the negative voltmeter lead onto pin number 4 (STO) of the test connector* **(see illustration).** *Connect a jumper wire from the test terminal (STI) to pin number 2 of the Diagnostic Test terminal.*

c) *Turn the ignition key ON (engine not running) and observe the needle sweeps on the voltmeter. For example code 223, the voltmeter will sweep twice indicating the first digit of the code (2). There will be a two second pause, then there will be two more distinct sweeps of the needle to indicate the second digit of the code number. After another two second pause, there will be three distinct sweeps to indicate the third digit in the code. Additional three-digit codes will be separated by a four second pause and then the indicated sweeps on the voltmeter. Be aware that the code sequence may continue into the continuous memory codes (read further).*

d) *The CHECK ENGINE light on the dash can be used in place of the voltmeter. The number of flashes indicates each digit instead of needle sweeps.*

21 **Interpreting the continuous memory codes:**

a) *After the KOEO codes are reported, there will be a short pause and any stored Continuous Memory codes will appear in order. Remember that the "Separator" code is 111. The computer will not enter the Continuous Memory mode without flashing the separator pulse code. The Continuous Memory codes are read the same as the initial codes or "Hard Codes." Record these codes onto a piece of paper and continue the test.*

22 **Perform the Engine Running (ER) tests.**

a) *Remove the jumper wires from the Diagnostic Test connector to start the test.*

b) *Run engine until it reaches normal operating temperature.*

c) *Turn the engine OFF for at least 10 seconds.*

d) *Install the jumper wire onto Diagnostic Test connector* **(see illustration 3.20b)** *and start the engine.*

e) *Observe that the voltmeter or CHECK ENGINE light will flash the engine identification code. This code indicates 1/2 the number of cylinders of the engine. For example, 2 flashes represent a 4 cylinder engine, or 3 flashes represent a six-cylinder engine.*

f) *Within 1 to 2 seconds of the ID code, turn the steering wheel at least 1/2 turn and release. This will store any power steering pressure switch trouble codes.*

g) *Depress the brake pedal and release.* **Note:** *Perform the steering wheel and brake pedal procedure in succession immediately (1 to 2 seconds) after the ID codes are flashed.*

h) *Observe all the codes and record them on a piece of paper. Be sure to count the sweeps or flashes very carefully as you jot them down.*

23 On some models the PCM will request a Dynamic Response check. This test quickly checks the operation of the TPS, MAF or MAP sensors in action. This will be indicated by a code 1 or a single sweep of the voltmeter needle (one flash on CHECK ENGINE light). This test will require the operator to simply full throttle ("goose") the accelerator pedal for one second. DO NOT throttle the accelerator pedal unless it is requested.

24 The next part of this test makes sure the system can advance the timing. This is called the Computed Timing test. After the last ER code has been displayed, the PCM will advance the ignition timing a fixed amount and hold it there for approximately 2 minutes. Use a timing light to check the amount of advance. The computed timing should equal the base timing plus 20 BTDC. The total advance should equal 27 to 33 degrees advance. If the timing is out of specification, have the system checked at a dealer service department. **Note:** *If it's necessary to adjust the base timing on (1993 and 1994) 3.0L engines, remember to remove the SPOUT from the connector as described in the ignition timing procedure in Chapter 5. This will remove the computer from the loop and give base timing. Base timing is not adjustable on other models.*

25 **Finally, perform the Wiggle Test.** This test can be used to recreate a possible intermittent fault in the harness wiring system:

a) *Use a jumper wire to ground the STI lead on the Diagnostic Test connector* **(see illustration 3.20b).**

b) *Turn the ignition key ON (engine not running).*

c) *Now deactivate the self-test mode (remove the jumper wire) and then immediately reactivate self test mode. Now the system has entered Continuous Monitor Test Mode.*

d) *Carefully wiggle, tap or remove any suspect wiring to a sensor or output actua-*

Chapter 6 Emissions control systems

tor. If a problem exists, a trouble code will be stored that indicates a problem with the circuit that governs the particular component. Record the codes that are indicated.

e) Next, enter Engine Running Continuous Monitor Test Mode to check for wiring problems only when the engine is running. Start first by deactivating the Diagnostic Test connector and turning the ignition key OFF. Now start the engine and allow it to idle.

f) Use a jumper wire to ground the STI lead on the Diagnostic Test connector **(see illustration 3.20a)**. Wait ten seconds and then deactivate the test mode and reactivate it again (install jumper wire). This will enter Engine Running Continuous Monitor Test Mode.

g) Carefully wiggle, tap or remove any suspect wiring to a sensor or output actuator. If a problem exists, a trouble code will be stored that indicates a problem with the circuit that governs the particular component. Record the codes that are indicated.

26 If necessary, perform the Cylinder Balance Test. This test must be performed by a dealer service department.

Clearing codes

To clear the codes from the PCM memory, start the KOEO self test diagnostic procedure **(see illustrations 3.20a and 3.20b)** and install the jumper wire into the Diagnostic Test connector. When the codes start to display themselves on the voltmeter or CHECK ENGINE light, remove the jumper wire from the Diagnostic Test connector. This will erase any stored codes within the system. **Note:** *You can also clear the codes by disconnecting the negative battery cable. However, this will erase stored operating parameters from the KAM (Keep Alive Memory) and cause the engine to run rough for a period of time (several miles) while the computer relearns the information.* **Caution:** *Do not disconnect the positive terminal to clear the codes from memory. If a spark should occur when reconnecting the battery, the PCM could be damaged.*

Diagnostic Trouble Codes (EEC-IV, 1993 and 1994 models)

Code	Description
111	System pass
112	Intake Air Temperature sensor circuit indicates circuit grounded/above 245 degrees F
113	Intake Air Temperature sensor circuit indicates open circuit/below -40 degrees F
114	Air Temperature sensor out-of-self-test range
116	Coolant Temperature sensor out-of-self test range
117	Temperature circuit below minimum voltage/indicates above 245 degrees F
118	Coolant Temperature sensor circuit above maximum voltage/indicates below -40 degrees F
121	Throttle Position sensor out-of-self test range
121	Indicates throttle position voltage inconsistent with MAF sensor (where applicable)
122	Throttle Position sensor below minimum voltage
123	Throttle Position sensor above maximum voltage
124	Throttle Position Sensor voltage higher than expected
125	Throttle Position Sensor voltage lower than expected
126	MAP/BARO sensor higher or lower than expected
128	MAP sensor vacuum hose damaged or disconnected
129	Insufficient Manifold Absolute Pressure/Mass Air Flow change during Dynamic Response Check
136	System indicates lean bank #2
137	System indicates rich bank #2 (KOER)
144	No exhaust sensor switching detected
157	Mass Air Flow (MAF) Sensor below minimum voltage
158	Mass Air Flow (MAF) Sensor above maximum voltage
159	Mass Air Flow Sensor out-of-self test range
167	Insufficient Throttle Position Sensor change during Dynamic Response Check
171	Fuel system at adaptive limits bank #1
172	System indicates lean bank #1
173	System indicates rich bank #1
175	Fuel system at adaptive limits bank #2
176	System indicates lean bank #2
177	System indicates rich bank #2
179	Adaptive Fuel lean limit reached at part throttle, system rich, bank #1
181	Adaptive Fuel rich limit reached at part throttle, bank #1
182	Adaptive fuel limit lean at idle
183	Adaptive fuel limit rich at idle
184	Mass Air Flow higher than expected
185	Mass Air Flow lower than expected
186	Injector Pulse-width higher or MAF lower than expected (without Baro Sensor)
186	Injector Pulse width higher than expected (with Baro Sensor)
187	Injector Pulse-width lower than expected (with Baro Sensor)
187	Injector Pulse width lower or MAF higher than expected (without Baro Sensor)
188	Adaptive Fuel lean limit reached, bank #2
189	Adaptive Fuel rich limit reached, bank #2
211	Profile Ignition Pick-up circuit failure
212	Ignition Module Circuit failure/SPOUT circuit grounded
213	SPOUT circuit open
214	Cylinder Identification (CID) circuit failure
224	SPOUT circuit open
225	Knock sensor not detected during dynamic response test (KOER)
311	Air System inoperative, bank #1 (with W/Dual HO2S)
312	Secondary Air Injection (AIR) misdirected
313	Secondary air Injection (AIR) not bypassed
327	EGR (EVP/PFE/DPFE) circuit below minimum voltage
328	EGR (EVP) closed valve voltage lower than expected
332	Insufficient EGR flow detected
334	EGR closed valve voltage higher than expected
335	EGR (PFE) Sensor voltage out-of-range
336	EGR circuit higher than expected
337	EGR Valve Pressure Transducer/Position Sensor circuit above maximum voltage
341	Octane adjust service pin open
411	Unable to control RPM during Low RPM Self-test
412	Unable to control RPM during High RPM Self-test
452	No input from Vehicle Speed Sensor
511	Read Only Memory test failed - replace PCM
512	Keep Alive Memory test failed
513	Internal voltage failure in PCM
519	Power steering pressure switch (PSP) circuit open states
522	Manual Lever Position (MLP) sensor circuit open/vehicle in gear
527	Neutral drive switch open
528	Clutch Pedal Position (CPP) switch circuit failure
536	Brake ON/Off (BOO) circuit failure/not activated during the KOER
538	Insufficient change in RPM/operator error in Dynamic Response Check
538	Invalid Cylinder Balance Test due to throttle movement during test
538	Invalid Cylinder Balance Test due to CID circuit failure
539	Air conditioning on during Self-test
542	Fuel Pump circuit open; PCM to motor

Diagnostic Trouble Codes (EEC-IV, 1993 and 1994 models) (continued)

- 543 .... Fuel Pump circuit open; Battery to PCM
- 551 .... Idle Air Control (IAC) circuit failure (KOEO)
- 552 .... Secondary Air Injection Bypass (AIRB) circuit failure
- 553 .... Secondary Air Injection Diverter (AIRB) circuit failure
- 556 .... Primary Fuel Pump circuit failure
- 558 .... EGR Vacuum Regulator circuit failure
- 565 .... Canister Purge circuit failure
- 566 .... 3-4 Shift Solenoid Circuit failure
- 569 .... Auxiliary Canister Purge (CANP2) circuit failure (KOEO)
- 617 .... 1-2 shift error
- 618 .... 2-3 shift error
- 619 .... 3-4 shift error
- 621 .... Shift Solenoid 1 (SS1) circuit failure (KOEO)
- 622 .... Shift Solenoid 2 (SS2) circuit failure (KOEO)
- 624 .... Electronic Pressure Control (EPC) circuit failure
- 625 .... Electronic Pressure Control (EPC) driver open in PCM
- 626 .... Coast Clutch Solenoid (CCS) circuit failure (KOEO)
- 627 .... Converter clutch solenoid circuit failure (E4OD)
- 628 .... Excessive converter clutch slippage
- 629 .... Torque Converter Clutch (TCC) solenoid circuit failure
- 631 .... Transmission Control Indicator Lamp (TCIL) circuit failure (KOEO)
- 632 .... Transmission Control Switch (TCS) circuit did not change states during KOER
- 634 .... Manual Lever Position (MLP) sensor voltage higher or lower than expected
- 636 .... Transmission Fluid Temp (TFT) higher or lower than expected
- 637 .... Transmission Fluid Temp (TFT) sensor circuit above maximum voltage/ -40°F (-40°C) indicated / circuit open
- 638 .... Transmission Fluid Temp (TFT) sensor circuit below minimum voltage/ 290°F (143°C) indicated / circuit shorted
- 639 .... Insufficient input from Transmission Speed Sensor (TSS)
- 654 .... Transmission Range (TR) sensor not indicating PARK during KOEO
- 655 .... Transmission Range (TR) sensor not indicating NEUTRAL during KOEO
- 656 .... Torque Converter Clutch continuous slip error
- 668 .... Transmission Range circuit voltage above maximum voltage
- 998 .... Hard fault present

Diagnostic Trouble Codes (EEC-V, 1995 and later)

Note: *"P0" preceding each code means the code is associated with the powertrain. A scan tool must be used to read these codes.*

- P0102 ... Mass air flow circuit low input
- P0103 ... Mass air flow circuit high input
- P0112 ... Intake Air Temperature circuit low
- P0113 ... Intake Air Temperature circuit high
- P0117 ... Engine Coolant Temp. circuit low
- P0118 ... Engine Coolant Temp. circuit high
- P0121 ... TP sensor voltage does not agree w/MAF sensor
- P0122 ... TP Sensor circuit low
- P0123 ... TP Sensor circuit high
- P0125 ... Coolant temp not high enough to enter closed loop operation
- P0126 ... Coolant temp not high enough for stable operation
- P0131 ... Upstream Oxygen sensor circuit low (Bank 1)
- P0132 ... Upstream Oxygen sensor circuit high (Bank 1)
- P0133 ... Upstream Oxygen sensor circuit slow response (Bank 1)
- P0135 ... Upstream Oxygen sensor heater malfunction (Bank 1)
- P0136 ... Downstream Oxygen sensor circuit malfunction (Bank 1)
- P0138 ... Downstream Oxygen sensor circuit high (Bank 1)
- P0141 ... Downstream Oxygen sensor heater malfunction (Bank 1)
- P0151 ... Upstream Oxygen sensor circuit low (Bank 2)
- P0152 ... Upstream Oxygen sensor circuit high (Bank 2)
- P0153 ... Upstream Oxygen sensor circuit slow response (Bank 2)
- P0155 ... Upstream Oxygen sensor heater malfunction (Bank 2)
- P0156 ... Downstream Oxygen sensor circuit malfunction (Bank 2)
- P0158 ... Downstream Oxygen sensor circuit high (Bank 2)
- P0161 ... Downstream Oxygen sensor heater malfunction (Bank 2)
- P0171 ... System adaptive fuel too lean (Bank 1)
- P0172 ... System adaptive fuel too rich (Bank 1)
- P0174 ... System adaptive fuel too lean (Bank 2)
- P0175 ... System adaptive fuel too rich (Bank 2)
- P0222 ... Throttle Position Sensor circuit low
- P0223 ... Throttle Position Sensor circuit high
- P0230 ... Fuel pump primary circuit malfunction
- P0231 ... Fuel pump secondary circuit low
- P0232 ... Fuel pump secondary circuit high
- P0300 ... Random misfire detected
- P0301 ... Misfire at cylinder 1 detected
- P0302 ... Misfire at cylinder 2 detected
- P0303 ... Misfire at cylinder 3 detected
- P0304 ... Misfire at cylinder 4 detected
- P0305 ... Misfire at cylinder 5 detected
- P0306 ... Misfire at cylinder 6 detected
- P0320 ... PIP input circuit malfunction
- P0340 ... Camshaft position sensor malfunction
- P0350 ... Ignition coil primary circuit malfunction
- P0351 ... Ignition coil A primary circuit malfunction
- P0352 ... Ignition coil B primary circuit malfunction
- P0400 ... EGR flow malfunction
- P0401 ... EGR flow insufficient
- P0402 ... EGR flow excessive
- P0420 ... Manifold catalyst efficiency low (Bank 1)
- P0430 ... Manifold catalyst efficiency low (Bank 2)
- P0443 ... Purge Control Valve permanently open
- P0500 ... Vehicle Speed sensor malfunction
- P0505 ... Idle air control malfunction
- P0603 ... Internal control module KAM error
- P0605 ... Internal control module ROM error
- P0703 ... BOO switch circuit malfunction
- P0704 ... Clutch Pedal Position Switch circuit malfunction
- P0707 ... Transmission range (TR) sensor circuit low
- P0708 ... Transmission range (TR) sensor circuit high
- P0712 ... Transmission fluid Temp. (TFT) sensor circuit low
- P0713 ... Transmission fluid Temp. (TFT) sensor circuit high
- P0715 ... Turbine shaft speed sensor circuit malfunction
- P0720 ... Output shaft speed sensor circuit malfunction
- P0741 ... Torque converter clutch mechanical system performance
- P0743 ... Torque converter clutch electrical system malfunction
- P0746 ... Electronic pressure control solenoid performance
- P0750 ... Shift Solenoid #1 circuit malfunction
- P0751 ... Shift Solenoid #1 performance
- P0755 ... Shift Solenoid #2 circuit malfunction
- P0756 ... Shift Solenoid #2 performance
- P0760 ... Shift Solenoid #3 circuit malfunction
- P0761 ... Shift Solenoid #3 performance
- P0781 ... 1 to 2 shift error
- P0782 ... 2 to 3 shift error
- P0783 ... 3 to 4 shift error
- P0784 ... 4 to 5 shift error

Chapter 6 Emissions control systems

4.1 The Power Train Control Module is located along the upper edge of the firewall in the engine compartment

5.2 Check resistance of the engine coolant temperature sensor

5.3 Check for power (VREF) to the engine coolant temperature sensor at the harness

4 Powertrain Control Module (PCM)

Refer to illustration 4.1

1 The Powertrain Control Module (PCM) is located on the passenger's side of the engine compartment, along the upper edge of the firewall **(see illustration)**.
2 Should it become necessary to remove or replace the PCM, disconnect the negative battery cable from the battery.
3 Loosen the bolt in the center of the connector and disconnect the PCM connector.
4 Remove the two nuts securing the PCM to the firewall and remove the PCM.
5 Installation is reversal of removal. **Note:** *The vehicle may have to be driven up to ten miles after PCM replacement to allow the PCM to relearn its adaptive strategy. During that time, the engine may run roughly.*

5 Information sensors

Caution: *Use only a high-impedance digital multi-meter when performing checks on the EEC system.*

Engine Coolant Temperature (ECT) sensor

Refer to illustrations 5.2, 5.3 and 5.4

General description

1 The coolant sensor is a thermistor (a resistor which varies the value of its voltage output in accordance with temperature changes). The change in the resistance values will directly affect the voltage signal from the coolant sensor. As the sensor temperature DECREASES, the resistance values will INCREASE. As the sensor temperature INCREASES, the resistance values will DECREASE. A failure in the coolant sensor circuit should set a Code 116, 117 or 118. These codes indicate a failure in the coolant temperature circuit. In most cases, this will involve a poor connection, a broken wire or a defective sensor.

Check

Note: *The coolant temperature sensor on the 2.3L four-cylinder engine is located in the water outlet pipe at the front of the engine. On V6 engines it's threaded into the lower intake manifold near the front, adjacent to the water outlet.*

2 Check the resistance value of the coolant temperature sensor while it is completely cold (50 to 65-degrees F). The resistance should equal about 40,000 ohms **(see illustration)**. Next, start the engine and warm it up until it reaches operating temperature (180 to 220-degrees F). The resistance should be lower, approximately 3,000 ohms.
3 If the resistance values on the sensor are correct, check the reference voltage (VREF) to the sensor from the PCM **(see illustration)**. It should be approximately 5.0 volts.

Replacement

Warning: *Wait until the engine is completely cool before replacing the sensor.*
4 Before installing the new sensor, wrap the threads with Teflon sealing tape to prevent leakage and thread corrosion **(see illustration)**. Remove the radiator cap to relieve any pressure that may still be present in the system, then reinstall it.
5 To remove the sensor, unplug the electrical connector, then carefully unscrew it. **Caution:** *Handle the coolant sensor with care. Damage to this sensor will affect the operation of the entire fuel injection system.* Install the sensor and tighten it securely. Check the coolant level and add some, if necessary (see Chapter 1).

Oxygen sensor

Refer to illustrations 5.8 and 5.10

General description and check

6 The heated oxygen sensors (HEGO), which are located in the exhaust manifolds, monitor the oxygen content of the exhaust gas stream. The oxygen content in the exhaust reacts with the oxygen sensor to produce a voltage output which varies from 0.1-volt (high oxygen, lean mixture) to 0.9-volts (low oxygen, rich mixture). The PCM constantly monitors this variable voltage output to determine the ratio of oxygen to fuel in the mixture. The PCM alters the air/fuel mixture ratio by controlling the pulse width (open time) of the fuel injectors. A mixture ratio of 14.7 parts air to 1 part fuel is the ideal mixture ratio for minimizing exhaust emissions, thus allowing the catalytic converter to operate at maximum efficiency. It is this ratio of 14.7 to 1 which the PCM and the oxygen sensor attempt to maintain at all times.
7 The oxygen sensor produces no voltage when it is below its normal operating temperature of about 600-degrees F. During this initial period before warm-up, the PCM operates in open loop mode.
8 Allow the engine to reach normal operating temperature and check that the oxygen sensor is producing a steady signal voltage

5.4 To prevent leakage, wrap the threads of the coolant temperature sensor with Teflon tape before installing it

Chapter 6 Emissions control systems

5.8 The oxygen sensor creates a very small voltage signal when the engine is warmed up. A quick oxygen sensor check is to immediately disconnect the sensor connector when it is warmed up and probe the signal wire (sensor side of connector) for a millivolt reading

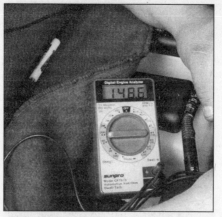

5.10 Working on the harness side, probe the electrical connector and check for battery voltage with the ignition key ON and the engine not running

5.23 With the connector attached and the ignition On, backprobe terminal TP SIG with the positive (+) probe of the voltmeter and SIG RTN with the negative (-) probe of the voltmeter (insert shows terminal locations inside the unplugged connector)

between 0.35 and 0.55-volts **(see illustration)**. *Warning: Be careful of burns from the exhaust system when disconnecting the oxygen sensor.*

9 A delay of two minutes or more between engine start-up and normal operation of the sensor, followed by a low or a high voltage signal or a short in the sensor circuit, will cause the PCM to also set a code. Codes that indicate problems in the oxygen sensor system are 136, 137, 144 and 171 through 183.

10 Also check to make sure the oxygen sensor heater is supplied with battery voltage **(see illustration)**.

11 When any of the above codes occur, the PCM operates in the open loop mode - that is, it controls fuel delivery in accordance with a programmed default value instead of feedback information from the oxygen sensor.

12 The proper operation of the oxygen sensor depends on four conditions:

a) **Electrical** - *The low voltages generated by the sensor depend upon good, clean connections which should be checked whenever a malfunction of the sensor is suspected or indicated.*

b) **Outside air supply** - *The sensor is designed to allow air circulation to the internal portion of the sensor. Whenever the sensor is removed and installed or replaced, make sure the air passages are not restricted.*

c) **Proper operating temperature** - *The PCM will not react to the sensor signal until the sensor reaches approximately 600-degrees F. This factor must be taken into consideration when evaluating the performance of the sensor.*

d) **Unleaded fuel** - *The use of unleaded fuel is essential for proper operation of the sensor. Make sure the fuel you are using is of this type.*

13 In addition to observing the above conditions, special care must be taken whenever the sensor is serviced:

a) *The oxygen sensor has a permanently-attached pigtail and electrical connector which should not be removed from the sensor. Damage or removal of the pigtail or electrical connector can adversely affect operation of the sensor.*

b) *Grease, dirt and other contaminants should be kept away from the electrical connector and the louvered end of the sensor.*

c) *Do not use cleaning solvents of any kind on the oxygen sensor.*

d) *Do not drop or roughly handle the sensor.*

e) *The silicone boot must be installed in the correct position to prevent the boot from being melted and to allow the sensor to operate properly.*

Replacement

Note: *Because it is installed in the exhaust manifold or pipe, which contracts when cool, the oxygen sensor may be very difficult to loosen when the engine is cold. Rather than risk damage to the sensor (assuming you are planning to reuse it in another manifold or pipe), start and run the engine for a minute or two, then shut it off. Be careful not to burn yourself during the following procedure.*

14 Disconnect the cable from the negative terminal of the battery.

15 Raise the vehicle and place it securely on jackstands.

16 Disconnect the electrical connector from the sensor.

17 Unscrew the sensor from the exhaust manifold.

18 Anti-seize compound must be used on the threads of the sensor to facilitate future removal. The threads of new sensors will already be coated with this compound, but if an old sensor is removed and reinstalled, recoat the threads.

19 Install the sensor and tighten it securely. **Caution:** *Excessive force may damage the threads.*

20 Reconnect the electrical connector of the pigtail lead to the main engine wiring harness.

21 Lower the vehicle and reconnect the cable to the negative terminal of the battery.

Throttle Position Sensor (TPS)

General description

Refer to illustration 5.23

22 The Throttle Position Sensor (TPS) is located on the end of the throttle shaft on the throttle body. By monitoring the output voltage from the TPS, the PCM can determine fuel delivery based on throttle valve angle (driver demand). A broken or loose TPS can cause intermittent bursts of fuel from the injector and an unstable idle because the PCM thinks the throttle is moving. Any problems in the TPS or circuit will set a code 122 through 125.

Check

23 To check the TPS, turn the ignition switch to ON (engine not running) and connect the probes of the voltmeter into the ground wire (SIG RTN) and signal wire (TP) terminals on the backside of the electrical connector **(see illustration)**. This test checks for the proper signal voltage from the TPS. **Note:** *Be careful when backprobing the electrical connector. Do not damage the wiring harness or pull on any connectors to make clean contact. Be sure the probes are placed in the correct position by referring to the illustration.*

24 The sensor should read 0.50 to 1.0-volt at idle. Open the throttle all the way and the sensor should increase voltage to 4.0 to 5.0-volts. If the TPS voltage readings are incorrect, replace it with a new unit.

25 Also, check the TPS reference voltage. With the ignition key ON (engine not running), connect the positive (+) probe of the voltmeter onto the voltage reference wire (VREF) and the negative probe to a good ground. There should be approximately 5.0 volts sent from the PCM to the TPS **(see illustration 5.23)**.

Chapter 6 Emissions control systems

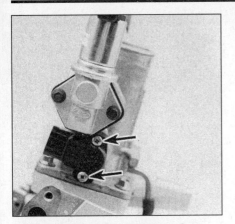

5.28 Remove the two screws from the TPS (arrows) and separate it from the throttle body

26 Also, check the resistance of the potentiometer within the TPS. Disconnect the TPS electrical connector and, working on the sensor side, connect the probes of the ohmmeter onto the reference voltage terminal (VREF) and the TPS signal terminal (TP). With the throttle valve fully closed, the TPS should read between 3.0 and 4.0 K-ohms.

27 Now open the throttle with one hand and check the resistance again. Slowly advance the throttle until fully open. The resistance should be approximately 350 ohms. The potentiometer should exhibit a smooth change in resistance as it travels from fully closed to wide open throttle. Any deviations indicate a possible worn or damaged TPS.

Replacement
Refer to illustration 5.28

28 The TPS is a non-adjustable unit. Remove the two retaining screws **(see illustration)** and separate the TPS from the throttle body. **Note:** *If you're replacing the TPS on a 2.3L four-cylinder engine, the job is much easier if you remove the throttle body from the upper intake manifold.*

29 Installation is the reverse of removal. Be sure to line up the slot in the TPS with the blade on the throttle shaft.

Mass Airflow Sensor (MAF)
General Information
Refer to illustrations 5.31 and 5.32

30 The Mass Airflow Sensor (MAF) is located on the air intake duct. This sensor uses a hot wire sensing element to measure the amount of air entering the engine. The air passing over the hot wire causes it to cool. Consequently this change in temperature can be converted into an analog voltage signal to the PCM which in turn calculates the required fuel injector pulse width.

Check
31 Check for power to the MAF sensor. Disconnect the MAF sensor electrical connector, work on the harness side and probe the connector to check for battery voltage **(see illustration)**.
32 Plug in the electrical connector and backprobe the MAF SIGNAL and MAF SIG RTN **(see illustration)** with the voltmeter and check for the voltage. The voltage should be 0.2 to 1.5 volts at idle.
33 Raise the engine rpm. The signal voltage from the MAF sensor should increase to about 2.0 volts at 60 mph. It is impossible to simulate these conditions in the driveway at home but it is necessary to observe the voltmeter for a fluctuation in voltage as the engine speed is raised. The vehicle will not be under load conditions but it should manage to vary slightly.
34 Disconnect the MAF harness connector and use an ohmmeter and probe the terminals MAF SIGNAL and MAF SIG RTN. If the hot wire element inside the sensor has been damaged it will be indicated by an open circuit (infinite resistance).
35 If the voltage readings are correct, check the wiring harness for open circuits or a damaged harness (see Chapter 12).

Replacement
36 Disconnect the electrical connector from the MAF sensor.
37 Remove the air cleaner outlet tube assembly (see Chapter 4).
38 Remove the MAF sensor retaining bolts and lift the MAF sensor out.
39 Installation is the reverse of removal.

Air conditioning clutch control
General Information
Refer to illustration 5.41
Note: *Refer to Chapter 12 for additional information on the location of the relays.*
40 During air conditioning operation, the PCM controls the application of the air conditioning compressor clutch. The PCM controls the air conditioning clutch control relay to delay clutch engagement after the air conditioning is turned ON to allow the idle air control valve to adjust the idle speed of the engine to compensate for the additional load. The PCM also controls the relay to disengage the clutch in the event of an excessively high or low pressure within the system or an overheating problem.

Check
41 The accompanying diagnostic chart outlines the testing procedures for the air conditioning clutch control **(see illustration on following page)**.
42 In most cases, if the air conditioning does not function, the problem is probably related to the air conditioning system relays and switches and not the PCM.
43 If the air conditioning is operating properly and idle is too low when the air conditioning compressor turns on or is too high when the air conditioning compressor turns off, check for an open circuit between the air conditioning control relay and the PCM.

Vehicle Speed Sensor (VSS) (1997 and earlier models only)
General description
Note: *1998 and later models are not equipped with a Vehicle speed Sensor. The PCM receives vehicle speed information from the antilock brake system rear wheel speed sensor.*
44 The Vehicle Speed Sensor (VSS) is located near the rear section of the transmission **(see illustration 3.1)**. This sensor is a

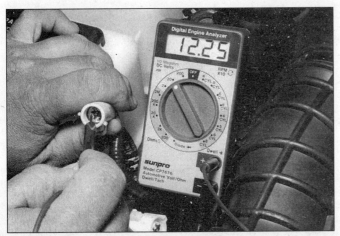

5.31 Probe the VPWR terminal on the harness side of the MAF sensor and check for battery voltage to the MAF sensor

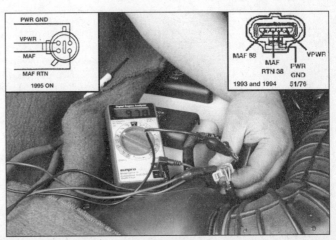

5.32 Use a voltmeter and probe MAF SIG and MAF SIG RTN for signal voltage

6-10   Chapter 6 Emissions control systems

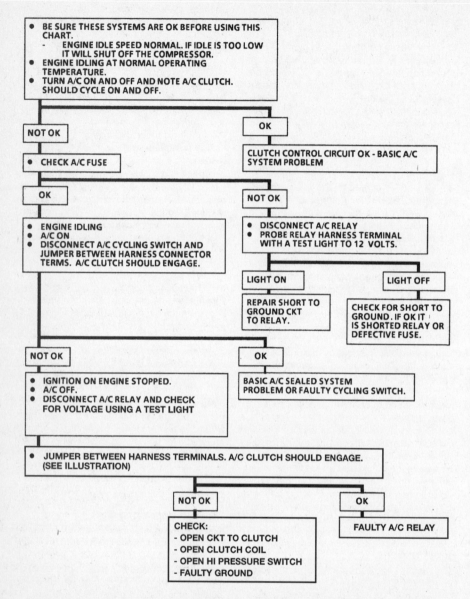

5.41 Diagnostic flow chart for checking the air conditioning clutch control system

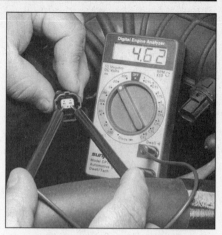

5.51 Disconnect the intake air temperature sensor and check for VREF (5.0 volts) at the harness connector

5.53 Measure the resistance across the intake air temperature sensor terminals. The resistance should be high for a low temperature and low for a high temperature

permanent magnetic variable-reluctance sensor that produces a pulsing voltage whenever vehicle speed is over 3 mph. These pulses are translated by the PCM and provided for other systems for fuel and transmission shift control. The VSS is part of the Transmission Converter Clutch (TCC) system. Any problems with the VSS will usually set a Code 452.

Check

45   To check the vehicle speed sensor, remove the electrical connector in the wiring harness near the sensor. Using a voltmeter, check for signal voltage to the sensor. The signal wire should have 10 volts or more available. If there is no voltage available, have the PCM diagnosed by a dealer service department or other qualified repair shop.

Replacement

46   To replace the VSS, disconnect the electrical connector from the VSS.
47   Remove the retaining bolt and lift the VSS from the transmission.
48   Installation is the reverse of removal.

Intake Air Temperature (IAT) Sensor

Refer to illustrations 5.51 and 5.53

General description

49   The Intake Air Temperature (IAT) sensor is located inside the air intake duct. This sensor acts as a resistor which changes value according to the temperature of the air entering the engine. Low temperatures produce a high resistance value (for example, at 68- degrees F the resistance is about 27K-ohms) while high temperatures produce low resistance values (at 160-degrees F the resistance is 7K-ohms. The PCM supplies approximately 5-volts (reference voltage) to the IAT sensor. The signal voltage back to the PCM will vary according to the temperature of the incoming air. Any problems with the IAT sensor will usually set a code 112, 113 or 114.

Check

50   To check the IAT sensor, disconnect the two-prong electrical connector and turn the ignition key ON but do not start the engine.
51   Measure the voltage (reference voltage) at the electrical connector. It should read approximately 5 volts **(see illustration)**.
52   If the voltage signal is not correct, have the PCM diagnosed by a dealer service department or other repair shop.
53   Measure the resistance across the sensor terminals **(see illustration)**. The resistance should be HIGH when the air temperature is LOW.
54   Next, start the engine and let it idle

Chapter 6 Emissions control systems

5.56 With the engine running, disconnect the IAC connector (arrow). The engine rpm should drop if the solenoid is working correctly

5.60 Remove the two screws securing the IAC solenoid to the throttle body

(cold). Wait a while and let the engine reach operating temperature. Turn the ignition OFF, disconnect the IAT sensor and measure the resistance across the terminals. The resistance should be LOW when the air temperature is HIGH. If the sensor does not exhibit this change in resistance, replace it with a new part.

Idle Air Control (IAC) solenoid

Refer to illustrations 5.56 and 5.60

General Information

55 The Idle Air Control solenoid controls engine idle speed by allowing a small amount of air to bypass the throttle plate. The PCM controls operation of the IAC solenoid.

Checking

56 With the engine running, disconnect the IAC solenoid. Engine rpm should drop or the engine might even stall, indicating that the IAC solenoid is functioning **(see illustration)**.
57 With the engine off, disconnect the IAC solenoid connector. Measure the resistance across the pins of the solenoid. It should read between 6.0 to 10.0 ohms with the positive lead on the VPWR pin and the negative lead on the IAC pin. You may get slightly lower readings if you reverse the leads due to a diode in parallel with the solenoid winding. If you get higher readings, the solenoid winding is probably open.
58 With the key on (engine not running) measure the voltage from the VPWR pin on the harness connector and ground. You should read at least 10.5 volts.

Replacement

59 Disconnect the electrical connector from the IAC solenoid.
60 Remove the two screws securing the IAC solenoid to the throttle body **(see illustration)**. **Note:** *If you're replacing the IAC valve on a four-cylinder engine, the job can be made easier by removing the throttle body from the upper intake manifold first.*
61 Clean off all traces of gasket from the IAC solenoid and throttle body. Be careful not to damage the gasket mating surfaces if you must scrape them. **Caution:** *Do not use solvent to clean the IAC solenoid.*
62 Installation is the reverse of removal.

Power steering pressure switch

General information

63 Turning the steering wheel increases power steering fluid pressure and engine load. The pressure switch will close before the load can cause an idle problem. A problem in the power steering pressure switch circuit will set a code 519 or 521.
64 A pressure switch that will not open or an open circuit from the PCM will cause timing to retard at idle and this will affect idle quality.
65 A pressure switch that will not close or an open circuit may cause the engine to die when the power steering system is used heavily.

Replacement

66 Any problems with the power steering pressure switch or circuit should be repaired by a dealer service department or other qualified repair shop.

Camshaft position sensor

General information

67 The camshaft position sensor (CMP) provides information that the PCM uses to determine Top Dead Center for cylinder number 1 so that it can sequence coil firing and fuel injection at the right time for all cylinders. Not all models require a camshaft position sensor. In many cases, only California models with manual transmissions are so equipped. **Note:** *The camshaft position sensor is sometimes referred to as the Cylinder ID (CID) sensor.*
68 On 1994 2.3L engines, the camshaft position sensor is located on the Synchronizer Assembly which is mounted on the left side of the block near the oil filter. on these models it's driven by the auxiliary shaft.
69 On 1995 and later 2.3L engines, the camshaft position sensor is mounted on the oil pump assembly behind the inner timing belt cover. A high point on the oil pump sprocket triggers the sensor.
70 On 3.0L and 4.0L engines, the camshaft position sensor is located behind the intake manifold on the top surface of the cylinder block where the distributor was on earlier models.

Replacement

Note: *On all models equipped with a camshaft position sensor (except 1995 and later 2.3L four-cylinder engines) the sensor is mounted on a synchronizer assembly, which is essentially a drive unit for the sensor. If you are simply replacing the cam position sensor it is not necessary to remove the synchronizer assembly from the engine - just remove the screws from the sensor, detach it from the synchronizer and install the new sensor. However, many engine repair procedures require removal of the synchronizer assembly, in which case it will be necessary to perform the following procedure to time the synchronizer.*

1994 2.3L four-cylinder engine

Note: *This procedure applies only to California models with a manual transmission. It requires a special tool, available at most auto parts stores, to properly time the synchronizer (cam position sensor).*

71 Position the engine at Top Dead Center (TDC) compression for cylinder number 1 (see Chapter 2, Part A). Disconnect the cable from the negative terminal of the battery.
72 Detach the electrical connector, remove the screws from the cam position sensor and detach the sensor from the synchronizer.
73 If you will be removing the synchronizer assembly, remove the bolt and clamp and withdraw the synchronizer from the engine.
74 Install synchronizer positioner, available at most auto parts stores, into the top of the synchronizer. If the synchronizer was removed, lubricate the gear with clean engine oil and install the assembly, hold-down clamp and bolt.
75 Rotate the auxiliary shaft sprocket until the synchronizer shutter is resting against the stop and the notch of the tool is engaged with the corresponding notch on the synchronizer bowl.
76 Tighten the synchronizer hold-down bolt.
77 Install the cam position sensor onto the synchronizer. Tighten the screws securely and plug in the electrical connector.

1995 and later four-cylinder engines

78 Disconnect the cable from the negative terminal of the battery.
79 Remove the two sensor mounting bolts and detach the sensor from the oil pump body.
80 Installation is the reverse of removal. Be sure to tighten the screws securely.

Chapter 6 Emissions control systems

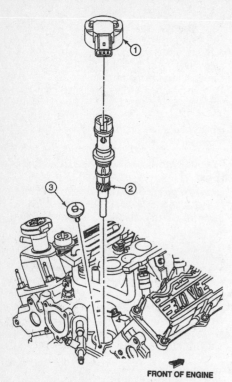

5.83 Camshaft Position Sensor and synchronizer installation details -V6 engine

1. Camshaft position sensor
2. Synchronizer assembly
3. Hold-down bolt

3.0L V6 engine and 1996 and later 4.0L V6 engines

Refer to illustrations 5.83, 5.84 and 5.86

Note: *This procedure requires the use of a special tool, available at most auto parts stores.*

81 Position the engine at Top Dead Center (TDC) compression for cylinder number 1 (see Chapter 2, Part B). The engine must remain in this position throughout the entire procedure. Disconnect the cable from the negative terminal of the battery.

82 Mark the relative position of the camshaft position sensor electrical connector so the assembly can be oriented properly upon installation (this is only necessary if the synchronizer assembly will be removed). Disconnect the electrical connector from the camshaft position sensor. Remove the screws and detach the sensor from the synchronizer assembly.

83 If you will be removing the synchronizer assembly, remove the bolt and withdraw the synchronizer from the engine **(see illustration)**.

84 Place a special positioning tool, available at most auto parts stores, onto the synchronizer assembly **(see illustration)**. Align the vane of the synchronizer with the radial slot in the special tool.

85 Turn the tool on the synchronizer until the boss on the tool is engaged with the

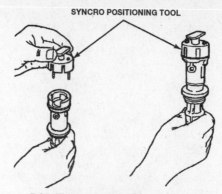

5.84 This special tool is required for timing the synchronizer assembly

notch on the synchronizer.

86 Lubricate the gear, thrust washer and lower bearing of the synchronizer assembly with clean engine oil. Insert the assembly into the engine, with the tool pointing about 75-degrees (3.0L) or 60-degrees (4.0L) counterclockwise from the engine's centerline **(see illustration)**.

87 Turn the tool clockwise a little so the synchronizer engages with the oil pump intermediate shaft. Push down on the synchronizer, turning the tool gently until the gear on the synchronizer engages with the gear on the camshaft.

88 On the 3.0L engine, the head of the tool should be pointing to the rear, approximately 33 to 45-degrees counterclockwise from the centerline of the engine when the synchronizer is fully inserted **(see illustration 5.86)**. **Note:** *No angle specification is available for the 4.0L engine.*

89 Install the hold-down bolt and tighten it securely. Remove the positioning tool.

90 Install the camshaft position sensor and tighten the screws securely. **Caution:** *Check the position of the electrical connector on the sensor to make sure it is aligned with the mark you made in Step 82. If it isn't oriented correctly, DO NOT rotate the synchronizer to reposition it - doing so will result in the fuel system being out of time with the engine, will could damage the engine (and at the very least cause driveability problems). If the connector is not oriented properly, repeat the synchronizer installation procedure.*

91 Plug in the electrical connector to the sensor and reconnect the cable to the negative terminal of the battery.

1994 and 1995 4.0L V6 engines

Refer to illustrations 5.92 and 5.95

92 Locate the TDC mark on the crankshaft damper. Check for an additional mark at 26-degrees After Top Dead Center (ATDC). If you can't find this mark, make one at exactly 34mm (1.34 inches, or 1-11/32 inches) counterclockwise from the TDC mark **(see illustration)**.

93 Disconnect the cable from the negative terminal of the battery. Disconnect the electrical connector, remove the hold-down bolt

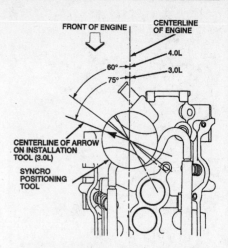

5.86 Synchronizer orientation details

and clamp and remove the camshaft position sensor.

94 Position the engine at TDC compression for cylinder number 1 (see Chapter 2, Part C).

95 Lubricate the drive gear and the O-ring with clean engine oil. Align the trailing edge of the sensor vane with the short mark at the left side of the sensor window **(see illustration)**.

96 Carefully insert the sensor and drive assembly into the engine. The sensor vane will rotate clockwise as the drive gear meshes with the gear on the camshaft.

97 Turn the sensor counterclockwise to make room for the hold-down clamp and bolt. Install the clamp and bolt, but don't tighten the bolt yet. Now turn the sensor back until it is parallel (lengthwise) with the centerline of the engine. Reconnect the electrical connector to the sensor.

98 Rotate the engine two revolutions and return the engine to TDC compression for cylinder number 1. This will take up any slack in the timing chain. Now continue to turn the engine until it is positioned at 26-degrees ATDC.

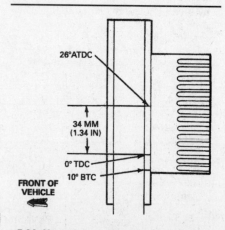

5.92 If one is not present, make a mark on the crankshaft damper exactly 34mm (1.34 inches) After Top Dead Center, which is 26-degrees ATDC

Chapter 6 Emissions control systems

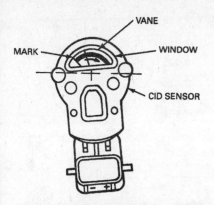

5.95 Line up the trailing edge of the sensor vane with the short mark at the left of the sensor window

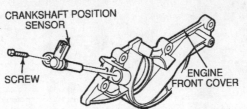

5.106 Crankshaft position sensor installation details - 2.3L four-cylinder engine

99 Reconnect the cable to the negative terminal of the battery, then turn the ignition key to the On position. Connect the positive (+) probe of a high-impedance (10-meg ohms) digital voltmeter to the backside of the center terminal of the sensor (the dark blue wire with the orange stripe). **Note:** *If the probes of your voltmeter are too large, you can use a pin or a straightened-out paper clip to backprobe the connector.* Connect the negative probe to a good ground.

100 Rotate the sensor counterclockwise until voltage reads on the voltmeter (it should be approximately battery voltage). Now turn the sensor clockwise until the voltmeter reads 0-volts. Finally, rotate the sensor slowly counterclockwise and stop exactly when the voltmeter reading changes from 0-volts to a positive voltage reading.

101 Detach the voltmeter and tighten the hold-down bolt securely.

Crankshaft position sensor

General information

102 The crankshaft position sensor (CKP) on 3.0L V6 engines equipped with distributors is located in the distributor. It consists of the distributor hall-effect switch and a rotary-vane cup. The passing of the open and closed portion of the vane past the hall switch results in a high/low voltage change that is seen by the PCM. The PCM then determines when to trigger the ignition module. On all other models the crankshaft position sensor is located at the front of the engine, in close proximity to the crankshaft damper. On these models the sensor detects the passing of teeth on the crankshaft damper. This data is used by the computer to calculate ignition firing and fuel injection timing.

Replacement

Four-cylinder engines

Refer to illustration 5.106

103 Disconnect the cable from the negative terminal of the battery. Drain the cooling system (see Chapter 1).

104 Remove the alternator (see Chapter 5).

105 Remove the lower radiator hose and the heater hose from the water pump inlet tube. Unscrew the two bolts and detach the water pump inlet tube from the water pump. **Note:** *Be sure to install a new O-ring on the tube during installation.*

106 Disconnect the electrical connector for the crankshaft position sensor. Unscrew the sensor mounting bolt and carefully pry the sensor from the engine front cover (see illustration).

107 Installation is the reverse of the removal procedure. Refill the cooling system (see Chapter 1), start the engine and check for leaks.

V6 engines

Refer to illustration 5.109

108 Detach the cable from the negative terminal of the battery.

109 Unplug the electrical connector from the sensor. Unscrew the two sensor mounting bolts and detach the sensor from the front cover (see illustration).

110 Installation is the reverse of the removal procedure.

Brake On/Off (BOO) switch

General Information

Refer to illustrations 5.111 and 5.113

111 The brake On/Off switch (BOO) tells the PCM when the brakes are being applied. The switch closes when brakes are applied and opens when the brakes are released. The BOO switch is located on the brake pedal assembly (see illustration).

112 The brake light circuit and bulbs are wired into the BOO circuit so it is important in diagnosing any driveability problems to make sure all the brake light bulbs are working properly (not burned out) or the driver may experience poor idle quality.

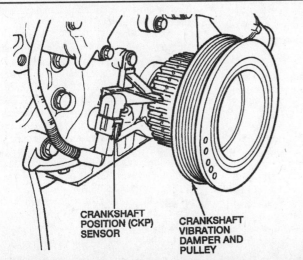

5.109 Crankshaft position sensor installation details - 4.0L V6 engine (3.0L V6 similar)

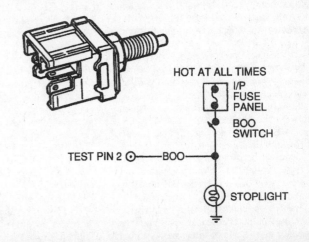

5.111 Since the brake light bulbs are wired into the BOO switch, check for any burned out bulbs that might cause the engine to idle poorly because of the open circuit (brake on signal)

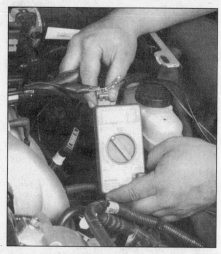

5.113 Checking for battery voltage to the BOO switch

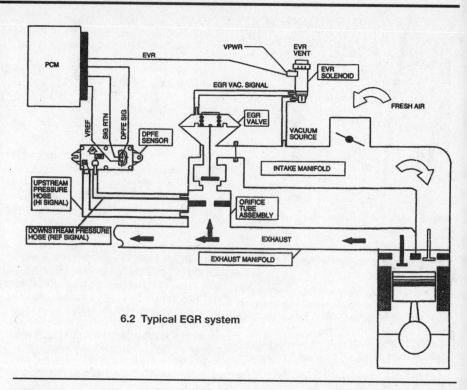

6.2 Typical EGR system

Check

113 Disconnect the electrical connector from the BOO switch and check for battery voltage to the BOO switch **(see illustration)**.
114 Also, check continuity from the BOO switch to the brake light bulbs. Change any burned out bulbs or damaged wire looms.

Replacement

115 Refer to Chapter 9 for the replacement procedure.

---

6 Exhaust Gas Recirculation (EGR) system

General description

Refer to illustration 6.2

1 The EGR system is used to lower NOx (oxides of nitrogen) emission levels caused by high combustion temperatures. The EGR recirculates a small amount of inert exhaust gases into the intake manifold. The additional mixture lowers the temperature of combustion thereby reducing the formation of NOx compounds.

2 All engines are equipped with the Differential Pressure Feedback EGR (DPFE) system **(see illustration)**. When the Powertrain Control Module (PCM) determines to operate the EGR, it drives the EGR vacuum regulator solenoid which applies a vacuum signal to the EGR valve. As the EGR valve opens it allows exhaust gases to enter the intake manifold to control combustion. The EGR gases must also pass through a metering orifice which creates a pressure differential that is proportional to the rate of EGR flow entering the intake manifold. The DPFE transducer picks up the differential pressure signal and provides the PCM with an analog feedback signal on the rate of EGR flow. The PCM uses the feedback signal to turn the EGR vacuum regulator solenoid on and off to achieve the desired EGR flow.

Check

3 Too much EGR flow tends to weaken combustion, causing the engine to run rough or stop. When EGR flow is excessive, the engine can stop after a cold start or at idle after deceleration, the vehicle can surge at cruising speeds or the idle may be rough. If the EGR valve remains constantly open, the engine may not idle at all.
4 Too little or no EGR flow allows combustion temperatures to get too high during acceleration and load conditions. This can cause spark knock (detonation), engine overheating or emission test failure.
5 The following checks will help you pinpoint problems in the EGR system.
Warning: *Where the procedure says to lift up on the EGR valve diaphragm, it's a good idea to wear a heat-resistant glove to prevent burns.*

EGR valve

6 The EGR valve is controlled by a normally-open solenoid which allows vacuum to pass when energized. The PCM energizes the solenoid to turn on the EGR. The PCM controls the EGR when three conditions are present: engine coolant is above 113-degrees F, the TPS is at part throttle and the MAF sensor is in its mid-range. A problem with the EGR system will set Code 328 through 337.
7 Make sure the vacuum hoses are in good condition and hooked up correctly.
8 To perform a leakage test, attach a vacuum pump to the EGR valve. Apply a vacuum of 5 to 6 in-Hg to the valve. The EGR valve should hold vacuum.
9 If access is possible, position your finger tip under the vacuum diaphragm and apply vacuum to the EGR valve. You should feel movement of the EGR diaphragm.
10 Remove the EGR valve (see Step 19) and clean the inlet and outlet ports with a wire brush or scraper. Do not sandblast the valve or clean it with gasoline or solvents. These liquids will destroy the EGR valve diaphragm.
11 If the specified conditions are not met, replace the EGR valve.

EGR control system

Refer to illustration 6.18

12 If a code is displayed, there are several possibilities for EGR failure. Engine coolant temperature sensor, TPS, MAF sensor, TCC system and the engine rpm govern the parameters the EGR system use for distinguishing the correct ON time.
13 All systems use an Electronic Vacuum Regulator to control the amount of exhaust gas through the EGR valve. The valve is normally open (engine at operating temperature), and vacuum not directed to the EGR valve is vented to the atmosphere. The PCM uses a controlled "pulse width" or electronic signal to turn the EGR ON and OFF (the "duty cycle"). The duty cycle should be zero percent (no EGR) when in Park or Neutral, when the TPS input is below the specified value or when Wide Open Throttle (WOT) is indicated.
14 To check the EGR vacuum regulator, disconnect the electrical connector to the EGR vacuum regulator, turn the ignition key ON (engine not running) and check for voltage to the solenoid connector. Battery voltage should be present.
15 Next, use an ohmmeter and check the resistance of the EGR vacuum regulator solenoid. It should be between 30 and 70 ohms.

Chapter 6 Emissions control systems

6-15

6.18 Check for reference voltage (VREF) to the DPFE sensor at the harness connector (arrow)

between 4.0-6.0 volts. If VREF is not within this range, the problem is most likely in the PCM. Don't overlook the possibility that the battery voltage is low.

Component replacement

EGR valve

Refer to illustrations 6.23a and 6.23b

Note: *When buying a new EGR valve, make sure that you get the correct one. Use the stamped code located on the top of the EGR valve when purchasing it.*

19 Disconnect the negative battery cable from the battery.
20 Remove the air cleaner outlet tube (see Chapter 4).
21 Disconnect the EGR valve vacuum hose.
22 Disconnect the EGR exhaust manifold tube.
23 Remove the EGR valve retainer bolts and EGR valve from intake manifold, or intake manifold EGR tube **(see illustrations)**.
24 Remove the EGR gasket, clean the manifold of any remaining gasket material and clean the EGR valve gasket surface if valve is to be reused.
25 Installation is the reverse of removal.

EGR tube

26 Remove the EGR valve.
27 Disconnect the EGR tube at the exhaust

16 Disconnect the vacuum hoses from the EGR regulator. Lightly blow air into the EGR regulator and verify that air does not flow through the regulator. Replace the regulator if this does not occur.
17 Disconnect the electrical connector from the EGR regulator. Using jumpers, ground one pin on the regulator and apply battery voltage to the other. Lightly blow air into the EGR regulator and verify that air flows through the regulator. Replace the regulator if this does not occur.
18 Check for reference voltage (VREF) to the DPFE sensor **(see illustration)**. With the ignition key on (engine not running), check for voltage on the harness side of the DPFE sensor electrical connector. VREF should be

manifold, on some models it is necessary to remove the EGR tube to manifold connector **(see illustrations 6.23a or 6.23b)**.
28 Installation is the reverse of removal.

EGR Vacuum Regulator Solenoid

29 Disconnect the battery ground cable.
30 Disconnect the electrical connector from the EGR vacuum regulator solenoid.
31 Disconnect the vacuum hoses from the EGR vacuum regulator solenoid.
32 Remove the nuts securing the solenoid and remove the solenoid.

Differential Pressure Feedback Sensor

33 Disconnect the battery ground cable.
34 Disconnect the electrical connector from the DPFE sensor.
35 Disconnect the vacuum hoses from the sensor.
36 Remove the bolts securing the sensor and remove it.

7 Evaporative Emissions Control System (EECS)

Refer to illustration 7.2

General description

1 This system is designed to trap and store fuel vapors that evaporate from the fuel

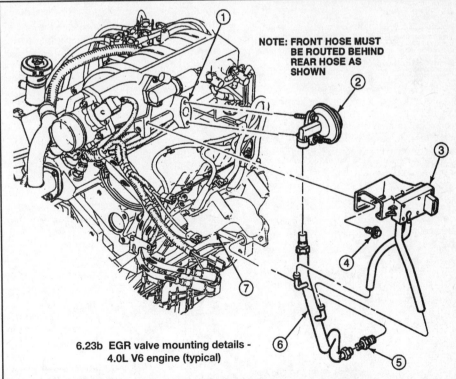

6.23a EGR valve mounting details - 2.3L four-cylinder engine

1	EGR valve	5	Exhaust manifold
2	Gasket		
3	Bolt	6	EGR tube
4	Intake manifold		

6.23b EGR valve mounting details - 4.0L V6 engine (typical)

1	Intake manifold	5	EGR valve tube to manifold connector
2	EGR valve	6	EGR tube
3	EGR transducer	7	Exhaust manifold
4	Bolt	8	Transmission kickdown cable

Chapter 6 Emissions control systems

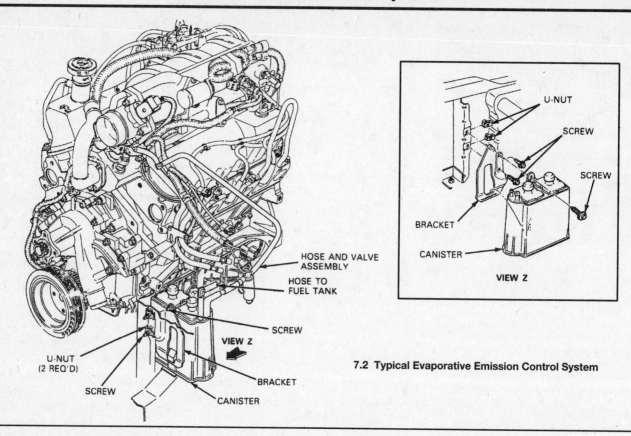

7.2 Typical Evaporative Emission Control System

tank, throttle body and intake manifold.

2   The Evaporative Emission Control System (EECS) consists of a charcoal-filled canister and the lines connecting the canister to the fuel tank, ported vacuum and intake manifold vacuum (see illustration).

3   Fuel vapors are transferred from the fuel tank, throttle body and intake manifold to a canister where they are stored when the engine is not operating. When the engine is running, the fuel vapors are purged from the canister by a purge control solenoid, which is PCM controlled, and consumed in the normal combustion process.

Check

4   Poor idle, stalling and poor driveability can be caused by an inoperative purge control solenoid, a damaged canister, split or cracked hoses or hoses connected to the wrong tubes.

5   Evidence of fuel loss or fuel odor can be caused by fuel leaking from fuel lines or the TBI, a cracked or damaged canister, an inoperative bowl vent valve, an inoperative purge valve, disconnected, misrouted, kinked, deteriorated or damaged vapor or control hoses or an improperly seated air cleaner or air cleaner gasket.

6   Inspect each hose attached to the canister for kinks, leaks and breaks along its entire length. Repair or replace as necessary.

7   Inspect the canister. If it is cracked or damaged, replace it.

8   Look for fuel leaking from the bottom of the canister. If fuel is leaking, replace the canister and check the hoses and hose routing.

9   Apply a short length of hose to the lower tube of the purge valve assembly and attempt to blow through it. Little or no air should pass into the canister (a small amount of air will pass because the canister has a constant-purge hole).

10   With a hand-held vacuum pump, apply vacuum through the control vacuum signal tube near the throttle body to the purge control solenoid diaphragm.

11   If the purge control solenoid does not hold vacuum for at least 20 seconds, the purge control solenoid is leaking and must be replaced.

12   If the diaphragm holds vacuum, apply battery voltage to the purge control solenoid and observe that vacuum (vapors) are allowed to pass through to the intake system.

Component replacement

13   Clearly label, then detach, all vacuum lines from the canister.

14   Loosen the canister mounting clamp bolt and pull the canister out.

15   Installation is the reverse of removal.

8   Positive Crankcase Ventilation (PCV) system

Refer to illustrations 8.1 and 8.2

1   The Positive Crankcase Ventilation (PCV) system reduces hydrocarbon emissions by scavenging crankcase vapors. It does this by circulating fresh air from the air cleaner through the crankcase, where it mixes with blow-by gases and is then rerouted through a PCV valve to the intake

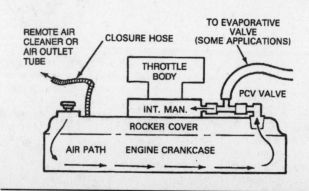

8.1 Gas flow in a typical PCV system

Chapter 6 Emissions control systems

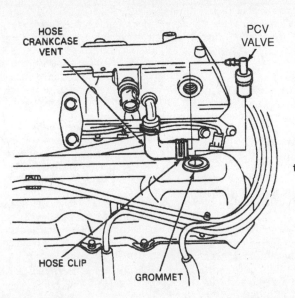

8.2 The PCV valve is mounted in the valve cover (on the driver's side on V6 engines). There is also a hose running from the air cleaner outlet tube to the oil filler neck on the passenger side valve cover on V6 engines

9.18 To remove the bimetal sensor, detach the hoses and pry off the retaining clip with a small screwdriver

manifold **(see illustration)**.

2 The main components of the PCV system are the PCV valve, a fresh-air filtered inlet and the vacuum hoses connecting these two components with the engine and the EECS system **(see illustration)**.

3 To maintain idle quality, the PCV valve restricts the flow when the intake manifold vacuum is high. If abnormal operating conditions arise, the system is designed to allow excessive amounts of blow-by gases to flow back through the crankcase vent tube into the air cleaner to be consumed by normal combustion.

4 Checking and replacement of the PCV valve and filter is covered in Chapter 1.

9 Inlet air temperature control system (1993 and 1994 models)

General description

1 The inlet air temperature control system provides heated intake air during warm-up, then maintains the inlet air temperature within a 70-degree F to 105-degree F operating range by mixing warm and cool air. This allows leaner fuel/air mixture settings for the EFI system, which reduces emissions and improves driveability.

2 Two fresh air inlets - one warm and one cold - are used. The balance between the two is controlled by the intake manifold vacuum, cold weather modulator and a bimetal sensor. A vacuum motor, which operates a heat duct valve in the air cleaner, is controlled by the vacuum switch **(see illustration)**.

3 When the underhood temperature is cold, warm air radiating off the exhaust manifold is routed by a shroud which fits over the manifold up through a hot air inlet tube and into the air cleaner (see Chapter 4). This provides warm air for the EFI, resulting in better driveability and faster warm-up. As the underhood temperature rises, a heat duct valve is gradually closed by a vacuum motor and the air cleaner draws air through a cold air duct instead. The result is a consistent intake air temperature.

4 A temperature vacuum switch mounted in the air cleaner cover monitors the temperature of the inlet air heated by the exhaust manifold. A bimetal disc in the temperature vacuum switch orients itself in one of two positions, depending on the temperature. One position allows vacuum through a hose to the motor; the other position blocks vacuum.

5 The vacuum motor itself is regulated by a Cold Weather Modulator (CWM), mounted in the side of the air cleaner housing assembly, between the temperature vacuum switch and the motor, which provides the motor with a range of graduated positions between fully open and fully closed.

Check

Note: *Make sure the engine is cold before beginning this test.*

6 Always check the vacuum source and the integrity of all vacuum hoses between the source and the vacuum motor before beginning the following test. Do not proceed until they're okay.

7 Apply the parking brake and block the wheels.

8 Remove components as necessary from the air intake duct in order to see the vacuum motor door (see Chapter 4).

9 Observe the vacuum motor door position - it should be open. If it isn't, it may be binding or sticking. Make sure it's not rusted in an open or closed position by attempting to move it by hand. If it's rusted, it can usually be freed by cleaning and oiling the hinge. If it fails to work properly after servicing, replace it.

10 If the vacuum motor door is okay but the motor still fails to operate correctly, check carefully for a leak in the hose leading to it. Check the vacuum source to and from the bimetal sensor and the cold weather modulator as well. If no leak is found, replace the vacuum motor (see Step 23).

11 Start the engine. If the duct door has moved or moves to the "heat on" (closed to fresh air) position, go to Step 15.

12 If the door stays in the "heat off" (closed to warm air) position, place a finger over the bimetal sensor bleed. The duct door must move rapidly to the "heat on" position. If the door doesn't move to the "heat on" position, stop the engine and replace the vacuum motor (see Step 23). Repeat this Step with the new vacuum motor.

13 With the engine off, allow the bimetal sensor and the Cold Weather Modulator to cool completely.

14 Restart the engine. The duct door should move to the "heat on" position. If the door doesn't move or moves only partially, replace the bimetal sensor (see Step 18).

15 Start and run the engine briefly (less than 15 seconds). The duct door should move to the "heat on" position.

16 Shut off the engine and watch the duct door. It should stay in the "heat on" position for at least two minutes.

17 If it doesn't stay in the "heat on" position for at least two minutes, replace the CWM.

Component replacement

Refer to illustrations 9.18 and 9.21

Bimetal sensor

18 Clearly label and detach both vacuum hoses from the bimetal sensor (one is coming from the vacuum source at the manifold and the other is going to the vacuum motor underneath the air cleaner housing) **(see illustration)**.

19 Remove the air cleaner housing cover (see Chapter 4).

20 Pry the sensor retaining clip off with a screwdriver.

9.21 The bimetal sensor is mounted inside the air cleaner housing

21 Remove the bimetal sensor **(see illustration)**.
22 Installation is the reverse of removal procedure.

Vacuum motor

23 Remove the air intake duct (see Chapter 1). Place the assembly on a workbench.
24 Detach the vacuum hose from the motor.
25 Drill out the vacuum motor retaining strap rivet.
26 Remove the motor.
27 Installation is the reverse of removal. Use a sheet metal screw of the appropriate size to replace the rivet.

10 Catalytic converter

General description

1 The catalytic converter is an emission-control device added to the exhaust system to reduce pollutants from the exhaust gas stream. A single-bed converter design is used in combination with a three-way (reduction) catalyst. The catalytic coating on the three-way catalyst contains platinum and rhodium, which lowers the levels of oxides of nitrogen (NOx) as well as hydrocarbons (HC) and carbon monoxide (CO).

Physical checks

2 The catalytic converter requires little if any maintenance or servicing at regular intervals. However, the system should be inspected whenever the vehicle is raised on a lift or if the exhaust system is checked or serviced.
3 Check all connections in the exhaust pipe assembly for looseness or damage. Also check all the clamps for damage, cracks, or missing fasteners. Check the rubber hangers for cracks.
4 The converter itself should be checked for damage or dents which could affect its performance and/or be hazardous to your health. At the same time the converter is inspected, check the metal protector plate under it and the heat insulator above it for damage or loose fasteners.

Functional checks

5 Potential converter problems can be associated with two situations: vehicle fails state smog certification check or exhibits low power. Both situations can be caused by a converter that has been overheated and is either non-functional or restrictive.
6 A non-functional converter is difficult to diagnose. If all other engine systems are operating properly, and the converter is relatively cold (this is a judgment call, no specifications exist, **do not** check by feeling converter) the converter is probably bad.
7 A restricted converter can be checked. If a performance issue is in question, proceed to backpressure check.

Backpressure check

8 Attach a vacuum gauge to a source of manifold vacuum and attach a tachometer to the engine.
9 Set the parking brake and put the transmission in Park.
10 Start the engine and warm it to operating temperature. Turn the engine off and wait for a few minutes (this allows any backpressure to escape).
11 Start the engine and observe the vacuum gauge. Vacuum should be 16 in-Hg or greater on an engine in a good state of tune.
12 Let the engine idle for a few minutes while watching the gauge. Vacuum should hold fairly steady.
13 Increase the engine speed and hold it for one minute at 2000 rpm while observing the vacuum gauge. The vacuum gauge should read continuously high vacuum at the end of one minute with no additional throttle required to maintain rpm. If the vacuum gauge dropped significantly or if more throttle was required to maintain rpm, converter restriction can be suspected. Turn the engine off and proceed to the next operation to isolate the problem further.
14 Let the exhaust system cool, then remove the exhaust pipe at the exhaust manifold(s).
15 Repeat the test. If vacuum is now steady and high in the engine-running test, the converter is restricted. Replace the converter. **Caution:** *Although rare, if a restricted muffler is suspected, reconnect the converter, disconnect the muffler and repeat the test to determine if the muffler is the cause.*

Replacement

16 Do not attempt to remove the catalytic converter until the complete exhaust system is cool. Raise the vehicle and support it securely on jackstands. Apply some penetrating oil to the clamp bolts and allow it to soak in.
17 Remove the bolts and the rubber hangers, then separate the converter from the exhaust pipes. Remove the old gaskets if they are stuck to the pipes.
18 Installation of the converter is the reverse of removal. Use new exhaust pipe gaskets and tighten the clamp bolts securely. Replace the rubber hangers with new ones if the originals are deteriorated. Start the engine and check carefully for exhaust leaks.

Chapter 7 Part A
Manual transmission

Contents

	Section
Extension housing oil seal (2WD models) - replacement	6
General information	1
Manual transmission - removal and installation	4
Shift lever - removal and installation	2
Speedometer pinion gear and seal - removal and installation	5
Transmission mount - check and replacement	3
Transmission overhaul - general information	7

Specifications

General
Transmission type... 5-speed synchromesh
Lubricant type... See Chapter 1

Torque specifications | Ft-lbs
Transmission-to-clutch housing bolt or nut .... 18 to 38
Mount-to-crossmember nut .... 65 to 85
Mount-to-transmission bolt .... 60 to 80
Crossmember-to-frame bracket nut and bolt .... 65 to 85

Chapter 7 Part A Manual transmission

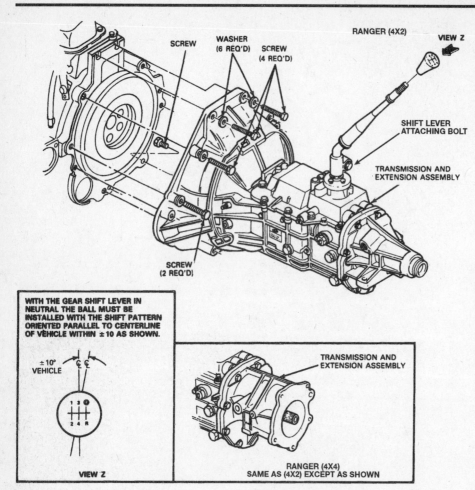

1.1 A typical 5-speed manual transmission

1 General information

Refer to illustration 1.1

All vehicles covered in this manual come equipped with a 5-speed manual transmission (see illustration) or an automatic transmission. All information on the manual transmission is included in this Part of Chapter 7. Information on the automatic transmission can be found in Part B of this Chapter.

Due to the complexity, unavailability of replacement parts and the special tools necessary, internal transmission repair procedures are not recommended for the home mechanic.

Depending on the expense involved in having a faulty transmission overhauled, it may be an advantage to consider replacing the unit with either a new or rebuilt one. Your local dealer or transmission shop should be able to supply you with information concerning cost, availability and exchange policy. Regardless of how you decide to remedy a transmission problem, you can still save a lot of money by removing and installing the unit yourself.

2 Shift lever - removal and installation

Refer to illustration 2.2.

1 Place the transmission in Neutral.
2 Carefully pull back on the carpeting and remove the shifter boot retaining bolts (see illustration).
3 Remove the bolt that secures the shift lever to the transmission (see illustration 1.1).
4 Pull the shift lever straight up and off the transmission.
5 Installation is the reverse of the removal Steps.

3 Transmission mount - check and replacement

Refer to illustrations 3.2, 3.3a and 3.3b

1 Insert a large screwdriver or pry bar into the space between the transmission extension housing and the frame crossmember and pry up.
2 The transmission should not move significantly away from the mount (see illustration). If it does, the mount should be replaced.
3 To replace the mount, remove the two nuts securing the mount to the frame crossmember and the two bolts securing the mount to the transmission extension housing (see illustrations).

2.2 Hold back the carpet and remove the shifter boot retaining bolts or screws

3.2 Pry up on the transmission mount and check for excessive looseness

Chapter 7 Part A Manual transmission

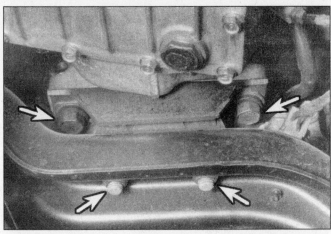

3.3a Remove the two bolts attaching the transmission mount to the frame crossmember and the two bolts securing the mount to the transmission extension housing

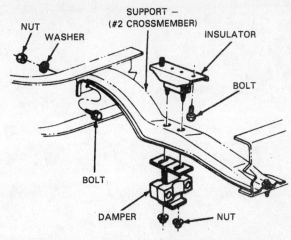

3.3b A typical transmission mount assembly - exploded view

4 Place a jack under the transmission with a piece of wood on top of it to protect the transmission case. Apply a slight amount of jack pressure and raise the transmission. Remove mount and install a new one.
5 Installation is the reverse of the removal Steps.

4 Manual transmission - removal and installation

Refer to illustration 4.8

1 Disconnect the negative cable from the battery.
2 Shift the transmission into neutral.
3 Remove the shift lever (see Section 2).
4 Drain the transmission lubricant only if you will be disassembling it. Otherwise, you are going to have a problem disposing of the lubricant yourself. (see Chapter 1).
5 Raise the vehicle and support it securely on jackstands.
6 Remove the driveshaft (see Chapter 8) and install a plug, on 2WD models, in the transmission rear extension housing to prevent lubricant leakage.
7 Disconnect the hydraulic fluid line from the clutch release cylinder (see Chapter 8) and plug the line to prevent fluid spillage.
8 Disconnect the speedometer cable from the transfer case or transmission **(see illustration)**.
9 Remove the starter motor (see Chapter 5).
10 Disconnect the electrical connections from the backup lamp and shift indicator switch.
11 Place a jack under the engine and protect the oil pan with a wood block. Apply a slight amount of jack pressure to support the rear of the engine.
12 On 4WD models, remove the transfer case (see Chapter 7, Part C).
13 On V6 engines, remove the exhaust system (see Chapter 4).

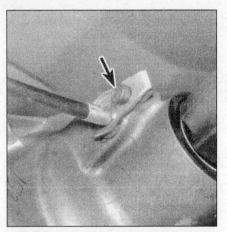

4.8 Remove the bolt and retainer securing the speedometer and disconnect the speedometer cable from the transmission or transfer case

14 Place a transmission jack under the transmission. Apply a slight amount of jack pressure to support the transmission. Remove the bolts, lockwashers and plain washers that secure the transmission to the engine **(see illustration 1.1)**.
15 Remove the bolts and nuts securing the transmission mount to the frame crossmember **(see illustration 3.3b)**.
16 Remove the nuts securing the frame crossmember to the frame side rails. **Note:** *If you do not remove the mount from the transmission, it will be necessary to raise the transmission enough to clear the mount bolts, then slide the crossmember toward the rear.*
17 Slowly lower the jack supporting the engine and carefully pull the transmission toward the rear and work it free from the locating dowels. Pull the transmission straight back until the input shaft clears the clutch assembly.
18 Lower the transmission jack and remove the transmission.
19 Installation is the reverse of the removal

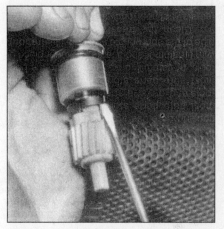

5.3 Pry the retaining clip from the pinion gear and slide the gear off the cable

Steps with the following additions:
a) Mount the transmission in a transmission jack and position it under the vehicle.
b) Start the transmission input shaft into the clutch disc. Align the shaft splines with the clutch disc splines and move the transmission forward.
c) Tighten the bolts and nuts to the torque listed in this Chapter's Specifications.

5 Speedometer pinion gear and seal - removal and installation

Refer to illustrations 5.3 and 5.4

1 Disconnect the speedometer cable from the transmission.
2 Pull the speedometer pinion gear straight out of the transfer case or extension housing.
3 Use a small screwdriver and remove the retaining clip from the pinion gear, then slide the gear off the cable **(see illustration)**.
4 If necessary, use a small screwdriver

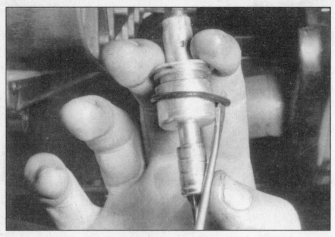

5.4 To replace the O-ring, pry off the old O-ring with a small screwdriver

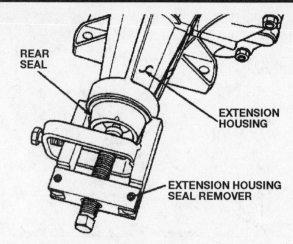

6.3 Use a special tool, or equivalent, to remove the oil seal from the extension housing

and remove the O-ring from the retaining groove **(see illustration)**. Discard the O-ring.
5 Lubricate the new O-ring with transmission lubricant and install it in the retaining groove. Make sure it's seated correctly.
6 Install the pinion gear on the cable and install the retaining clip. Make sure the clip is properly seated in the groove.
7 Install the speedometer pinion gear and cable in the transfer case or rear extension housing and secure it with the bolt and bracket.

6 Extension housing oil seal (2WD models) - replacement

Refer to illustration 6.3.
1 Raise the vehicle and support it securely on jackstands.
2 Remove the driveshaft (see Chapter 8). Some transmission lubricant may drain out when the driveshaft slip joint is withdrawn from the extension housing.
3 To remove the oil seal from the end of the extension housing, use a special tool, available at most auto parts stores, and withdraw the oil seal **(see illustration)**. If the special tool is not available and there is access, use a thin blade screwdriver or chisel and pry the oil seal out.
4 Inspect the sealing surface on the driveshaft slip joint for scoring or burrs that may damage the new oil seal. If the yoke is damaged in any way, replace it.
5 Inspect the extension housing counterbore for burrs. If found, remove them with emery cloth or medium grit wet-and-dry sandpaper. Use a clean cloth dipped in solvent and remove any sanding residue from the counterbore.
6 Apply silicone sealant to the inside diameter of the oil seal and apply grease to the end of the rubber boot portion.
7 Using a seal driver, large socket or section of pipe, install a new oil seal and make sure it is completely seated in the counterbore.
8 Install the driveshaft (see Chapter 8).
9 Lower the vehicle and check the transmission lubricant level. Top up if necessary (see Chapter 1).

7 Transmission overhaul - general information

Overhauling a manual transmission is a difficult job for the do-it-yourselfer. It involves the disassembly and reassembly of many small parts. Numerous clearances must be precisely measured and, if necessary, changed with select-fit spacers and snap-rings. As a result, if transmission problems arise, it can be removed and installed by a competent do-it-yourselfer, but overhaul should be left to a transmission repair shop. Rebuilt transmissions may be available. Check with your dealer's parts department and auto parts stores. At any rate, the time and money involved in an overhaul is almost sure to exceed the cost of a rebuilt unit.

Nevertheless, it's not impossible for an inexperienced mechanic to rebuild a transmission if the special tools are available and the job is done in a deliberate step-by-step manner so nothing is overlooked.

The tools necessary for an overhaul include internal and external snap-ring pliers, bearing puller, slide hammer, set of pin punches, dial indicator and possibly a hydraulic press. In addition, a large, sturdy workbench and a large vise or transmission stand will be required.

During disassembly of the transmission, make careful notes of how each piece comes off, where it fits in relation to other pieces and what holds it in place.

Before taking the transmission apart for repair, it will help if you have some idea what area of the transmission is malfunctioning. Certain problems can be closely tied to specific areas in the transmission, which can make component examination and replacement easier. Refer to the *Troubleshooting* Section at the front of this manual for information regarding possible sources of trouble.

Chapter 7 Part B
Automatic transmission

Contents

	Section		Section
Automatic transmission - removal and installation	8	Neutral start switch - removal, installation and adjustment	9
Diagnosis - general	2	Shift cable - adjustment	3
Extension housing oil seal - replacement	7	Shift cable - replacement	4
General information	1	Transmission mount - check and replacement	11
Kickdown cable - adjustment	5	Vacuum diaphragm - removal and installation	10
Kickdown cable - replacement	6		

Specifications

General

Transmission type
  1993 and 1994 .... A4LD
  1995 on
    Models with four-cylinder and 3.0L V6 engines 4R44E
    Models with 4.0L V6 engine .... 4R55E or 5R55E
Fluid type and capacity .... See Chapter 1

Torque specifications   **Ft-lbs** (unless otherwise indicated)
Transmission-to-engine bolts .... 28 to 38
Torque converter-to-driveplate nuts .... 20 to 34
Neutral start switch .... 84 to 120 in-lbs

1 General information

Ford Rangers covered in this manual come equipped with a five-speed manual transmission or an automatic transmission. All information on the automatic transmission is included in this Part of Chapter 7. Information on the manual transmission can be found in Part A of this Chapter.

1993 and 1994 models with automatic transmissions come equipped with the A4LD automatic transmission for all engines. This is a 4-speed transmission with automatic overdrive and a lockup torque converter.

1995 and later models use a 4R44E with the four-cylinder and 3.0L V6 engine. The 4.0L V6 engine uses the 4R55E on 1995 and 1996 models and the 5R55E on 1997 models. These electronically controlled transmissions are also 4-speeds.

Due to the complexity of the automatic transmissions covered in this manual and the need for specialized equipment to perform most service operations, this Chapter contains only general diagnosis, routine maintenance, adjustment and removal and installation procedures.

If the transmission requires major repair work, it should be taken to a dealer service department or an automotive or transmission repair shop. You can, however, remove and install the transmission yourself and save the expense, even if the repair work is done by a transmission shop.

2 Diagnosis - general

Note: *Automatic transmission malfunctions may be caused by five general conditions: poor engine performance, improper adjustments, hydraulic malfunctions, mechanical malfunctions or malfunctions in the computer or its signal network. Diagnosis of these problems should always begin with a check of the easily repaired items: fluid level and condition (see Chapter 1) and shift linkage adjustment. Next, perform a road test to determine if the problem has been corrected or if more diagnosis is necessary. If the problem persists after the preliminary tests and corrections are completed, additional diagnosis should be done by a dealer service department or transmission repair shop. Refer to the Troubleshooting Section at the front of this manual for information on symptoms of transmission problems.*

Preliminary checks

1 Drive the vehicle to warm the transmission to normal operating temperature.
2 Check the fluid level as described in Chapter 1:
  a) *If the fluid level is unusually low, add enough fluid to bring the level within the designated area of the dipstick, then check for external leaks (see below).*

b) If the fluid level is abnormally high, drain off the excess, then check the drained fluid for contamination by coolant. The presence of engine coolant in the automatic transmission fluid indicates that a failure has occurred in the internal radiator walls that separate the coolant from the transmission fluid (see Chapter 3).

c) If the fluid is foaming, drain it and refill the transmission, then check for coolant in the fluid or a high fluid level.

3 Check the engine idle speed. **Note:** *If the engine is malfunctioning, do not proceed with the preliminary checks until it has been repaired and runs normally.*

4 Inspect the shift linkage (see Section 3). Make sure that it's properly adjusted and that the linkage operates smoothly.

Fluid leak diagnosis

5 Most fluid leaks are easy to locate visually. Repair usually consists of replacing a seal or gasket. If a leak is difficult to find, the following procedure may help.

6 Identify the fluid. Make sure it's transmission fluid and not engine oil or brake fluid (automatic transmission fluid is a deep red color when new and may turn light brown after the vehicle has been driven some distance).

7 Try to pinpoint the source of the leak. Drive the vehicle several miles, then park it over a large sheet of cardboard. After a minute or two, you should be able to locate the leak by determining the source of the fluid dripping onto the cardboard.

8 Make a careful visual inspection of the suspected component and the area immediately around it. Pay particular attention to gasket mating surfaces. A mirror is often helpful for finding leaks in areas that are hard to see.

9 If the leak still cannot be found, clean the suspected area thoroughly with a degreaser or solvent, then dry it.

10 Drive the vehicle for several miles at normal operating temperature and varying speeds. After driving the vehicle, visually inspect the suspected component again.

11 Once the leak has been located, the cause must be determined before it can be properly repaired. If a gasket is replaced but the sealing flange is bent, the new gasket will not stop the leak. The bent flange must be straightened.

12 Before attempting to repair a leak, check to make sure that the following conditions are corrected or they may cause another leak. **Note:** *Some of the following conditions cannot be fixed without highly specialized tools and expertise. Such problems must be referred to a transmission repair shop or a dealer service department.*

Gasket leaks

13 Check the pan periodically. Make sure the bolts are tight, no bolts are missing, the gasket is in good condition and the pan is flat (dents in the pan may indicate damage to the valve body inside).

14 If the pan gasket is leaking, the fluid level or the fluid pressure may be too high, the vent may be plugged, the pan bolts may be too tight, the pan sealing flange may be warped, the sealing surface of the transmission housing may be damaged, the gasket may be damaged or the transmission casting may be cracked or porous. If sealant instead of gasket material has been used to form a seal between the pan and the transmission housing, it may be the wrong sealant.

Seal leaks

15 If a transmission seal is leaking, the fluid level or pressure may be too high, the vent may be plugged, the seal bore may be damaged, the seal itself may be damaged or improperly installed, the surface of the shaft protruding through the seal may be damaged or a loose bearing may be causing excessive shaft movement.

16 Make sure the dipstick tube seal is in good condition and the tube is properly seated. Periodically check the area around the speedometer gear or sensor for leakage. If transmission fluid is evident, check the O-ring for damage.

Case leaks

17 If the case itself appears to be leaking, the casting is porous and will have to be repaired or replaced.

18 Make sure the oil cooler hose fittings are tight and in good condition.

Fluid comes out vent pipe or fill tube

19 If this condition occurs, the transmission is overfilled, there is coolant in the fluid, the case is porous, the dipstick is incorrect, the vent is plugged or the drain back holes are plugged.

3 Shift cable - adjustment

Refer to illustration 3.3

1 Raise the vehicle and support it securely on jackstands.

2 Have an assistant position the column shift selector lever in the Drive Overdrive position and hold in place during adjustment. If working by yourself, hang a three-pound weight on the gear selector lever.

3 Working under the vehicle, pull down on the lock tab on the shift cable and remove the fitting from the manual shift lever ball stud with a screwdriver **(see illustration)**.

4 Position the transmission manual shift lever in the Drive Overdrive position by moving the bellcrank lever all the way to the rear (counterclockwise), then forward three detents (clockwise).

5 Connect the cable end fitting to the manual lever ball stud.

6 Push the lock tab all the way down to lock the cable in the correctly adjusted position.

7 If used, remove the weight from the gear

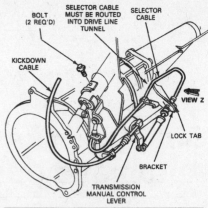

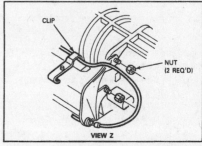

3.3 Shift cable details

selector lever.

8 After adjustment, check for Park engagement. The control lever must move to the right when engaged in Park. Check the control lever in all detent positions with the engine running to ensure correct detent/transmission action and adjust and readjust if necessary.

9 Remove the jackstands and lower the vehicle to the ground.

10 Test drive the vehicle slowly at first to make sure the adjustment is correct.

4 Shift cable - replacement

1 Use a screwdriver and pry the end fitting from the steering column lever ball stud.

2 Remove the nuts securing the shift cable bracket to the steering column bracket and remove the bracket.

3 Raise the vehicle and support it securely on jackstands.

4 Working under the vehicle, pull down on the lock tab on the shift cable and remove the fitting from the manual lever ball stud with a screwdriver **(see illustration 3.3)**.

5 Bend back the retaining tab and disengage the cable from the transmission bracket. Slide the cable down from the transmission bracket and remove the cable.

6 Installation is the reverse of the removal procedure with the following additions:

a) *Adjust the shift linkage (see Section 3).*
b) *Remove the jackstands and lower the vehicle to the ground.*

Chapter 7 Part B Automatic transmission

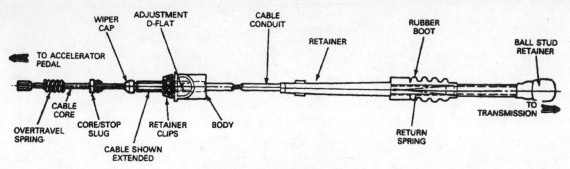

5.1 Details of the kickdown cable

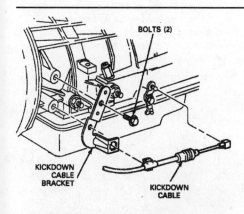

6.2 The kickdown cable connects at one end to the transmission ball stud...

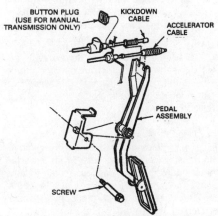

6.4 ...and at the other end to the accelerator cable

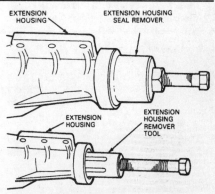

7.3a These special tools are available to remove the oil seal and bushing from the transmission extension housing

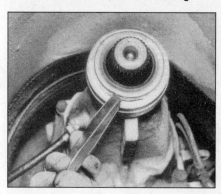

7.3b Be extremely careful not to damage the sealing surface on the housing when prying out the rear seal

5 Kickdown cable - adjustment

Refer to illustration 5.1

Note: *The kickdown cable for the automatic transmission is self adjusting. If there are problems with proper kickdown when depressing the accelerator pedal to the floor, check for aftermarket carpets or pads that may be preventing the throttle from opening fully.*

1   From under the hood, depress the D-flat while pulling the cable conduit out of the cable body **(see illustration)**.
2   Have an assistant depress the accelerator pedal firmly to the floor. This adjusts the kickdown cable.

6 Kickdown cable - replacement

Refer to illustrations 6.2 and 6.4

1   Note the cable routing carefully before removal.
2   Disconnect the kickdown cable from the ball stud on the transmission **(see illustration)**.
3   Collapse the retaining tabs on the cable conduit. Remove the cable from the transmission bracket.
4   Disconnect the kickdown cable from the top of the accelerator pedal **(see illustration)**.
5   Collapse the retaining clips at the dash panel and push the cable through the hole into the engine compartment.
6   Readjust the cable by depressing the D-flat while pulling the cable conduit back from the cable body.
7   Install the cable through the dash panel hole into the passenger compartment. Make sure the cable retaining tabs are fully seated.
8   Attach the cable end to the top of the accelerator cable.
9   Route the cable exactly the same as the original.
10   Install the cable conduit to the transmission mounting bracket. Make sure the cable retaining tabs are fully seated.
11   Connect the kickdown cable end to the ball stud on the transmission kickdown lever.
12   Adjust the cable (see Section 5).

7 Extension housing oil seal - replacement

Refer to illustrations 7.3a, 7.3b, 7.7 and 7.8

1   Raise the vehicle and support it securely on jackstands.
2   Remove the driveshaft (see Chapter 8). Some transmission fluid may drain out when the driveshaft universal joint is withdrawn from the extension housing.
3   To remove the oil seal from the end of the extension housing, use a special tool made for this purpose, available at most auto parts stores that carry tools, and withdraw the oil seal **(see illustration)**. If the special tool is not available, and there is access, use a thin blade screwdriver or chisel to remove the oil seal **(see illustration)**.
4   The extension housing bushing can also be replaced. If the bushing is worn or damaged, removed it with the specified Ford tool or equivalent **(see illustration 7.3a)**. Makeshift tools may damage the bushing.
5   Inspect the oil seal contact surface on the universal joint yoke for scoring or burrs that may damage the new oil seal. If the yoke is damaged in any way, replace it.
6   Inspect the extension housing counterbore for burrs. If found, remove them with emery cloth or medium grit wet-and-dry sandpaper. Use a clean cloth dipped in sol-

7B-4 Chapter 7 Part B Automatic transmission

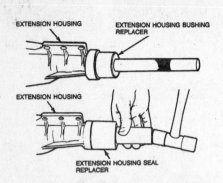

7.7 These special tools are available to install the oil seal and bushing from the transmission extension housing

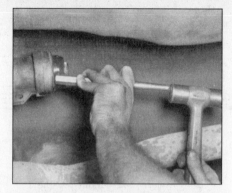

7.8 Installing the new extension housing oil seal with a punch and hammer - large socket or length of pipe of the proper diameter may also be used to install the oil seal

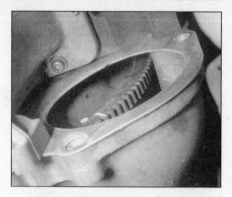

8.8 Mark one of the torque converter studs with white paint and make a corresponding mark on the driveplate so the torque converter and driveplate can be reassembled in the same relative positions

vent and remove any sanding residue from the counterbore.

7   If the extension housing bushing was removed, install a new one with a special tool made for this purpose, available at most auto parts stores that carry tools. **(see illustration).**
8   Install a new extension housing oil seal and make sure it is completely seated in the counterbore **(see illustration).**
9   Install the driveshaft (see Chapter 8).
10 Lower the vehicle and check the transmission fluid level. Top up if necessary (see Chapter 1).

8 Automatic transmission - removal and installation

Refer to illustrations 8.8, 8.11, 8.12, 8.13, 8.14, 8.17, 8.19 and 8.23

1   Disconnect the negative cable from the battery.
2   Remove the transmission fluid level dipstick.
3   Raise the vehicle and support it securely on jackstands.
4   On 4WD models, remove the transfer case (see Chapter 7 Part C). **Caution:** *Remove the Vehicle Speed Sensor (VSS) from the transfer case.*

5   Drain the transmission fluid only if you will disassembling the transmission, then reinstall the pan. (see Chapter 1) Otherwise, you are going to have a problem disposing of the lubricant yourself.
6   Unbolt the starter and tie it up out of the way (see Chapter 5).
7   On vehicles with 3.0L V6 engine, remove the transmission access cover from the converter housing to gain access to the torque converter nuts. The starter mounting hole provides access to the torque converter nuts on vehicles with four-cylinder and 4.0L V6 engines.
8   Mark the torque converter and one of the studs with white paint so they can be installed in the same position **(see illustration).**
9   Remove the four torque converter-to-driveplate nuts. Turn the crankshaft for access to each nut. **Caution:** *Turn the crankshaft in a clockwise direction only (as viewed from the front). You can remove the spark plugs to make this easier.*
10 On 2WD models, remove the driveshaft (see Chapter 8). Tie a plastic bag over the end of the extension housing to prevent the entry of dirt and to catch any residual transmission fluid.
11 On 2WD models, remove the bolt secur-

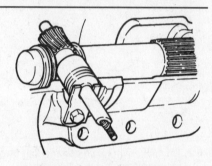

8.11 On 2WD models, remove the speedometer gear retainer bolt and remove the speedometer cable from the transmission

ing the speedometer cable retaining bracket and remove the bracket and bolt as an assembly **(see illustration).** Disconnect the speedometer cable from the transmission.
12 Disconnect the shift cable from the transmission manual shift lever **(see illustration).**
13 On 1993 and 1994 models, disconnect the kickdown cable from the transmission kickdown lever **(see illustration).**
14 On 1993 and 1994 models, detach the electrical connectors for the neutral safety switch and torque converter lockup solenoid

8.12 Disconnect the shift cable from the transmission manual shift lever . . .

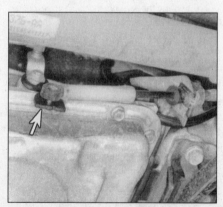

8.13 . . . and the kickdown cable from the transmission downshift lever

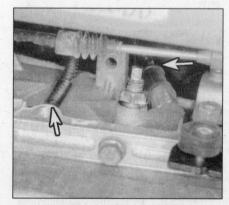

8.14 Disconnect the electrical connectors for the shift solenoids and neutral start switch

Chapter 7 Part B Automatic transmission

8.17 Detach the transmission mount from the frame crossmember

8.19 Detach the front crossmember from the frame rail on each side of the vehicle

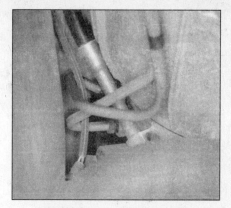

8.23 Disconnect the fluid cooler lines and remove the filler tube

from the transmission **(see illustration)**.
15 On 1995 and later models, disconnect the transmission range (TR) sensor connector and the 16-pin transmission connector.
16 Place a transmission jack beneath the transmission and apply jack pressure to slightly raise the transmission. Use safety chains to help steady the transmission on the jack.
17 Remove the bolts and nuts securing the rear mount and insulator to the frame crossmember **(see illustration)**.
18 Remove the crossmember-to-frame side support mounting bolts and remove the crossmember insulator, support and damper.
19 Remove the bolts and nuts securing the crossmember to the frame rails **(see illustration)**, raise the transmission slightly and remove the crossmember.
20 Support the engine with a jack. Use a block of wood under the oil pan to spread the load.
21 On V6 engines, remove the catalytic converter (see Chapter 4).
22 Slightly lower the transmission. **Caution:** *Do not allow the transmission to hang free.*
23 Disconnect the transmission oil cooler lines from the transmission and plug them to prevent the entry of dirt **(see illustration)**. Use a flare nut wrench to avoid rounding off the nuts.
24 Remove the lower converter housing-to-engine bolts.
25 Remove the transmission fluid filler (dipstick) tube **(see illustration 8.23)**.
26 Make sure the transmission is securely mounted on the transmission jack. Remove the two upper converter housing-to-engine bolts.
27 On 1995 and later models, remove the four side converter housing-to-engine bolts.
28 Carefully move the transmission to the rear to disengage it from the engine block dowel pins and make sure the torque converter is detached from the driveplate. Secure the torque converter to the transmission so it won't fall out during removal.
29 Installation is the reverse of the removal Steps with the following additions:

a) Install the converter to the transmission, making sure the converter hub is fully engaged in the pump gear.
b) Rotate the converter to align the bolt drive lugs and the drain plug with the holes in the driveplate.
c) With the transmission secured to the jack, raise it into position. Be sure to keep it level so the torque converter does not slide forward and disengage from the pump gear.
d) Turn the torque converter to line up the studs with the holes in the driveplate. The white paint mark on the torque converter and the stud made in Step 5 must line up.
e) Move the transmission forward carefully until the dowel pins and the torque converter are engaged.
f) When installing the driveplate-to-converter nuts, position the driveplate so the pilot hole is in the six o'clock position. First install one nut through the pilot hole and tighten it, then install the remaining nuts. Tighten the nuts to the torque listed in this Chapter's Specifications.
g) Adjust the shift cable (see Section 3).
h) Lower the vehicle.
i) Fill the transmission with the specified fluid (see Chapter 1), run the engine and check for fluid leaks.

9 Neutral start switch - removal, installation and adjustment

Refer to illustration 9.3

1 Disconnect the negative cable from the battery.
2 Disconnect the electrical connector from the switch.
3 Carefully remove the switch **(see illustration)**. **Caution:** *It is easy to crush or puncture the walls of the switch. A special tool designed to remove the switch without damaging it is available at most auto parts stores.*
4 Install the switch and tighten to the torque listed in this Chapter's Specifications. If available, use the same special tool used for removal to avoid damaging the switch.
5 Install the electrical connector.
6 Connect the negative cable to the battery.
7 Check that the engine will start only when the shift selector is in the Neutral or Park positions.

10 Vacuum diaphragm - removal and installation

Refer to illustrations 10.2 and 10.5
Note: *This procedure applies only to 1993 and 1994 Models with the A4LD transmission.*

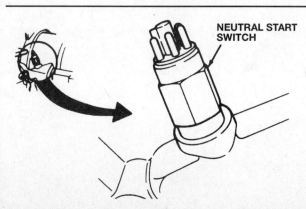

9.3 Neutral start switch - the switch is very fragile and the special Ford tool is recommended for removal and installation

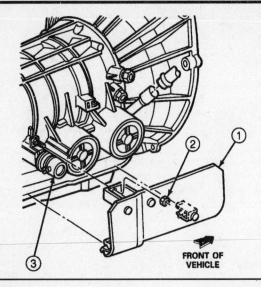

10.2 Details of the vacuum diaphragm and heat shield
1 Heat shield
2 Nut
3 Vacuum diaphragm

10.5 Remove the retaining bolt and clamp, then pull the vacuum diaphragm out of the transmission

1 Raise the vehicle and support it securely on jackstands.
2 Remove the heat shield covering the diaphragm **(see illustration)**.
3 Carefully pry the heat shield from the oil pan, then slide it forward to gain access to the vacuum diaphragm.
4 Disconnect the hose from the vacuum diaphragm.
5 Remove the bolt and the retaining clamp and pull the diaphragm out of the transmission case **(see illustration)**.
6 Remove the control rod from the transmission case.
7 Installation is the reverse of the removal procedure, but be sure to install a new O-ring on the vacuum diaphragm and tighten the clamp bolt securely.

11 Transmission mount - check and replacement

1 Insert a large screwdriver or pry bar into the space between the transmission extension housing and the crossmember and try to pry the transmission up slightly **(see illustration 8.17)**.
2 The transmission should not move away from the mount much at all.
3 To replace the mount, remove the nuts attaching the mount to the crossmember and the bolts attaching the mount to the transmission.
4 Raise the transmission slightly with a jack and remove the mount, noting which holes are used in the crossmember for proper alignment during installation.
5 Installation is the reverse of the removal procedure. Be sure to tighten the nuts/bolts securely.

Chapter 7 Part C
Transfer case

Contents

	Section		Section
Electronic shift control module - removal and installation	5	Shift lever - removal and installation	3
Electronic shift controls - removal and installation	4	Transfer case electronic shift motor - general information, diagnosis and replacement	10
Front output shaft oil seal - removal and installation	6		
General information	1	Transfer case overhaul - general information	9
Linkage adjustment - manual shift transfer case	2	Transfer case - removal and installation	8
Rear output shaft oil seal - removal and installation	7		

Specifications

Lubricant type .... See Chapter 1

Torque specifications .... **Ft-lbs** (unless otherwise indicated)

Cam plate bolt A	70 to 90
Cam plate bolt B	31 to 42
Breather vent	72 to 168 in-lbs
Upper shift control and heat shield bolts	27 to 37
Transfer case-to-transmission bolts	25 to 35
Yoke nut	150 to 180
Skid plate-to-frame bolt	22 to 30
Speedometer cable bolt	20 to 25 in-lbs
Bearing retainer bolts	26 to 32
Shift motor bolts	71 to 97 in-lbs

1 General information

Refer to illustration 1.1a and 1.1b.

Ford Rangers are equipped with the Borg Warner 13-54 Manual or Electronic Shift transfer case. Manual selection is accomplished with a shift lever that provides neutral, 2-high, 4-high and 4-low **(see illustration)**. Electronic selection is accomplished with a switch on the instrument panel **(see illustration)**.

The 13-54 chain-driven transfer case transfers power from the transmission to the rear axle and, when activated, also to the front driveaxle. The unit is lubricated by a positive displacement oil pump that channels oil through drilled holes in the rear output shaft. The pump turns with the rear output shaft and allows towing of the vehicle at maximum legal road speeds for extended distances without disconnecting the front or rear driveshafts.

Although each of the systems differ internally, the external components on the transfer case are similar.

Due to the complexity of the transfer cases covered in this manual and the need for specialized equipment to perform most service operations, this Chapter contains only general diagnosis, routine maintenance, adjustment and removal and installation procedures.

If the transfer case requires major repair work, it should be taken to a dealer service department or an automotive or transmission repair shop. You can, however, remove and install the transfer case yourself and save the expense, even if the repair work is done by a transmission shop.

2 Linkage adjustment - manual shift transfer case

Refer to illustration 2.2

1 The linkage should be adjusted if the transfer case won't engage properly or when-

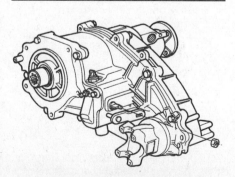

1.1a The manually controlled Borg-Warner 13-54 transfer case

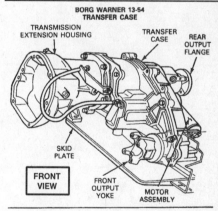

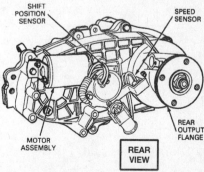

1.1b The electronically controlled Borg-Warner 13-54 transfer case

Chapter 7 Part C Transfer case

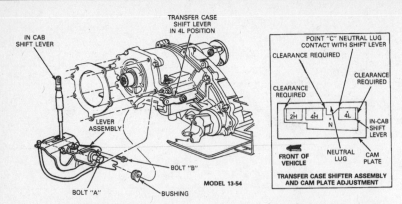

2.2 Shift linkage adjustment points (manual shift transfer case)

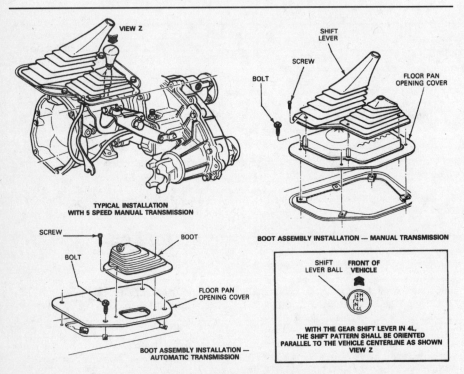

3.3 Manual control transfer case shift lever assembly

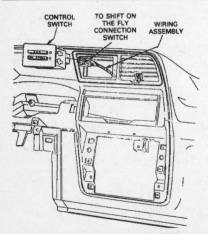

4.4 Lift the connector latch and pull the electrical connector from the electronic shift control switch

ever the transfer case control assembly is removed from the vehicle.

2 Lift up the shift boot so the upper surface of the cam plate is visible **(see illustration)**.

3 Loosen bolts A and B one turn each.

4 Move the shift lever on the side of the transfer case to the 4L position (down).

5 Pivot the cam plate clockwise around bolt A until the bottom chamfered corner of the neutral lug just touches the forward right edge of the shift lever on the transfer case.

6 While holding the cam plate from turning, tighten bolt A, then bolt B to the torque listed in this Chapter's Specifications.

7 Move the shift lever inside the vehicle to all positions and check for positive engagement. In the 2H, 4H and 4L positions, there should be clearance between the cam plate and the shift lever on the side of the transfer case.

8 When the linkage is adjusted properly, install the shift lever boot.

3 Shift lever - removal and installation

Refer to illustration 3.3

Removal

1 Place the transfer case shift lever in the 4L position.

2 **Note:** *Don't perform Step 2 unless the shift ball, boot or lever is to be replaced.* Remove the plastic insert from the shift ball. Warm the ball with a heat gun or portable hair dryer to 140 to 180 degrees F and knock the ball off the lever with a piece of wood and a hammer. Be careful not to damage the finish on the shift lever.

3 Remove the rubber boot and the floor pan cover **(see illustration)**.

4 Disconnect the vent hose from the control lever.

5 Unscrew the shift lever from the control lever.

6 Remove the large and small housing bolts securing the shifter to the extension housing. Remove the control lever and bushing.

Installation

7 Adjust the linkage (see Section 2).

8 Install the vent assembly so the white mark on the housing is indexed into the notch in the shifter, if applicable. **Note:** *The upper end of the vent hose should be 3/4-inch above the top of the shifter and positioned just under the vehicle floor.*

9 Install the floor pan cover and the rubber boot.

10 If the shift ball was removed, warm it with a heat gun or portable hair dryer to 140 to 180-degrees F and carefully tap the ball onto the lever with a piece of wood and a hammer. Install the plastic insert into the shift ball. **Note:** *The shift ball must be tapped onto the lever until all of the knurled portion of the shaft is covered.*

11 Check the transfer case for proper operation.

4 Electronic shift controls - removal and installation

Refer to illustration 4.4

1 Disconnect the negative cable from the battery.

2 Remove the instrument cluster trim panel (see Chapter 12).

3 Remove the screws securing the shift control assembly and partially pull the assembly out from the instrument panel.

4 Disconnect the electrical connector from the switch assembly and remove the assembly **(see illustration)**.

5 Installation is the reverse of the removal Steps.

Chapter 7 Part C Transfer case   7C-3

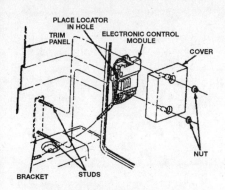

5.2a The shift control module is located behind the driver's seat (standard model shown)

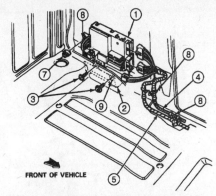

5.2b Location of the shift control module (Supercab models)

1 Shift control module
2 Shift module bracket
3 Screw
4 Wiring harness
5 Wiring harness
6 Wiring harness
7 Nut and washer assembly
8 Stud

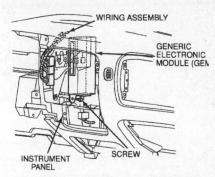

5.7 Location of the Generic Electronic Module (GEM) on 1995 and later models

5 Electronic shift control module - removal and installation

Note: *1995 and later models are equipped with a Generic Electronic Module (GEM). This module incorporates the functions of several different modules into one and offers diagnostic codes to locate any multitude of failures (see Chapter 6).*

1 Disconnect the negative cable from the battery.

1993 through 1994 models

Refer to illustrations 5.2a and 5.2b.

2 Working inside the vehicle, remove the two nuts securing the control module and cover to the trim panel behind the driver's seat. Remove the two bolts securing the control module and the bracket **(see illustrations)**.
3 Pull the control module part way out and disconnect the electrical connectors from the module. Remove the control module.
4 Installation is the reverse of removal.

1995 and later models

Refer to illustration 5.7.

5 Remove the radio (see Chapter 12).
6 Remove the radio trim panel (see Chapter 12).
7 Remove the retaining screw, and slide the GEM off the retaining tabs and toward the front of the vehicle **(see illustration)**.
8 Disconnect the electrical connectors from the GEM and remove the unit from the dash.
9 Installation is the reverse of removal.

6 Front output shaft oil seal - removal and installation

Refer to illustration 6.4

Removal

1 Raise the vehicle and support it securely on jackstands.
2 Drain the transfer case lubricant (see Chapter 1).
3 Remove the front driveshaft from the axle input yoke (see Chapter 8). Tie it up out of the way.
4 Use a 30 mm thin-walled socket and remove the front output shaft nut, washer, rubber seal and yoke **(see illustration)**.
5 Remove the dust seal from the yoke opening of the transfer case.
6 Carefully pry out the oil seal with a large screwdriver.

Installation

7 Inspect the oil seal contact surface on the housing for scoring or burrs that may damage the new oil seal. If found, remove them with emery cloth or medium grit wet-and-dry sandpaper. Use a clean cloth dipped in solvent and remove any sanding residue from the counterbore.
8 Apply multipurpose grease to the oil seal. Position the seal into the front output shaft housing bore and make sure the oil seal is not cocked in the bore.
9 To drive the front oil seal into the output housing bore use a special tool and driver, available at most auto parts stores, or carefully drive the seal into the bore with a hammer and a large socket or piece of pipe of the appropriate size.
10 Clean the transfer case front output female splines and apply multi-purpose grease to them. Insert the front driveshaft splined shaft into the female splines.
11 Install the front yoke onto the splines, then install the rubber seal, steel washer and nut. Tighten the nut to the torque listed in this Chapter's Specifications.
12 Connect the front driveshaft to the axle input yoke (see Chapter 8).
13 Refill the transfer case lubricant (see Chapter 1).
14 Remove the jackstands and lower the vehicle.

7 Rear output shaft oil seal - removal and installation

Removal

1 Raise the vehicle and support it securely on jackstands.
2 Remove the rear driveshaft from the transfer case output shaft yoke and wire the driveshaft out of the way (see Chapter 8).
3 Use a 30 mm thin-walled socket to remove the nut securing the flange **(see illustration 6.4)**.
4 Remove the steel washer and rubber

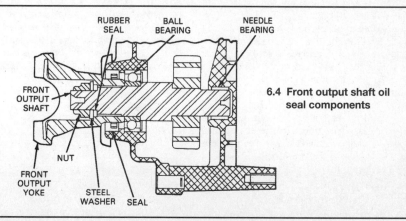

6.4 Front output shaft oil seal components

7C-4 Chapter 7 Part C Transfer case

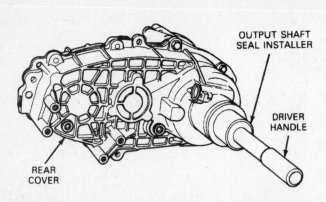

7.8 A large pipe or socket can be used to install the seal if the special tool isn't available

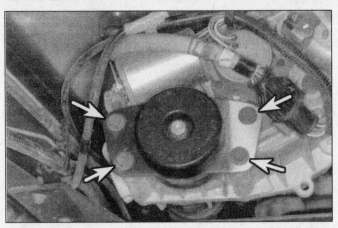

8.3 Some models are equipped with a transfer case damper - this must be removed to remove the transfer case

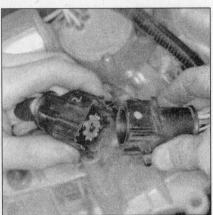

8.5 Disconnect the shift motor connector (electronic shift models)

8.6 Disconnect the speed sensor connector (if equipped) and speedometer cable

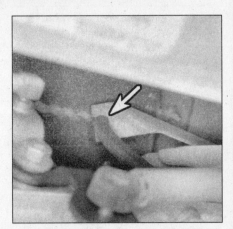

8.8 Disconnect the vent hose from its bracket (electronic shift models)

seal, then remove the flange.
5   Carefully pry and pull on the outer curved lip of the oil seal and remove it from transfer case.

Installation

Refer to illustration 7.8

6   Inspect the oil seal contact surface on the housing for scoring or burrs that may damage the new oil seal. If found, remove them with emery cloth or medium grit wet-and-dry sandpaper. Use a clean cloth dipped in solvent and remove any sanding residue from the counterbore.
7   Apply multi-purpose grease to the oil seal. Position the seal into the front output shaft housing bore and make sure the oil seal is not cocked in the bore.
8   To drive the front oil seal into the output housing bore use a special tool available at most auto parts stores that carry special tools,and driver a special tool available at most auto parts stores that carry special tools, or carefully drive the seal into the bore with a hammer and a large socket or piece of pipe of the appropriate size **(see illustration)**.
9   Clean the transfer case rear output female splines and apply multi-purpose grease to them.

10 Install the output flange onto the output shaft splines.
11 Install the rubber seal, steel washer and nut. Tighten the nut to the torque listed in this Chapter's Specifications.
12 Connect the front driveshaft to the axle output flange (see Chapter 8).
13 Remove the jackstands and lower the vehicle.

8 Transfer case - removal and installation

Refer to illustrations 8.3, 8.5, 8.6, 8.8, 8.12, 8.14a and 8.14b

1   Raise the vehicle and support it securely on jackstands.
2   On models so equipped, remove the four bolts securing the skid plate to the frame and remove it.
3   On models so equipped, remove the damper from the transfer case **(see illustration)**.
4   Drain the transfer case lubricant only if you will disassembling it. Otherwise, you are going to have a problem disposing of the lubricant yourself. (see Chapter 1).

5   On electronically controlled models, squeeze the locking tabs (the part labeled PUSH) toward each other, then pull the electrical connector apart. This may take some effort, but it shouldn't be necessary to use tools to separate the connector halves. Disconnect the electrical connector from the transfer case motor and from the connector mounting bracket **(see illustration)**.
6   Disconnect the speed sensor electrical connector (if equipped) and speedometer cable from the transfer case **(see illustration)**.
7   Disconnect both driveshafts (see Chapter 8).
8   Disconnect the vent hose from the shift lever bracket (manual shift) or mounting bracket (electronic shift) **(see illustration)**. **Warning:** *The catalytic converter is next to the vent hose and is extremely hot after the engine has been run. Be careful when working around the converter and if possible allow the converter to cool down several hours prior to working around it.*
9   On manually controlled models, remove the nut from the shift lever and remove the shift lever.
10 On manually controlled models, remove the large bolt and the small bolt retaining the

Chapter 7 Part C Transfer case

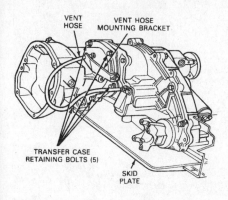

8.12 The transfer case is secured by five bolts

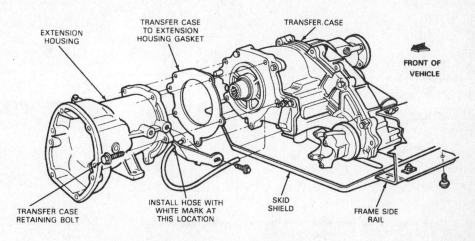

8.14a Transfer case to extension housing gasket location

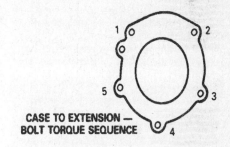

8.14b Tighten the transfer case mounting bolts in the sequence shown

shifter to the extension housing. Pull on the control lever unit until the bushing slides off the transfer case shift lever pin.

11 Place a transmission jack beneath the transfer case and apply jack pressure to slightly raise the transfer case. Use safety chains to steady the transfer case on the jack.

12 Remove the five bolts securing the transfer case to the transmission extension housing **(see illustration)**.

13 Slide the transfer case rearward and off the transmission output shaft, then lower the transfer case from the vehicle. Remove and discard the gasket between the transfer case and the extension housing.

14 Installation is the reverse of the removal Steps with the following additions:

a) *Install a new gasket between the transfer case and the extension housing* **(see illustration)**.

b) *Slide the transfer case onto the transmission output shaft, making sure the splines align, until the transfer case seats over the dowel pin.*

c) *Install the five bolts securing the transfer case and tighten to the torque listed in this Chapter's Specifications and in the torque sequence shown* **(see illustration)**. **Note:** *On manually controlled models, always tighten the large bolt securing the shifter to the extension housing before tightening the small bolt.*

d) *Install the vent assembly so the white marking on the hose is positioned in the notch in the shift lever bracket (manual shift) or mounting bracket (electronic shift).* **Note:** *The upper end of the vent hose should be 3/4 inch above the top of the shifter on manually controlled models.*

e) *Connect the rear driveshaft to the transfer case output flange. Tighten the bolts to the torque listed in the Chapter 8 Specifications.*

f) *Refill the transfer case with the proper type of lubricant (see Chapter 1).*

9 Transfer case overhaul - general information

Overhauling a transfer case is a difficult job for the do-it-yourselfer, not unlike overhauling a transmission. It involves the disassembly and reassembly of many small parts. Numerous clearances must be precisely measured and, if necessary, changed with select-fit spacers and snap-rings. As a result, if transfer case problems arise, it can be removed and installed by a competent do-it-yourselfer, but overhaul should be left to a competent repair shop. Rebuilt transfer cases may be available. Check with your dealer parts department and auto parts stores. At any rate, the time and money involved in an overhaul is almost sure to exceed the cost of a rebuilt unit.

Nevertheless, it's not impossible for an inexperienced mechanic to rebuild a transfer case if the special tools are available and the job is done in a deliberate step-by-step manner so nothing is overlooked.

The tools necessary for an overhaul include internal and external snap-ring pliers, bearing puller, slide hammer, set of pin punches, dial indicator and possibly a hydraulic press. In addition, a large, sturdy workbench and a large vise will be required.

During disassembly, make careful notes of how each piece comes off, where it fits in relation to other pieces and what holds it in place.

Before taking the transfer case apart for repair, it will help if you have some idea what area of the transfer case is malfunctioning. Certain problems can be closely tied to specific areas in the transfer case, which can make component examination and replacement easier. Refer to the *Troubleshooting* Section at the front of this manual for information regarding possible sources of trouble.

10 Transfer case electronic shift motor – general information, diagnosis and replacement

General information and diagnosis

1 Some models are equipped with an electronic shift motor which shifts the transfer case from 2-high to 4-high or 4-low in response to an electrical signal from a button on the dash.

2 If the shift motor should malfunction while the vehicle is in 4-low, top speed will be restricted. Should this happen to you, the following procedure will show you how to manually shift the transfer case into 2-high.

3 First, remove the shift motor (go to Step 6). It isn't necessary to unplug any connectors; just remove the motor mounting bolts and pull off the motor. To test the motor, have an assistant press the button on the dash (key on) while you watch the motor.

a) *If the motor runs, the problem is in the transfer case.*

Chapter 7 Part C Transfer case

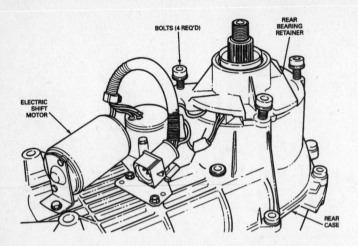

10.13 Remove the four bolts from the rear bearing retainer and remove the retainer

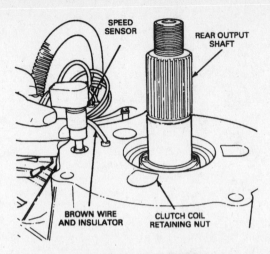

10.14 Remove the speed sensor from the transfer case (1991 through 1997 models)

b) *If the motor does not run, check for voltage at the shift motor (see Chapter 12, if necessary). If there's voltage, replace the motor. If there's no voltage, the problem lies elsewhere in the shift motor circuit (check the fuse first).*

4   To shift the transfer case into 2-High, rotate the shift shaft on the transfer case clockwise until it stops. Pliers may help rotate the recessed shaft.
5   Reattach the shift motor (see below).

Replacement

6   Raise the vehicle and place it securely on jackstands.

1991 through 1997 models

Refer to illustrations 10.13, 10.14 and 10.16

7   Unplug the electrical connector for the shift motor harness.
8   Remove the two wire harness bracket bolts and detach the bracket from the transfer case.
9   Bend a paper clip into a hook and remove the locking sleeve from the center terminal of the electrical connector by pulling it up from the bottom of the connector with the hook.
10   Remove the brown wire from the No. 1 (center) terminal of the shift motor side of the electrical connector. If necessary, remove the green wire (speed sensor) from the No. 4 terminal and the blue wire from the No. 5 terminal (this isn't necessary on all models).
11   Remove the rear driveshaft (see Chapter 8).
12   Remove the rear output shaft yoke (see Section 7).
13   Remove the four (No. 50 Torx) bolts from the rear bearing retainer **(see illustration)**. Remove the bearing retainer by inserting a prybar between the bosses (next to the electrical connections) and prying it off.

Remove all old RTV gasket sealant from the mating surfaces of the case and bearing retainer.
14   Remove the speed sensor from the transfer case **(see illustration)**.
15   Note the position of the triangular shaft protruding from the rear of the transfer case. It must be in this same position when installing the motor.
16   Remove the four shift motor retaining bolts (one on the motor bracket, three on the motor) and remove the shift motor **(see illustration)**.
17   Installation is the reverse of removal, with the following additions:
   a) *Coat the bearing retainer with a new coat of RTV gasket sealant before installing it. Be sure not to pinch the brown wire between the retainer and the transfer case. Tighten the bearing retainer bolts to the torque listed in this Chapter's Specifications.*
   b) *Using soft-jaw pliers, rotate the triangular shift shaft so that it's aligned with the triangular slot in the motor.*
   c) *Apply a thin coat of RTV sealant to the motor base.*
   d) *Slightly loosen the two nuts that attach the slotted motor bracket to the motor (so that the bracket doesn't misalign the motor during installation).*
   e) *Tighten the shift motor bolts to the torque listed in this Chapter's Specifications. Tighten the two shift motor bracket-to-shift motor nuts securely.*

1998 and later models

18   Unplug the electrical connector for the shift motor harness.
19   Remove the two wire harness bracket bolts and detach the harness bracket.
20   Remove the brown wire from the back of the shift motor side of the electrical connector.

21   Remove the motor bracket bolt.
22   Note the position of the triangular shaft protruding from the rear of the transfer case. It must be in this same position when installing the motor.
23   Remove the three motor retaining bolts and remove the motor.
24   Installation is the reverse of removal, with the following additions:
   a) *Using soft-jaw pliers, rotate the triangular shift shaft so that it's aligned with the triangular slot in the motor.*
   b) *Apply a thin coat of RTV sealant to the motor base.*
   b) *Tighten the shift motor bolts to the torque listed in this Chapter's Specifications.*

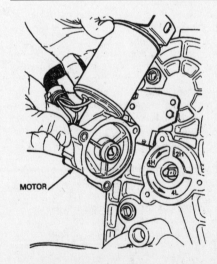

10.16 Remove the four shift motor retaining bolts and remove the shift motor

Chapter 8
Clutch and drivetrain

Contents

	Section		Section
Axleshaft and joint assemblies (front) - removal, component replacement and installation............................	20	Clutch/starter interlock switch - removal and installation.............	11
Axleshafts, bearings and oil seals (rear) - removal and installation...	18	Driveaxle boot replacement and CV joint overhaul (1998 and later 4WD models)	22
Center bearing - replacement	15	Driveshaft and universal joints - description and check	12
Clutch - description and check	2	Driveshafts - removal and installation...	13
Clutch components - removal, inspection and installation............	3	Front axle assembly (4WD models) - removal and installation	19
Clutch hydraulic line quick-disconnect fittings - general information...	5	General information...	1
		Pilot bearing - inspection and replacement	4
Clutch hydraulic system - bleeding...	9	Rear axle - description and check	16
Clutch master cylinder and reservoir - removal and installation....	8	Rear axle assembly - removal and installation...	17
Clutch pedal - removal and installation...	10	Right slip yoke and stub shaft assembly, carrier, carrier oil seals and bearings - removal and installation...	21
Clutch release bearing - removal, inspection and installation	6	Universal joints - replacement...	14
Clutch release cylinder - removal and installation	7		

Specifications

Clutch
Fluid type	See Chapter 1
Type	Single dry plate, diaphragm spring
Actuation	Hydraulic
Driveshaft type	One-piece with single or double cardan type joints

Rear axle
Type	Integral carrier
Ring gear size	8.8-inch
Lubricant type	See Chapter 1

Front axle
Type	Dana Model 35 IFS
Hubs	Automatic locking type
Lubricant type	See Chapter 1

Torque specifications
Ft-lbs

Clutch
Pressure plate-to-flywheel bolts...	15 to 24
Clutch master cylinder-to-firewall bolts...	15 to 20
Release cylinder-to-clutch housing bolts	13 to 19

Rear driveshaft
Driveshaft-to-transfer case bolts (4WD models only).............................	12 to 16
Driveshaft-to-rear axle bolts	
With double-cardan U-joint	70 to 95
With single-cardan U-joint	8 to 15

Front driveshaft
Driveshaft-to-transfer case flange bolts	70 to 95
Driveshaft-to-differential yoke bolts	10 to 15

Rear axle
Cover bolts	See Chapter 1
Pinion shaft lock-bolt...	15 to 30
Leaf spring U-bolt nuts...	See Chapter 10
Rear shock absorber-to-axle bracket bolt or nut	See Chapter 10
Brake backing plate nuts	See Chapter 9

8-2 Chapter 8 Clutch and drivetrain

Front axle

Pivot bolt	120 to 150
Pivot bracket-to-frame nut	70 to 92
Axle stud	190 to 230
Lower balljoint nut	95 to 110
Upper balljoint nut	85 to 100
Carrier to axle arm bolts	35 to 53
Carrier shear bolt	75 to 95
Front shock absorber-to-radius arm nut	See Chapter 10
Front spring seat nut	See Chapter 10
Front radius arm bracket bolts	See Chapter 10

1 General information

The information in this Chapter deals with the components that transmit power to the wheels, except for the transmission and transfer case, which are dealt with in Chapter 7. For the purposes of this Chapter, these components are grouped into three categories; clutch, driveshaft and axles. Separate Sections within this Chapter offer general descriptions and checking procedures for components in each of the three groups.

Since nearly all the procedures covered in this Chapter involve working under the vehicle, make sure it's securely supported on sturdy jackstands or on a hoist where the vehicle can be easily raised and lowered.

2 Clutch - description and check

Refer to illustration 2.5

1 All models with a manual transmission use a single dry plate, diaphragm spring type clutch. The clutch disc has a splined hub which allows it to slide along the splines of the transmission input shaft. The clutch and pressure plate are held in contact by spring pressure exerted by the diaphragm in the pressure plate.

2 The clutch release system is operated by hydraulic pressure. The hydraulic release system consists of the clutch pedal, a master cylinder and fluid reservoir, the hydraulic line, an internal release cylinder and the clutch release (or throwout) bearing.

3 The internal release cylinder, mounted concentrically on the transmission input shaft, pushes directly on the release bearing. This eliminates the need for a release lever.

4 Terminology can be a problem when discussing the clutch components because common names are in some cases different from those used by the manufacturer. For example, the driven plate is also called the clutch plate or disc, the clutch release bearing is sometimes called a throwout bearing, the release cylinder is sometimes called the operating or slave cylinder.

5 Other than to replace components with obvious damage, some preliminary checks should be performed to diagnose clutch problems.

a) The first check should be of the fluid level in the clutch master cylinder. If the fluid level is low, add fluid as necessary and inspect the hydraulic system for leaks. If the master cylinder reservoir has run dry, bleed the system as described in Section 9 and retest the clutch operation.

b) To check "clutch spin down time," run the engine at normal idle speed with the transmission in Neutral (clutch pedal up - engaged). Disengage the clutch (pedal down), wait several seconds and shift the transmission into Reverse. No grinding noise should be heard. A grinding noise would most likely indicate a problem in the pressure plate or the clutch disc.

c) To check for complete clutch release, run the engine (with the parking brake applied to prevent movement) and hold the clutch pedal approximately 1/2-inch from the floor. Shift the transmission between 1st gear and Reverse several times. If the shift is hard or the transmission grinds, component failure is indicated. Check the release bearing travel **(see illustration)**. With the clutch pedal depressed completely, the release cylinder should extend substantially. If it doesn't, check the fluid level in the clutch master cylinder (see Chapter 1). Bleed the system (see Section 9).

d) Visually inspect the pivot bushing at the top of the clutch pedal to make sure there is no binding or excessive play.

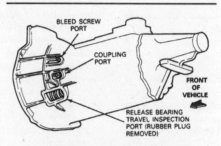

2.5 Release cylinder and bearing travel can be checked through the inspection port

3 Clutch components - removal, inspection and installation

Warning: *Dust produced by clutch wear and deposited on clutch components may contain asbestos, which is hazardous to your health. DO NOT blow it out with compressed air and DO NOT inhale it. DO NOT use gasoline or petroleum-based solvents to remove the dust. Brake system cleaner should be used to flush the dust into a drain pan. After the clutch components are wiped clean with a rag, dispose of the contaminated rags and cleaner in a covered, marked container.*

Removal

Refer to illustrations 3.3, 3.5a and 3.5b

1 Access to the clutch components is normally accomplished by removing the transmission (and transfer case on 4WD models), leaving the engine in the vehicle. IÒf, of course, the engine is being removed for major overhaul, then check the clutch for wear and replace worn components as necessary. However, the relatively low cost of the clutch components compared to the time and trouble spent gaining access to them warrants their replacement anytime the engine or transmission is removed, unless they are new or in near perfect condition. The following procedures are based on the assumption the engine will stay in place.

2 Referring to Chapter 7 Part A and Part C, remove the transmission from the vehicle. Support the engine while the transmission is out. Preferably, an engine hoist should be used to support it from above. However, if a jack is used underneath the engine, make sure a piece of wood is positioned between the jack and oil pan to spread the load. **Caution:** *The pick-up for the oil pump is very close to the bottom of the oil pan. If the pan is bent or distorted in any way, engine oil starvation could occur.*

3 To support the clutch disc during removal, install a clutch alignment tool through the clutch disc hub **(see illustration)**.

3.3 A clutch alignment tool can be purchased at most auto parts stores and eliminates all guesswork when centering the clutch disc in the pressure plate

Chapter 8 Clutch and drivetrain

3.5a Loosen the pressure plate bolts (arrows) one turn at a time in rotation until they are free

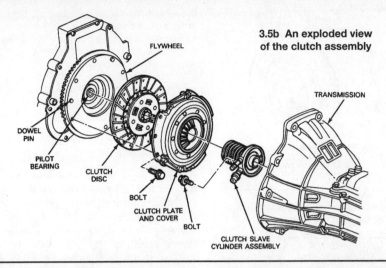

3.5b An exploded view of the clutch assembly

4 Carefully inspect the flywheel and pressure plate for indexing marks. The marks are usually an X, an O or a white letter. If they cannot be found, scribe marks yourself so the pressure plate and the flywheel will be in the same alignment during installation.

5 Turning each bolt only 1/4-turn at a time, loosen the pressure plate-to-flywheel bolts **(see illustration)**. Work in a criss-cross pattern until all spring pressure is relieved evenly. Then hold the pressure plate securely and completely remove the bolts, followed by the pressure plate and clutch disc **(see illustration)**.

Inspection

Refer to illustrations 3.9 and 3.11

6 Ordinarily, when a problem occurs in the clutch, it can be attributed to wear of the clutch driven plate assembly (clutch disc). However, all components should be inspected at this time.

7 Inspect the flywheel for cracks, heat checking, grooves and other obvious defects. If the imperfections are slight, a machine shop can machine the surface flat and smooth, which is highly recommended

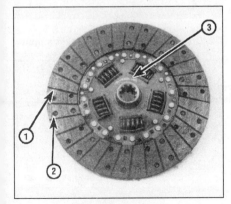

3.9 The clutch disc

1 **Lining** - this will wear down in use
2 **Rivets** - these secure the lining and will damage the flywheel or pressure plate if allowed to contact the surfaces
3 **Marks** - "Flywheel side" or similar

regardless of the surface appearance. Refer to Chapter 2 for the flywheel removal and installation procedure.

8 Inspect the pilot bearing (see Section 4).
9 Inspect the lining on the clutch disc. There should be at least 1/16-inch of lining above the rivet heads. Check for loose rivets, distortion, cracks, broken springs and other obvious damage **(see illustration)**. As mentioned above, ordinarily the clutch disc is routinely replaced, so if in doubt about the condition, replace it with a new one.
10 The release bearing should also be replaced along with the clutch disc (see Section 6).
11 Check the machined surfaces and the diaphragm spring fingers of the pressure plate **(see illustration)**. If the surface is grooved or otherwise damaged, replace the pressure plate. Also check for obvious damage, distortion, cracking, etc. Light glazing can be removed with medium grit emery cloth. If a new pressure plate is required, new and factory-rebuilt units are available.

Installation

12 Before installation, clean the flywheel and pressure plate machined surfaces with lacquer thinner or acetone. It's important that no oil or grease is on these surfaces or the lining of the clutch disc. Handle the parts only with clean hands.
13 Position the clutch disc and pressure plate against the flywheel with the clutch held in place with an alignment tool **(see illustration 3.3)**. Make sure it's installed properly (most replacement clutch plates will be marked "flywheel side" or something similar - if not marked, install the clutch disc with the flat side of the hub toward the flywheel).
14 Tighten the pressure plate-to-flywheel bolts only finger tight, working around the pressure plate.
15 Center the clutch disc by ensuring the alignment tool extends through the splined hub and into the pilot bearing in the crankshaft. Wiggle the tool up, down or side-to-side as needed to bottom the tool in the pilot bearing. Tighten the pressure plate-to-

flywheel bolts a little at a time, working in a criss-cross pattern to prevent distorting the cover. After all of the bolts are snug, tighten them to the torque listed in this Chapter's Specifications. Remove the alignment tool.
16 If removed, install the clutch release bearing as described in Section 6.
17 Install the transmission and all components removed previously. Tighten all fasteners to the torques listed in this Chapter's Specifications.

4 Pilot bearing - inspection and replacement

Refer to illustrations 4.1 and 4.5

1 The clutch pilot bearing is a needle roller type bearing which is pressed into the rear of

NORMAL FINGER WEAR

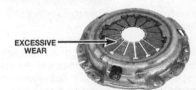

EXCESSIVE WEAR

EXCESSIVE FINGER WEAR

BROKEN OR BENT FINGERS

3.11 Replace the pressure plate if excessive or abnormal wear is noted

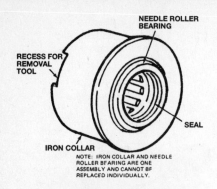

4.1 The pilot bearing seal must face out when the bearing is installed in the crankshaft

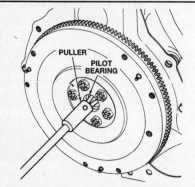

4.5 One method of removing the pilot bearing requires an internal puller connected to a slide hammer

the crankshaft **(see illustration)**. It is greased at the factory and does not require additional lubrication. Its primary purpose is to support the front of the transmission input shaft. The pilot bearing should be inspected whenever the clutch components are removed from the engine. Due to its inaccessibility, if you are in doubt as to its condition, replace it with a new one. **Note:** *If the engine has been removed from the vehicle, disregard the following steps which do not apply.*

2 Remove the transmission (see Chapter 7 Part A).
3 Remove the clutch components (see Section 3).
4 Inspect for any excessive wear, scoring, lack of grease, dryness or obvious damage. If any of these conditions are noted, the bearing should be replaced. A flashlight will be helpful to direct light into the recess.
5 Removal can be accomplished with a special puller and slide hammer **(see illustration)**, but an alternative method also works.
6 Find a solid steel bar which is slightly smaller in diameter than the bearing. Alternatives to a solid bar would be a wood dowel or a socket with a bolt fixed in place to make it solid.
7 Check the bar for fit - it should just slip into the bearing with very little clearance.
8 Pack the bearing and the area behind it (in the crankshaft recess) with heavy grease. Pack it tightly to eliminate as much air as possible.

9 Insert the bar into the bearing bore and strike the bar sharply with a hammer which will force the grease to the back side of the bearing and push it out. Remove the bearing and clean all grease from the crankshaft recess.
10 To install the new bearing, lightly lubricate the outside surface with lithium-based grease, then drive it into the recess with a soft-face hammer. The seal must face out.
11 Install the clutch components, transmission and all other components removed previously, tightening all fasteners properly.

5 Clutch hydraulic line quick-disconnect fittings - general information

Refer to illustration 5.2

1 The hydraulic lines are equipped with quick-disconnect fittings and a special tool available at most auto parts stores that carry special tools, is required for separation
2 Depress the white retainer within the quick-disconnect fitting with the special tool while pulling on the hydraulic line, then disconnect the male connector **(see illustration)**. To install, push the male connector into the female connector until it locks into place.
3 There should be no loss of hydraulic fluid during separation of the fittings so there is no need to bleed the system after a fitting has been disconnected unless shifting is hard or there's a lack of clutch reserve travel.

6 Clutch release bearing - removal, inspection and installation

Warning: *Dust produced by clutch wear and deposited on clutch components may contain asbestos, which is hazardous to your health. DO NOT blow it out with compressed air and DO NOT inhale it. DO NOT use gasoline or petroleum-based solvents to remove the dust. Brake system cleaner should be used to flush the dust into a drain pan. After the clutch components are wiped clean with a rag, dispose of the contaminated rags and cleaner in a covered, marked container.*

Removal and inspection

Refer to illustration 6.2

1 Remove the transmission (see Chapter 7 Part A).
2 Within the clutch housing, twist the release bearing and carrier assembly until you feel resistance. Keep turning the assembly and the preload spring will push the release bearing off the release cylinder **(see illustration)**.
3 Hold the center of the bearing and rotate the outer portion while applying pressure. If the bearing doesn't turn smoothly or it's noisy, replace it with a new one. Wipe the bearing with a clean rag and inspect it for damage, wear and cracks. Don't immerse the bearing in solvent - it's sealed for life and to do so would ruin it.

Installation

4 Apply a light coat of high-temperature lithium based grease to the face of the release bearing, where it contacts the pressure plate diaphragm fingers.
5 Lubricate the inner bore of the bearing and bearing carrier with lithium-based grease. Don't use petroleum-based grease.
6 Push the release bearing and carrier onto the slave cylinder until it bottoms out **(see illustration 6.2)**.
7 Install the transmission (see Chapter 7 Part A).

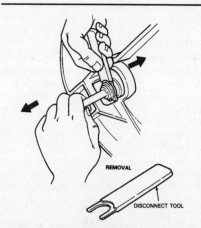

5.2 Quick-disconnect hydraulic line fitting

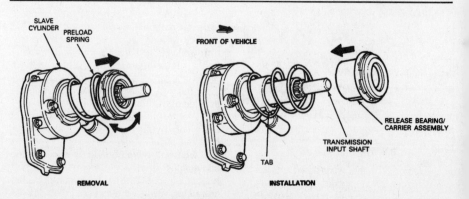

6.2 Release bearing details

Chapter 8 Clutch and drivetrain   8-5

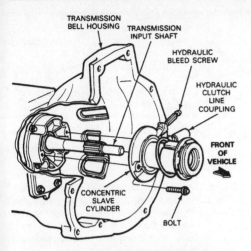

7.4 Clutch release cylinder removal

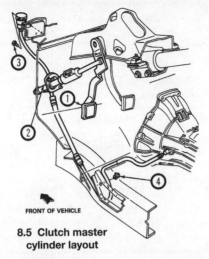

8.5 Clutch master cylinder layout

1. Clutch pedal
2. Master cylinder
3. Reservoir bolt
4. Clip

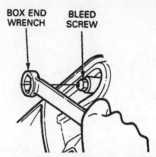

9.5 The bleeder valve is accessible through a port on the left side of the clutch housing

7 Clutch release cylinder - removal and installation

Refer to illustration 7.4

Note: *Prior to removing the release cylinder, disconnect the master cylinder pushrod from the clutch pedal. If not disconnected, permanent damage to the master cylinder will occur if the clutch pedal is depressed while the release cylinder is disconnected.*

1 Pry off the retainer bushing and disconnect the master cylinder pushrod from the clutch pedal.
2 Disconnect the hydraulic quick-disconnect fitting from the slave cylinder (see Section 5).
3 Remove the transmission (see Chapter 7 Part A).
4 Remove the bolts securing the release cylinder to the transmission and slide the release cylinder assembly off the transmission input shaft **(see illustration)**.
5 Installation is the reverse of the removal Steps with the following additions:
 a) *Install the release cylinder onto the input shaft with the hydraulic line facing toward the left side of the transmission.*
 b) *Install the release cylinder mounting bolts and tighten them to the torque listed in this Chapter's Specifications.*
 c) *If necessary, bleed the clutch hydraulic system (see Section 9).*

8 Clutch master cylinder and reservoir - removal and installation

Removal

Refer to illustration 8.5

1 Disconnect the negative cable from the battery.
2 Working inside the vehicle, remove the spring clip or retainer bushing and disconnect the pushrod from the top of the clutch pedal.
3 Remove the clutch/starter interlock switch from the pushrod (see Section 11).
4 Disconnect the hydraulic line from the slave cylinder (see Section 5).
5 Within the engine compartment, remove the bolts securing the master cylinder reservoir to the firewall **(see illustration)**. Be careful not to spill any of the fluid.
6 Working in the same area, remove the bolts securing the master cylinder to the firewall. Be careful not to spill any of the fluid.
7 Remove the master cylinder and reservoir from the engine compartment.

Installation

8 Install the master cylinder through the firewall and make sure the pushrod is positioned on the correct side of the clutch pedal so it can be installed onto the pedal's pivot pin.
9 Attach the master cylinder to the firewall and install the mounting bolts. Tighten the bolts to the torque listed in this Chapter's Specifications.
10 Attach the master cylinder reservoir to the firewall and install the mounting bolts, tightening them securely.
11 Connect the hydraulic line to the slave cylinder (see Section 5).
12 Working inside the vehicle, connect the pushrod to the clutch pedal and install the clip or retainer bushing.
13 Install the clutch/starter interlock switch (see Section 11).
14 Connect the negative cable to the battery.
15 Fill the clutch master cylinder reservoir with brake fluid conforming to DOT 3 specifications and bleed the clutch system (see Section 9).

9 Clutch hydraulic system - bleeding

Refer to illustration 9.5

1 The hydraulic system should be bled to remove all air whenever air enters the system. This occurs if the fluid level has been allowed to fall so low that air has been drawn into the master cylinder. Under normal circumstances, air should not enter the system when the quick-disconnect hydraulic line fittings are disconnected. The procedure is very similar to bleeding a brake system, but depends mainly on gravity, rather than the pumping action of the pedal, for the bleeding effect.
2 Fill the master cylinder to the top with new brake fluid conforming to DOT 3 specifications. **Caution:** *Do not re-use any of the fluid coming from the system during the bleeding operation and don't use fluid which has been inside an open container for an extended period of time.*
3 Raise the vehicle and place it securely on jackstands to gain access to the bleeder valve, which is located on the left side of the clutch housing.
4 Remove the dust cap which fits over the bleeder valve and push a length of clear plastic hose over the valve. Place the other end of the hose into a clear container.
5 Open the bleeder valve **(see illustration)**. Fluid will run from the master cylinder, down the hydraulic line, into the release cylinder and out through the clear plastic tube. Let the fluid run out until it is free of bubbles. **Note:** *Don't let the fluid level drop too low in the master cylinder, or air will be drawn into the hydraulic line and the whole process will have to be started over.*
6 Close the bleeder valve.
7 Have an assistant slowly depress the clutch pedal and hold it. Open the bleeder valve on the release cylinder, allowing fluid to flow through the clear plastic hose. Close the bleeder valve when the flow stops. Once closed, have the assistant release the pedal.
8 Slowly press and release the pedal five times, waiting for two seconds each time the pedal is released.
9 Fill the fluid reservoir to the step.
10 The clutch should now be completely bled. If it isn't (indicated by failure to disengage completely), repeat Steps 5 through 9.
11 Continue this process until all air is evacuated from the system, indicated by a solid stream of fluid being ejected from the bleeder valve each time with no air bubbles in the hose or container.
12 Install the dust cap and lower the vehi-

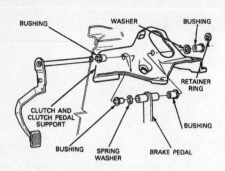

10.5 An exploded view of the clutch pedal

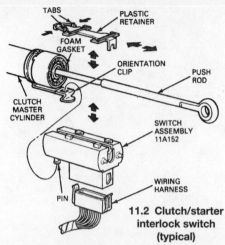

11.2 Clutch/starter interlock switch (typical)

cle. Check carefully for proper operation before placing the vehicle in normal service. Check the fluid level.

10 Clutch pedal - removal and installation

Refer to illustration 10.5

1 Disconnect the negative cable from the battery.
2 Disconnect the barbed end of the clutch/starter interlock switch rod from the clutch pedal. Remove the lockpin and remove the master cylinder pushrod from the clutch pedal. Remove the plastic bushing.
3 Remove the plastic side panel on the driver's sidewall.
4 Unbolt the parking brake and move it aside.
5 Pry the retainer from the clip on the pedal support bracket **(see illustration)**.
6 Slide the pedal out of the bracket.
7 Installation is the reverse of the removal Steps.

11 Clutch/starter interlock switch - removal and installation

Refer to illustration 11.2

1 Disconnect the negative cable from the battery.
2 Disconnect the electrical connector from the switch **(see illustration)**.
3 Rotate the switch so the plastic retainer is visible. Squeeze the retainer tabs together and slide the retainer rearward off the switch.
4 Detach the switch from the master cylinder pushrod and take it out.
5 Installation is the reverse of the removal Steps. Be sure the switch is securely seated on the master cylinder pushrod.

12 Driveshaft and universal joints - description and check

1 The driveshaft is a tube running between the transfer case or transmission and the rear axle or front axle. Universal joints are located at either end of the driveshaft(s) and permit power to be transmitted to the rear (and, on 4WD models the front) wheels at varying angles. The driveshaft on long wheelbase models is actually two shafts connected by a universal joint and supported in the middle by a bearing attached to the frame.
2 Some driveshafts feature a splined slip yoke at one end, which fits into the transmission extension housing. Others incorporate a slip yoke as part of the driveshaft. This arrangement allows the length of the driveshaft to change, which compensates for suspension movement.
3 On models with a slip yoke at the front of the driveshaft, an oil seal is used to prevent leakage of fluid and to keep dirt and contaminants from entering the transmission. If leakage is evident at the front of the driveshaft, replace the oil seal, referring to the procedures in Chapter 7.
4 The driveshaft assembly(ies) requires very little service. The universal joints on models not equipped with grease fittings are lubricated for life and must be replaced if problems develop. The driveshaft(s) must be removed from the vehicle for this procedure.
5 Since the driveshaft is a balanced unit, it's important that no undercoating, mud, etc. be allowed to stay on it. When the vehicle is raised for service it's a good idea to clean the driveshaft(s) and inspect it for any obvious damage. Also check that the small weights used to originally balance the driveshaft(s) are in place and securely attached. Whenever a driveshaft is removed it's important that it be reinstalled in the same relative position to preserve the balance.
6 Problems with the driveshaft(s) are usually indicated by a noise or vibration while driving the vehicle. A road test should verify if the problem is the driveshaft(s) or another vehicle component:

a) On an open road, free of traffic, drive the vehicle and note the engine speed (rpm) at which the problem is most evident.
b) With this noted, drive the vehicle again, this time manually keeping the transmission in first, then second, then third gear ranges and running the engine up to the engine speed noted.
c) If the noise or vibration decreased or was eliminated, visually inspect the driveshaft(s) for damage, material on the shaft which would effect balance, missing weights and damaged universal joints. Another possibility for this condition would be tires which are out-of-balance or a bent or damaged wheel(s).
d) If the noise or vibration occurs at the same engine speed regardless of which gear the transmission is in, the driveshaft is not at fault.

7 To check for worn universal joints:

a) On an open road, free of traffic, drive the vehicle slowly until the transmission is in High gear. Let off on the accelerator, allowing the vehicle to coast, then accelerate. A clunking or knocking noise will indicate worn universal joints.
b) Drive the vehicle at a speed of about 10 to 15 mph and then place the transmission in Neutral, allowing the vehicle to coast. Listen for abnormal driveline noises.
c) Raise the vehicle and support it securely on jackstands. With the transmission in Neutral, manually turn the driveshaft(s), watching the universal joints for excessive play.

13 Driveshafts - removal and installation

Rear driveshaft (one-piece)

Refer to illustrations 13.3 and 13.4

1 Disconnect the negative cable from the battery.
2 Raise the vehicle and support it securely on jackstands. Place the transmission in Neutral with the parking brake off.
3 Look for the yellow paint marks made at the factory on the axle companion flange and the driveshaft flange **(see illustration)**. If these aren't visible, use a scribe or white paint to place marks on the driveshaft and the differential flange in line with each other. This is to make sure the driveshaft is reinstalled in the same position to preserve the balance.

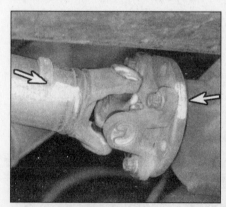

13.3 The rear driveshaft comes from the factory with yellow paint marks that indicate the correct alignment of the driveshaft with the companion flange

Chapter 8 Clutch and drivetrain

13.4 On 4WD models, make alignment marks on the driveshaft and transfer case flanges

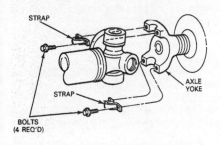

13.18 Make alignment marks on the front driveshaft and differential yoke - wrap the universal joint with tape after unbolting it so the bearings won't fall off

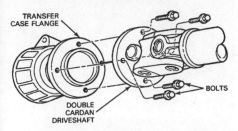

13.17 Make alignment marks on the front driveshaft and transfer case flanges, then unbolt the driveshaft from the transfer case

4   On 4WD models, repeat Step 3 for the front end of the driveshaft **(see illustration)**.
5   Remove the bolts and lower the rear of the driveshaft. On 4WD models, remove the bolts and detach the front end from the transfer case. On 2WD models, slide the front end out of the transmission.
6   Installation is the reverse of the removal Steps with the following additions:
  a) On 2WD models, slide the front of the driveshaft into the transmission.
  b) On 4WD models, position the front end of the driveshaft on the transfer case with the alignment marks lined up.
  c) Raise the rear of the driveshaft into position, checking to be sure the marks are in alignment. If not, make sure the transmission is still in Neutral, then turn the rear wheels to match the pinion flange and the driveshaft.

Rear driveshaft (two-piece)

Note: *Ranger SuperCab models are equipped with a two-piece driveshaft.*

7   Disconnect the negative cable from the battery.
8   Raise the vehicle and support it securely on jackstands. Place the transmission in Neutral with the parking brake off.
9   Look for the yellow paint marks made at the factory on the axle companion flange and the driveshaft flange. If these aren't visible, use a scribe or white paint to place marks on the driveshaft and the differential flange in line with each other. This is to make sure the driveshaft is reinstalled in the same position to preserve the balance.
10 Remove bolts that hold the driveshaft center bearing bracket to the frame crossmember. Make note of any spacers and save them for installation. Support the driveshaft with wire or rope. Do not allow it to hang free.
11 Remove the bolts securing the driveshaft to the rear axle circular flange.
12 On 2WD models, disconnect the driveshaft from the rear axle companion flange. Pull the driveshaft toward the rear of the vehicle until the slip yoke clears the extension housing. Plug the extension housing to prevent lubricant from leaking out.
13 On 4WD models, disconnect the driveshaft from the transfer case circular flange, and remove the driveshaft.
14 Installation is the reverse of removal with the following additions:
  a) Lubricate the slip yoke splines with chassis grease.
  b) Inspect the housing seal for damage and replace it if necessary.
  c) Do not allow the driveshaft slip yoke to bottom out on the output shaft.
  d) Make sure that the center bearing bracket is aligned at right angles or "square" with the driveshaft. If there were spacers under the bracket, install them again.
  e) Line up the paint marks on the rear yoke and the rear axle companion flange.
  f) Torque all bolts to the specifications shown in the beginning of this Chapter.

Front driveshaft (4WD models)

Refer to illustrations 13.17 and 13.18

15 Using a scribe or white paint, place marks on the driveshaft and differential flanges in line with each other **(see illustration 13.3)**. This is to make sure the driveshaft is reinstalled in the same position to preserve the balance.
16 Make alignment marks on the driveshaft and transfer case flanges.
17 Unbolt the flange that secures the driveshaft double cardan joint to the transfer case **(see illustration)**.
18 Remove the bolts and straps that secure the front end of the driveshaft to the differential yoke **(see illustration)**.
19 Wrap tape around the universal joint bearings at the axle end of the driveshaft so they won't fall off.
20 Installation is the reverse of the removal Steps with the following additions:
  a) Raise the front of the driveshaft into position, checking to be sure the marks are in alignment. If not, make sure the transmission is still in Neutral, then turn the front wheels to line up the marks.
  b) Remove the tape securing the bearing cups and install the straps and bolts. Tighten the bolts to the torque listed in this Chapter's Specifications.

14 Universal joints - replacement

Note: *A press or large vise will be required for this procedure. It may be a good idea to take the driveshaft to a repair or machine shop where the universal joints can be replaced for you, normally at a reasonable charge.*

1   Remove the driveshaft as outlined in the previous Section.

Single cardan U-joint

Refer to illustrations 14.3, 14.5a and 14.5b

2   Mark the position of driveshaft components in relation to the driveshaft tube so that all components can be reassembled in the same relationship.
3   Using a small pair of pliers, remove the snap-rings from the spider **(see illustration)**.
4   Supporting the driveshaft, place it in position on either an arbor press or on a workbench equipped with a vise.
5   Place a piece of pipe or a large socket with the same inside diameter over one of the bearing caps. Position a socket which is of slightly smaller diameter than the cap on the

14.3 Remove the universal joint snap-rings with small pliers

14.5a The universal joint bearing cups can be pressed out with a vise and sockets

14.5b The bearing cup can be extracted with locking pliers

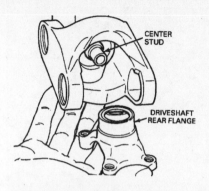

14.20 Remove the rubber seal from the center stud

opposite bearing cap **(see illustration)** and use the vise or press to force the cap out (inside the pipe or large socket), stopping just before it comes completely out of the yoke. Use the vise or large pliers to work the cap the rest of the way out **(see illustration)**.

6 Transfer the sockets to the other side and press the opposite bearing cap out in the same manner.

7 Clean all parts that will be reused. **Caution:** *A new U-joint must be replaced as an assembly. Do not use old parts if new ones are included with the new U-joint.*

8 Pack the new universal joint bearings with chassis grease. Ordinarily, specific instructions for lubrication will be included with the universal joint servicing kit and should be followed carefully.

9 Position the spider in the yoke and partially install one bearing cap in the yoke. If the replacement spider is equipped with a grease fitting, be sure it's offset in the proper direction (toward the driveshaft).

10 Start the spider into the bearing cap and partially install the other cap. Align the spider and press the bearing caps into position, being careful not to damage the dust seals.

11 Install the snap-rings. If difficulty is encountered in seating the snap-rings, strike the driveshaft yoke sharply with a brass or plastic hammer. This will spring the yoke ears slightly and allow the snap-rings to seat in the groove. **Caution:** *Do not strike the bearings.*

12 Install the grease fitting and fill the joint with grease. Be careful not to overfill the joint, as this could blow out the grease seals.

13 Install the driveshaft. Tighten the flange bolts to the specified torque listed in this Chapter's Specifications.

Double cardan U-joint

Refer to illustration 14.20

Note: *The double cardan joint must be replaced as an assembly - don't attempt to replace half of it, even if only one cross-and-roller assembly is worn.*

14 Mark the position of the spiders, the driveshaft center yoke, and the driveshaft centering socket yoke in relation to the stud yoke welded to the front of the driveshaft tube. All components must be reassembled in the same relationship.

15 Using a small pair of pliers, remove the snap-rings from the spider **(see illustration 14.3)**.

16 Supporting the driveshaft, place it in position on either an arbor press or on a workbench equipped with a vise.

17 Place a piece of pipe or a large socket with the same inside diameter over one of the bearing caps. Position a socket which is of slightly smaller diameter than the cap on the opposite bearing cap **(see illustrations 14.5a and 14.5b)** and use the vise or press to force the cap out (inside the pipe or large socket), stopping just before it comes completely out of the yoke. Use the vise or large pliers to work the cap the rest of the way out.

18 Transfer the sockets to the other side and press the opposite bearing cap out in the same manner.

19 Remove the spider from the center yoke.

20 Pull the centering socket yoke off the center stud. Remove the rubber seal from the centering ball stud **(see illustration)**.

21 Repeat steps 15 through 19 to remove the other spider from the center yoke.

22 Clean all parts that will be reused. **Caution:** *A new U-joint must be replaced as an assembly. Do not use old parts if new ones are included with the new U-joint.*

23 Pack the new universal joint bearings with Ford Premium Long-Life Grease XG-1-C, or equivalent grease. Ordinarily, specific instructions for lubrication will be included with the universal joint servicing kit and should be followed carefully.

24 Position the spider in the driveshaft yoke and partially install one bearing cap in the yoke. If the replacement spider is equipped with a grease fitting, be sure it's offset in the proper direction (toward the driveshaft).

25 Start the spider into the bearing cap and then partially install the other cap. Align the spider and press the bearing caps into position, being careful not to damage the dust seals.

26 Install the snap-rings. If difficulty is encountered in seating the snap-rings, strike the driveshaft yoke sharply with a brass or plastic hammer. This will spring the yoke ears slightly and allow the snap-rings to seat in the groove. **Caution:** *Do not strike the bearings.*

27 Install the grease fitting (if equipped) and fill the joint with grease. Be careful not to overfill the joint, as this could blow out the grease seals.

28 Pack the socket relief and the ball with Ford Premium Long-Life Grease XG-1-C, or equivalent grease.

29 Position the driveshaft center yoke over the spider ends and press in the bearing. Install the snap rings.

30 Install a new seal on the centering ball stud. Position the driveshaft centering socket yoke on the stud.

31 Start the front spider into the bearing cap and then partially install the other cap. Align the spider and press the bearing caps into position, being careful not to damage the dust seals. Install the snap-rings.

32 Install the grease fitting (if equipped) and fill the joint with grease. Be careful not to overfill the joint, as this could blow out the grease seals.

33 Install the driveshaft. Tighten the flange bolts to the torque listed in this Chapter's Specifications.

15 Center bearing - replacement

Refer to illustration 15.1

Note: *A center bearing is used to support the two-piece driveshaft on long wheelbase models.*

1 Unbolt the center bearing mount **(see illustration)**. Save any spacers so that they can be reinstalled. Spin the bearing by hand. If it does not spin freely, replace it.

2 Inspect the rubber insulator. Replace it if there is evidence of hardening, cracking or deterioration before removing the shaft.

3 Remove the driveshaft as outlined in Section 13.

4 Separate the driveshaft from the coupling shaft. Make sure to mark the driveshaft and coupling shaft.

5 Remove the nut securing the half-round yoke to the coupling shaft and remove the yoke.

6 Installation is the reverse of these steps with the following additions:

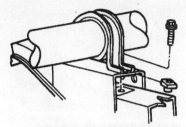

CENTER BEARING INSTALLATION RANGER SUPERCAB ONLY

15.1 Center bearing details (long-wheelbase models only)

a) Make sure the yoke and coupling shaft have the same orientation as originally installed.
b) Torque the nut to the specification shown in the Specifications in this Chapter.
c) Make sure the driveshaft and coupling shaft have the same orientation as originally installed.

16 Rear axle - description and check

Refer to illustration 16.1

Description

1 The rear axle assembly is a hypoid (the centerline of the pinion gear is below the centerline of the ring gear) semi-floating type **(see illustration)**. The differential carrier is a casting with a pressed steel cover. The axle tubes are made of steel, pressed and welded into the carrier.

2 An optional Traction-Lok limited slip rear axle is also available. This differential allows for normal operation until one wheel loses traction. The unit utilizes multi-disc clutch packs and a speed sensitive engagement mechanism which locks both axleshafts together, applying equal rotational power to both wheels.

3 In order to undertake certain operations, particularly replacement of the axleshafts, it's important to know the axle identification number. It's located on a small metal tag near one of the cover bolts.

4 Many times a problem is suspected in the rear axle area when, in fact, it lies elsewhere. For this reason, a thorough check should be performed before assuming a rear axle problem.

5 The following noises are those commonly associated with rear axle diagnosis procedures:

a) Road noise is often mistaken for mechanical faults. Driving the vehicle on different surfaces will show whether the road surface is the cause of the noise. Road noise will remain the same if the vehicle is under power or coasting.
b) Tire noise is sometimes mistaken for mechanical problems. Tires which are worn or low on air pressure are particularly susceptible to emitting vibrations and noises. Tire noise will remain about the same during varying driving situations, where rear axle noise will change during coasting, acceleration, etc.
c) Engine and transmission noise can be deceiving because it will travel along the driveline. To isolate engine and transmission noises, make a note of the engine speed at which the noise is most pronounced. Stop the vehicle and place the transmission in Neutral and run the engine to the same speed. If the noise is the same, the rear axle is not at fault.

6 Overhaul and general repair of the rear axle is beyond the scope of the home mechanic due to the many special tools and critical measurements required. Thus, the procedures listed here will involve axleshaft removal and installation, axleshaft oil seal replacement, axleshaft bearing replacement and removal of the entire unit for repair or replacement.

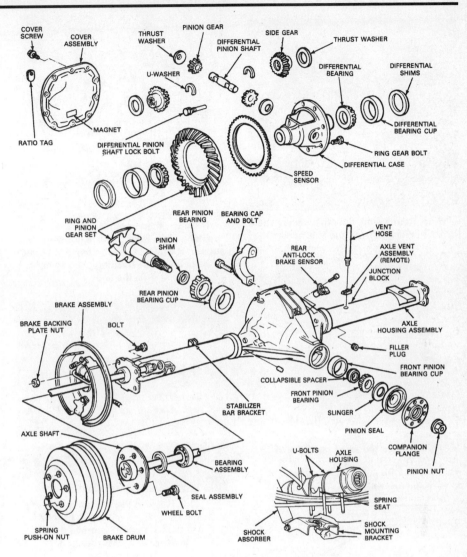

16.1 An exploded view of a typical rear axle assembly

17 Rear axle assembly - removal and installation

Refer to illustrations 17.6a, 17.6b, 17.7, 17.8, 17.9, 17.10a, 17.10b, 17.11 and 17.13

1 Loosen the lug nuts on the rear wheels.
2 Raise the rear of the vehicle, support it securely on jackstands and remove the rear wheels.
3 Remove the axleshafts (see Section 20).
4 Unbolt the brake line junction block from the axle housing and detach the brake lines from the clips on the axle housing. Disconnect the rear anti-lock electrical connector from the sensor on the axle housing (see Chapter 9).
5 Remove the rear brake drums (see Chapter 9).

Chapter 8 Clutch and drivetrain

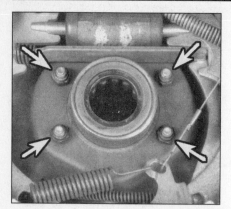

17.6a Remove the nuts that secure the brake backing plate to the axle housing . . .

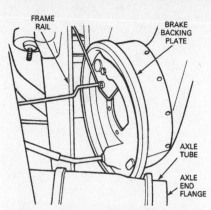

17.6b . . . then carefully separate the backing plate from the axle housing and support it with wire - don't bend the brake line or allow the backing plate to hang on the line

17.7 The vent hose is secured to the axle by a crimp-type clamp - you'll need to replace it with a screw-type clamp

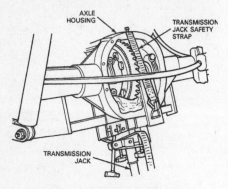

17.8 Place the axle housing on a transmission jack and secure it firmly - the axle housing is heavy and can cause injury if it falls

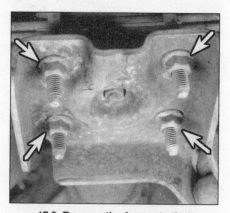

17.9 Remove the four nuts that secure the shock mounting bracket to the axle - don't remove the lower shock absorber bolt

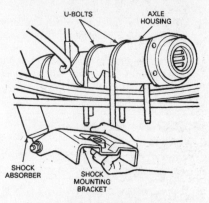

17.10a Pull the bracket down off the U-bolts . . .

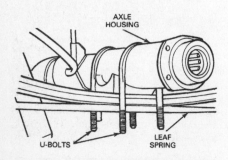

17.10b . . . then remove the U-bolts from the axle housing and spring

17.11 Unbolt the stabilizer bar retainer from the axle housing

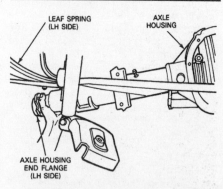

17.13 Tilt the axle housing a shown to clear the spring, then lower the jack and take it out from under the vehicle

6 Remove the nuts and bolts that secure the brake backing plate to the axle (see illustration). Separate the backing plates from the axle and tie them up out of the way (see illustration).
7 Detach the vent hose from the axle housing (see illustration).
8 Place a transmission jack beneath the axle housing. Secure the axle housing to the jack with a safety strap or chain (see illustration). **Note:** *You'll need to tilt the axle housing on the jack to remove it. Secure the axle so it can't fall off the jack, but position the strap or chain so the axle can be tilted.*

9 Remove the U-bolt nuts that secure the shock mounting bracket to the axle (see illustration). Don't detach the shock absorber from the bracket.
10 Lower the bracket away from the spring, then remove the U-bolts (see illustrations).
11 Unbolt the stabilizer bar bracket bolts from the axle housing (see illustration). Move the stabilizer bar out of the way.

12 Jack the axle up off the spring. Move it toward the passenger side of the vehicle until the driver's side end of the axle shaft clears the spring.
13 Tilt the driver's side of the axle housing down (see illustration). Lower the axle, move it toward the driver's side and remove it from beneath the vehicle.
14 Installation is the reverse of the removal Steps. Be sure to tighten the fasteners to the torque values listed in this Chapter's Specifications. Tighten the wheel lug nuts to the torque listed in the Chapter 1 Specifications.

Chapter 8 Clutch and drivetrain

18.6a The pinion shaft lock-bolt fits in a recess in the differential housing

18.6b Remove the bolt with a thin-wall socket . . .

18.6c . . . then pull the pinion shaft out - on Traction-Lok models, the spacer spring (arrow) need not be removed

18.7a Remove the C-lock from the end of each axleshaft - the differential lubricant is very slippery and makes the C-locks hard to hold with pliers, so be careful not to drop them

18.7b There's a rubber O-ring in the C-lock groove in the end of each axleshaft

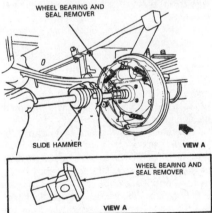

18.8 Lift and pull the axleshaft out of the axle housing - be careful not to damage the oil seal

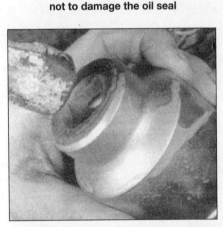

18.11 The wheel bearing and seal can be installed with a large socket or piece of pipe of the correct diameter if the special tool isn't available

18 Axleshafts, bearings and oil seals (rear) - removal and installation

Axleshaft removal

Refer to illustrations 18.6a, 18.6b, 18.6c, 18.7a, 18.7b and 18.8

1 Loosen the lug nuts on the rear wheels.
2 Raise the rear of the vehicle, support it securely on jackstands and remove the rear wheels.
3 Remove the rear brake drums (see Chapter 9).
4 Clean all dirt from the area surrounding the carrier cover.
5 Drain the differential lubricant (see Chapter 1).
6 Remove the differential shaft lock-bolt and differential pinion shaft **(see illustration)**. On Traction-Lok differentials, it isn't necessary to remove the preload spring **(see illustrations)**. **Note:** *The pinion gears may be left in place. Once the axleshafts are removed, reinstall the pinion shaft and lock-bolt.*
7 Push the axleshafts toward the center of the vehicle and remove the C-locks from the button end of the axleshafts **(see illustration)**. Don't damage the rubber O-ring which fits in the axle shaft groove under the C-lock **(see illustration)**.

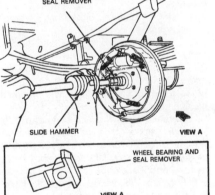

18.9 Remove the wheel bearing and seal with a puller - a special tool is shown here, which can be purchased or rented at most auto parts stores

8 Carefully remove the axleshafts from the housing **(see illustration)**. Don't damage the oil seal at the end of the housing.

Oil seal and bearing replacement

Refer to illustrations 18.9 and 18.11

9 Insert a rear axle bearing remover attached to a slide hammer into the bore of the bearing and position it behind the bearing outer race. Using the slide hammer withdraw the bearing and oil seal as a unit **(see illustration)**. These tools are available at most auto parts stores.
10 Thoroughly clean the rear axle housing bore with a rag and solvent.
11 Lubricate the new bearing with rear axle lubricant and install the bearing into the housing bore. Use an axle bearing driver tool or a large socket that matches the outer bearing race **(see illustration)**. Tap the bearing in until it's into the full depth of its recess. **Caution:** *Do not tap on the inner bearing race while installing*

18.14 The O-ring in the C-lock groove goes toward the wheel end of the axleshaft

18.15 Align the lock-bolt hole in the pinion lock shaft with the hole in the differential housing

the bearing, as it will be damaged.

12 Apply multi-purpose grease between the lips of the axleshaft seal.

13 Install a new axleshaft seal using an axle seal installation tool, available at most auto parts stores, or a large socket that matches the seal diameter. **Caution:** *If the seal is installed incorrectly or gets cocked during installation it will be damaged, leading to component failure and an oil leak.*

Axleshaft installation

Refer to illustrations 18.14 and 18.15

14 Make sure the rubber O-ring is in the groove in the end of the axleshaft. Push the O-ring toward the outer end of the groove (toward the wheel) **(see illustration)**. Carefully insert the axle into the housing and install the C-lock on the button end of the axleshaft splines. Be sure the O-ring is in the axleshaft groove **(see illustration 19.7b)**. Push the shaft out until the C-lock seats in the counterbore of the differential side gear.

15 Position the differential pinion shaft through the case and pinion gears, aligning the hole in the shaft with the lock-bolt hole **(see illustration)**. Apply thread locking compound to the lock-bolt, install the bolt and tighten it to the torque listed in this Chapter's Specifications.

16 Clean the gasket mounting surface on the rear axle housing and cover of all old gasket material residue with lacquer thinner or gasket remover. Apply a continuous bead of silicone sealant. Run the bead to the inside of the bolt holes and install the cover and bolts.

17 Tighten the bolts in a criss-cross pattern to the torque listed in this Chapter's Specifications.

18 Refill the axle with the correct quantity and grade of lubricant (see Chapter 1).

19 Install the brake drums (see Chapter 9).

20 Install the wheels and lug nuts.

21 Lower the vehicle and tighten the lug nuts to the value listed in the Chapter 1 Specifications.

19 Front axle assembly (4WD models) - removal and installation

1993 through 1997 models

Refer to illustration 19.7

1 Raise the front of the vehicle and install jackstands under the radius arm brackets.

2 Disconnect the driveshaft from the front axle yoke (see Section 13).

3 Remove the front brake calipers (see Chapter 9). **Caution:** *Tie the calipers up with wire to keep any strain off the flexible brake lines.*

4 Detach the tie-rod ends from the steering linkage (see Chapter 10).

5 Remove the stabilizer bar and its connecting links (see Chapter 10).

6 Position a floor jack under the axle arm and slightly compress the front coil spring. Remove the nut securing the lower portion of the spring to the axle beam.

7 Carefully lower the jack and remove the coil spring, spacer, seat and stud **(see illustration)**. **Caution:** *The axle arm assembly must be supported on the jack throughout spring removal and installation. Do not let the arm assembly hang suspended by the brake hose.*

8 Detach the shock absorber from the radius arm bracket.

9 Detach the radius arm and bracket from the axle arm and remove them from the vehicle.

10 Remove the pivot bolt securing the right axle arm assembly to the frame crossmember. Remove the clamps securing the axleshaft boot to the axleshaft slip yoke and axleshaft. Slide the rubber boot over the stub shaft.

11 Once the right driveaxle is disconnected from the slip yoke, lower the floor jack and remove the right axle arm assembly.

12 Position the jack under the differential housing and remove the bolt securing the left axle arm to the frame crossmember. Lower and remove the left axle arm assembly.

13 Installation is the reverse of the removal procedure with the following addition.

14 Tighten all fasteners to the torque listed in this Chapter's Specifications and related Chapters' Specifications.

1998 and later models

Removal

Refer to illustrations 19.21 and 19.22

Note: *A special retainer ring/seal installer tool is required to reinstall the driveaxle retainer ring. Be sure this tool is on hand before beginning this procedure.*

15 Loosen the wheel lug nuts, raise the vehicle and support it securely on jackstands. Remove the wheel(s).

16 Remove the hub lock (see Section 13 in Chapter 10)

17 Remove the brake caliper (see Chapter 9), then disconnect the ABS sensor, if equipped, and secure the sensor and wire harness aside.

18 Using snap-ring pliers, remove the plastic driveaxle retainer ring and discard it. **Caution:** *You MUST replace the plastic driveaxle retainer ring every time it's removed. Using an old retainer ring can damage the hub and/or the driveaxle.*

19 Support the lower control arm with a floor jack. Make sure the control arm is securely supported at its outer end.

20 Remove the upper control arm balljoint-to-steering knuckle nut and pinch bolt and separate the upper arm from the knuckle (see Chapter 10). **Warning:** *The jack must remain in this position throughout the entire procedure.*

21 To separate the driveaxle assembly from the steering knuckle, simply swing the knuckle out, compress the outer CV joint and pull the driveaxle assembly out of the knuckle. It should come right out. If the driveaxle splines are "frozen" to the hub and bearing assembly, knock the driveaxle loose with a brass hammer, or press the driveaxle out of the hub with a two-jaw puller, using a prybar or large screwdriver to keep the hub from turning **(see illustration)**. Once the driveaxle is loose from the hub splines, pull out on the steering knuckle assembly and guide the outer CV joint out of the hub.

22 Using a prybar or slide hammer with CV joint adapter, pry the inner CV joint assembly from the front differential **(see illustration)**. Be careful not to damage the front differential. Suspend the axle with a piece of wire - don't let it hang, or damage to the outer CV joint may occur.

23 Remove the driveaxle from the vehicle.

Installation

24 Installation is basically the reverse of removal, except that you MUST install a NEW retainer ring as follows. Place the new retainer ring on the special retainer ring/seal installer tool (make sure that the steel band on the retainer faces out), slide the installer tool onto the end of the axleshaft splines, install the snap-ring sleeve on the seal

Chapter 8 Clutch and drivetrain

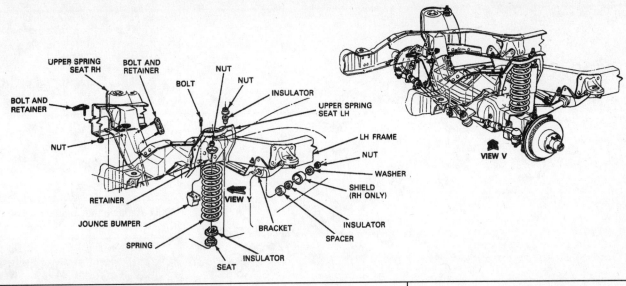

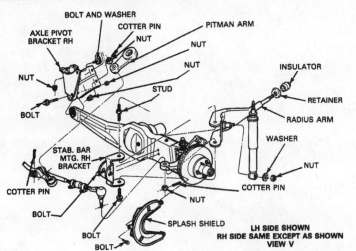

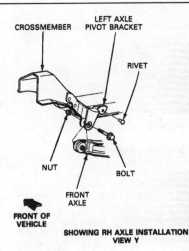

19.7 The front axle assembly (4WD models) - exploded view

installer and install the seal remover guide on the snap-ring sleeve **(see illustrations)**. (The seal remover guide serves as a pressing handle when used with the seal installer tool and the snap-ring sleeve). Pull *out* (toward the steering knuckle) on the driveaxle assembly and simultaneously press *in* with the seal remover guide until the retainer ring seats. You should hear an audible "click" when the retainer snaps into place. Verify that the retainer ring has been correctly seated by pushing and pulling firmly on the driveaxle assembly and on the seal remover guide. **Caution:** *If the retainer ring is not fully seated, the driveaxle seal will leak, the hub and bearing assembly will be contaminated and the vehicle will not operate in 4WD.*

25 Installation is otherwise the reverse of removal.

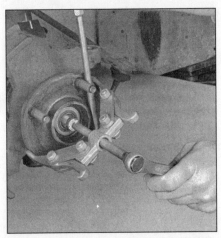

19.21a If the driveaxle does not come out easily, use a puller to press it out

19.22 Using a prybar or slide hammer with CV joint adapter, disengage the inner CV joint assembly from the front differential

Chapter 8 Clutch and drivetrain

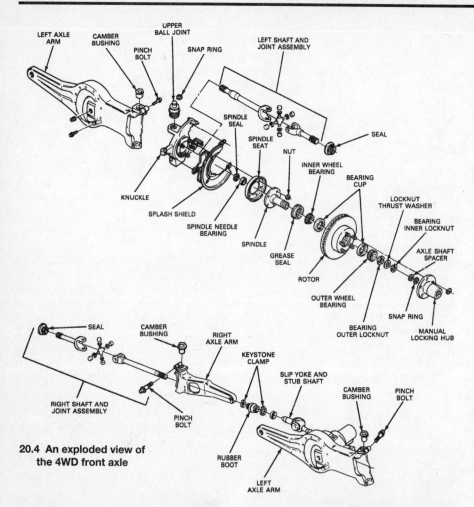

20.4 An exploded view of the 4WD front axle

20.5a Remove the spindle securing nuts

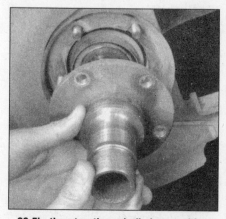

20.5b then tap the spindle loose with a soft face hammer and remove it

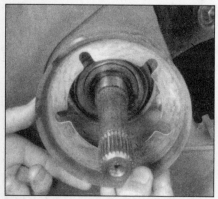

20.6 Take the spindle seat off

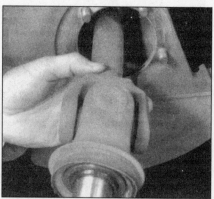

20.7a Take out the left shaft and joint assembly

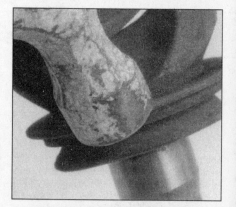

20.7b If necessary, tap the slinger off with a hammer - a press is needed to reinstall it

20 Axleshaft and joint assemblies (front) - removal, component replacement and installation

Removal

Refer to illustrations 20.4, 20.5a, 20.5b, 20.6, 20.7a, 20.7b and 20.8

1 Loosen the lug nuts on the front wheels.
2 Raise the front of the vehicle, support it securely on jackstands and remove the front wheels.
3 Remove the front brake calipers (see Chapter 9). Hang the calipers out of the way with pieces of wire. Don't disconnect the brake hoses.
4 Remove the locking hubs, wheel bearings and locknuts (see illustration). Refer to Chapter 1, Section 29, for the hub/brake disc/wheel bearing removal procedure.
5 Remove the nuts securing the spindle to the steering knuckle. Tap the spindle with a plastic or soft faced hammer to jar it free, then remove it (see illustrations).
6 Remove the spindle seat (see illustration).
7 On the left side of the vehicle, pull the shaft and joint assembly out of the carrier (see illustration). If necessary, tap the slinger off the shaft with a hammer (see illustration). Note: *Don't remove the slinger unless necessary. You'll need a press to install it.*
8 Working on the right side, remove the

Chapter 8 Clutch and drivetrain

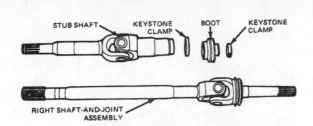

20.8 The right shaft and joint assembly - exploded view

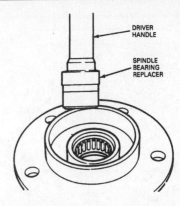

20.12 Install the new bearing in the bore

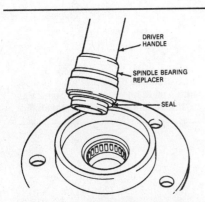

20.13 Drive in a new seal with its lip toward the driver

20.14a Mark the shaft and yoke with paint so they can be reassembled in the same relative positions

20.14b Push the snap-rings out of the joint with a screwdriver

20.15 Position sockets in a vise to push the bearings out of the joint - the small socket is smaller than the bearing; the large socket is large enough so the bearing will fit into it

metal clamps from the shaft and joint assembly and the stub shaft. Slide the rubber boot onto the stub shaft and pull the shaft and joint assembly from the splines of the stub shaft **(see illustration)**.

Spindle bearing and oil seal replacement

Refer to illustrations 20.12 and 20.13

9　Place the spindle in a padded vise so the vise grips the spindle's second step.
10　Remove the bearing and oil seal with a slide hammer and puller attachment.
11　Clean all dirt and grease from the spindle bore. Make sure there are no nicks or burrs in the bearing bore.
12　Install the new bearing in the bore with the manufacturer's mark facing out. Install the bearing with a spindle bearing replacer and driver handle, or equivalent available at most auto parts stores that carry special tools. **(see illustration)**.
13　Use the same tools to install the seal **(see illustration)**. The lip of the seal faces the installer. After installation, coat the lip of the seal with high-temperature lubricant, available at most auto parts stores.

Shaft and joint assembly universal joint replacement

Refer to illustrations 20.14a, 20.14b and 20.15

14　Remove the internal snap-rings from the U-joint with a screwdriver **(see illustrations)**.
15　The remainder of the procedure is the same as that for the driveshaft universal joints, which is described in Section 14 (see illustration).

16　If the slinger was removed from the spindle, have it pressed on by a machine shop.
17　On the right side of the carrier, install the rubber boot and new metal clamps on the stub shaft slip yoke. There is no blind spline to ensure correct alignment, so pay special attention to the yoke ears and make sure they are in phase. Be sure the assembly is fully seated and crimp the metal clamp.
18　On the left side of the carrier, slide the shaft and joint assembly through the knuckle and engage the splines on the shaft in the carrier.
19　Install the splash shield. Install the spindle on the steering knuckle, then install and tighten the spindle nuts to the torque listed in Chapter 10 Specifications.
20　Install the front wheel bearings and locking hubs (see Chapter 1).
21　Install the brake caliper (see Chapter 9).
22　Install the front wheels and lower the vehicle.
23　Tighten the lug nuts to the torque listed in Chapter 1 Specifications.

21　Right slip yoke and stub shaft assembly, carrier, carrier oil seals and bearings - removal and installation

Removal

Refer to illustrations 21.7, 21.9a, 21.9b, 21.10, 21.11, 21.12 and 21.14

1　Loosen the lug nuts on the front wheels.
2　Raise the front of the vehicle, support it securely on jackstands and remove the front wheels.
3　Detach the front driveshaft from the axle yoke (see Section 13). Tie the driveshaft up

Chapter 8 Clutch and drivetrain

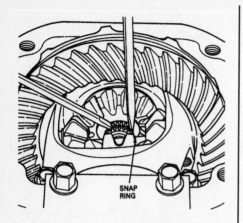

21.7 Push the C-clip off the shaft with a pair of screwdrivers

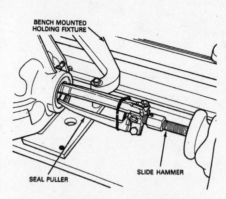

21.9a Remove the passenger's side oil seal with a slide hammer and a puller that hooks to the inside of the seal

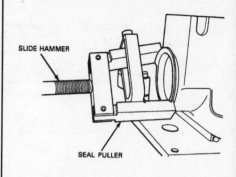

21.9b Remove the driver's side oil seal with a slide hammer and a puller that hooks to the outside of the seal

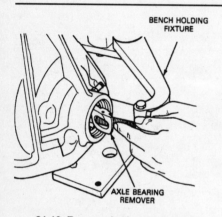

21.10 Remove both caged needle bearings with a slide hammer and puller

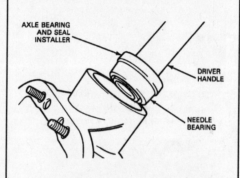

21.11 Drive in the bearings with the manufacturer's marking facing toward the driver tool (out)

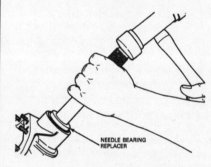

21.12 After the bearings are installed, drive in the seals

out of the way.
4 Remove the spindles and the left and right shaft and joint assemblies (see Section 20).
5 Place a transmission jack under the carrier and unbolt the carrier from the left axle arm. Separate the carrier from the arm, let the lubricant drain and remove the carrier.
6 Place the carrier in a vise or on a workbench.
7 Turn the shaft assembly so the open side of the C-clip is exposed, then push the C-clip off with a pair of screwdrivers **(see illustration)**.
8 Pull the slip yoke and shaft assembly out of the carrier.
9 Remove the oil seals with a slide hammer and puller **(see illustrations)**.
10 Remove the caged needle bearings with a slide hammer and puller attachment **(see illustration)**.

Installation

11 Clean and inspect the bearing bore for nicks and burrs before installing a new caged bearing. Position the bearing with the manufacturer's marks facing out, then drive the bearing in until it is fully seated in the bore

(see illustration).
12 Coat the oil seal with high-temperature lubricant and drive it into the carrier housing **(see illustration)**.
13 Install the slip yoke and shaft assembly into the carrier so the C-clip groove in the shaft is visible.
14 Push the C-clip into the groove with a screwdriver **(see illustration)**. Make sure it is completely seated in the groove. **Caution:** *Don't push on the center of the C-clip with the screwdriver, as it may be damaged.*
15 Clean all traces of gasket sealant from the mating surfaces with lacquer thinner or gasket remover. Apply RTV sealant in an unbroken 1/4-inch wide bead, inboard of the bolt holes.
16 Position the carrier on the transmission jack and install it in position on the support arm. Use the guide pins for alignment. Install and tighten the bolts in a clockwise or counterclockwise pattern to the torque listed in this Chapter's Specifications.
17 Install the shear bolt securing the carrier to the axle arm and tighten to the torque listed in this Chapter's Specifications.
18 Install the shaft and joint assemblies and spindles.

19 Connect the driveshaft to the yoke (see Section 13).
20 Fill the carrier with lubricant (see Chapter 1).
21 Install the front wheels and lower the vehicle.

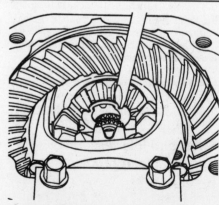

21.14 Push the C-clip into the groove with a screwdriver - don't push on the center section of the C-clip or it may be damaged

Chapter 8 Clutch and drivetrain

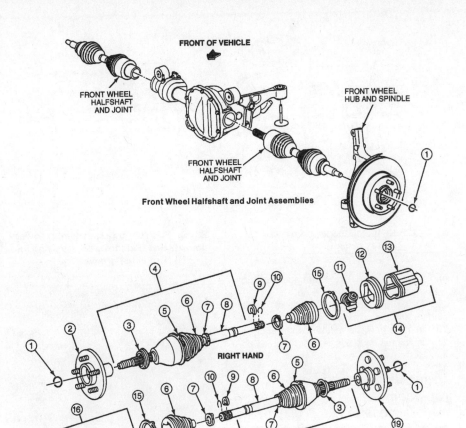

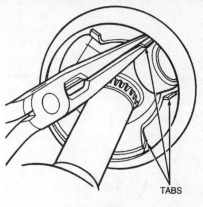

22.5 Bend the retaining tabs down slightly . . .

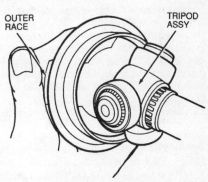

22.6 . . . and remove the outer race from the tripod assembly

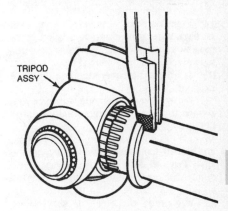

22.7 Move the stop ring down the axleshaft

22.1 An exploded view of the driveaxle assembly (later 4WD models)

1 Retainer rings	7 Boot clamp	14 Inner CV joint
2 Front wheel hub	8 Axleshaft	15 Boot clamp
3 Excluder seal	9 Stop ring	16 Inner CV joint
4 Driveaxle and outer joint assembly	10 Circlip	17 Circlip
	11 Tripod assembly	18 Housing
5 Boot clamp	12 Tri-lobe insert	19 Front wheel hub
6 Boot	13 Housing	

22 Tighten the lug nuts to the torque listed in Chapter 1 Specifications.

22 Driveaxle boot replacement and CV joint overhaul (1998 and later 4WD models)

Inner CV joint and boot replacement

Disassembly

Refer to illustrations 22.1, 22.5, 22.6, 22.7 and 22.8

1 1998 and later 4WD models are equipped with Constant Velocity (CV) joint type driveaxle assemblies **(see illustration)**. The inboard tripod design CV joint and driveaxle boot are serviceable units. The outer CV joint is not serviceable and must be replaced as an assembly. However, the outer driveaxle boot is serviceable with a similar procedure as the inner boot.

2 Remove the driveaxle from the vehicle (see Section 19).

3 Mount the driveaxle in a vise. The jaws of the vise should be lined with wood or rags to prevent damage to the axleshaft.

4 Cut the boot clamps from the inner boot and discard them, then move the boot toward the center of the shaft..

5 Bend the retaining tabs slightly to allow for tripod removal **(see illustration)**.

6 Remove the tripod assembly from the outer race **(see illustration)**.

7 Move the inner (exposed) stop ring down the shaft about 1/2-inch **(see illustration)**.

8 Move the tripod down the shaft towards the inner stop-ring until the circlip is visible

Chapter 8 Clutch and drivetrain

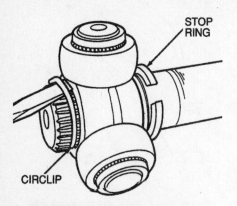

22.8 Remove the circlip and pull the tripod assembly off the axleshaft

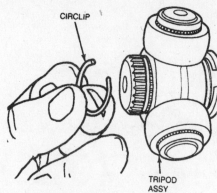

22.15a Install the tripod circlip

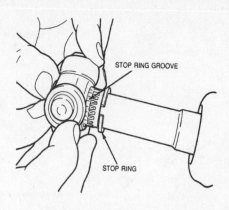

22.15b Push the tripod assembly toward the axle end and move the stop ring into the groove

on the end of the driveaxle. Remove the circlip and remove the tripod assembly from the driveaxle **(see illustration)**.
9 No further disassembly of the tripod is possible. Inspect the tripod rollers, roller bearings and races carefully for damage, worn spots and smooth operation. Damaged or worn tripods cannot be rebuilt and must be replaced.
10 Remove the inner stop ring from the axleshaft and remove the old boot.

Reassembly

Refer to illustrations 22.15a and 22.15b

11 Slide a new clamp and the inner CV joint boot on the axleshaft.
12 Install a new inner stop ring past the second ring groove about 1/2-inch.
13 Install the tripod assembly on the driveaxle with the chamfered side inward.
14 Push the tripod assembly down the axle far enough to allow circlip installation. Install the new circlip and push the tripod assembly towards the axle end until the tripod seats on the circlip and the inner stop ring groove is exposed.
15 Move the inner stop ring to its groove to secure the tripod assembly **(see illustrations)**.
16 Fill the outer race with CV joint grease and spread some on the inside of the boot as well. The left axle tripods use about 6.5 ounces of grease and the right axles use about 7 ounces. Push the tripod assembly into outer race and bend the six retaining tabs back to their original shape.
17 Wipe any excess grease from the axle boot groove on the outer race. Seat the small diameter of the boot in the recessed area on the axleshaft and install the clamp.
18 Equalize the pressure in the boot by inserting a dull screwdriver between the boot and the outer race. Don't damage the boot with the tool.
19 Install the boot clamp. A pair of special clamp-crimping pliers are required.
20 Install a new clip on the stub axle.
21 Install the driveaxle as described in Section 20.

Outer CV joint and boot replacement

Refer to illustration 22.27

Note: *The outer CV joint is a non-serviceable item and is permanently retained to the driveaxle. If any damage or excessive wear occurs to the axle or the outer CV joint, the driveaxle and outer CV joint assembly must be replaced. Complete rebuilt driveaxle assemblies are available at most auto parts stores. If after disassembly and inspection the CV joint is deemed serviceable, replace the boot as described below.*
22 Remove the inner CV joint and boot.
23 Cut the boot clamps and remove them from the outer CV joint.
24 Detach the boot from the CV joint and slide it off the axleshaft.
25 Check the grease for contamination by rubbing a small amount between your fingers. If the grease feels gritty, the CV joint is contaminated and must be replaced.
26 If the grease is not contaminated, remove all traces of grease from the CV joint using solvent and a soft bristle brush. Thoroughly dry the CV joint using compressed air, if available. All traces of solvent must be removed or damage to the CV joint may occur. **Warning:** *Wear eye protection when using compressed air!*
27 Inspect the cage, balls and races for pitting, score marks, cracks and other signs of wear or damage **(see illustration)**. Shiny, polished spots are normal and will not

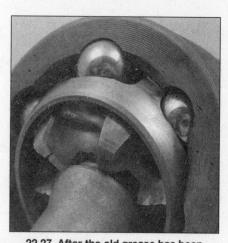

22.27 After the old grease has been completely removed and the solvent has dried, rotate the outer joint housing through its full range of motion and inspect the bearing surfaces for wear or damage; if the balls, race or cage appear damaged, replace the driveaxle shaft and outer joint assembly

adversely affect CV joint performance.
28 Install the new outer boot and clamps onto the driveaxle shaft. Completely pack the CV joint with the grease provided in the boot kit. Force the grease into the joint and around the balls and races. Spread the remaining grease evenly inside the boot. Total grease fill is approximately 6.0 ounces.
29 Position the outer boot on the CV joint and using the appropriate boot clamp pliers, tighten the clamps.
30 Assemble the inner CV joint and boot and install the driveaxle.

Chapter 9 Brakes

Contents

	Section		Section
Anti-lock Brake System (ABS) - general information	2	Brake system check	See Chapter 1
Brake caliper - removal, overhaul and installation	4	General information	1
Brake disc - inspection, removal and installation	5	Master cylinder - removal, overhaul and installation	8
Brake fluid level check	See Chapter 1	Parking brake - adjustment	11
Brake hoses and lines - inspection and replacement	9	Parking brake cables - replacement	12
Brake hydraulic system - bleeding	10	Power brake booster - check, removal, installation and adjustment	13
Brake light switch - removal and installation	14	Wheel cylinder - removal, overhaul and installation	7
Brake pads - replacement	3		
Brake shoes (rear) - replacement	6		

Specifications

Brake fluid type .... See Chapter 1

Disc brakes
Minimum pad lining thickness .... 1/8 inch
Front brake disc
  Standard thickness
    1993 and 1994 models .... Not specified
    1995 on .... 1.023 inches
  Minimum thickness .... Refer to marks stamped on disc
  Runout limit
    Measured on vehicle .... 0.010 inch
    After machining .... 0.003 inch
  Thickness variation (parallelism) .... 0.00035 inch

Drum brakes
Drum wear limit .... Refer to marks stamped on drum
Minimum lining thickness .... 1/16 inch above rivet heads

Torque specifications
Ft-lbs (unless otherwise indicated)

Backing plate-to-axle housing nuts	25 to 35
Wheel cylinder-to-backing plate bolts	
1993 and 1994 models	108 to 156 in-lbs
1995 on	106 to 159 in-lbs
Brake booster-to-firewall nuts	
1993 and 1994 models	13 to 25
1995 on	16 to 21
Brake hose -to-caliper banjo bolt	29
Caliper mounting bolts (pins)	38 to 48
Caliper mounting bracket-to-spindle bolts	73 to 97*
Master cylinder-to-brake booster nuts	
1993 and 1994	13 to 25
1995 on	16 to 21
RABS valve mounting bolt	14 to 19
ABS rear sensor mounting bolt	25 to 29
Front wheel sensor mounting bolt	68 to 92 in-lbs

*Use new bolts or apply non-permanent thread locking agent to the threads of the old bolts.

1 General information

Refer to illustration 1.3

General description

All models covered by this manual are equipped with hydraulically operated, power-assisted brake systems. All front brake systems are disc type, while the rear brakes are drum type. Some models are equipped with an Anti-lock Brake System (ABS), which is described in Section 2.

All brakes are self-adjusting. The front disc brakes automatically compensate for pad wear, while the rear drum brakes incorporate an adjustment mechanism which is activated as the brakes are applied.

The hydraulic system is a split design, meaning there are separate circuits for the front and rear brakes **(see illustration)**. If one circuit fails, the other circuit will remain functional and a warning indicator will light up on the dashboard, showing that a failure has occurred.

Master cylinder

The master cylinder is located under the hood, mounted to the power brake booster, and is best recognized by the large fluid reservoir on top. The removable plastic reservoir is partitioned to prevent total fluid loss in the event of a front or rear brake hydraulic system failure.

The master cylinder is designed for the "split system" mentioned earlier and has separate primary and secondary piston assemblies, the piston nearest the firewall being the primary piston, which applies hydraulic pressure to the front brakes.

Power brake booster

The power brake booster, utilizing engine manifold vacuum and atmospheric pressure to provide assistance to the hydraulically operated brakes, is mounted on the firewall in the engine compartment.

Parking brake

The parking brake mechanically operates the rear brakes only.

The parking brake cables pull on a lever attached to the brake shoe assembly, causing the shoes to expand against the drum.

Precautions

There are some general cautions and warnings involving the brake system on this vehicle:

a) Use only brake fluid conforming to DOT 3 specifications.
b) The brake pads and linings may contain asbestos fibers which are hazardous to your health if inhaled. Whenever you work on brake system components, clean all parts with brake system cleaner. Do not allow the fine dust to become airborne. Also, wear an approved filtering mask

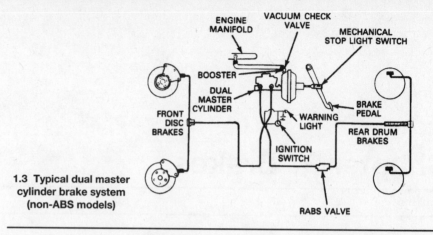

1.3 Typical dual master cylinder brake system (non-ABS models)

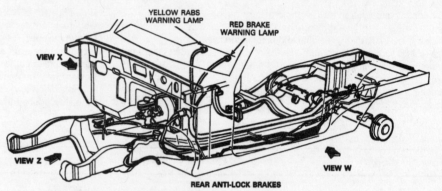

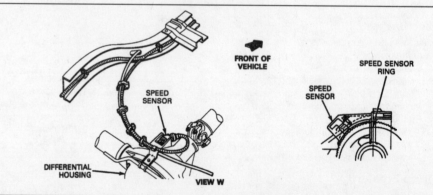

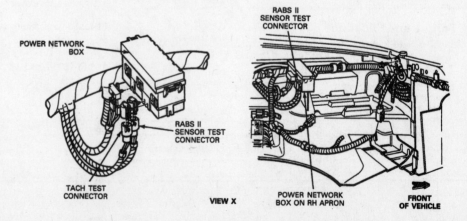

2.2 Rear Anti-lock Brake System (RABS) component layout

Chapter 9 Brakes

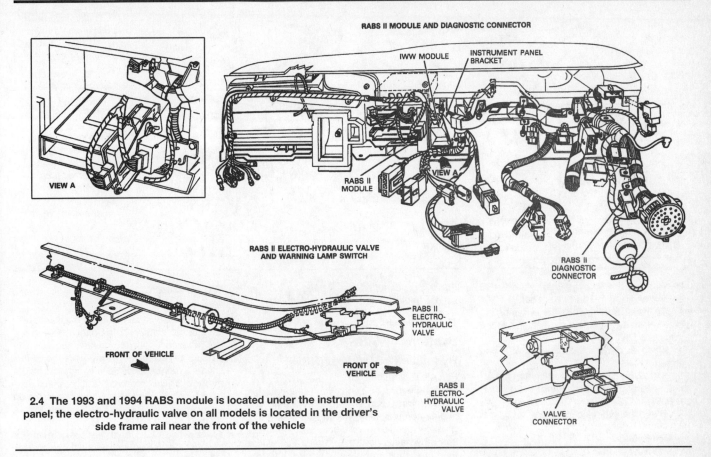

2.4 The 1993 and 1994 RABS module is located under the instrument panel; the electro-hydraulic valve on all models is located in the driver's side frame rail near the front of the vehicle

c) *Safety should be paramount whenever any servicing of the brake components is performed. Do not use parts or fasteners which are not in perfect condition, and be sure that all clearances and torque specifications are adhered to. If you are at all unsure about a certain procedure, seek professional advice. Upon completion of any brake system work, test the brakes carefully in a controlled area before putting the vehicle into normal service.*

If a problem is suspected in the brake system, don't drive the vehicle until it's fixed.

2 Anti-lock Brake System (ABS) - general information

General information

1 Some models are equipped with an Anti-lock Brake System (ABS). Early models and later 2WD models are equipped with rear wheel anti-lock brake systems while later 4WD models are equipped with 4-wheel anti-lock brake systems. The ABS system is designed to maintain vehicle steerability, directional stability and optimum deceleration under severe braking conditions and on most road surfaces. It does so by monitoring the rotational speed of each wheel and controlling the brake line pressure to each wheel during braking. This prevents the wheel from locking-up and provides maximum vehicle controllability.

Rear wheel Anti-lock Brake System (RABS)

Refer to illustration 2.2

2 The Rear Anti-lock Brake System (RABS) consists of a computer module, anti-lock brake valve, an exciter ring and a sensor **(see illustration)**.

3 Disconnect the cable from the negative battery terminal.

RABS module

1993 and 1994 models

Refer to illustration 2.4

4 Working under the instrument panel, press on the plastic tab on the electrical connector and disconnect it from the RABS module **(see illustration)**.

5 Remove the screws securing the RABS module to the instrument panel bracket and remove it.

6 Installation is the reverse of the removal steps.

1995 and later models

Refer to illustration 2.8

7 Remove the center finish panel from the instrument panel (see Chapter 11).

8 Slide the module's upper tab free of the instrument panel brace, then disconnect the electrical connectors and take it out **(see illustration)**.

9 Installation is the reverse of the removal steps.

RABS valve

10 Disconnect the two brake lines from the RABS valve **(see illustration 2.4)**. Plug the ends of the lines to prevent the loss of hydraulic fluid and the entry of foreign matter and moisture.

11 Disconnect the electrical connector from the RABS valve.

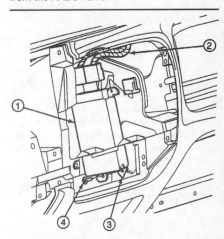

2.8 The 1995 and later RABS module is secured under the center of the instrument panel by a screw and J-nut

1 RABS module
2 Wiring harness
3 J-nut
4 Screw

Chapter 9 Brakes

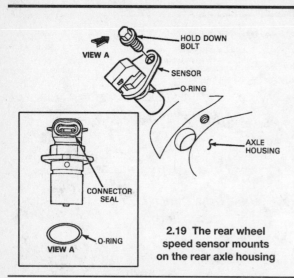

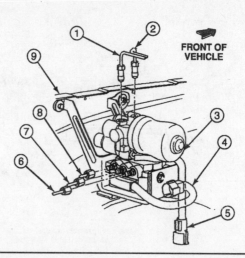

2.19 The rear wheel speed sensor mounts on the rear axle housing

2.28 Hydraulic control unit details on 1995 and later models with four-wheel ABS

1. Primary brake tube
2. Secondary brake tube
3. Pump motor
4. Solenoid control wiring
5. Pump motor wiring
6. Front brake tube (right side)
7. Front brake tube (left side)
8. Rear brake tube
9. Inner fender panel (left side)

12 Remove the screw securing the RABS valve to the frame rail and take it out.
13 To install, position the RABS valve on the frame rail. Install the screw and tighten it to the torque listed in this Chapter's Specifications.
14 Connect the electrical connector to the valve.
15 Install the brake lines and tighten the fittings securely. **Caution:** *Do not overtighten the fittings.*
16 Bleed the hydraulic system.

Speed sensor

Refer to illustration 2.19

17 Disconnect the electrical connector from the sensor located on the rear axle housing.
18 Prior to removing the sensor, thoroughly clean the rear axle housing surrounding the sensor.
19 Unbolt the sensor and remove it from the rear axle housing **(see illustration)**. Do not allow any dirt to fall into the interior of the rear axle housing.
20 Thoroughly clean the sensor mounting surface on the rear axle housing.
21 Inspect and clean the magnetized sensor pole piece. Remove any small metal particles that could cause erratic system operation.
22 If installing a new sensor, lightly lubricate the O-ring seal with clean engine oil.
23 If installing an old sensor, remove the old O-ring seal, install a new one and lubricate it with clean engine oil.
24 Firmly grasp the sides of the sensor (do not apply pressure to the electrical connector) and push it into the mounting hole.
25 Align the mounting flange bolt hole with that on the rear axle housing and install the hold-down bolt. Tighten the bolt to the torque listed in this Chapter's Specifications.
26 Connect the electrical connector to the sensor.

Exciter ring

27 To service the exciter ring the rear axle must be disassembled and the ring gear removed with a press. It is recommended that this procedure be performed by a dealer service department or other repair shop.

Four wheel Anti-lock Brake System (4WABS)

Hydraulic Control Unit (HCU)

Refer to illustration 2.28

28 The hydraulic control unit is located in the left front corner (driver's side) of the engine compartment. It consists of a brake pressure control valve block, a pump motor and a hydraulic control unit reservoir with a fluid level indicator assembly **(see illustration)**.
29 During normal braking conditions, brake hydraulic fluid from the master cylinder enters the hydraulic control unit through two inlet ports and passes through four normally open inlet valves, one to each wheel.
30 When the anti-lock brake control module senses that a wheel is about to lock up, the anti-lock brake control module closes the appropriate inlet. This prevents any more fluid from entering the affected brake. If the module determines that the wheel is still decelerating, the module opens the outlet valve, which bleeds off pressure in the affected brake.

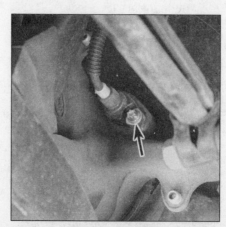

2.32 ABS front wheel sensor (arrow)

Wheel sensors

Refer to illustration 2.32

31 The ABS system uses four "variable-reluctance" sensors to monitor wheel speed ("reluctance" is a term used to indicate the amount of resistance to the passage of flux lines - lines of force in a magnetic field - through a given material). Each sensor contains a small inductive coil that generates an electromagnetic field. When paired with a toothed sensor ring which interrupts this field as the wheels turn, each sensor generates a low-voltage analog (continuous) signal. This voltage signal, which rises and falls in proportion to wheel rotation speed, is continuously sampled (monitored) by the control module, converted into digital data inside the module and processed (interpreted).
32 The front wheel sensors **(see illustration)** are mounted in the steering knuckle in close proximity to the toothed sensor rings, which are pressed onto the wheel hubs. The rear wheel sensors are mounted in the rear axle housing and the sensor ring is part of the ring gear in the rear axle.

Brake control module

Refer to illustration 2.33

33 The brake control module **(see illustration)**, which is mounted at the right front of

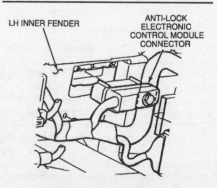

2.33 ABS control module

Chapter 9 Brakes

the battery on the radiator support, is the "brain" of the ABS system. The module (referred to as an Electronic Control Unit or ECU) constantly monitors the incoming analog voltage signals from the four ABS wheel sensors, converts these signals to digital form, processes this digital data by comparing it to the map (program), makes decisions, converts these (digital) decisions to analog form and sends them to the hydraulic control unit, which opens and closes the front and/or rear circuits as necessary.

34 The module also has a self-diagnostic capability which operates during both normal driving as well as ABS system operation. If a malfunction occurs, a red "BRAKE" warning indicator or an amber "CHECK ANTI-LOCK BRAKES" warning indicator will light up on the dash.

a) If the red BRAKE light glows, the brake fluid level in the master cylinder reservoir has fallen below the level established by the fluid level switch. Top up the reservoir and verify that the light goes out.
b) If the amber CHECK ANTI-LOCK BRAKES light glows, the ABS has been turned off because of a symptom detected by the module. Normal power-assisted braking is still operational, but the wheels can now lock up if you're involved in a panic-stop situation. A diagnostic code is also stored in the module when a warning indicator light comes on; when retrieved by a service technician, the code indicates the area or component where the problem is located. Once the problem is fixed, the code is cleared. These procedures, however, are beyond the scope of the home mechanic.

Relays

35 The main relay and pump motor relay are located in a relay box mounted near the master cylinder on the left side of the engine compartment. Refer to Chapter 12 for the locations and designations for the relays.

Diagnosis and repair

Warning: *If a dashboard warning light comes on and stays on while the vehicle is in operation, the ABS system requires immediate attention!*

36 Although a special electronic ABS diagnostic tester is necessary to properly diagnose the system, the home mechanic can perform a few preliminary checks before taking the vehicle to a dealer who is equipped with this tester.

a) Check the brake fluid level in the reservoir.
b) Verify that the control module electrical connector is securely connected.
c) Check the electrical connectors at the hydraulic control unit.
d) Check the fuses.
e) Follow the wiring harness to each wheel and check that all connections are secure and that the wiring is not damaged.

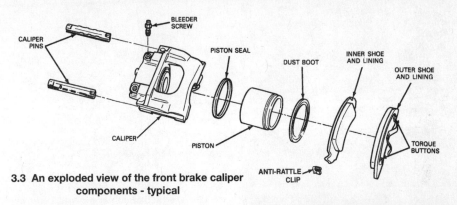

3.3 An exploded view of the front brake caliper components - typical

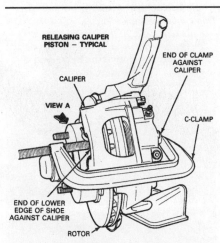

3.6a Use a large C-clamp and push the piston back into the bore just enough to allow the caliper to slide off the brake disc easily - note the one end of the C-clamp is on the flat area of the inner side of the caliper and the other end (screw end) is pressing on the outer pad

If the above preliminary checks do not rectify the problem, the vehicle should be diagnosed by a dealer service department or other qualified repair shop. Due to the rather complex nature of this system, all actual repair work must be done by the dealer service department or repair shop.

3 Brake pads - replacement

Refer to illustrations 3.3 and 3.6a through 3.6l

Warning 1: *Disc brake pads must be replaced on both front wheels at the same time - never replace the pads on only one wheel. Also, the dust created by the brake system may contain asbestos, which is harmful to your health. Never blow it out with compressed air and don't inhale any of it. An approved filtering mask should be worn when working on the brakes. Do not, under any circumstances, use petroleum-based solvents to clean brake parts. Use brake system cleaner only!*

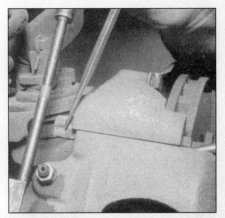

3.6b On 1994 and earlier models, use pliers and squeeze the caliper retaining pin while prying the other end until the tabs on the pin enter the spindle groove . . .

Warning 2: *1995 and later models use a special lubricating gel on certain parts of the brake caliper. This gel is non-toxic under most conditions. If it is heated to a certain point, however, it can give off toxic fumes which can cause health problems or even death. Never apply heat to these calipers and don't smoke around them, either (the gel might get on your cigarette). Always wash your hands thoroughly after working on the brakes.*

1 Remove the cover from the brake fluid reservoir and siphon out about 1/2 of the brake fluid.
2 Loosen the wheel lug nuts, raise the vehicle and support it securely on jackstands.
3 Remove the wheels. Work on one brake assembly at a time, using the assembled brake for reference if necessary **(see illustration)**.
4 Inspect the brake disc carefully as outlined in Section 5.
5 If machining is necessary, follow the information in that Section to remove the disc, at which time the pads can be removed from the caliper as well.
6 Follow the accompanying photos, beginning with **illustration 3.6a**, for the front brake pad replacement procedure. Be sure to stay in order and read the caption under each illustration.

3.6c ... then drive out the pin with a punch and hammer

3.6f This spring clip (arrow) holds the outside brake pad to the caliper on 1994 and earlier models - push down on the pad, releasing the locking tabs and slide the pad out of the caliper, then remove the inner pad and the anti-rattle clips. On 1995 and later models, detach the pads from the caliper mount

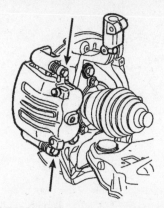

3.6d On 1995 and later models, remove the two caliper mounting bolts. Note: *Don't wipe the lubricating gel from these pins or the pin boots. At the time of publication there was no substitute for this gel* **(also read the Warning at the beginning of Section 3)**

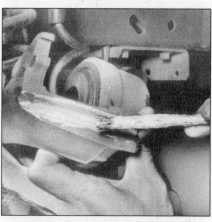

3.6g Prior to installing the caliper, lightly lubricate the V-grooves where the caliper slides into the anchor plate with disc brake caliper slide rail grease (1994 and earlier models)

3.6e If the caliper isn't going to be removed for service, suspend it with a length of wire to relieve any strain on the brake hose

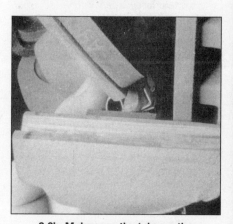

3.6h Make sure the tabs on the spring clip are positioned correctly and the clip is fully seated

7 When reinstalling the caliper on 1995 and later models, be sure to tighten the caliper bolts to the torque listed in this Chapter's Specifications.

8 After the job has been completed, firmly depress the brake pedal a few times to bring the pads into contact with the disc.

9 Check the brake fluid level and add some, if necessary, to bring it to the appropriate level (see Chapter 1).

4 Brake caliper - removal, overhaul and installation

Warning 1: *Dust created by the brake system may contain asbestos, which is harmful to your health. Never blow it out with compressed air and don't inhale any of it. An approved filtering mask should be worn when working on the brakes. Do not, under any circumstances, use petroleum-based solvents to clean brake parts. Use brake system cleaner only!*

Warning 2: *1995 and later models use a special lubricating gel on certain parts of the brake caliper. This gel is non-toxic under most conditions. If it is heated to a certain point, however, it can give off toxic fumes which can cause health problems or even death. Never apply heat to these calipers and don't smoke around them, either (the gel might get on your cigarette). Always wash your hands thoroughly after working on the brakes.*

Warning 3: *If the vehicle is equipped with an Anti-lock Brake System, make sure you plug the brake hose immediately after disconnecting it from the brake caliper, to prevent the fluid from draining out of the line and air entering the Hydraulic Control Unit (HCU). The HCU on the four-wheel ABS system cannot be bled without a very expensive tool.*

Note: *If an overhaul is indicated (usually because of fluid leakage) explore all options before beginning the job. New and factory-rebuilt calipers are available on an exchange basis, which makes this job quite easy. If it is decided to rebuild the calipers, make sure that a rebuild kit is available before proceeding. Always rebuild the calipers in pairs - never rebuild just one of them.*

Removal

1 Apply the parking brake and block the wheels opposite the end being worked on. Loosen the wheel lug nuts, raise the vehicle and support it securely on jackstands. Remove the wheel.

2 Unscrew the brake hose banjo bolt and detach the hose from the caliper. **Caution:** *Plug the brake hose immediately to prevent excessive fluid loss and, on models with a four-wheel Anti-lock Brake System, to keep air from getting into the Hydraulic Control Unit (HCU). If air gets into the HCU, you will not be able to bleed the brakes properly at home, since bleeding the brakes on these models requires a very expensive special tool.* **Note:** *If the caliper will not be completely removed from the vehicle - as for pad inspection or disc removal - leave the hose connected and suspend the caliper with a length of wire. This will save the trouble of bleeding the brake system.*

Chapter 9 Brakes

3.6i Insert a new anti-rattle clip on the lower end of the inner pad

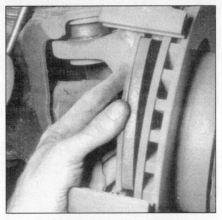

3.6j Compress the anti-rattle clip and slide the upper end of the pad into position

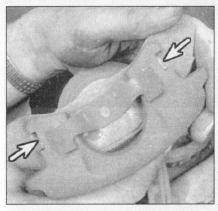

3.6k On 1994 and earlier models, push the outer pad into position on the caliper ears, making sure the torque buttons on the pad seat fully in the retention notches (arrows)

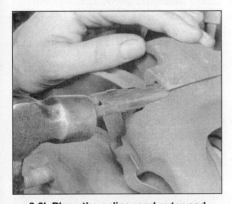

3.6l Place the caliper and outer pad assembly over the disc and inner pad, then drive the caliper pins into their grooves or, on 1995 and later models, install the mounting bolts and tighten them to the torque listed in this Chapter's Specifications

4.5 With the caliper padded to catch the piston, use compressed air to force the piston out of its bore - make sure your hands and fingers are not between the piston and the caliper!

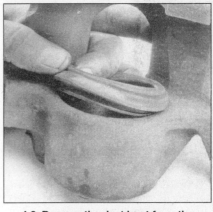

4.6 Remove the dust boot from the caliper bore groove

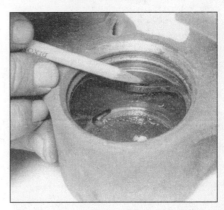

4.7 To remove the seal from the caliper bore, use a plastic or wooden tool, such as a pencil

3 Refer to the first six steps in Section 3 to separate the caliper from the steering knuckle - it's part of the brake pad replacement procedure.

Overhaul

Refer to illustrations 4.5, 4.6, 4.7, 4.17, 4.18, 4.19a, 4.19b, 4.20 and 4.21

4 Clean the exterior of the caliper with brake system cleaner. Never use gasoline, kerosene or other petroleum-based cleaning solvents. Place the caliper on a clean workbench.

5 Position a wood block or rags in the center of the caliper as a cushion, then use compressed air to remove the piston from the caliper **(see illustration)**. **Note:** *1995 and later models have two pistons.* Use only enough air to ease the piston out of the bore. If the piston is blown out, even with the cushion in place, it may be damaged. **Warning:** *Never place your fingers in front of the piston in an attempt to catch or protect it when applying compressed air, as serious injury could occur.*

6 Pull the dust boot(s) out of the caliper bore **(see illustration)**.

7 Using a wooden or plastic tool, remove the piston seal(s) from the caliper bore **(see illustration)**. Metal tools may cause bore damage.

8 Carefully examine the piston for nicks, burrs, cracks, loss of plating, corrosion or any signs of damage. If surface defects are present, the parts must be replaced.

9 Check the caliper bore in a similar way. Light polishing with crocus cloth is permissible to remove light corrosion and stains.

10 Remove the bleeder valve and rubber cap.

11 On 1995 and later models, inspect the caliper pins (bolts) for corrosion and damage. Don't wipe the lubricating gel from the pins or pin boots. At the time of publication there was no suitable substitute for this gel.

12 If the pins (bolts) are in need of addition lubrication, Ford recommends replacing them along with the boots, which come pre-filled with the special lubricating gel. Petroleum or mineral-based greases can destroy the caliper pin boot.

13 Use brake system cleaner to clean all the parts. **Warning:** *Do not, under any circumstances, use petroleum-based solvents to clean brake parts. Allow the parts to dry, preferably using compressed air to blow out all passages. Make sure the compressed air is filtered, as a harmful lubricant residue or moisture may be present in unfiltered systems.*

14 Examine carefully the bore of the caliper. Replace the caliper if the bore is excessively pitted, damaged or scored.

15 Check the fit of the piston in the bore by sliding it into the caliper. The piston should

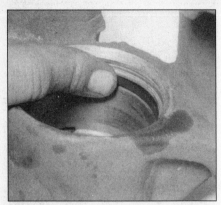

4.17 Push the new seal into the groove with your fingers, then check to see that it isn't twisted or kinked

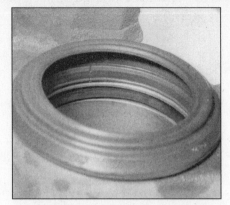

4.18 Install the dust boot in the upper groove in the caliper bore, making sure it's completely seated (1994 and earlier models)

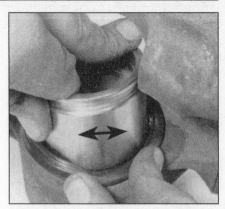

4.19a Lubricate the piston and bore with clean brake fluid, insert the piston into the dust boot (NOT the bore) at an angle, then, using a rotating motion, work the piston completely into the dust boot . . .

4.19b . . . and push it straight into the caliper as far as possible by hand (1994 and earlier models)

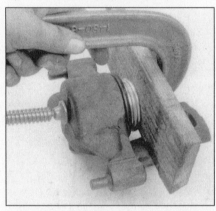

4.20 Use a C-clamp and a block of wood to bottom the piston in the caliper bore - make sure it goes in perfectly straight, or the sides of the piston may be damaged, rendering it useless

4.21 Install the lip of the dust boot in the groove on the caliper piston

move easily (don't install it yet).
16 Thread the bleeder valve into the caliper and tighten it securely. Install the rubber cap.
17 Lubricate the new piston seal(s) and caliper bore(s) with clean brake fluid. Position the seal(s) in the caliper bore groove(s), making sure it doesn't twist **(see illustration)**.

1994 and earlier models

18 Fit the new dust boot in the caliper bore upper groove, making sure it's seated **(see illustration)**.
19 Lubricate the caliper piston with clean brake fluid. Push the piston into the caliper, using a turning motion to roll the lip of the dust boot over the piston **(see illustrations)**. Push the piston into the caliper by hand as far as possible.
20 Using a C-clamp and a block of wood, push the piston all the way to the bottom of the bore. Work slowly, keeping an eye on the side of the piston, making sure it enters the bore perfectly straight with no resistance **(see illustration)**.
21 Seat the lip of the dust boot in the groove on the piston **(see illustration)**.

1995 and later models

22 Lubricate the caliper pistons with clean brake fluid. Push the pistons into the caliper bores by hand as far as possible.
23 Using a C-clamp and a block of wood, push the pistons about two-thirds of the way into the bores. Work slowly, keeping an eye on the side of the pistons, making sure they enter the bore perfectly straight with no resistance.
24 Install the dust boots on the pistons and seat the boots in the caliper. Now push the pistons to the bottom of the bores.

Installation

25 Refer to Section 3 for the caliper installation procedure, as it is part of the brake pad replacement procedure.
26 Connect the brake hose to the caliper, using new sealing washers. Tighten the banjo bolt to the torque listed in this Chapter's Specifications.
27 Bleed the brakes as outlined in Section 10. This is not necessary if the banjo bolt was not loosened or removed (if the caliper was removed for access to other components, for example).
28 Install the wheel and lower the vehicle. Tighten the lug nuts to the torque listed in the Chapter 1 Specifications. Pump the brake pedal several times to bring the pads into contact with the disc.
29 Test the operation of the brakes before placing the vehicle into normal service.

5 Brake disc - inspection, removal and installation

Inspection

Refer to illustrations 5.4a, 5.4b, 5.5a and 5.5b
Note: *This procedure applies to front and rear disc brake assemblies.*
1 Loosen the wheel lug nuts, raise the vehicle and support it securely on jackstands. Remove the wheel.
2 Remove the brake caliper as outlined in Section 4. It isn't necessary to disconnect the brake hose for this procedure. After removing the caliper, suspend the it out of the way with a piece of wire. Don't let the caliper hang by the hose and don't stretch or twist the hose.
3 Visually check the disc surface for score marks and other damage. Light scratches and shallow grooves are normal after use and may not always be detrimental to brake operation, but deep score marks - over 0.015-

Chapter 9 Brakes

5.4a To check disc runout, mount a dial indicator as shown and rotate the disc

5.4b Using a swirling motion, remove the glaze from the disc with sandpaper or emery cloth

5.5a The minimum thickness limit is cast into the inside of the disc

inch - require disc removal and refinishing by an automotive machine shop. Be sure to check both sides of the disc. If pulsating has been noticed during application of the brakes, suspect disc runout.

4 To check disc runout, place a dial indicator at a point about 1/2-inch from the outer edge of the disc **(see illustration)**. Set the indicator to zero and turn the disc. The indicator reading should not exceed the specified allowable runout limit. If it does, the disc should be refinished by an automotive machine shop. **Note:** *Professionals recommend resurfacing of brake discs regardless of the dial indicator reading (to produce a smooth, flat surface that will eliminate brake pedal pulsations and other undesirable symptoms related to questionable discs). At the very least, if you elect not to have the discs resurfaced, deglaze the surface with emery cloth or sandpaper (use a swirling motion to ensure a non-directional finish)* **(see illustration)**.

5 The disc must not be machined to a thickness less than the specified minimum refinish thickness. The minimum wear (or discard) thickness is cast into the inside of the disc **(see illustration)**. The disc thickness can be checked with a micrometer **(see illustration)**.

Removal

6 If you're working on a 2WD model, refer to Chapter 1, *Front wheel bearing check, repack and adjustment* for the disc removal procedure. If you're working on a 4WD model refer to Chapter 1, *Front hub lock, spindle bearing and wheel bearing maintenance*.

Installation

7 Install the disc by reversing the removal procedure.
8 Install the caliper (see Section 4).
9 Install the wheel and lug nuts, lower the vehicle to the ground and tighten the lug nuts to the torque listed in this Chapter's Specifications. Depress the brake pedal a few times to bring the brake pads into contact with the disc. Bleeding of the system will not be necessary unless the brake hose was disconnected from the caliper. Check the operation of the brakes carefully before placing the vehicle into normal service.

6 Brake shoes (rear) - replacement

Refer to illustrations 6.4a through 6.4aa and 6.5

Warning: *Drum brake shoes must be replaced on both wheels at the same time - never replace the shoes on only one wheel. Also, the dust created by the brake system may contain asbestos, which is harmful to your health. Never blow it out with compressed air and don't inhale any of it. An approved filtering mask should be worn when working on the brakes. Do not, under any circumstances, use petroleum-based solvents to clean brake parts. Use brake system cleaner only!*

Caution: *Whenever the brake shoes are replaced, the retractor and hold-down springs should also be replaced. Due to the continuous heating/cooling cycle that the springs are subjected to, they lose their tension over a period of time and may allow the shoes to drag on the drum and wear at a much faster rate than normal.*

1 Loosen the wheel lug nuts, raise the rear of the vehicle and support it securely on jackstands. Block the front wheels to keep the vehicle from rolling.
2 Release the parking brake.
3 Remove the wheel. **Note:** *All four rear brake shoes must be replaced at the same time, but to avoid mixing up parts, work on only one brake assembly at a time.*
4 Follow the accompanying photos **(illustrations 6.4a through 6.4aa)** for the inspection and replacement of the brake shoes. Be sure to stay in order and read the caption under each illustration. **Note:** *If the brake drum cannot be easily pulled off, pry the rubber plug from the backing plate inspection hole and insert a screwdriver and a brake

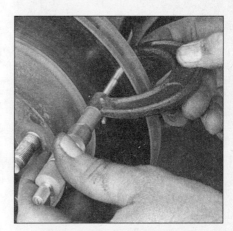

5.5b Use a micrometer to measure disc thickness at several points, about 1/2-inch from the edge

adjusting tool to lift the adjusting lever and rotate the adjusting screw. This will cause the brake shoes to pull together. Spray the assembly with penetrating oil and allow the oil to soak in if the mechanism is difficult to turn. The drum should now come off.*

5 Before reinstalling the drum it should be checked for cracks, score marks, deep scratches and hard spots, which will appear as small discolored areas. If the hard spots cannot be removed with fine emery cloth or if any of the other conditions listed above exist, the drum must be taken to an automotive machine shop to have it resurfaced. **Note:** *Professionals recommend resurfacing the drums whenever a brake job is performed. Resurfacing will eliminate the possibility of out-of-round drums. If the drums are worn so much that they can't be resurfaced without exceeding the maximum allowable diameter, which is stamped into the drum* **(see illustration)**, *then new ones will be required. At the very least, if you elect not to have the drums resurfaced, remove the glazing from the surface with medium-grit emery cloth using a swirling motion.*

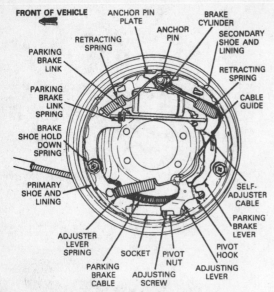

6.4a Components of the 1993 and 1994 rear brake assembly (left side shown)

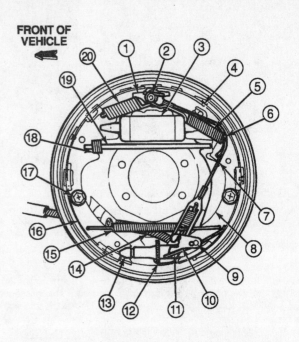

6.4b Components of the 1995 and later model rear brake assembly (left side shown)

1	Anchor pin guide plate	11	Adjusting lever return spring
2	Anchor pin	12	Adjusting wheel
3	Wheel cylinder	13	Adjuster screw
4	Secondary (rear) brake shoe	14	Parking brake cable
5	Long retracting spring	15	Adjusting screw spring
6	Cable guide	16	Primary (front) brake shoe
7	Adjusting cable	17	Shoe hold-down spring
8	Parking brake lever	18	Parking brake link spring
9	Adjusting lever pin	19	Parking brake lever link
10	Adjusting lever	20	Short retracting spring

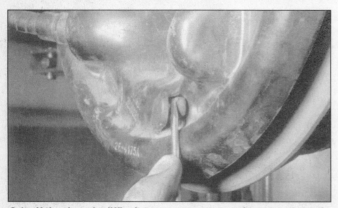

6.4c If the drum is difficult to remove, you may have to retract the brake shoes: Remove the rubber plug from the backing plate . . .

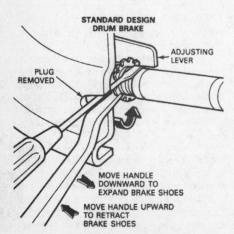

6.4d . . . use a narrow bladed screwdriver to push the adjusting lever away from the adjusting screw star wheel so you can turn the star wheel down with a brake tool (turn the tool handle up) to retract the shoes . . .

6.4e . . . and slide the drum off the shoes

6.4f Use a brake spring removal tool to remove the shoe retracting springs

Chapter 9 Brakes

6.4g Pull up on the adjusting cable and disconnect the cable eye from the anchor pin

6.4h Remove the anchor pin plate

6.4i Remove the shoe retaining springs and pins (one on each shoe)

6.4j Pull the shoes apart and remove the adjusting screw

6.4k Remove the primary shoe, then slide out the parking brake strut and spring

6.4l Remove the adjuster pawl

6.4m Pull the secondary shoe away from the backing plate

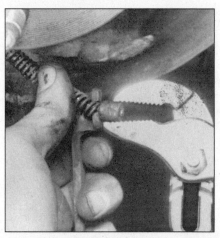

6.4n Separate the parking brake cable and spring from the actuating lever

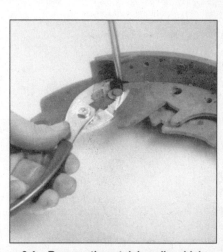

6.4o Remove the retaining clip which holds the parking brake actuating lever to the brake shoes

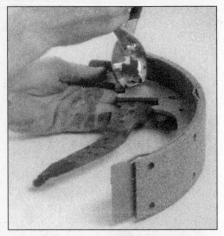

6.4p Install the parking brake actuator lever on the new brake shoe and install the retaining clip

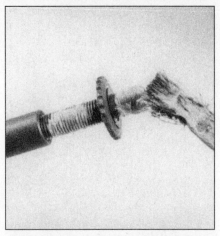

6.4q Lubricate the threads and end of the adjusting screw assembly with high-temperature brake grease

6.4r Lightly coat the shoe guide pads on the backing plate with high-temperature brake grease

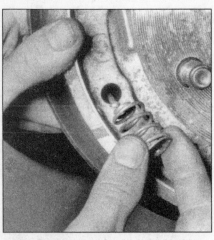

6.4s Position the shoes on the backing plate, insert the retaining pins through the backing plate and shoes and put the springs over them

6.4t Install the retaining spring caps

6.4u Make sure the slots in the wheel cylinder pushrods and the parking brake strut properly engage the brake shoes

6.4v Install the adjusting screw with the long end pointing towards the front of the vehicle

6.4w Install the adjusting pawl

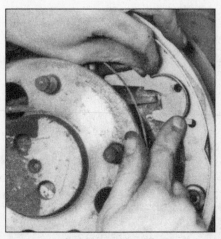

6.4x Install the anchor pin plate, cable guide and cable

Chapter 9 Brakes 9-13

6.4y Install the shoe guide and adjusting cable eye to the anchor pin, then install the shoe retracting springs

6.4z On 1993 and 1994 models, connect the cable and spring to the lever, then install the drum and adjust the brake shoe-to-drum clearance (adjust the shoes so they rub slightly as the drum is turned, then back-off the adjuster until they don't rub)

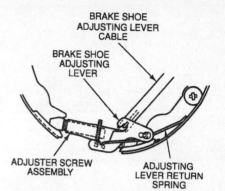

6.4aa On 1995 and later models, hook the adjuster cable into the lever from the side nearest the backing plate, then install the adjuster lever and return spring on the secondary shoe. The adjusting lever should contact the adjusting wheel just below the center of the adjusting wheel. Install the drum and adjust the brake shoe -to-drum clearance (adjust the shoes so they rub slightly as the drum is turned, then back-off the adjuster until they don't rub)

6   Install the brake drum on the hub. Turn the adjuster wheel until the brake shoes just drag on the drum when the drum is rotated, then back-off the adjuster wheel until the shoes don't drag.

7   Mount the wheel, install the lug nuts, then lower the vehicle.

8   Make a number of forward and reverse stops to adjust the brakes until satisfactory pedal action is obtained.

7   Wheel cylinder - removal, overhaul and installation

Note: *If an overhaul is indicated (usually because of fluid leakage or sticky operation) explore all options before beginning the job. New wheel cylinders are available, which makes this job quite easy. If it's decided to rebuild the wheel cylinder, make sure that a rebuild kit is available before proceeding. Never overhaul only one wheel cylinder - always rebuild both of them at the same time.*

Removal

Refer to illustration 7.4

1   Raise the rear of the vehicle and support it securely on jackstands. Block the front

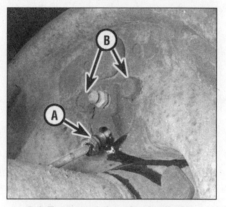

7.4 To remove the wheel cylinder, disconnect the brake line fitting (A) and remove the two mounting bolts (B)

wheels to keep the vehicle from rolling.

2   Remove the brake shoe assembly (see Section 6).

3   Remove all dirt and foreign material from around the wheel cylinder.

4   Disconnect the brake line **(see illustra-**

tion). Don't pull the brake line away from the wheel cylinder.

5   Remove the wheel cylinder mounting bolts.

6   Detach the wheel cylinder from the brake backing plate and place it on a clean workbench. Immediately plug the brake line to prevent fluid loss and contamination.

Overhaul

Refer to illustration 7.7

7   Remove the bleeder screw, cups, pistons, boots and spring assembly from the wheel cylinder body **(see illustration)**.

8   Clean the wheel cylinder with brake fluid, denatured alcohol or brake system cleaner. **Warning:** *Do not, under any circumstances, use petroleum-based solvents to clean brake parts!*

6.5 The maximum permissible diameter specification is cast into the brake drum

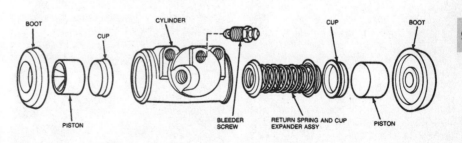

7.7 An exploded view of a typical wheel cylinder assembly

Chapter 9 Brakes

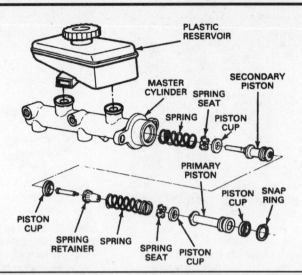

8.7a An exploded view of the master cylinder assembly

8.7b Use a Phillips head screwdriver to push the primary piston into the cylinder, then remove the snap-ring

9 Use compressed air to remove excess fluid from the wheel cylinder and to blow out the passages. Make sure the compressed air is filtered and unlubricated.
10 Check the cylinder bore for corrosion and score marks. Crocus cloth can be used to remove light corrosion and stains, but the cylinder must be replaced with a new one if the defects cannot be removed easily, or if the bore is scored.
11 Lubricate the new cups, pistons and cylinder bore with brake fluid.
12 Assemble the brake cylinder components. Make sure the cup lips face in.

Installation

13 Place the wheel cylinder in position, install the mounting bolts and tighten them to the torque listed in this Chapter's Specifications.
14 Connect the brake line and install the brake shoe assembly.
15 Bleed the brakes (see Section 10).

8 Master cylinder - removal, overhaul and installation

Warning: *If the vehicle is equipped with a four-wheel Anti-lock Brake System (ABS), do not attempt to remove or overhaul the master cylinder. Have the master cylinder removed, rebuilt and installed at a dealer service department or other qualified repair shop. Removing a master cylinder from a four-wheel ABS system can allow air to get into the ABS Hydraulic Control Unit (HCU), which requires a special tool and bleeding procedure which makes it impossible to perform at home.*
Note: *Before deciding to overhaul the master cylinder, check on the availability and cost of a new or factory rebuilt unit and also the availability of a rebuild kit.*

Removal

1 Place rags under the brake line fittings and prepare caps or plastic bags to cover the ends of the lines once they are disconnected.

Caution: *Brake fluid will damage paint. Cover all body parts and be careful not to spill fluid during this procedure.*
2 Unscrew the tube nuts at the ends of the brake lines where they enter the master cylinder. To prevent rounding off the flats on these nuts, a flare-nut wrench, which wraps around the fitting, should be used.
3 Pull the brake lines away from the master cylinder slightly and plug the ends to prevent contamination.
4 Disconnect the brake warning light electrical connector, remove the two master cylinder mounting nuts, and detach the master cylinder from the vehicle.
5 Remove the reservoir cap, then discard any fluid remaining in the reservoir.

Overhaul

Refer to illustrations 8.7a, 8.7b, 8.8, 8.9, 8.10, 8.14 and 8.19
6 Mount the master cylinder in a vise with the vise jaws clamping on the mounting flange.
7 Remove the primary piston snap-ring by depressing the piston and extracting the ring with a pair of snap-ring pliers **(see illustrations)**.
8 Remove the primary piston assembly from the cylinder bore **(see illustration)**.

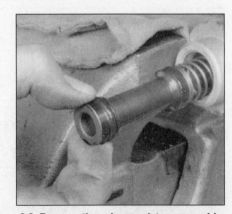

8.8 Remove the primary piston assembly from the cylinder

9 Remove the secondary piston assembly from the cylinder bore. It may be necessary to remove the master cylinder from the vise and invert it, carefully tapping it against a block of wood to expel the piston **(see illustration)**.
10 If fluid has been leaking past the reservoir grommets, pry the reservoir from the cylinder body with a screwdriver **(see illustration)**. Remove the grommets. Clean the master cylinder body and components with brake system cleaner. **Warning:** *DO NOT use petroleum-based solvents to clean brake parts - use brake system cleaner only.*
11 Inspect the cylinder bore for corrosion and damage. If any corrosion or damage is found, replace the master cylinder with a new one, as abrasives cannot be used on the bore.
12 Lubricate the new reservoir grommets with silicone grease and press them into the master cylinder body. Make sure they're properly seated. **Note:** *If silicone grease is not available, use clean brake fluid.*
13 Lay the reservoir on a hard surface and press the master cylinder body onto the reservoir, using a rocking motion.
14 Lubricate the cylinder bore and primary and secondary piston assemblies with clean brake fluid. Insert the secondary piston assembly into the cylinder **(see illustration)**.
15 Install the primary piston assembly in

8.9 Tap the master cylinder against a block of wood to eject the secondary piston assembly

Chapter 9 Brakes

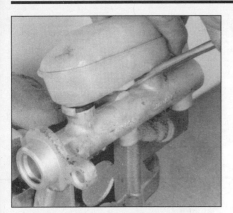

8.10 If you must remove the fluid reservoir to replace leaking seals or a broken reservoir, gently pry it off with a screwdriver or small prybar

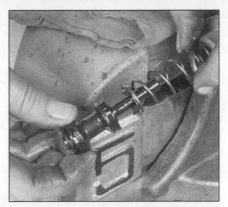

8.14 Coat the secondary piston with clean brake fluid and install it in the master cylinder, spring end first

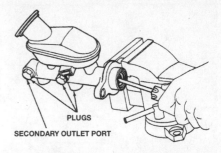

8.19 When bench bleeding the master cylinder, start with the secondary outlet port

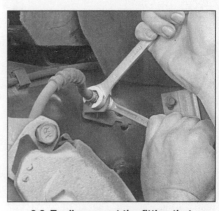

9.2 To disconnect the fitting that attaches the flexible brake hose to the metal brake line at the bracket in the wheel well, use a backup wrench on the hose fitting to ensure that the metal line doesn't get twisted

the cylinder bore, depress it and install the snap-ring. If equipped with a stop bolt, install it now, using a new sealing washer and tightening it securely.

16 Inspect the reservoir cap and diaphragm for cracks and deformation. Replace any damaged parts with new ones and attach the diaphragm to the cap.

17 **Note:** *Whenever the master cylinder is removed, the complete hydraulic system must be bled. The time required to bleed the system can be reduced if the master cylinder is filled with fluid and bench bled (refer to Steps 18 through 22) before the master cylinder is installed on the vehicle.*

18 Insert threaded plugs of the correct size into the cylinder outlet holes and fill the reservoirs with brake fluid. The master cylinder should be supported in such a manner that brake fluid will not spill during the bench bleeding procedure.

19 Loosen one plug at a time, starting with the secondary outlet port first, and push the piston assembly into the bore to force air from the master cylinder **(see illustration)**. To prevent air from being drawn back into the cylinder, the appropriate plug must be replaced before allowing the piston to return to its original position.

20 Stroke the piston three or four times for each outlet to ensure that all air has been expelled.

21 Since high pressure is not involved in the bench bleeding procedure, an alternative to the removal and replacement of the plugs with each stroke of the piston assembly is available. Before pushing in on the piston assembly, remove one of the plugs completely. Before releasing the piston, however, instead of replacing the plug, simply put your finger tightly over the hole to keep air from being drawn back into the master cylinder. Wait several seconds for the brake fluid to be drawn from the reservoir to the piston bore, then repeat the procedure. When you push down on the piston it will force your finger off the hole, allowing the air inside to be expelled. When only brake fluid is being ejected from the hole, replace the plug and

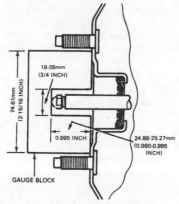

8.23 Check the master cylinder-to-booster pushrod clearance

go on to the other port.

22 Refill the master cylinder reservoirs and install the diaphragm and cap assembly.

Installation

Refer to illustration 8.23

23 Using a hand-held vacuum pump, apply a vacuum of about 20-in. Hg to the power booster and check the distance from the outer end of the booster pushrod to the front face of the brake booster assembly **(see illustration)**. Turn the pushrod adjusting screw in or out as required to obtain the correct length.

24 Carefully install the master cylinder by reversing the removal steps, then bleed the brakes (see Section 10).

9   Brake hoses and lines - inspection and replacement

Caution: *If the vehicle is equipped with a four-wheel Anti-lock Brake System (ABS), make sure you plug the brake line immediately after disconnecting it from the brake hose, to prevent the fluid from draining out of the line and air from entering the HCU. The HCU on an ABS system cannot be bled without a very expensive tool.*

Inspection

1 About every six months, with the vehicle raised and supported securely on jackstands, the rubber hoses which connect the steel brake lines with the front and rear brake assemblies should be inspected for cracks, chafing of the outer cover, leaks, blisters and other damage. These are important and vulnerable parts of the brake system and inspection should be complete. A light and mirror will be helpful for a thorough check. If a hose exhibits any of the above conditions, replace it with a new one.

Replacement

Flexible hose

Refer to illustrations 9.2 and 9.3

2 Using a flare nut wrench, disconnect the brake line from the hose fitting, being careful not to bend the frame bracket or brake line. Hold the fitting on the hose with a wrench to prevent the metal line from twisting and the frame bracket from bending **(see illustration)**.

3 Remove the large retaining clip **(see illustration)** and detach the hose from the bracket and the body. **Caution:** *Plug the metal brake line immediately. This is espe-*

9.3 Once the fitting has been unscrewed, remove this large retainer clip and separate the hose from the bracket

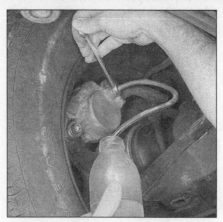

10.8 When bleeding the brakes, a hose is connected to the bleed screw at the caliper or wheel cylinder and then submerged in brake fluid - air will be seen as bubbles in the tube and container (all air must be expelled before moving to the next wheel)

cially important on models with four wheel ABS, to prevent air from getting into the Hydraulic Control Unit (HCU). If air gets into the HCU, you will not be able to bleed the brakes properly at home - the vehicle will have to be towed to a dealer service department or other repair shop equipped with the necessary tool.

4 Remove the banjo bolt from the caliper and discard the sealing washers.
5 Connect the hose to the caliper, using new sealing washers. Tighten the banjo bolt to the torque listed in this Chapter's Specifications.
6 Without twisting the hose, connect the other end of the line to the bracket on the chassis.
7 Connect the metal brake line to the hose fitting by hand, then, using a flare nut wrench, tighten the fitting securely. Be sure to use a wrench on the hose fitting to prevent the bracket from bending or the metal line from twisting.
8 When the brake hose installation is complete, there should be no kinks in the hose. Make sure the hose doesn't contact any part of the suspension. Check this by turning the wheels to the extreme left and right positions. If the hose makes contact, remove it and correct the installation as necessary.

Metal brake line

9 When replacing brake lines be sure to use the correct parts. Don't use copper tubing for any brake system components. Purchase steel brake lines from a dealer or auto parts store.
10 Prefabricated brake line, with the tube ends already flared and fittings installed, is available at auto parts stores and dealers. These lines are also sometimes bent to the proper shapes.
11 When installing the new line make sure it's securely supported in the brackets and has plenty of clearance between moving or hot components.
12 After installation, check the master cylinder fluid level and add fluid as necessary. Bleed the brake system as outlined in the next Section and test the brakes carefully before driving the vehicle in traffic.

10 Brake hydraulic system - bleeding

Refer to illustration 10.8
Warning: *Wear eye protection when bleeding the brake system. If the fluid comes in contact with your eyes, immediately rinse them with water and seek medical attention.*
Note: *Bleeding the hydraulic system is necessary to remove any air that manages to find its way into the system when it's been opened during removal and installation of a hose, line, caliper or master cylinder.*

Conventional brakes (non-ABS)

1 It will probably be necessary to bleed the system at all four brakes if air has entered the system due to low fluid level, or if the brake lines have been disconnected at the master cylinder.
2 If a brake line was disconnected only at a wheel, then only that caliper or wheel cylinder must be bled.
3 If a brake line is disconnected at a fitting located between the master cylinder and any of the brakes, that part of the system served by the disconnected line must be bled.
4 Remove any residual vacuum from the brake power booster by applying the brake several times with the engine off.
5 Remove the master cylinder reservoir cover and fill the reservoir with brake fluid. Reinstall the cover. **Note:** *Check the fluid level often during the bleeding operation and add fluid as necessary to prevent the fluid level from falling low enough to allow air bubbles into the master cylinder.*
6 Have an assistant on hand, as well as a supply of new brake fluid, an empty clear plastic container, a length of 3/16-inch plastic, rubber or vinyl tubing to fit over the bleeder valve and a wrench to open and close the bleeder valve.
7 Beginning at the right rear wheel, loosen the bleeder valve slightly, then tighten it to a point where it is snug but can still be loosened quickly and easily.
8 Place one end of the tubing over the bleeder valve and submerge the other end in brake fluid in the container **(see illustration)**.
9 Have the assistant pump the brakes slowly a few times to get pressure in the system, then hold the pedal firmly depressed.
10 While the pedal is held depressed, open the bleeder valve just enough to allow a flow of fluid to leave the valve. Watch for air bubbles to exit the submerged end of the tube. When the fluid flow slows after a couple of seconds, close the valve and have your assistant release the pedal.
11 Repeat Steps 9 and 10 until no more air is seen leaving the tube, then tighten the bleeder valve and proceed to the left rear wheel, the right front wheel and the left front wheel, in that order, and perform the same procedure. Be sure to check the fluid in the master cylinder reservoir frequently.
12 Never use old brake fluid. It contains moisture which will deteriorate the brake system components.
13 Refill the master cylinder with fluid at the end of the operation.
14 Check the operation of the brakes. The pedal should feel solid when depressed, with no sponginess. If necessary, repeat the entire process. **Warning:** *Do not operate the vehicle if you are in doubt about the effectiveness of the brake system.*

Anti-lock brake system (ABS)

Rear-wheel Anti-lock Brake System (RABS)

15 Place several rags or newspapers underneath the master cylinder. Using a flare nut wrench, loosen the brake line fittings at the master cylinder. Have an assistant slowly depress the brake pedal and hold it in the depressed position, then tighten the fittings. Repeat this step until the fluid flowing from the fittings is free of air bubbles. This will bleed out any air trapped in the master cylinder.
16 Working underneath the vehicle (raise and support it on jackstands, if necessary) attach a hose to the bleeder valve on the rear-wheel anti-lock brake valve (located on the inside of the driver's side frame rail, behind the front wheel) and place the other end of the hose in a container partially filled with clean brake fluid.
17 Open the bleeder valve and have your assistant slowly depress the brake pedal, holding it in the depressed position, then close the bleeder valve. Repeat this step until the fluid flowing from the valve is free of air bubbles.

Chapter 9 Brakes

18 The remainder of the bleeding procedure is the same as that for the conventional braking system. Follow Steps 7 through 14, making sure to check the fluid level in the master cylinder often.

Four-wheel Anti-lock Brake System

19 Four-wheel ABS-equipped models cannot be bled at home if air gets into the master cylinder and/or the Hydraulic Control Unit (HCU). The first step in the bleeding procedure for these two components requires a special anti-lock test adapter which must be plugged into the control module. Any attempt to bleed the master cylinder and HCU without this special device will trap air in the HCU, which will result in a spongy brake pedal.

20 However, as long as no air has gotten into the master cylinder or the HCU, the brake lines and the calipers can be bled in the conventional manner. Refer to Steps 1 through 14 above.

11 Parking brake - adjustment

The parking brake system on all models is completely self-adjusting. No means of manual adjustment is provided.

12 Parking brake cables - replacement

Warning: *Cable tension must be released before removing any parts of the parking brake system, then reset after installation.*
Refer to illustration 12.3

1 This procedure requires a steel pin to be inserted in the parking brake control. Use a 5/32-inch drill bit or fabricate the pin from a piece of stiff wire and bend a loop in one end so you can pull the pin out when you're done with it. You'll also need an assistant to pull on the parking brake cable while you insert the pin.
2 Release the parking brake.
3 Locate the lockout pin hole in the parking brake control **(see illustration)**.
4 Have an assistant, working beneath the vehicle, pull the front parking brake cable rearward. Insert the pin into the hole in the parking brake control assembly, then have your assistant let go of the cable.
5 To reset cable tension, remove the pin and operate the parking brake pedal several times.

Front cable

6 Release cable tension as described in Steps 1 through 4.
7 Raise the vehicle and support it securely on jackstands. Disconnect the rear end of the front cable from the parking brake rear cable **(see illustration 12.3)**.
8 From inside the vehicle, disconnect the cable from the pedal assembly **(see illustration 12.3, view Y)**.
9 Detach the cable from the mounting bracket on the frame. Remove the cable-to-frame bolt (if equipped) and take the cable out.
10 Installation is the reverse of the removal steps. Be sure the retainer fingers on the cable secure it to the front mounting bracket.
11 After installation, reset cable tension as described in Step 5.

Rear cable

12 Release cable tension as described in Steps 1 through 4.
13 Raise the vehicle and support it securely on jackstands. Remove the hubcap, the wheel and tire, and brake drum.
14 Disconnect the front cable from the rear cable **(see illustration 12.3, view W)**.
15 Disconnect the end of the right rear cable from the equalizer, then slip the cable out of the bracket.
16 On the wheel side of the backing plate, compress the retainer fingers so the retainer passes through the hole in the backing plate.
17 Lift the cable out of the slot in the parking brake lever attached to the secondary brake shoe and remove the cable through the backing plate hole.
18 Installation is the reverse of removal.
19 After installation, reset cable tension as described in Step 5.

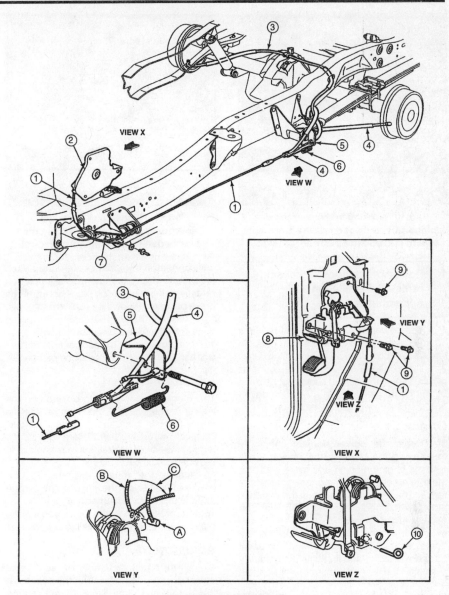

12.3 Parking brake details

1	Front cable	7	Front cable bracket	A	Front cable before installation
2	Control assembly	8	Release handle	B	Front cable installed in pivot hole
3	Left rear cable	9	Control assembly mounting bolts	C	Front cable rotated into installed position
4	Right rear cable	10	5/32-inch steel pin or drill bit		
5	Rear cable bracket				
6	Equalizer spring				

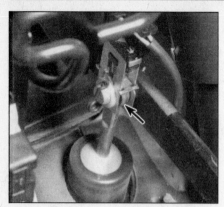

13.10 Working in the passenger compartment, disconnect the brake light switch and booster pushrod (arrow) . . .

13.11 . . . then pull down the floorboard padding and remove the four booster mounting nuts (arrows)

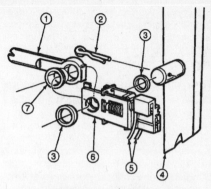

14.2 Installation details of a typical brake light switch assembly

1. Pushrod
2. Hairpin clip
3. Pushrod spacer
4. Brake pedal
5. Wiring harness
6. Brake light switch
7. Pushrod bushing

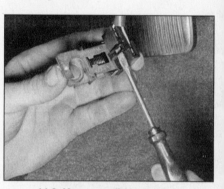

14.3 Use a small screwdriver to disengage the clip inside the electrical connector

13 Power brake booster - check, removal, installation and adjustment

1 The power brake booster unit requires no special maintenance apart from periodic inspection of the vacuum hose and the case.
2 Dismantling of the brake booster requires special tools and should not be done by the home mechanic. If a problem develops, install a new or factory rebuilt unit.

Check

3 Begin the power booster check by depressing the brake pedal several times with the engine running and make sure that there is no change in the pedal reserve distance. The reserve distance is the distance between the pedal and the floor when the pedal is fully depressed.
4 Turn off the engine and depress the brake pedal several times slowly. If the pedal goes down farther the first time but gradually rises after the second or third depression, the booster is airtight.
5 Now, depress the pedal and start the engine. If the pedal goes down slightly, operation is normal. Release the brake pedal and let the engine run for a couple of minutes.
6 Depress the brake pedal, then stop the engine with the pedal still depressed. If there is no change in the reserve distance after holding the pedal for about 30-seconds, the booster is airtight.
7 If the pedal feels "hard" when the engine is running, the booster isn't operating properly or there is a vacuum leak in the hose to the booster.

Removal

Refer to illustrations 13.10 and 13.11

8 Remove the nuts attaching the master cylinder to the booster (see Section 8) and carefully pull the master cylinder forward until it clears the mounting studs. Use caution so as not to bend or kink the brake lines.
9 Detach the manifold vacuum hose from the booster check valve.
10 Working in the passenger compartment under the steering column, unplug the electrical connector from the brake light switch (see illustration), then remove the pushrod retaining clip and nylon washer from the brake pedal pin. Slide the pushrod off the pin.
11 Remove the nuts attaching the brake booster to the firewall (see illustration).
12 Carefully detach the booster from the firewall and lift it out of the engine compartment.

Installation

13 Place the booster into position on the firewall and tighten the mounting nuts to the torque listed in this Chapter's Specifications. Connect the pushrod and brake light switch to the brake pedal. Install the retaining clip in the brake pedal pin.
14 Install the master cylinder to the booster, tightening the nuts to the torque listed in this Chapter's Specifications.
15 Carefully check the operation of the brakes before driving the vehicle in traffic.

Adjustment

16 Some boosters feature an adjustable pushrod. They are matched to the booster at the factory and most likely will not require adjustment, but if a misadjusted pushrod is suspected, a gauge can be fabricated out of heavy gauge sheet metal (see illustration 8.23).
17 Some common symptoms caused by a misadjusted pushrod include dragging brakes (if the pushrod is too long) or excessive brake pedal travel accompanied by a groaning sound from the brake booster (if the pushrod is too short).
18 To check the pushrod length, unbolt the master cylinder from the booster and position it to one side. It isn't necessary to disconnect the hydraulic lines, but be careful not to bend them.
19 Apply vacuum to the booster with a hand-held vacuum pump, then place the pushrod gauge against the end of the pushrod, exerting a force of approximately five pounds to seat the pushrod in the power unit (see illustration 8.23). The rod should just touch the cutout on the gauge. If it doesn't, adjust it by holding the knurled portion of the pushrod with a pair of pliers and turning the end with a wrench.
20 When the adjustment is complete, reinstall the master cylinder and check for proper brake operation before driving the vehicle in traffic.

14 Brake light switch - removal and installation

Removal

Refer to illustrations 14.2 and 14.3

1 Remove the under dash panel.
2 Locate the brake light switch assembly (see illustration) near the top of the brake pedal and disconnect the switch from the brake pedal by removing the retaining clip.
3 Use a small screwdriver to unlock the electrical connector, then unplug the connector from the brake light switch (see illustration).

Installation

4 Install the switch to the electrical connector by snapping the clip into place.
5 Reconnect the assembly to the brake pedal.
6 Install the under dash panel.
7 Check the brake lights for proper operation.

Chapter 10
Suspension and steering systems

Contents

	Section
Axle pivot bracket (1993 through 1997 models) - removal and installation	10
Axle pivot bushing (1993 through 1997 models) - removal and installation	9
Balljoints (1993 through 1997 models) - removal and installation	4
Coil spring (1993 through 1997 models) - removal and installation	5
Coil spring (1998 and later 2WD models) - removal and installation	18
Drag link (1993 through 1997 models) - removal and installation	25
Front axle arm (1993 through 1997 models) - removal and installation	8
Front end alignment - general information	33
Front hub and bearing assembly (1998 and later models) - removal and installation	13
Front shock absorber (1988 and later models) - removal and installation	12
Front shock absorber (1993 through 1997 models) - inspection, removal and installation	2
Front stabilizer (1993 through 1997 models) - removal and installation	11
Front stabilizer bar (1998 and later models) - removal and installation	14
Front wheel spindle (1993 through 1997 models) - removal and installation	3
Front wheel spindle (1998 and later models) - removal and installation	15
General information	1
Leaf spring - removal and installation	21
Lower control arm (1998 and later models) - removal, inspection and installation	17
Power steering line quick disconnect fittings	30
Power steering pump - removal and installation	29
Power steering system - bleeding	31
Radius arm (1993 through 1997 models) - removal and installation	6
Radius arm insulators (1993 through 1997 models) - replacement	7
Rear shock absorbers - inspection, removal and installation	20
Rear stabilizer bar - removal and installation	22
Steering connecting rod (1993 through 1997 models) - removal and installation	26
Steering gear - removal and installation	28
Steering system - general information	23
Steering wheel - removal and installation	24
Tie-rod ends - removal and installation	27
Torsion bar (1998 and later 4WD models) - removal and installation	16
Upper control arm (1998 and later models) - removal, inspection and installation	19
Wheels and tires - general information	32

Specifications

General
Power steering fluid type and capacity .... See Chapter 1

Torque specifications
Ft-lbs (unless otherwise indicated)

Front suspension (1993 through 1997)
2WD models
Shock absorber-to-radius arm nut	39 to 53
Shock absorber upper nut	25 to 34
Stabilizer bar-to-mounting bracket bolts	22 to 30
Coil spring retainer nut	70 to 100
Coil spring retainer nut and bolt	
1993 and 1994	191 to 231
1995-on	188 to 254
Stabilizer bar end nuts and bolts	30 to 41
Radius arm-to-frame nut	
1993 and 1994	81 to 120
1995-on	82 to 113
Radius arm-to-axle arm bolt	
1993 and 1994	191 to 231
1995-on	188 to 254
Radius arm bracket-to-frame bolts	
1993 and 1994	77 to 110
1995-on	82 to 113

Torque specifications   Ft-lbs (unless otherwise indicated)

Front suspension (1993 through 1997)
2WD models (continued)
Axle pivot bolt
- 1993 and 1994 .... 120 to 150
- 1995-on .... 111 to 148

Axle pivot bracket nuts/bolts
- 1993 and 1994 .... 77 to 110
- 1995-on .... 72 to 97

Jounce bumper bolt .... 18 to 26
Lower balljoint stud nut .... 89 to 133
Balljoint pinch bolt
- 1993 and 1994 .... 48 to 65
- 1995-on .... 50 to 68

4WD models
Shock absorber-to-radius arm nut .... 39 to 53
Shock absorber upper nut .... 25 to 35
Spring retainer nut .... 70 to 100
Stabilizer bar bracket bolts
- 1993 and 1994 .... 35 to 50
- 1995-on .... 22 to 30

Stabilizer bar link nuts .... 30 to 40
Radius arm-to-rear bracket nut .... 83 to 113
Radius arm front bracket front bolts .... 15 to 27
Radius arm front bracket lower bolt
- 1993 and 1994 models .... 190 to 230
- 1995-on .... 190 to 255

Radius arm-to-front bracket upper stud
- 1993 and 1994 models .... 190 to 230
- 1995-on .... 190 to 255

Axle pivot bolt
- 1993 and 1994 models .... 120 to 150
- 1995-on .... 111 to 148

Axle pivot bracket bolts
- Left bracket (right axle arm) 155
- Right bracket (left axle arm)
  - 1993 and 1994 models .... 77 to 110
  - 1995-on .... 83 to 113

Jounce bumper bolt .... 18 to 26
Balljoint pinch bolt .... 65 to 85

Front suspension (1998 on)
Shock absorber-to-lower control arm mounting nuts .... 15 to 21
Shock absorber upper mounting nut .... 30 to 41
Hub and bearing-to-steering knuckle mounting bolts (4WD) .... 74 to 96
Upper control arm-to-spindle pinch bolt/nut .... 35 to 46
Lower balljoint castellated nut .... 83 to 113
Lower control arm pivot bolts/nuts .... 110 to 148
Upper control arm pivot bolts .... 83 to 113
Front stabilizer bar link nuts .... 15 to 21
Front stabilizer bar bushing clamp bolts .... 25 to 34

Rear suspension
Rear leaf spring U-bolt nut
- 1993 and 1994 .... 65 to 80
- 1995-on .... 65 to 87

Shock-to-lower bracket nut .... 39 to 53
Shock-to-upper bracket nut .... 39 to 53
Shackle-to-spring nut and bolt
- 1993 and 1994 .... 74 to 115
- 1995-on .... 65 to 87

Spring shackle-to-rear bracket bolt
- 1993 and 1994 .... 74 to 115
- 1995-on .... 65 to 87

Spring-to-front bracket bolt
- 1993 and 1994 .... 56 to 76
- 1995-on .... 65 to 87

Stabilizer-to-axle bracket
  1993 and 1994 models (U-bolt nuts)................................... 30 to 42
  1995-on (bolts) .... 25 to 34
Stabilizer-to-link nut
  1993 and 1994 .... 40 to 60
  1995-on .... 55 to 59
Stabilizer link-to-frame bolt
  1993 and 1994 .... 40 to 60
  1995-on .... 55 to 59

Steering linkage
Drag link-to-Pitman arm nut (1993 through 1997)........................ 51 to 73
Drag link-to-steering connecting rod nut (1993 through 1997) ..... 51 to 73
Tie-rod adjusting sleeve nut (1993 through 1997) .... 30 to 42
Tie-rod-to-spindle nut (1993 through 1997) .... 51 to 73
Pitman arm-to-steering gear nut (1993 through 1997).................. 170 to 228
Tie-rod end-to-spindle nut (1998 on).. 44 to 59

Steering system
Flex coupling-to-steering gear input shaft bolt (1993 and 1994) ...... 27 to 32
Steering column-to-gear bolt (1995 on) .... 30 to 40
Steering gear-to-frame bolts
  1993 through 1997 .... 53 to 59
  1998 on .... 102 to 127
Power steering pump-to-bracket bolts
  2.3L/2.5L engines .... 32 to 38
  3.0L engine .... 30 to 40
  4.0L engine .... 15 to 21
Steering wheel bolt .... 25 to 34

1 General information

Refer to illustrations 1.2a and 1.2b

Warning: *On models so equipped, whenever working in the vicinity of the front grille/bumper, steering wheel, steering column or other components of the airbag system, the system should be disarmed. To do this, perform the following steps:*

a) *Turn the ignition switch to Off.*
b) *Detach the cable from the negative battery terminal, then detach the positive cable. Wait two minutes for the electronic module backup power supply to be depleted.*

To enable the system

a) *Turn the ignition switch to the Off position.*
b) *Connect the positive battery cable first, then connect the negative cable.*

Front suspension

1997 and earlier models

The front suspension on 2WD models is a twin I-beam type, which is composed of coil springs, I-beam axle arms, radius arms, upper and lower balljoints and spindles, tie-rods, shock absorbers and an optional stabilizer bar. The 4WD model is basically the same except the front driveline system is composed of a two-piece driveaxle assembly.

The front suspension consists of two independent axle arm assemblies **(see illustrations)**. One end of the assembly is anchored to the frame and the other is supported by the coil spring and radius arm. The spindle is connected to the axle by upper and lower balljoints. The balljoints are constructed of a lubricated-for-life special bearing material. Lubrication points are found on the tie-rods and steering linkage. Movement of the spindles is controlled by the tie-rods and the steering linkage.

Two adjustments can be performed on the axle assembly. Camber is adjusted by removing and replacing an adapter between the upper balljoint stud and the spindle on 2WD models. 4WD models require replacing the camber adapter on the upper balljoint stud. Adapters are available in 0-degree, 1/2-degree, 1-degree and 1-1/2-degree increments. Toe-in adjustment is accomplished on both models by turning the tie-rod adjusting sleeve.

1998 and later models

The front suspension on 1998 and later models **(see illustration 14.8)** consists of upper and lower control arms, shock absorbers, coil springs (2WD models) or torsion bars (4WD models) and a stabilizer bar. The inner ends of the control arms are attached to the frame; the outer ends are attached to the spindle. The upper control arms pivot on eccentric bolts. The lower control arms pivot on two separate, non-adjustable bolts. The shock absorbers and, on 2WD models, the coil springs are mounted between the lower control arms and the frame. On 4WD models, the torsion bars are mounted between the lower control arms and the frame. The front stabilizer bar is attached to the frame with mounting clamps. The balljoints are an integral part of the control arms and cannot be replaced separately; if a balljoint is damaged or worn out, the control arm must be replaced.

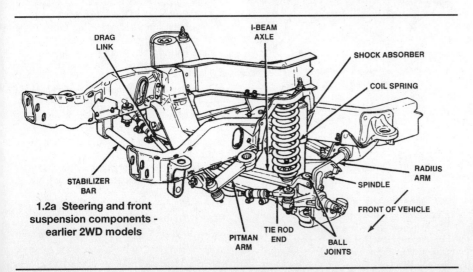

1.2a Steering and front suspension components - earlier 2WD models

Chapter 10 Suspension and steering systems

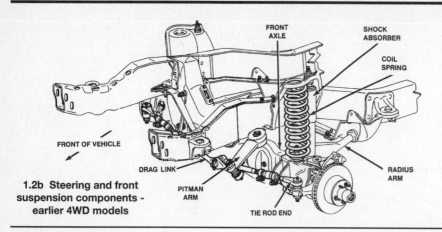

1.2b Steering and front suspension components - earlier 4WD models

2.6 Hold the shock absorber shaft with an open end wrench while turning the top retainer nut (1993 through 1997 models)

2.7 The shock absorber lower mounting stud is attached to the radius arm (earlier models)

Rear suspension

The rear suspension uses shock absorbers and semi-elliptical leaf springs. The forward end of each spring is attached to the bracket on the frame side rail (**see illustration 13.4**). The rear of each spring is shackled to a bracket on the frame rail. The rear shock absorbers are direct, double acting units with staggered mounting positions. The right shock is mounted forward of the axle and the left mounted behind it.

Since most procedures that are dealt with in this Chapter involve jacking up the vehicle and working underneath it, a good pair of jackstands will be needed. A hydraulic floor jack is the preferred type of jack to lift the vehicle, and it can also be used to support certain components during various operations. **Warning 1:** *Never, under any circumstances, rely on a jack to support the vehicle while working under it.*
Warning 2: *Whenever any of the suspension or steering fasteners are loosened or removed they must be inspected and, if necessary, replaced with new ones of the same part number or of original equipment quality and design.*

Torque specifications must be followed for proper reassembly and component retention.

2 Front shock absorber - (1993 through 1997 models) inspection, removal and installation

Inspection

1 The common test of shock damping is simply to bounce the corners of the vehicle several times and observe whether or not the vehicle stops bouncing once you let go if it. A slight rebound and settling indicates good damping, but if the vehicle continues to bounce several times, the shock absorbers must be replaced.
2 If the shock absorbers stand up to the bounce test, visually inspect the shock body for signs of fluid leakage, punctures or deep dents in the metal of the body. Replace any shock absorber which is leaking or damaged, even if it passed the bounce test in Step 1.
3 After removing the shock absorber, pull the piston rod out and push it back in several times to check for smooth operation throughout the travel of the piston rod. Replace the shock absorber if it gives any signs of hard or soft spots in the piston rod travel.
4 Prior to installing the new shock absorbers, pump the piston rod fully in and out several times to lubricate the seals and fill the hydraulic sections of the unit.

Removal and installation

Refer to illustrations 2.6 and 2.7
Caution: *The low pressure gas shock absorbers are pressurized to 135 PSI with nitrogen gas. Do not attempt to open, puncture or apply heat to the shock absorbers.*
5 Loosen the front wheel lug nuts on the side to be dismantled. Raise the front of the vehicle, support it securely on jackstands, block the rear wheels and set the parking brake. Remove the front wheel.
6 Remove the top nut with a deep socket while holding the shaft with an open end wrench (**see illustration**). Lift off the washer.
7 Remove the bolt and nut securing the shock absorber to the radius arm (**see illustration**).
8 To remove the shock absorber, slightly compress the shock and remove it from its brackets.
9 Installation is the reverse of the removal Steps with the following additions.

a) *Install and tighten the new nuts and bolts to the torque listed in this Chapter's Specifications.*

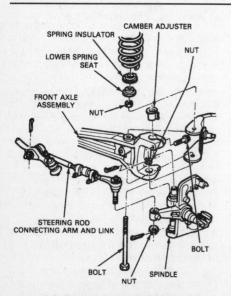

3.3 Typical earlier 2WD front spindle installation details

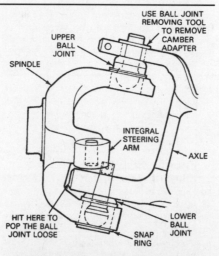

3.7 Separate the upper balljoint with a special tool, available at most auto parts stores, then strike the edge of the front spindle with a hammer to break it loose from the balljoint studs (earlier models)

Chapter 10 Suspension and steering systems

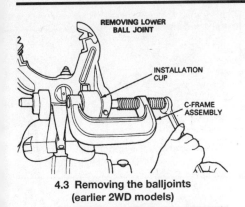

4.3 Removing the balljoints (earlier 2WD models)

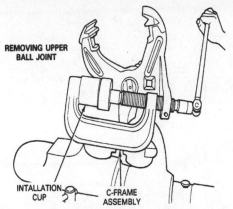

4.4 Removing the upper balljoint (earlier 2WD models)

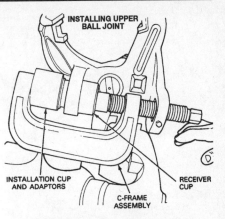

4.5 Note the additional receiver cup required to install the upper balljoint - install the upper balljoint first (earlier 2WD models)

b) Install the wheel and lug nuts, lower the vehicle and tighten the lug nuts to the torque listed in Chapter 1.

3 Front wheel spindle (1993 through 1997 models) - removal and installation

Refer to illustrations 3.3 and 3.7

2WD models

Removal

1 Remove the front wheel and brake disc (see Chapter 1).
2 Remove the brake dust shield.
3 Remove the cotter pin and nut and disconnect the tie-rod end from the spindle **(see illustration)**.
4 Remove the cotter pin from the lower balljoint stud nut, then remove the nut.
5 Remove the clamp bolt from the upper balljoint.
6 Use a special balljoint removal tool (or equivalent) to remove the camber adjuster from the upper balljoint.
7 Strike the inside of the spindle near the balljoint to break the spindle loose from the balljoint studs **(see illustration)**. **Caution:** *Don't use a fork-type separator to detach the balljoint. This will damage the balljoint seal.*
8 Remove the spindle together with the balljoints.

Installation

9 Before installation, check that the upper and lower balljoint seals were not damaged during spindle removal and that they are positioned correctly. Replace them if necessary.
10 Position the spindle and balljoints in the axle arm.
11 Install the camber adjuster on the upper balljoint, making sure it's aligned correctly.
12 Tighten the lower balljoint nut to the torque listed in this Chapter's Specifications, then tighten further until a cotter pin hole lines up. Install a new cotter pin and bend it to secure the nut.
13 Install the clamp bolt on the upper balljoint and tighten it to the torque listed in this Chapter's Specifications.
14 Install the dust shield.
15 Install the brake disc and the front wheel (see Chapter 1).

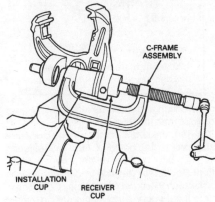

4.7 Installing the lower balljoint (earlier 2WD models)

4WD models

16 Refer to Chapter 8 for the axleshaft removal procedure, then refer to the 2WD spindle removal and installation procedure described above.

4 Balljoints (1993 through 1997 models) - removal and installation

Warning: *Do not heat the axle or balljoint to aid removal since the temper may be removed from the component(s), leading to premature failure.*

2WD models

Refer to illustrations 4.3, 4.4, 4.5 and 4.7

1 Remove the spindle (see Section 3).
2 Remove the snap-ring from the lower balljoint. **Note:** *Remove the lower balljoint first.*
3 Install a special C-frame assembly tool, available at most auto parts stores, and the appropriate size receiving cup on the lower balljoint **(see illustration)**. Tighten the tool and press the balljoint out of the spindle. **Note:** *If the C-frame assembly tool and receiver tool are not available or will not remove the balljoint, take the spindle assembly to a dealership service department or machine shop and have the balljoints pressed out.*

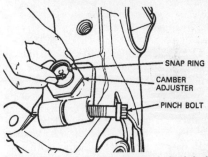

4.12 Remove the snap-ring and pinch bolt (earlier models)

4 Use the same tool setup used in Step 3 on the upper balljoint and press the balljoint out **(see illustration)**.
5 Install the upper balljoint with the C-frame assembly, balljoint receiver cup and installation cup **(see illustration)**. **Note:** *Always install the upper balljoint first since the special tool must pass through the lower balljoint receptacle in the axle.*
6 Turn the screw in the C-frame clockwise and press the balljoint into the axle until it's completely seated. **Caution:** *Don't heat the axle or balljoint to aid in installation since the temper may be removed from the component(s) leading to premature failure.*
7 Install the lower balljoint with the C-frame assembly, balljoint receiver cup and installation cup **(see illustration)**.
8 Turn the screw in the C-frame clockwise and press the balljoint into the axle until it is completely seated.
9 Install the lower balljoint snap-ring.

4WD models

Refer to illustrations 4.12, 4.13, 4.18, 4.19 and 4.22

10 Remove the spindle, axleshaft and joint assembly (see Chapter 8).
11 Remove the cotter pin and nut securing the tie-rod (see Section 19).
12 Remove the snap-ring from the upper balljoint, then remove the pinch bolt **(see illustration)**.

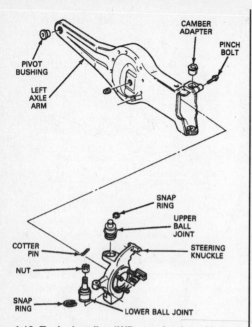

4.13 Typical earlier 4WD steering knuckle and balljoint installation details

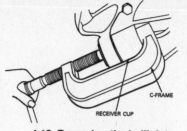

4.18 Removing the balljoints (earlier 4WD models)

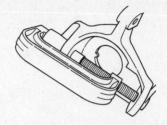

4.19 Removing the upper balljoint (earlier 4WD models)

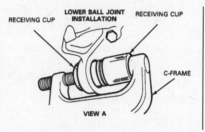
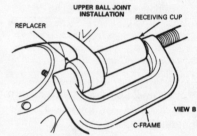

4.22 Note the additional receiver cup required to install the upper balljoint; install the lower balljoint first (earlier 4WD models)

13 Remove the cotter pin from the lower balljoint, then loosen the nut to the end of the stud (but no farther) **(see illustration)**.
14 Hit the inside of the steering knuckle near each balljoint with a hammer. This will break the axle arm loose from the balljoint studs. Remove the steering knuckle from the axle arm.
15 Remove the camber adjuster sleeve **(see illustration 4.12)**. Mark the position of the slot in the camber adjuster. It should be installed in its original position to maintain the correct alignment. **Note:** *If the camber adjuster is difficult to remove, use a Pitman arm puller. These are available at most auto parts stores and can often be rented for a reasonable fee.*
16 Remove the nut from the lower balljoint.
17 Place the steering knuckle in a vise. Remove the snap ring (if equipped) from the lower balljoint socket. **Note:** *Always remove the lower balljoint first.*
18 Install the C-frame assembly tool, forcing screw and balljoint remover) on the lower balljoint **(see illustration)**. Tighten the special tool and press the lower balljoint out of the steering knuckle. **Note:** *If the C-frame assembly tool and receiver tool are not available or won't remove the balljoint, take the knuckle assembly to a dealership service department and have the balljoints pressed out.*
19 Use the same tool setup used in Step 18 on the upper balljoint and press the balljoint out **(see illustration)**.
20 Clean the steering knuckle balljoint bores.
21 Insert the lower balljoint into the knuckle as straight as possible. **Note:** *Always install the lower balljoint first since the special tool must pass through the upper balljoint receptacle in the axle.*
22 Position the C-frame assembly, balljoint receiver cup and installation cup on the lower balljoint **(see illustration)**.
23 Turn the screw in the C-frame clockwise and press the balljoint into the steering knuckle until it is completely seated. **Caution:** *Don't heat the axle or balljoint to ease installation since the temper may be removed from the component(s) leading to premature failure. If the balljoint won't go in all the way, realign the C-frame and receiver cup.*
24 Install the lower balljoint snap-ring (if equipped).
25 Position the C-frame assembly, balljoint receiver cup and installation cup on the upper balljoint **(see illustration 4.22)**.
26 Turn the screw in the C-frame clockwise and press the balljoint into the axle until it is completely seated.
27 Install the camber adjuster into the axle arm. Place the slot in the original position noted during removal. **Note:** *Install the camber adjuster with the arrow pointing toward the outside for positive camber or with the arrow pointing toward the inside of the vehicle for negative camber. Zero camber bushings do not have an arrow and may be rotated in either direction as long as the lugs on the yoke engage the slots in the bushing.* **Caution:** *The following tightening sequence must be followed exactly when securing the steering knuckle. Excessive spindle turning effort may result in reduced steering returnability if this procedure is not followed.*
28 Install the steering knuckle in the axle arm. Do not disrupt the camber adjuster during installation.
29 Install a new nut on the lower balljoint stud and tighten to the torque listed in this Chapter's Specifications, then advance the nut until a cotter pin slot lines up. Install a new cotter pin and bend the ends over completely.
30 Install the snap-ring on the upper balljoint stud. Install the pinch bolt and tighten to the torque listed in this Chapter's Specifications. **Note:** *The camber adjuster will position itself in the knuckle during adjustment. DO NOT try to change its position.*
31 Attach the tie-rod end to the knuckle. Tighten the nut to the torque listed in this Chapter's Specifications, install a new cotter pin and bend the ends over completely.
32 The remainder of installation is the reverse of the removal Steps.
33 Have the front end alignment checked by a dealership service department or an alignment shop.

5 Coil spring (earlier models) - removal and installation

Removal

Refer to illustrations 5.2, 5.5a, 5.5b and 5.5c

1 Loosen the front wheel lug nuts on the side to be dismantled. Raise the front of the vehicle, support it securely on jackstands, block the rear wheels and set the parking brake. Remove the front wheel.
2 Place a floor jack under the front axle and raise it just enough to support the weight of the axle **(see illustration)**.
3 Remove the bolt and nut securing the shock absorber to the radius arm **(see illustration 2.8)**.
4 Remove the brake caliper assembly and suspend it with a length of wire to relieve any strain on the brake hose (see Chapter 9).
5 At the lower end of the spring, remove the retaining nut and the retainer securing the spring to the front axle **(see illustration)**. **Note:** *The nut is attached to a stud on 4WD models and on 2WD models it is attached to a bolt that runs through the axle* **(see illustrations)**.

Chapter 10 Suspension and steering systems

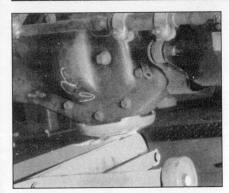

5.2 With the vehicle supported on jackstands, use a floor jack to raise and lower the front axle (earlier models)

5.5a Remove the lower spring retainer nut (earlier models)

6 On models so equipped, remove the through bolt securing the stabilizer bar to the front axle.

7 Lower the floor jack slightly, the front axle should now be free to allow spring removal.

8 If necessary, use a pry bar and lift the spring up and over the bolt or stud that passes through the lower spring seat. Rotate the spring so the upper built-in spring seat retainer is cleared, then remove the spring.

Installation

9 Install the spring lower seat and insulator onto the front axle.

10 Push the front axle down to allow installation of the spring. Install the upper end of the spring into the upper seat and rotate the spring into place.

11 If necessary, use a pry bar and lift the lower end of the spring up and over the bolt or stud on the axle and into place on the lower spring seat and insulator.

12 Apply slight pressure on the floor jack and lift the front axle until the spring is correctly seated.

13 Install the lower spring retainer and new nut. Tighten the nut to the torque listed in this Chapter's Specifications.

14 Install the brake caliper assembly (see Chapter 9).

15 Install the bolt and new nut securing the shock absorber to the radius arm. Tighten the nut to the torque listed in this Chapter's Specifications.

16 Install the wheel and lug nuts, lower the vehicle and tighten the lug nuts to the torque listed in Chapter 1.

6 Radius arm (1993 through 1997 models) - removal and installation

Removal

Refer to illustrations 6.2, 6.3 and 6.4

1 Remove the coil spring assembly (see Section 5).

2 On 2WD models, perform the following:

  a) Remove the spring lower seat from the radius arm.

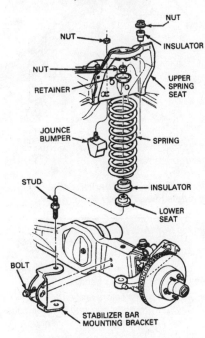

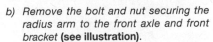

5.5c Front spring and related components (earlier 4WD models)

  b) Remove the bolt and nut securing the radius arm to the front axle and front bracket (see illustration).

3 On 4WD models, perform the following:

  a) Remove the spring lower seat and stud.
  b) Remove the bolts securing the radius arm to the front axle bracket (see illustration).

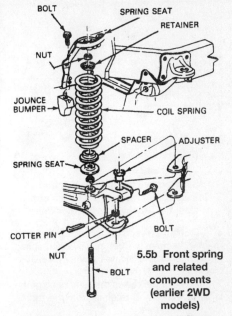

5.5b Front spring and related components (earlier 2WD models)

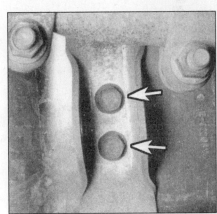

6.2 Remove the two bolts (arrows) securing the radius arm to the front bracket (earlier models)

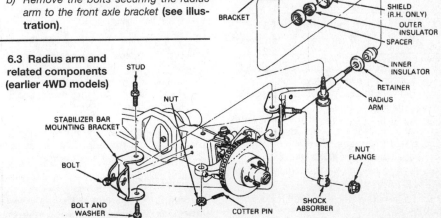

6.3 Radius arm and related components (earlier 4WD models)

6.4 Remove the nut, washer and insulator and remove the radius arm from the bracket (earlier models)

4   From the rear side of the radius arm bracket, remove the nut, rear washer, shield (passenger side only) and insulator **(see illustration)**.

5   Remove the radius arm and remove the inner insulator and retainer from the radius arm threaded end.

Installation

6   Install the front end of the radius arm onto the front axle.

7   On 2WD models, from underneath the axle, install the attaching bolt and a new nut. Tighten the nut only finger tight at this time.

8   On 4WD models, position the front end of the radius arm onto the bracket and axle. Install the bolts and stud (and washer on the left axle only) in the bracket. Tighten the bolts and stud only finger tight at this time.

9   Install the rear retainer and insulator onto the threaded end of the radius arm.

10   Install the radius arm into the rear bracket and install the rear insulator, washer and new nut. Tighten the nut to the torque listed in this Chapter's Specifications.

11   Tighten the bolts and nuts installed in Step 7 and Step 8 to the torque listed in this Chapter's Specifications.

12   On 2WD models, install the spring lower seat onto the radius arm.

13   On 4WD models, install the spring lower seat and insulator onto the left radius arm.

14   Install the spring assembly (see Section 6).

7   Radius arm insulators (1993 through 1997 models - replacement

1   Remove the front spring (see Section 6).
2   Loosen the axle arm pivot bolt.
3   Remove the shock absorber upper nut and compress the shock.
4   From the rear side of the radius arm bracket, remove the nut, rear washer, insulator, shield (right side of V6 engine models only) and spacer **(see illustration 6.3)**.
5   Raise the front axle arm with a floor jack until the radius arm is level.
6   Push the radius arm forward until it is free of the radius arm bracket. **Note:** *If necessary on 4WD models, detach the driveshaft*

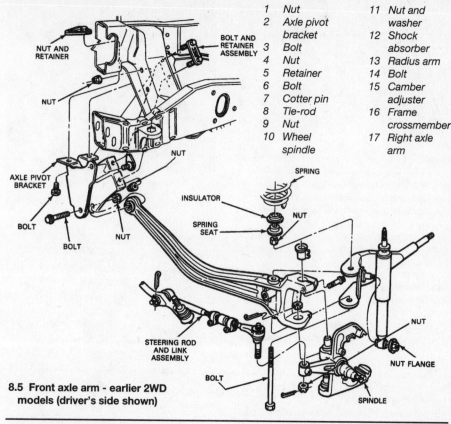

1	Nut
2	Axle pivot bracket
3	Bolt
4	Nut
5	Retainer
6	Bolt
7	Cotter pin
8	Tie-rod
9	Nut
10	Wheel spindle
11	Nut and washer
12	Shock absorber
13	Radius arm
14	Bolt
15	Camber adjuster
16	Frame crossmember
17	Right axle arm

8.5 Front axle arm - earlier 2WD models (driver's side shown)

from the front axle flange.

7   Remove the front insulator from the radius arm threaded end.

8   Installation is the reverse of the removal Steps with the following additions:

a) Install new insulators and nut.
b) Tighten the bolts and nut to the torque listed in this Chapter's Specifications.

8   Front axle arm (1993 through 1997 models) - removal and installation

Refer to illustration 8.5

Note: *Front axle arm removal and installation for 4WD models is covered in Chapter 8.*

Removal

1   Raise the front of the vehicle, support it securely on jackstands, block the rear wheels and set the parking brake. Position the front wheels in the straight ahead position.
2   Remove the coil spring (see Section 5) and front spindle (see Section 3).
3   Remove the radius arm (see Section 6).
4   Remove the front stabilizer bar (see Section 11).
5   Remove the bolt and nut securing the axle arm to the frame pivot bracket **(see illustration)**. Remove the axle arm.

Installation

6   Install the axle arm to the frame pivot bracket with a new retaining bolt and nut. Tighten the nut finger-tight.

7   Install the front stabilizer bar (see Section 11).
8   Install the front axle arm onto the radius arm (see Section 7).
9   Install the front spindle (see Section 3) and the coil spring (see Section 6).
10   Install the axle arm retaining bolt and nut and tighten them slightly. Don't torque them to specifications yet.
11   Install the wheel and lug nuts. Lower the vehicle and tighten the lug nuts to torque listed in Chapter 1.
12   Tighten the axle arm retaining bolt and nut to the torque listed in this Chapter's Specifications.
13   Have the front end alignment checked by a dealership service department or an alignment shop.

9   Axle pivot bushing (1993 through 1997 models) - removal and installation

Removal

2WD models

Refer to illustration 9.4

1   Remove the front spring (see Section 6).
2   To remove the left axle pivot bushing, remove the retaining bolt and nut, then pull the pivot end of the axle down until the bushing is exposed.
3   To remove the right axle pivot bushing, remove the axle arm from the frame (see Section 8).

Chapter 10 Suspension and steering systems

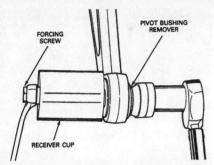

9.4 Removing the pivot bushing (2WD models)

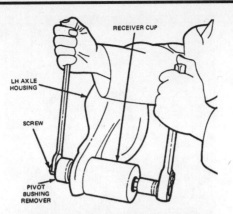

9.6 Removing the pivot bushing (earlier 4WD models)

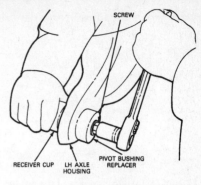

9.7a Installing the pivot bushing (earlier 2WD models)

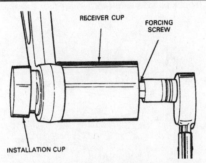

9.7b Installing the pivot bushing (earlier 4WD models)

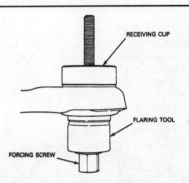

9.8 Flaring the bushing to prevent it from moving within the axle

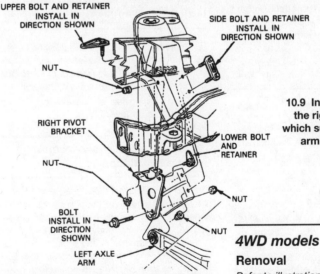

10.9 Installation details of the right pivot bracket, which supports the left axle arm (earlier models)

4   Install a special forcing tool, bushing remover and receiver cup, available at most auto parts stores, or equivalent. Position the spacer(s) between the walls of the axle, then turn the forcing screw and press the bushing out of the axle (see illustration).

4WD models

Refer to illustration 9.6

5   Remove the front axle assembly (see Chapter 8).
6   Install a special forcing tool, bushing remover and receiver cup. Turn the forcing screw and press the bushing out of the axle (see illustration).

Installation

Refer to illustrations 9.7a, 9.7b and 9.8

7   Insert the bushing in the axle receptacle. Install the receiver cup, forcing tool and bushing installer, available at most auto parts stores that carry special tools, onto the axle. Turn the forcing screw and press the bushing into the axle (see illustrations).
8   The new bushing must be flared to prevent movement within the axle. Install a forcing screw, receiving cup, and flaring tool, available at most auto parts stores that carry special tools Turn the forcing screw and flare the lip of the bushing (see illustration).
9   The remainder of installation is the reverse of the removal Steps.

10 Axle pivot bracket (1993 through 1997 models) - removal and installation

2WD models

Removal

1   Loosen the front wheel lug nuts on the side to be dismantled. Raise the front of the vehicle, support it securely on jackstands, block the rear wheels and set the parking brake. Remove the front wheel.
2   Remove the front spring (see Section 5), the radius arm (see Section 6), the wheel spindle (see Section 3) and the front axle arm (see Section 8).
3   Remove the bolts and nuts securing the axle pivot bracket to the frame and remove the bracket from the frame crossmember (see illustration 8.5).

Installation

4   Position the front axle bracket onto the frame crossmember. Install new bolts from within the frame crossmember and out through the bracket, then install new nuts. **Note:** *The nuts have an undercut to clear the knurled section of the bolts. If standard nuts are used, a hardened washer 0.20-inch thick must be installed under each nut.* After all of the bolts and nuts have been installed, tighten them to the torque listed in this Chapter's Specifications.
5   The remainder of installation is the reverse of the removal Steps.

4WD models

Removal

Refer to illustrations 10.9 and 10.10

6   Loosen the wheel lug nuts on the side to be dismantled, raise the front of the vehicle and support it securely on jackstands. Block the rear wheels and apply the parking brake. Remove the wheel.
7   Remove the front spring (see Section 5).
8   Place a jack beneath the axle arm near the pivot bracket to support it.
9   On the right axle pivot bracket (which supports the left axle arm), remove the nuts, upper bolts, side bolts and retainers. Discard the lower bolt and retainer. Remove the axle pivot bracket from the frame crossmember (see illustration).

Chapter 10 Suspension and steering systems

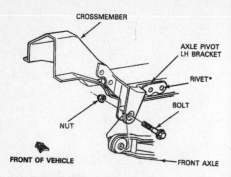

10.10 Installation details of the left axle pivot bracket, which supports the right axle arm (earlier models)

11.2a Unbolt the stabilizer bar and link from the axle arm (arrow) (earlier models)

11.2b Unbolt the stabilizer bar-to-frame mounting brackets (earlier models)

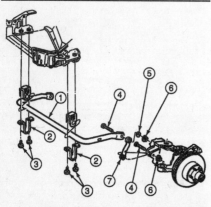

11.2c Front stabilizer bar installation details (earlier 4WD models)

1 Stabilizer bar
2 Bracket
3 Bolts
4 Bolts
5 Washer
6 Nut
7 Link (install with arrow pointing down)

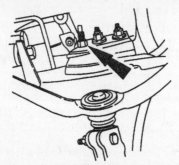

12.3 To detach the upper end of the shock absorber, remove this nut (arrow), (2WD model shown, 4WD models similar)

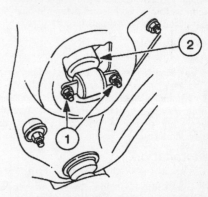

12.4 To detach the lower end of the shock absorber, remove these two nuts (arrows) (2WD model shown, 4WD models similar)

10 On the left axle pivot bracket (which supports the right axle arm), use a 9/16-inch drill bit and drill out the rivets securing the bracket to the frame. Remove the axle bracket from the frame crossmember **(see illustration)**.

Installation

11 Install the left axle pivot bracket onto the frame crossmember and align the 9/16 inch holes of the bracket and frame. Install 9/16-12x1-1/2-inch Grade 8 bolts. **Caution:** *Be sure to use Grade 8 bolts, not a lesser grade!* Install the washer and nuts onto the bolts and tighten to the torque listed in this Chapter's Specifications.

12 To install the right axle pivot bracket, perform the following:

a) Install the right axle pivot bracket onto the frame crossmember. Install all bolts in the direction shown in the illustration **(see illustration 10.9)**. The side bolt heads must face the engine oil pan to give maximum clearance. Install new nuts and tighten to the torque listed in this Chapter's Specifications.

b) Use a 9/16-inch drill bit and drill out the lower mounting hole bracket and frame crossmember.

c) Install a 9/16-inch Grade 8 replacement bolt with two flat washers and a new retaining nut. Tighten the nut to the torque listed in this Chapter's Specifications.

13 The remainder of installation is the reverse of the removal Steps.

11 Front stabilizer bar - removal and installation

Refer to illustrations 11.2a, 11.2b and 11.2c

1 Raise the front of the vehicle, support it securely on jackstands, block the rear wheels and set the parking brake. Position the front wheels in the straight ahead position.
2 Detach the stabilizer bar from the axle arms, then from the frame **(see illustrations)**.
3 Remove the stabilizer bar assembly.
4 Installation is the reverse of the removal Steps. Tighten the bolts and nuts to the torque listed in this Chapter's Specifications.

12 Front shock absorber (1998 and later models) - removal and installation

Refer to illustrations 12.3 and 12.4

1 Loosen the front wheel lug nuts, raise the vehicle and support it securely on jackstands. Remove the front wheels.

2 Remove the front disc brake water shield (see Chapter 9).
3 Remove the upper shock absorber mounting nut **(see illustration)**, washer and bushing.
4 Remove the two lower shock absorber mounting nuts **(see illustration)** and remove the shock.
5 Installation is the reverse of removal. Be sure to tighten the upper and lower shock mounting nuts to the torque listed in this Chapter's Specifications.

13 Front hub and bearing assembly (1998 and later 4WD models) - removal and installation

Note: *The hub and bearing assembly is a sealed unit and is neither adjustable nor serviceable. If it's defective, it must be replaced. Also, a number of special tools are needed to remove the hub lock and the hub and bearing assembly, to disassemble and reassemble the hub and bearing, and to install the hub and bearing and hub lock. Before undertaking this procedure, carefully read through the following steps and then obtain the tools you'll need from an auto parts store or dealer service department.*

Chapter 10 Suspension and steering systems

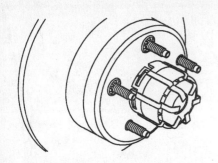

13.3a Install the six hub-lock removal clips as shown . . .

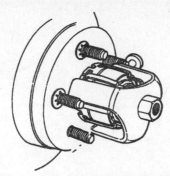

13.3b . . . and install the hub-lock puller over them

13.8 To detach the hub and bearing assembly from the steering knuckle, remove these three bolts (arrows)

Removal

Refer to illustrations 13.3a, 13.3b, 13.5 and 13.8

Note: *To remove the hub lock on these models, you'll need six special hub-lock removal clips, a hub-lock puller and a slide hammer designed to screw into the puller. Before proceeding, check on availability of these special tools.*

1 Loosen the front wheel lug nuts, raise the vehicle and support it securely on jackstands. Remove the wheel.

2 Remove the brake caliper and the brake disc (see Chapter 9).

3 Install the hub-lock removal clips and the hub-lock puller **(see illustrations)**. Align the puller with the three slots in the hub lock and hand-tighten the set screw on the side of the puller. Screw the slide hammer into the puller and remove the hub lock assembly.

4 Unbolt and remove the brake disc splash shield from the steering knuckle.

5 Remove the driveaxle retainer ring (see Section 19 in Chapter 8).

6 Detach the clips for the hub-lock system vacuum line, disconnect the vacuum line from the steering knuckle and set the vacuum line aside.

7 On models with ABS, disconnect the ABS electrical wire harness clips, remove the ABS sensor retaining bolt, remove the sensor from the knuckle and set the sensor and wire harness safely aside.

8 Remove the hub assembly-to-steering knuckle bolts **(see illustration)**.

9 Tap the hub assembly from side-to-side to break it loose from the steering knuckle. Pull the hub assembly off the end of the driveaxle. Wrap the end of the driveaxle with a rag to prevent damaging it. If the hub is stuck on the splines on the end of the driveaxle, use a puller to free it.

Disassembly

Note: *To disassemble the hub and bearing assembly, you'll need a special needle bearing puller and a slide hammer designed to screw into the puller (the same slide hammer tool that's used with the hub-lock puller). Before proceeding, consult with auto parts stores or your dealer service department regarding the availability of these special tools.*

10 Remove and discard the old O-ring(s) from the hub and bearing assembly (there are two on 1998 models, one on 1999 models). Always use new O-rings when installing the hub and bearing assembly. **Caution:** *You MUST replace the two hub and bearing O-rings every time the hub and bearing assembly is removed. Using an old O-ring can result in a vacuum leak, which could cause a loss of four-wheel-drive and/or contaminate the hub-lock vacuum system.*

11 Place the hub and bearing assembly in a bench vise.

12 Remove the old thrust washer and discard it. Always use a new thrust washer when installing the hub and bearing assembly. **Caution:** *You MUST replace the thrust washer every time the hub and bearing assembly is removed. Using an old thrust washer can damage the wheel hub and/or the driveaxle.*

13 Install the special needle bearing puller, screw the slide hammer into the puller and remove the caged needle bearing.

14 **Caution:** *Do NOT unscrew the big nut on the hub and bearing assembly! This nut is the bearing retaining nut. The bearing preload is preset at the factory. If you unscrew the bearing retaining nut, the bearing preload cannot be readjusted. The hub and bearing assembly MUST be replaced.*

Inspection

15 Wipe off the hub and bearing assembly so that you can inspect the condition of the internal parts. **Caution:** *Do NOT immerse the hub and bearing assembly in any kind of solvent! Doing so will ruin the sealed bearing inside and you will have to replace the hub and bearing assembly.*

16 Inspect the bearing preload nut and bore slots to ensure that the slots are aligned. If they're not aligned, or if the slots are damaged, replace the hub and bearing assembly.

17 Inspect the axleshaft bore for scoring and excessive wear. If it's scored or worn, replace the hub and bearing assembly. If there are metal particles in the axleshaft bore, the vacuum system must be flushed out and tested.

18 Inspect the needle bearing bore for scoring, bluing and excessive wear. If it's scored, blued or worn, replace the hub and bearing assembly.

19 Inspect the sealed hub bearing for excessive looseness or binding. If it's loose or binding, replace the hub and bearing assembly.

Reassembly

Note: *To reassemble the hub and bearing assembly, you'll need a special needle bearing installation tool and a driver designed to screw into the installation tool. Before proceeding, consult with your local Ford dealer or auto parts store regarding the availability of these special tools.*

20 Install the caged needle bearing into the hub and bearing assembly as follows: Place the caged needle bearing in position with the stamped part number side of the bearing race facing up. Position the bearing installation tool on the caged needle bearing and screw the driver handle into the installation tool. Using a hammer, tap the driver until the installation tool seats against the wheel hub and bearing. To verify that the caged needle bearing is correctly seated, measure the distance between the face of the hub and bearing assembly and the face of the needle bearing. It should be within the range of 0.35433 to 0.3937 inch.

21 Lubricate the caged needle bearings with wheel bearing grease. Install a new thrust washer. Make sure that all eight locking tabs on the thrust washer are correctly seated in the groove in the hub and bearing assembly. **Caution:** *If the thrust washer locking tabs are not fully seated in the groove in the hub and bearing assembly, the thrust washer can come loose from the hub and bearing, allowing the bearing preload nut to back off and destroying the wheel hub bearing.*

Installation

Note: *To install the hub and bearing assembly, you'll need a special main seal replacer tool, a main seal depth setting tool, a driveaxle seal installer tool, a seal remover guide and a snap-ring sleeve. Before proceeding, consult with auto parts stores or a dealer service department regarding the availability of these special tools.*

22 Wipe off and inspect the steering knuckle main seal. If it's damaged or worn, the wheel end of the vacuum system will leak and the vehicle will not operate in 4WD. If the main seal looks marginal, now is the time to

replace it. Knock out the old main seal with a hammer and punch. Clean the seal bore thoroughly and blow out the steering knuckle vacuum chamber and vacuum hose. Inspect the bore for burrs, nicks and other damage. If necessary, clean up the bore with crocus or emery cloth. Apply a thin coat of multi-purpose grease to the seal bore, place the new main seal in the special seal replacer, install the seal replacer tool and the depth setting tool on the steering knuckle, then drive the seal into position **(see illustration)** until it seats against the knuckle. Lubricate the sealing surface of the new seal **(see illustration)** with a liberal amount of wheel bearing grease. Lubricate the driveaxle splines with multi-purpose grease.

23 Install the hub and bearing assembly in the steering knuckle, install the three hub and bearing mounting bolts and tighten them to the torque listed in this Chapter's Specifications.

24 Install a new retainer ring on the driveaxle (see Section 19 in Chapter 8).

25 Reattach the vacuum line clips and, if equipped, the ABS harness clips to the frame.

26 If the vehicle is equipped with ABS, reconnect the ABS harness electrical connector.

27 Before reconnecting the vacuum line to the steering knuckle, check the wheel-end for vacuum leaks as follows: Hook up a hand-operated vacuum pump/gauge to the wheel-end of the vacuum line and apply 15 in-Hg of vacuum (you may have to pump rapidly for as much as 10 seconds). The line must not leak more than 0.50 in-Hg in 30 seconds. If the line holds vacuum satisfactorily, proceed to the next Step. If the line is leaking more than 0.50 in-Hg over 30 seconds, carefully inspect the line and connections between the wheel end and the pulse vacuum hub-lock solenoid (on the transfer case) for leaks. Repair or replace any leaking lines and/or connections.

28 Reconnect the vacuum line to the steering knuckle.

29 Install the brake disc splash shield.

30 Place the hub lock in position on the wheel, then use a soft-tipped rubber mallet to tap it all the way on. Make sure that the hub-lock locking arms are fully seated in the hub groove **(see illustration)**.

31 Install the brake disc and caliper (see Chapter 9).

32 Install the wheel, remove the vehicle from all jacks and stands and lower the vehicle. Tighten the wheel lug nuts to the torque listed in the Chapter 1 Specifications.

14 Front wheel spindle (1998 and later models) - removal and installation

Refer to illustration 14.8

Note: *On 4WD models, the spindle is referred to as the steering knuckle.*

1 Loosen the front wheel lug nuts, position the front wheels in the straight ahead position, raise the vehicle and support it securely on jackstands. Remove the front wheel.

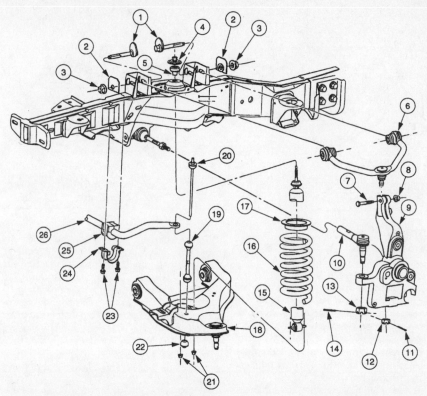

14.8 An exploded view of the front suspension assembly (later 2WD models; 4WD models are similar except that they use torsion bars instead of coil springs to make room for the driveaxles)

1 Upper control arm cam bolts and nuts	8 Spindle-to-upper control arm pinch bolt nut	18 Lower control arm
2 Upper control arm cam assemblies	9 Spindle	19 Stabilizer bar link
3 Upper control arm cam assembly nuts	10 Tie-rod end	20 Stabilizer bar stud and bushing assembly
4 Upper shock absorber nut and washer	11 Cotter pin	21 Shock absorber lower mounting nuts
5 Upper shock absorber bushing	12 Lower balljoint castellated nut	22 Stabilizer bar nut and washer
6 Upper control arm	13 Tie-rod end castellated nut	23 Stabilizer bar bracket bolts
7 Spindle-to-upper control arm pinch bolt	14 Cotter pin	24 Stabilizer bar bracket
	15 Shock absorber	25 Stabilizer bar bushing
	16 Coil spring	26 Stabilizer bar
	17 Front spring insulator	

2 Remove the disc brake caliper, brake disc and, on 2WD models, the disc brake splash shield (see Chapter 9). Wire the caliper to the underbody to prevent damage to the brake hose. On 2WD models with ABS, remove the ABS sensor.

3 On 2WD models, place a jack under the lower arm at the balljoint area. Raise the jack until it supports the spring load on the lower arm. The jack must remain in this position throughout the remainder of this procedure. **Warning:** *Failure to place a jack under the lower control arm could result in serious injury.*

4 On 4WD models, remove the hub lock assembly and disconnect the driveaxle from the hub and bearing assembly (see Section 13).

5 On 4WD models, remove the disc brake splash shield. If the vehicle is equipped with ABS, remove the ABS sensor from the steering knuckle.

6 On 4WD models, unload the torsion bar (see Section 16), then support the lower control arm with a jack.

7 Remove the upper shock absorber mounting nut, washer and bushing **(see illustration 12.3)**.

8 Remove the upper balljoint pinch bolt nut and pinch bolt **(see illustration)**.

9 Disconnect the lower shock absorber mounting nuts **(see illustration 12.4)** and remove the shock absorber.

10 Disconnect the tie-rod end from the spindle (see Section 27).

11 On 2WD models, compress the coil spring with a coil spring compressor (see Section 18).

12 Remove the lower balljoint cotter pin and loosen the castellated nut, install a small puller and separate the lower control arm

Chapter 10 Suspension and steering systems

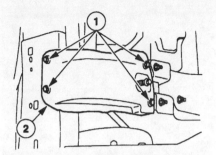

16.2 To detach the torsion bar cover plate (2), remove these four bolts (1)

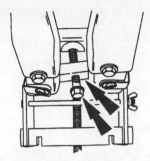

16.6 As you unscrew the adjustment bolt (lower arrow) from the torsion bar nut (upper arrow), count the number of turns to remove the bolt

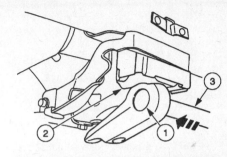

16.8 Pull down the torsion bar adjuster (1), remove the insulator (2) and mark the relationship of the torsion bar (3) to the torsion bar adjuster

balljoint from the spindle.
13 Remove the spindle/steering knuckle.
Caution: *To prevent damage to the CV joints on 4WD models, support the driveaxle assembly while the steering knuckle is removed.*
14 Installation is the reverse of removal with the following additions:
  a) Be sure to tighten all fasteners to the torque values listed in this Chapter's Specifications.
  b) Have the front-end alignment checked by a dealer service department or an alignment shop.

15 Front stabilizer bar (1998 and later models) - removal and installation

1 Position the front wheels in the straight-ahead position. Raise the vehicle and support it securely on jackstands.
2 To disconnect the stabilizer bar from the lower control arms, remove the nuts from the links that attach the bar to the arms **(see illustration 14.8)**.
3 To detach the stabilizer bar from the frame, remove the four stabilizer bar bracket bolts (two per bracket) **(see illustration 14.8)**.
4 Remove the stabilizer bar assembly.
5 Inspect all rubber bushings for cracks, tears and other damage. Replace any bushings that are worn, damaged or deteriorated.
6 Installation is the reverse of removal. Be sure to tighten the bolts and nuts to the torque listed in this Chapter's Specifications.
Note: *The stabilizer-to-frame bracket bolts are self-tapping. If they can no longer be tightened to the correct torque, obtain a kit with special fasteners from your dealer.*

16 Torsion bar (1998 and later 4WD models) - removal and installation

Removal

Refer to illustrations 16.2, 16.6 and 16.8
1 Position the front wheels in the straight-ahead position. Raise the vehicle and support it securely on jackstands.

2 Remove the torsion bar cover plate **(see illustration)** from the frame.
3 Before removing or adjusting the torsion bar, measure the distance that the torsion bar adjustment bolt protrudes from the adjustment nut. Jot down this dimension; you'll need it to readjust the torsion bar during installation.
4 Install the torsion bar tool, or a suitable aftermarket puller. **Caution:** *If you use an aftermarket puller, make sure that it has an adjustable bridge that keeps the jaws from spreading apart under pressure.*
5 Tighten the torsion bar tool until the torsion bar adjuster lifts off the adjustment bolt.
6 Unscrew the adjustment bolt from the torsion bar nut **(see illustration)** while counting the number of turns to remove the bolt. Record this number on a piece of paper for installation purposes. Remove the adjustment bolt and nut. Discard the old adjustment bolt. **Caution:** *Do NOT reuse the old adjustment bolt. The bolt is coated with a special adhesive to prevent it from backing out. When the bolt is backed out, this adhesive is removed from the threads. Using an old bolt without this adhesive could allow it to back out, resulting in the front end lowering substantially.*
7 Loosen the torsion bar tool until there's no tension on the torsion bar.
8 Mark the relationship of the torsion bar to the torsion bar adjuster **(see illustration)**.
9 Lower the torsion bar adjuster, remove the torsion bar insulator and pull the torsion bar out of the lower control arm.
10 Remove the torsion bar adjuster from the torsion bar.

Installation

11 Installation is the reverse of removal, with the following points:
  a) Be sure to use a new adjustment bolt, factory-coated with the special adhesive.
  b) When installing the new adjustment bolt, screw it in the same number of turns it took to remove the old one, then measure the height of the bolt from the adjustment nut and compare this measurement to the measurement you made before removing the old adjustment bolt.
  c) After lowering the vehicle to the ground, turn the torsion bar adjustment bolt in or out to adjust the ride height so the vehicle is level from side-to-side, with the vehicle sitting on its suspension. Also, roll the vehicle back and forth and jounce the suspension before adjusting and in between adjustments.
  d) Have the front-end alignment checked by a dealer service department or an alignment shop.

17 Lower control arm (1998 and later models) - removal, inspection and installation

Removal

Refer to illustration 17.4
1 Loosen the wheel lug nuts. Raise the front of the vehicle and support it securely on jackstands. Remove the front wheel.
2 On 2WD models, remove the brake disc and dust shield (see Chapter 9).
3 Remove the shock absorber (see Section 12).
3 Disconnect the stabilizer bar from the lower control arm (see Section 15).
4 On 2WD models, compress the coil spring **(see illustration)**.
5 On 4WD models, remove the torsion bar

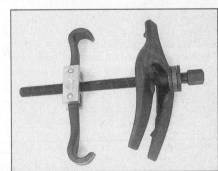

17.4 A typical aftermarket internal spring compressor tool: The hooked arms grip the upper coils of the spring, the plate is inserted below the lower coil, and the nut on the threaded rod is turned to compress the spring

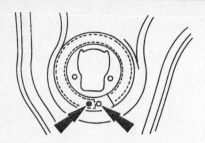

17.11 On 2WD models, make sure that the end of the coil spring covers the first hole (left arrow) in the lower control arm, but is not visible through the second hole (right arrow)

20.3 Remove the shock absorber lower bolt and nut (arrow) . . .

20.4 . . . and the upper nut (arrow)

(see Section 16).
6   Position the floor jack under the lower control arm to give support during removal.
7   Remove and discard the cotter pin from the lower control arm balljoint stud, loosen the castellated nut on the stud and install a suitable small puller. Separate the lower control arm balljoint from the spindle/steering knuckle. On 2WD models, carefully remove the compressed coil spring and set it and the spring compressor aside.
8   Loosen the two control arm pivot nuts and bolts and remove the lower control arm. **Caution:** *To prevent damage to the CV joints on 4WD models, support the steering knuckle and driveaxle assembly while the control arm is removed*

Inspection

9   Inspect the bushings for cracks and tears. If they're damaged or worn, take the control arm to an automotive machine shop to have new bushings installed. This procedure requires a number of specialized tools, so it's not worth tackling at home. If the balljoint is worn out the entire control arm must be replaced.
10 Inspect the lower control arm balljoint and boot seal for damage. If the balljoint or the boot is damaged, replace the lower control arm. The balljoint cannot be serviced.

Installation

Refer to illustration 17.11
11 Installation is the reverse of removal, with the following additions:
a) During assembly, tighten the pivot bolts and nuts finger-tight only. Do not tighten them to the specified torque until the vehicle is lowered to the ground.
b) On 2WD models, make sure that the end of the coil spring covers the first hole in the lower control arm, but is not visible through the second hole **(see illustration)**.
c) After the vehicle is back on the ground, tighten the lower control arm pivot bolts and nuts to the torque listed in this Chapter's Specifications.
d) Have the front-end alignment checked by a dealer service department or an alignment shop.

18 Coil spring (1998 and later 2WD models) - removal and installation

1   Removing the coil spring is an integral part of removing the lower control arm (see Steps 1 through 7 in Section 17).
2   To remove the coil spring, you will need a suitable coil spring compressor **(see illustration 17.4)**. These tools are available at auto parts stores.

19 Upper control arm (1998 and later models) - removal, inspection and installation

Note: *The upper control arm and balljoint cannot be separated and must be replaced as a single unit.*

Removal

1   Loosen the wheel lug nuts, raise the vehicle, support it securely on jackstands and remove the wheel.
2   On 2WD models, remove the brake disc and dust shield (see Chapter 9).
3   Place a floor jack under the lower control arm and raise the control arm slightly. The jack must remain in this position during the entire operation.
4   On 2WD models, mark the position of the cams on the upper control arm cam bolts **(see illustration 14.8)**.
5   Remove the pinch bolt nut and pinch bolt that secure the upper balljoint stud to the spindle/steering knuckle **(see illustration 14.8)**.
6   Before separating the upper control arm from the steering knuckle on 4WD models, support the knuckle to prevent it from tilting out (away from the vehicle) to protect the driveaxle assembly.
7   Using a puller, separate the upper control arm from the spindle/steering knuckle.
8   Remove the upper control arm cam bolts and detach the arm from the frame.

Inspection

9   Inspect the control arm bushings for cracks and tears. If they're damaged or worn, take the control arm to an automotive machine shop to have new bushings installed. This procedure requires a number of specialized tools, so it's not worth tackling at home. If the balljoint is worn out, the control arm must be replaced.

Installation

10 Installation is the reverse of removal, with the following additions:
a) On 2WD models, be sure to align the marks you made on the upper control arm adjustment cams before tightening the pivot nuts and bolts. Don't tighten the pivot nuts and bolts until the vehicle is on the ground.
b) On 4WD models, replace all four alignment plates with new alignment cams. Tighten the pivot nuts and bolts to the torque listed in this Chapter's Specifications.
c) After a 2WD model is lowered to the ground, tighten the pivot nuts and bolts to the torque listed in this Chapter's Specifications. Tighten the forward bolt and nut first, then the rear bolt and nut.

20 Rear shock absorbers - inspection, removal and installation

Inspection

1   Refer to Section 3 for inspection procedures.

Removal and installation

Refer to illustrations 12.3 and 12.4
2   Place a floor jack under the axle adjacent to shock absorber being removed. Apply just enough pressure with the jack to take the load off the shock absorber.
3   Remove the nut and bolt securing the lower end of the shock absorber to the spring plate **(see illustration)**.
4   Remove the nut securing the top of the shock absorber to the upper mounting bracket on the frame **(see illustration)**.
5   Installation is the reverse of the removal

Chapter 10 Suspension and steering systems

Steps. Tighten the nuts and bolts to the torque listed in this Chapter's Specifications.

21 Leaf spring - removal and installation

Refer to illustration 13.4

Removal

1 Raise the rear of the vehicle until the weight is off the rear springs but the tires are still touching the ground.
2 Support the vehicle securely on jackstands and support the rear axle with a jack. DO NOT get under a vehicle that's supported only by a jack, even if the tires are still installed.
3 Remove the nut and bolt securing the lower end of the shock absorber to the spring plate **(see illustration 12.3)**. Compress the shock up and out of the way.
4 Remove the nuts from the U-bolts and remove the U-bolts and the spring plate from the spring **(see illustration)**.
5 Remove the bolts and nuts securing the shackle assembly at the rear of the spring. Let the spring pivot down and rest on the floor.
6 Remove the spring hanger bolt and nut at the front of the spring. Remove the spring.
7 Inspect the spring eye bushings for wear or distortion. If worn or damaged have them replaced by a dealership service department or properly equipped shop.

Installation

8 Place the spring in the rear shackle. Install the bolt and nut and tighten them finger-tight.
9 Install the spring in the front bracket. Tighten the bolt and nut finger-tight.
10 Position the rear shackle on the frame. Install the bolt and nut and tighten them finger-tight.
11 Position the axle on the spring and install the spring seat. Make sure the spring tie-bolt is positioned in the hole in the spring, then install the U-bolts and nuts. Tighten the nuts finger-tight.
12 Lower the vehicle to the ground so its weight is resting on the tires. Tighten the spring bracket bolt and nut, shackle bolts and nuts and U-bolt nuts to the torque listed in this Chapter's Specifications.

22 Rear stabilizer bar - removal and installation

Refer to illustration 14.1

Removal

1 Remove the nut and washer securing the rear stabilizer bar ends to the link at each end **(see illustration)**.
2 Unbolt the retainers from the rear axle housing.

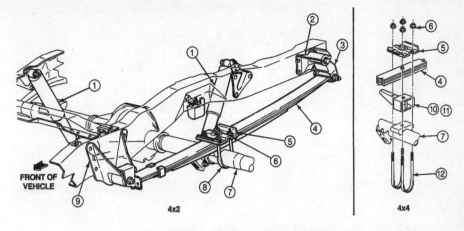

21.4 Rear suspension components

1 Shock absorber
2 Shackle bracket
3 Shackle
4 Spring
5 Spring cap and plate
6 Nut
7 Rear axle
8 U-bolt (2WD models)
9 Front hanger bracket
10 Spring spacer
11 Spring spacer
12 U-bolt (4WD models)

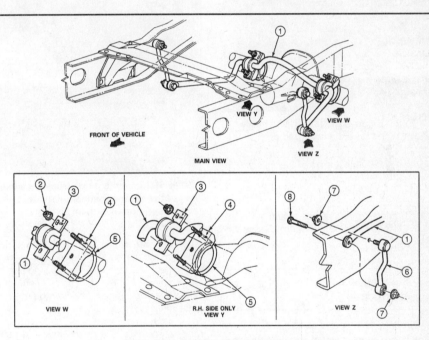

22.1 Rear stabilizer bar components

1 Stabilizer bar
2 Nut
3 Retainer
4 Mounting bracket
5 U-bolt
6 Link
7 Nut and washer
8 Bolt

3 Remove the mounting brackets, retainer and the stabilizer bar from the vehicle.
4 Inspect the rubber isolators on the stabilizer bar and replace if necessary.

Installation

5 Position the stabilizer bar onto the rear axle assembly. Position the retainer with the UP mark facing up.
6 Install the stabilizer bar and retainer onto the mounting brackets. Make sure the UP mark on the retainer is facing up toward the truck bed.
7 Install the U-bolts and nuts. Tighten the nuts finger-tight.
8 Move the stabilizer bar ends up into

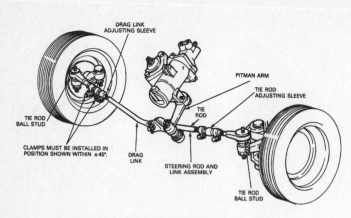

23.1 Steering system components (earlier models)

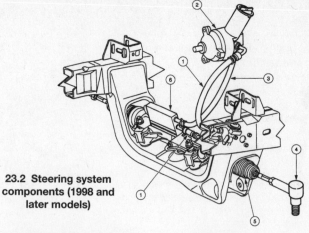

23.2 Steering system components (1998 and later models)

1 Power steering pressure hose
2 Power steering pump
3 Power steering return line hose
4 Tie-rod end
5 Rack-and-pinion steering gear
6 Power steering fluid cooler

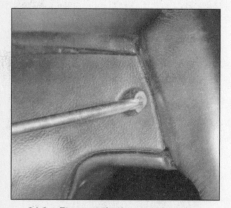

24.2a Remove the horn pad screws and take the horn pad off

position and connect them to the links. Install the bolt, washer and nut on each end. Tighten the bolts and nuts to the torque listed in this Chapter's Specifications.

9 Tighten the retainer bolts and link nuts to the torque listed in this Chapter's Specifications.

23 Steering system - general information

Refer to illustrations 23.1 and 23.2

The steering system on 1993 through 1997 models consists of a Pitman arm, drag link, steering connecting rod and tie-rods (see illustration). The Pitman arm transfers the steering gear movements through the drag link and steering connecting rod to the tie-rods at each end. The tie-rods move the spindles (or knuckle) and front wheels to the desired steering movement. The tie-rods are equipped with an adjusting sleeve for setting the toe-in. 1993 through 1997 models are equipped with the Ford Integral Power Steering gear that consists of a belt-driven Ford C-II pump and associated lines and hoses. The power steering pump reservoir fluid level should be checked periodically (see Chapter 1). The steering wheel operates the

24.2b Remove the plugs from the sides of the steering wheel on models equipped with an airbag

steering shaft, which actuates the steering gear through universal joints and the intermediate shaft. Looseness in the steering can be caused by wear in the steering shaft universal joints, the steering gear, the tie-rod ends and loose retaining bolts.

The steering system on 1998 and later models consists of a power-assisted rack-and-pinion type steering gear which is connected to the spindle/steering knuckles by tie-rod ends (see illustration). The power steering system is similar to earlier systems, except that it uses a power steering fluid cooler.

24 Steering wheel - removal and installation

Warning 1: *Some models have airbags. Always disconnect the negative battery cable, then the positive battery cable and wait two minutes before working in the vicinity of the impact sensors, steering column or*

instrument panel to avoid the possibility of accidental deployment of the airbag, which could cause personal injury (see Chapter 12).
Warning 2: *On models equipped with a driver's side airbag, put the wheels in the straight ahead position and lock the steering column. The steering column must not be rotated while the steering wheel is removed.*

Removal

Refer to illustrations 24.2a, 24.2b, 24.3a, 124.3b, 24.4, 24.5 and 24.7

1 Disconnect the negative cable from the battery (also disconnect the positive cable if equipped with an airbag).

2 On non-airbag systems, remove the screws securing the horn pad to the steering wheel (see illustration). On airbag systems, pry the two plugs covering the driver's side airbag module screws on the sides of the steering wheel (see illustration).

3 On non-airbag systems, partially pull the horn pad away from the steering wheel, unplug the electrical connector then remove the horn pad. On models with airbags, remove the two screws from the side of the steering wheel, lift the airbag off and detach the electrical connector (see illustrations). Store the airbag in a safe place until needed.
Warning: *Whenever handling an airbag module, always keep the airbag opening pointed away from your body. Never place the airbag module on a bench or other surface with the airbag opening facing the surface. Always place the airbag module in a safe location with the airbag opening facing up.*

4 Loosen the steering wheel bolt two or three turns, but don't remove it yet (see illustration).

5 Use a steering wheel puller to remove the steering wheel. On airbag equipped models, use a two-jaw puller (Ford tool no. T77F-4220-B1 or equivalent). Attach it to the areas on the steering wheel marked Pull (see illus-

Chapter 10 Suspension and steering systems

24.3a Remove the airbag module retaining screw from each side of the steering wheel . . .

24.3b . . . lift the module up and disconnect the airbag electrical connector

24.4 Loosen the steering wheel bolt (arrow) two or three turns

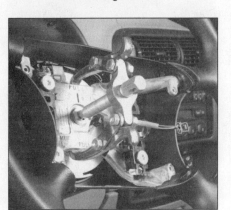

24.5 Position the puller against the steering wheel bolt, then tighten the puller to free the steering wheel - DO NOT pound on the wheel to loosen it

24.7 There should be alignment marks on the steering wheel and shaft (arrow); if not, make your own

24.8 Make sure the horn spring is in position before installing the wheel

tration). **Caution:** *Don't hammer on the shaft to remove the steering wheel. Once the steering wheel is loose, remove the bolt and discard it. Use a new bolt on installation.*

6 Lift off the damper (if equipped).

7 Look for alignment marks on the steering wheel and column **(see illustration)**. If they aren't visible, use white paint and make your own marks. These marks will be used during installation to align the wheel and column. Remove the steering wheel fro the column. On airbag systems, guide the sliding contact wires through the steering wheel.

Installation

Refer to illustration 24.8

8 Make sure the horn spring is in position **(see illustration)**. Align the mark on the steering wheel hub with the mark on the steering shaft and slip the wheel onto the shaft. Install the new mounting bolt and tighten it to the torque listed in this Chapter's Specifications.

9 The remainder of installation is the reverse of the removal Steps. **Caution:** *On airbag systems, make sure the airbag wiring does not get pinched between the airbag sliding contact and the steering wheel.*

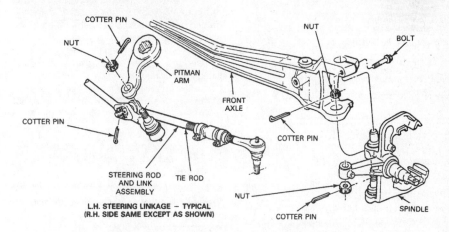

25.2a Steering linkage components (2WD models)

25 Drag link (1993 through 1997 models) - removal and installation

Refer to illustrations 25.2a, 25.2b and 25.2c

Removal

1 Raise the front of the vehicle, support it securely on jackstands, block the rear wheels and set the parking brake. Position the front wheels in the straight ahead position.

2 Remove the cotter pins and nuts securing the drag link to the steering connecting arm and to the Pitman arm **(see illustrations)**.

3 Loosen the clamp bolts on the tie-rod adjusting sleeve.

4 Unscrew the adjusting sleeve. Count and record the number of turns it takes to back the sleeve off the drag link.

5 Remove the drag link.

25.2b Steering linkage components (4WD models)

1 Tie-rod
2 Cotter pin
3 Nut
4 Front axle arm
5 Pitman arm
6 Steering drag link

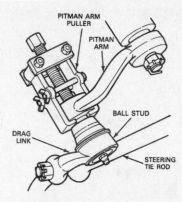

25.2c Disconnect the steering linkage joints with a Pitman arm puller

27.5 Remove the tie-rod end using a Pitman arm puller (earlier models)

Installation

6 Loosely install the drag link ball stud into the Pitman arm. Position the steering connecting arm ball end into the drag link. Make sure the front wheels and steering wheel are in the straight ahead position. Make sure the ball ends are seated in the taper to prevent them from rotating while tightening the nuts.
7 Install the drag link onto the tie-rod adjusting sleeve and turn it the same number of turns noted during removal in Step 4. Tighten the clamp bolts on the tie-rod adjusting sleeve to the torque listed in this Chapter's Specifications.
8 Install new nuts on the studs of the Pitman arm and steering connecting arm and tighten to the torque listed in this Chapter's Specifications. Install new cotter pins and bend the ends over completely.
9 Install the wheel and lug nuts. Lower the vehicle and tighten the lug nuts to torque listed in Chapter 1.
10 Have the front end alignment checked by a dealership service department or an alignment shop.

26 Steering connecting rod (1993 through 1997 models) - removal and installation

Removal

1 Raise the front of the vehicle, support it securely on jackstands, block the rear wheels and set the parking brake. Position the front wheels in the straight ahead position.
2 Remove the cotter pin and nut from the ball end of the steering connecting rod (see illustration 25.2a or 25.2b).
3 Use a Pitman arm puller and remove the ball end from the drag link (see illustration 25.2c).
4 Loosen the clamp bolts on the tie-rod adjusting sleeve.
5 Unscrew the connecting rod from the tie-rod adjusting sleeve. Count and record the number of turns it takes to remove the connecting rod from the tie-rod adjusting sleeve.
6 Remove the steering connecting rod.

Installation

7 Install the steering connecting rod onto the tie-rod and turn it the same number of turns noted during removal in Step 5. Tighten the clamp bolts on the tie-rod adjusting sleeve to the torque listed in this Chapter's Specifications.
8 Install the steering connecting rod end into the drag link. Make sure the front wheels and steering wheel are in the straight ahead position. Make sure the ball end is seated in the taper to prevent them from rotating while tightening the nut.
9 Install a new nut on the stud and tighten to the torque listed in this Chapter's Specifications. Install a new cotter pin and bend the ends over completely.
10 Install the wheel and lug nuts. Lower the vehicle and tighten the lug nuts to torque listed in Chapter 1.
11 Have the alignment checked by a dealership service department or an alignment shop.

27 Tie-rod ends - removal and installation

Warning: *Whenever any of the suspension or steering fasteners are loosened or removed discard the originals and don't re-use them. They must be replaced with new fasteners of the same part number or of original equipment quality and design. Torque specifications must be followed for proper reassembly and component retention.*

Removal

Refer to illustration 27.5

1 Loosen the front wheel lug nuts on the side to be dismantled. Raise the front of the vehicle, support it securely on jackstands, block the rear wheels and set the parking brake.
2 Remove the front wheel.
3 Remove the cotter pin and loosen the nut on the tie-rod end stud. Discard the cotter pin.
4 On 1993 through 1997 models, loosen the clamp bolts on the tie-rod adjusting sleeve. On 1998 and later models, loosen the tie-rod end jam nut and back it off several turns.
5 Disconnect the tie-rod from the steering spindle with a Pitman arm puller (see illustrations). **Note:** *On 1993 through 1997 4WD models the tie-rod end is inserted in from the top of the spindle. On 2WD models the tie-rod is inserted from the bottom of the spindle (see illustrations 25.2a and 25.2b).*
6 On 1993 through 1997 models, unscrew the tie-rod end from the adjusting sleeve. Count and record the number of turns it takes to back the tie-rod end off the sleeve. On 1998 and later models, if you're going to install a new tie-rod end, count and record the number of turns it takes to back the tie-rod end off the connecting rod. If you're going to re-use the same tie-rod end on a 1998 or later model, simply apply a paint mark on the threads adjacent to the tie-rod end and unscrew the tie-rod end from the connecting rod.

Installation

7 On 1993 through 1997 models, install the tie-rod end into the adjusting sleeve the

Chapter 10 Suspension and steering systems

28.4 The power steering gear line fittings are accessible through the wheel well (earlier models)

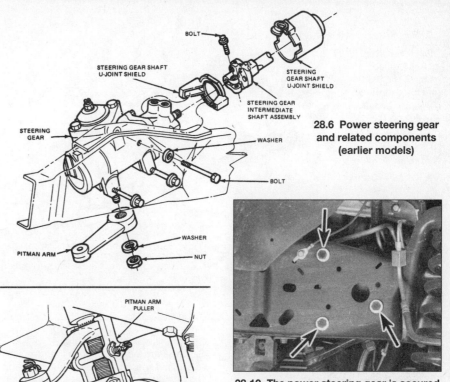

28.6 Power steering gear and related components (earlier models)

28.9 Remove the Pitman arm from the gear with a Pitman arm puller (earlier models)

28.10 The power steering gear is secured to the frame member by three bolts (arrows) (earlier models)

same number of turns noted during removal in Step 6. On 1998 and later models, if you're installing the old tie-rod end, simply screw the tie-rod into the connecting rod until the tie-rod end is adjacent to the paint mark you made in Step 6. If you're installing a new tie-rod end on a 1998 or later model, screw it in the same number of turns noted during removal.

8 Install the tie-rod onto the steering spindle. Make sure the front wheels and steering wheel are in the straight ahead position. Make sure the tie-rod stud is seated in the taper to prevent it from rotating while tightening the nut.
9 Install a new nut on the stud and tighten it to the torque listed in this Chapter's Specifications. Install a new cotter pin and bend the ends over completely.
10 On 1993 through 1997 models, tighten the clamp bolts on the tie-rod adjusting sleeve to the torque listed in this Chapter's Specifications. Make sure the tie-rod is positioned correctly in the same position as when it was removed. On 1998 and later models, tighten the jam nut securely.
11 Install the wheel and lug nuts. Lower the vehicle and tighten the lug nuts to the torque listed in Chapter 1.
12 Have the front end alignment checked by a dealership service department or an alignment shop.

28 Steering gear - removal and installation

Warning 1: *Some models have airbags. Always disconnect the negative battery cable, then the positive battery cable and wait two minutes before working in the vicinity of the impact sensors, steering column or instrument panel to avoid the possibility of accidental deployment of the airbag, which could cause personal injury (see Chapter 12).*

Warning 2: *On models equipped with airbags, DO NOT rotate the steering shaft while the steering gear is removed from the vehicle or damage to the airbag coil will occur. To prevent the shaft from turning, turn the ignition key to the Lock position before beginning*

work or run the seat belt through the steering wheel and clip the seat belt into place.
Warning 3: *Whenever any of the suspension or steering fasteners are loosened or removed discard the originals and don't re-use them. They must be replaced with new fasteners of the same part number or of original equipment quality and design. Torque specifications must be followed for proper reassembly and component retention.*

1993 through 1997 models

Removal

Refer to illustrations 28.4, 28.6, 28.9 and 28.10

1 Position the front wheels in the straight-ahead position.
2 Disconnect the cable from the negative battery terminal (also disconnect the positive cable if equipped with an airbag).
3 Turn the ignition key to the Lock position to lock the steering wheel.
4 Place a pan under the steering gear. Remove the power steering hose fittings from the steering gear and cap the ends and the ports to prevent excessive fluid loss and contamination **(see illustration)**.
5 Remove the upper and lower U-joint shaft coupling shield from the steering gear.

6 Remove the bolt securing the flex coupling to the steering gear **(see illustration)**.
7 Loosen the front wheel lug nuts. Raise the front of the vehicle, support it securely on jackstands, block the rear wheels and set the parking brake. Remove both front wheels.
8 Remove the large nut and washer securing the Pitman arm to the sector shaft.
9 Using a Pitman arm puller (Ford tool no. T64P-3590-F or equivalent), remove the Pitman arm from the sector shaft **(see illustration)**. **Caution:** *Don't hammer on the end of the puller as the steering gear will be damaged.*
10 Loosen all three bolts securing the steering gear box to the frame **(see illustration)**. Support the steering gear as you remove the bolts and washers and remove the steering gear from the vehicle.

Installation

11 Make sure the steering gear is centered. Turn the input shaft to full lock in one direction, then count and record the number of turns required to rotate it to the full lock position in the opposite direction. Turn the input shaft back one-half the number of turns to center the assembly.
12 Check that the front wheels are in the straight ahead position.
13 Install the steering gear input shaft lower shield onto the steering gear box. Install the upper shield onto the intermediate shaft.
14 Position the flat on the gear input shaft so it is facing down. Position the flex coupling on the steering gear input shaft, aligning the flat with the flat on the input shaft.

10-20 Chapter 10 Suspension and steering systems

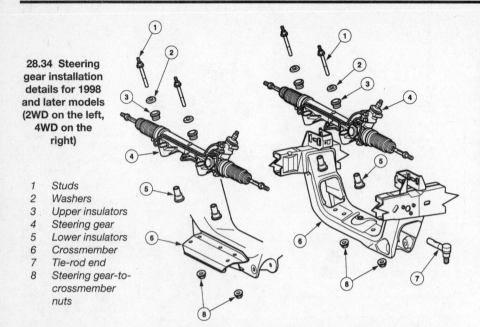

28.34 Steering gear installation details for 1998 and later models (2WD on the left, 4WD on the right)

1. Studs
2. Washers
3. Upper insulators
4. Steering gear
5. Lower insulators
6. Crossmember
7. Tie-rod end
8. Steering gear-to-crossmember nuts

29.6 This special tool is needed to remove the power steering pump pulley

15 Install all three bolts securing the steering gear box to the frame. Tighten the bolts to the torque listed in this Chapter's Specifications.
16 Align the two blocked teeth on the Pitman arm with the four missing teeth on the steering gear sector shaft and install the Pitman arm. Install the washer and nut and tighten the nut to the torque listed in this Chapter's Specifications.
17 Install the bolt securing the flex coupling to the steering gear input shaft. Tighten the bolt to the torque listed in this Chapter's Specifications.
18 Move the flex coupling shield into place on the steering gear input shield.
19 Install the front wheel(s) and lug nuts.
20 Lower the vehicle and tighten the lug nuts to the torque listed in Chapter 1 Specifications.
21 Connect the pressure and return lines to the steering gear assembly and tighten to the torque listed in this Chapter's Specifications.
22 Connect the battery cables.
23 Fill the fluid reservoir with the specified fluid and refer to Section 23 for the power steering bleeding procedure.

1998 and later models

Removal

Refer to illustration 28.34

24 Set the front wheels to the straight-ahead position. Raise and securely support the vehicle on jackstands.
25 Disconnect the cable from the negative battery terminal.
26 Secure the steering wheel in the straight-ahead position, but do not lock the column.
27 On 2WD models, disconnect the hoses from the power steering fluid cooler, drain the fluid into a suitable container and remove the cooler.
28 On 4WD models, remove the air deflector retaining screws, pull down on the deflector to disengage the retaining pins.
29 Separate the tie-rod ends from the spindle/steering knuckle (see Section 27).
30 On 4WD models, remove the front stabilizer bar (see Section 15).
31 On 4WD models, remove the power steering fluid cooler-to-crossmember nuts, then disconnect the fluid return line hose, allow the fluid to drain into a suitable container, disconnect the return hose and remove the cooler.
32 Disconnect the power steering pressure and return lines at the steering gear. Plug the lines and the steering gear inlet and outlet ports to prevent the entry of dirt or moisture.
33 Remove the pinch bolt retaining the lower steering column intermediate shaft to the steering gear input shaft. Turn the ignition switch to the LOCK position, then disconnect the intermediate shaft from the steering gear input shaft. **Caution:** *Do not rotate the steering wheel while the intermediate shaft is disconnected from the steering column or damage to the airbag sliding contact will result.*
34 Remove the nuts and mounting stud, nut, washer and stop assemblies **(see illustration)**.
35 On 2WD models, remove the steering gear assembly.
36 On 4WD models, remove the insulator bushings from the steering gear.
37 On 4WD models, rotate the steering gear control valve housing forward, turn the steering gear input shaft clockwise until it stops, move the steering gear as far to the right as possible, move the left tie-rod forward to clear the crossmember, then pull out the steering gear.

Installation

38 Installation is the reverse of removal **(see illustration)**, with the following additions:
  a) *Replace any insulators that are cracked, torn or deteriorated.*
  b) *Be sure to tighten all fasteners to the torque listed in this Chapter's Specifications.*
  c) *Using a special seal installation tool, install new seals on the power steering fluid pressure and return hose fittings.*
  d) *Have the vehicle's alignment checked when you're done.*

29 Power steering pump - removal and installation

Refer to illustrations 29.6, 29.7a, 29.7b, 29.7c and 29.10
Note: *Refer to Section 22 for information on quick disconnect fittings used on the power steering system.*

Removal

1 Disconnect the cable from the negative battery terminal.
2 Remove the drivebelt from the power steering pulley (see Chapter 1).
3 Loosen the front wheel lug nuts. Raise the front of the vehicle, support it securely on jackstands, block the rear wheels and set the parking brake. Remove both front wheels.
4 Disconnect the return hose at the pump and drain the fluid into a container.
5 Disconnect the pressure hose from the pump.
6 Remove the drivebelt pulley as follows:
  a) *Install a pump pulley removal tool, available at most auto parts stores, onto the pulley* **(see illustration)**.
  b) *Hold onto the pulley and rotate the tool nut counterclockwise to remove the pulley from the pump. Do not apply excessive force on the pulley shaft as it may damage the internal parts of the pump.*
7 Remove the bolts securing the pump to the mounting bracket **(see illustrations)**.
8 Remove the pump and pulley through the mounting bracket.

Installation

9 Place the pump in the bracket and install the mounting bolts. Tighten the bolts to the torque listed in this Chapter's Specifications.

Chapter 10 Suspension and steering systems

29.7a Typical power steering pump mounting details (2.3L four-cylinder engine shown; 2.5L engine similar)

1 Air conditioning compressor bracket
2 Bolt
3 Bolt
4 Bolt
5 Screw
6 Power steering pulley
7 Idler pulley
8 Bolt
9 Power steering pump
10 Air conditioning compressor

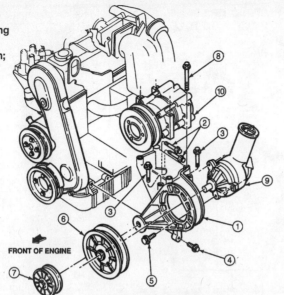

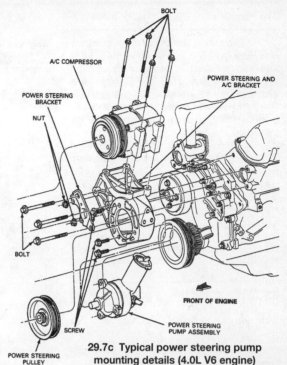

29.7c Typical power steering pump mounting details (4.0L V6 engine)

29.7b Typical power steering pump mounting details (3.0L V6 engine)

1 Power steering pump
2 Air conditioning compressor mounting bracket
3 Stud
4 Bolt
5 Bolt
6 Power steering pulley
7 Drivebelt tensioner
8 Bolt
9 Idler pulley
10 Bolt
11 Air conditioning compressor
12 Water pump pulley

29.10 This tool is used to install the power steering pump pulley

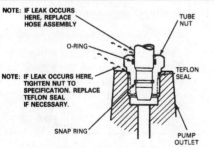

30.1 Quick disconnect fitting details

12 Install the pressure and return hoses to the proper fittings on the pump (see Section 22).
13 Fill the pump reservoir with the specified fluid and bleed the system (see Section 23).

30 Power steering line quick disconnect fittings

Refer to illustrations 30.1 and 30.3

1 If a fitting leaks, note whether the leak is between the tubing and the flare nut or between the flare nut and the pump **(see illustration)**. If it's between the tubing and the flare nut, replace the hose as an assembly. If it's between the flare nut and the pump, repair the leak (see Steps 2 and 3 below).
2 Make sure the nut is tightened to the torque listed in this Chapter's Specifications.
3 If the fitting still leaks, unscrew the nut

30.3 To install a new Teflon washer, stretch it with a center punch or similar tool until it fits over the flare nut - it will slowly shrink to its original size

and check the Teflon washer. Replace the washer if its condition is in doubt. **Note:** *It may be necessary to stretch the washer with a center punch or similar tapered tool so it will fit over the flare nut* **(see illustration)**. *The washer will slowly return to its original size after being stretched.*
4 The O-ring in the fitting can't be replaced. If it's causing the leak, the hose must be replaced as an assembly.
5 If a quick-disconnect fitting disconnects while the pump is in operation, replace the hose as an assembly.

10 Install the pulley onto the pump using a special pulley installation tool, available at most auto parts stores **(see illustration)**. Press the pulley onto the shaft until the hub is flush with the end of the shaft (within 0.010-inch).
11 Install the pump drivebelt (see Chapter 1).

31 Power steering system - bleeding

1 The power steering system must be bled whenever a line is disconnected. Bubbles can be seen in power steering fluid which has air in it and the fluid will often have a tan or milky appearance. On later models, low fluid level can cause air to mix with the fluid, resulting in a noisy pump as well as foaming of the fluid.

2 Open the hood and check the fluid level in the reservoir, adding the specified fluid necessary to bring it up to the proper level (see Chapter 1).

3 Start the engine and slowly turn the steering wheel several times from left-to-right and back again. Do not turn the wheel completely from lock-to-lock. Check the fluid level, topping it up as necessary until it remains steady and no more bubbles are visible.

32 Wheels and tires - general information

Refer to illustration 32.1

All vehicles covered by this manual are equipped with metric-sized fiberglass or steel-belted radial tires **(see illustration)**. Use of other size or type of tires may affect the ride and handling of the vehicle. Don't mix different types of tires, such as radials and bias belted, on the same vehicle as handling may be seriously affected. It's recommended that tires be replaced in pairs on the same axle, but if only one tire is being replaced, be sure it's the same size, structure and tread design as the other.

Because tire pressure has a substantial effect on handling and wear, the pressure on all tires should be checked at least once a month or before any extended trips (see Chapter 1).

Wheels must be replaced if they are bent, dented, leak air, have elongated bolt holes, are heavily rusted, out of vertical symmetry or if the lug nuts won't stay tight. Wheel repairs that use welding or peening are not recommended.

Tire and wheel balance is important to the overall handling, braking and performance of the vehicle. Unbalanced wheels can adversely affect handling and ride characteristics as well as tire life. Whenever a tire is installed on a wheel, the tire and wheel should be balanced by a shop with the proper equipment.

33 Front end alignment - general information

Refer to illustrations 33.3 and 33.4

A front end alignment refers to the adjustments made to the front wheels so they are in proper angular relationship to the suspension and the ground. Front wheels that are out of proper alignment not only affect steering control, but also increase tire wear.

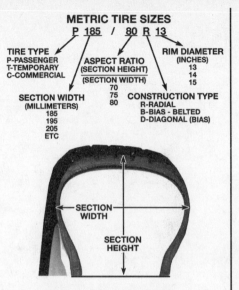

32.1 Metric tire size code

The only front end adjustments possible on these vehicles are caster and toe-in.

Getting the proper front wheel alignment is a very exacting process, one in which complicated and expensive machines are necessary to perform the job properly. Because of this, you should have a technician with the proper equipment perform these tasks. We will, however, use this space to give you a basic idea of what is involved with front end alignment so you can better understand the process and deal intelligently with the shop that does the work.

Toe-in is the turning in of the front wheels **(see illustration)**. The purpose of a toe specification is to ensure parallel rolling of the front wheels. In a vehicle with zero toe-in, the distance between the front edges of the wheels will be the same as the distance between the rear edges of the wheels. The actual amount of toe-in is normally only a fraction of an inch. Toe-in adjustment is controlled by the tie-rod end position on the tie-rod. Incorrect toe-in will cause the tires to wear improperly by making them scrub against the road surface.

Caster is the tilting of the top of the front steering axis from the vertical **(see illustration)**. A tilt toward the rear is positive caster and a tilt toward the front is negative caster. On 1993 through 1997 models, this angle is adjusted by adding or subtracting shims at the rear of the lower suspension arms. On 1998 and later models, caster is adjusted by means of eccentric cams on the upper control arm pivot bolts.

Camber is the tilting of the front wheels from vertical when viewed from the front of the vehicle **(see illustration 25.4)**. On 1993 through 1997 models, It is adjusted by means of adapters that fit into the upper balljoint hole in the spindle (2WD) or axle arm (4WD). On 1998 and later models, camber is adjusted by means of eccentric cams on the upper control arm pivot bolts.

33.3 Toe-in is the only normally adjusted alignment setting

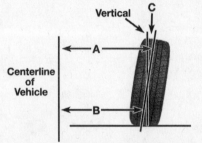

CAMBER ANGLE (FRONT VIEW)

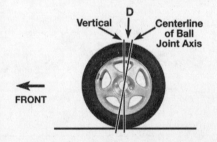

CASTER ANGLE (SIDE VIEW)

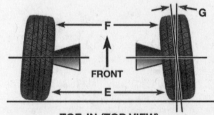

TOE-IN (TOP VIEW)

33.4 Camber can be adjusted by changing adapters. Caster (the tilting of the top of the front steering axis from vertical) affects the self-return characteristics of the steering system; it is not a routine adjustment, but should be checked whenever the front axle housing or front suspension arms are changed

Chapter 11 Body

Contents

Section		Section	
Body - maintenance	2	Hood latch control cable - removal and installation	9
Body repair - major damage	6	Inner door handle and latch assembly - removal and installation	13
Body repair - minor damage	5		
Console - removal and installation	24	Instrument cluster bezel - removal and installation	26
Door - removal, installation and alignment	10	Instrument panel - removal and installation	27
Door hinge - removal and installation	11	Mirrors (exterior) - removal and installation	31
Door latch striker - removal, installation and adjustment	12	Outer door handle - removal and installation	14
Door lock cylinder - removal and installation	15	Radiator grille - removal and installation	21
Door trim panel and watershield - removal and installation	16	Rear bumper - removal and installation	20
Front bumper and valance - removal and installation	19	Seat belt check	30
Front door window glass - replacement and adjustment	17	Seats - removal and installation	29
Front door window regulator - replacement	18	Steering column cover and shroud - removal and installation	25
Front fender - removal and installation	22		
General information	1	Upholstery and carpets - maintenance	4
Glove box - removal and installation	28	Vinyl trim - maintenance	3
Hinges and locks - maintenance	7	Windshield and fixed glass - removal and installation	23
Hood - removal, installation and adjustment	8		

Specifications

Torque specifications   Ft-lbs
Front bumper
  Mounting nuts .... 65 to 87
  Fog lamp bracket nuts .... 57 to 77
  Valance screws .... 23 to 34
Rear bumper bolts and nuts .... 84 to 112

---

1 General information

Warning: *Some models are equipped with airbags. Always disconnect the negative battery cable, then the positive battery cable and wait two minutes before working in the vicinity of the impact sensors, steering column or instrument panel to avoid the possibility of accidental deployment of the airbag, which could cause personal injury (see Chapter 12 for more information on the airbag system).*

These models feature a welded body that is attached to a separate frame. Certain components are particularly vulnerable to accident damage and can be unbolted and repaired or replaced. Among these parts are the body moldings, bumpers, hood, doors, tailgate and all glass.

Only general body maintenance procedures and body panel repair procedures within the scope of the do-it-yourselfer are included in this Chapter.

2 Body - maintenance

1 The condition of your vehicle's body is very important, because the resale value depends a great deal on it. It's much more difficult to repair a damaged body than it is to repair mechanical components. The hidden areas of the body, such as the fenderwells, the frame and the engine compartment, are equally important, although they don't require as frequent attention as the rest of the body.
2 Once a year, or every 12,000 miles, it's a good idea to have the underside of the body steam cleaned. All traces of dirt and oil will be removed and the area can then be inspected carefully for rust, damaged brake lines, frayed electrical wires, damaged cables and other problems. The front suspension components should be greased after completion of this job.
3 At the same time, clean the engine and the engine compartment with a steam cleaner or water soluble degreaser.
4 The fenderwells should be given close attention, since undercoating can peel away and stones and dirt thrown up by the tires can cause the paint to chip and flake, allowing rust to set in. If rust is found, clean down to the bare metal and apply an anti-rust paint.
5 The body should be washed about once a week (or when dirty). Wet the vehicle thoroughly to soften the dirt, then wash it down with a soft sponge and plenty of clean soapy water. If the surplus dirt is not washed off very carefully, it can wear down the paint.
6 Spots of tar or asphalt thrown up from the road should be removed with a cloth soaked in solvent.
7 Once every six months, wax the body and chrome trim. If a chrome cleaner is used to remove rust from any of the vehicle's plated parts, remember that the cleaner also removes part of the chrome, so use it sparingly.

3 Vinyl trim - maintenance

Don't clean vinyl trim with detergents, caustic soap or petroleum-based cleaners. Plain soap and water works just fine, with a soft brush to clean dirt that may be ingrained. Wash the vinyl as frequently as the rest of the vehicle.

After cleaning, application of a high quality rubber and vinyl protectant will help prevent oxidation and cracks. The protectant can also be applied to weatherstripping, vacuum lines and rubber hoses, which often fail as a result of chemical degradation, and to the tires.

4 Upholstery and carpets - maintenance

1 Every three months remove the carpets or mats and clean the interior of the vehicle (more frequently if necessary). Vacuum the upholstery and carpets to remove loose dirt and dust.

These photos illustrate a method of repairing simple dents. They are intended to supplement *Body repair - minor damage* in this Chapter and should not be used as the sole instructions for body repair on these vehicles.

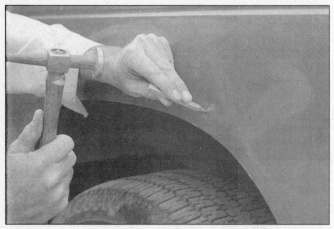

1 If you can't access the backside of the body panel to hammer out the dent, pull it out with a slide-hammer-type dent puller. In the deepest portion of the dent or along the crease line, drill or punch hole(s) at least one inch apart . . .

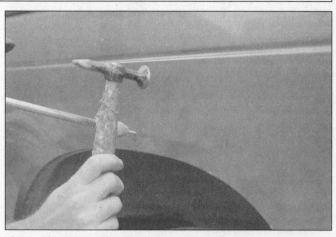

2 . . . then screw the slide-hammer into the hole and operate it. Tap with a hammer near the edge of the dent to help 'pop' the metal back to its original shape. When you're finished, the dent area should be close to its original contour and about 1/8-inch below the surface of the surrounding metal

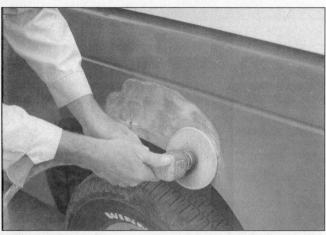

3 Using coarse-grit sandpaper, remove the paint down to the bare metal. Hand sanding works fine, but the disc sander shown here makes the job faster. Use finer (about 320-grit) sandpaper to feather-edge the paint at least one inch around the dent area

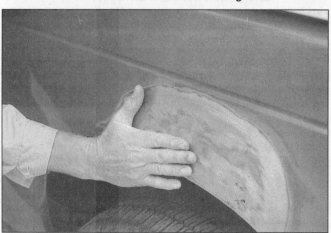

4 When the paint is removed, touch will probably be more helpful than sight for telling if the metal is straight. Hammer down the high spots or raise the low spots as necessary. Clean the repair area with wax/silicone remover

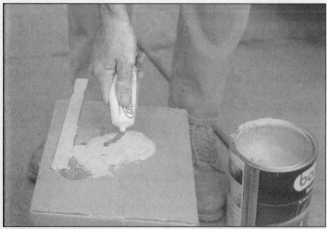

5 Following label instructions, mix up a batch of plastic filler and hardener. The ratio of filler to hardener is critical, and, if you mix it incorrectly, it will either not cure properly or cure too quickly (you won't have time to file and sand it into shape)

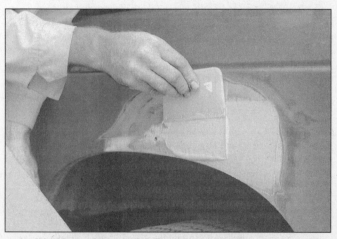

6 Working quickly so the filler doesn't harden, use a plastic applicator to press the body filler firmly into the metal, assuring it bonds completely. Work the filler until it matches the original contour and is slightly above the surrounding metal

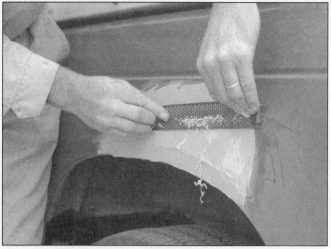

7 Let the filler harden until you can just dent it with your fingernail. Use a body file or Surform tool (shown here) to rough-shape the filler

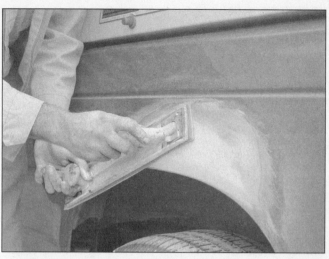

8 Use coarse-grit sandpaper and a sanding board or block to work the filler down until it's smooth and even. Work down to finer grits of sandpaper - always using a board or block - ending up with 360 or 400 grit

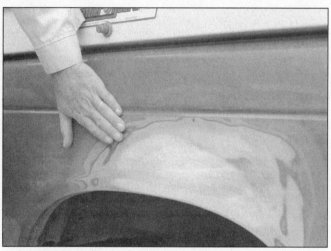

9 You shouldn't be able to feel any ridge at the transition from the filler to the bare metal or from the bare metal to the old paint. As soon as the repair is flat and uniform, remove the dust and mask off the adjacent panels or trim pieces

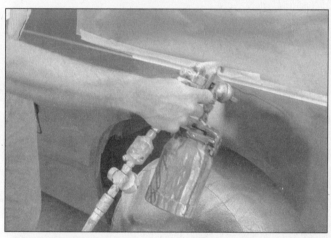

10 Apply several layers of primer to the area. Don't spray the primer on too heavy, so it sags or runs, and make sure each coat is dry before you spray on the next one. A professional-type spray gun is being used here, but aerosol spray primer is available inexpensively from auto parts stores

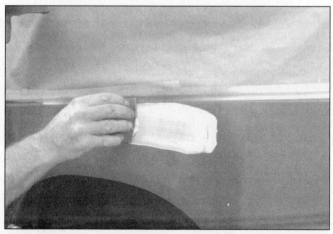

11 The primer will help reveal imperfections or scratches. Fill these with glazing compound. Follow the label instructions and sand it with 360 or 400-grit sandpaper until it's smooth. Repeat the glazing, sanding and respraying until the primer reveals a perfectly smooth surface

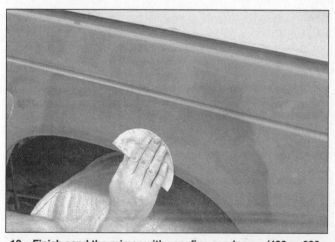

12 Finish sand the primer with very fine sandpaper (400 or 600-grit) to remove the primer overspray. Clean the area with water and allow it to dry. Use a tack rag to remove any dust, then apply the finish coat. Don't attempt to rub out or wax the repair area until the paint has dried completely (at least two weeks)

2 Leather upholstery requires special care. Stains should be removed with warm water and a very mild soap solution. Use a clean, damp cloth to remove the soap, then wipe again with a dry cloth. Never use alcohol, gasoline, nail polish remover or thinner to clean leather upholstery.

3 After cleaning, regularly treat leather upholstery with a leather wax. Never use car wax on leather upholstery.

4 In areas where the interior of the vehicle is subject to bright sunlight, cover leather seats with a sheet if the vehicle is to be left out for any length of time.

5 Body repair - minor damage

See photo sequence

Repair of minor scratches

1 If the scratch is superficial and does not penetrate to the metal of the body, repair is very simple. Lightly rub the scratched area with a fine rubbing compound to remove loose paint and built up wax. Rinse the area with clean water.

2 Apply touch-up paint to the scratch, using a small brush. Continue to apply thin layers of paint until the surface of the paint in the scratch is level with the surrounding paint. Allow the new paint at least two weeks to harden, then blend it into the surrounding paint by rubbing with a very fine rubbing compound. Finally, apply a coat of wax to the scratch area.

3 If the scratch has penetrated the paint and exposed the metal of the body, causing the metal to rust, a different repair technique is required. Remove all loose rust from the bottom of the scratch with a pocket knife, then apply rust inhibiting paint to prevent the formation of rust in the future. Using a rubber or nylon applicator, coat the scratched area with glaze-type filler. If required, the filler can be mixed with thinner to provide a very thin paste, which is ideal for filling narrow scratches. Before the glaze filler in the scratch hardens, wrap a piece of smooth cotton cloth around the tip of a finger. Dip the cloth in thinner and then quickly wipe it along the surface of the scratch. This will ensure that the surface of the filler is slightly hollow. The scratch can now be painted over as described earlier in this Section.

Repair of dents

Warning: *Some models are equipped with airbags. Always disconnect the negative battery cable, then the positive battery cable and wait two minutes before working in the vicinity of the impact sensors, steering column or instrument panel to avoid the possibility of accidental deployment of the airbag, which could cause personal injury. See Chapter 12 for more information on the airbag system.*

4 When repairing dents, the first job is to pull the dent out until the affected area is as close as possible to its original shape. There is no point in trying to restore the original shape completely as the metal in the damaged area will have stretched on impact and cannot be restored to its original contours. It is better to bring the level of the dent up to a point about 1/8-inch below the level of the surrounding metal. In cases where the dent is very shallow, it is not worth trying to pull it back out at all.

5 If the back side of the dent is accessible, it can be hammered out gently from behind using a soft-face hammer. While doing this, hold a block of wood firmly against the opposite side of the metal to absorb the hammer blows and prevent the metal from being stretched.

6 If the dent is in a section of the body which has double layers, or some other factor makes it inaccessible from behind, a different technique is required. Drill several small holes through the metal inside the damaged area, particularly in the deeper sections. Screw long, self tapping screws into the holes just enough for them to get a good grip in the metal. Now the dent can be pulled out by pulling on the protruding heads of the screws with locking pliers.

7 The next stage of repair is the removal of the paint from the damaged area and from an inch or so of the surrounding metal. This is easily done with a wire brush or sanding disk in a drill motor, although it can be done just as effectively by hand with sandpaper. To complete the preparation for filling, score the surface of the bare metal with a screwdriver or the tang of a file or drill small holes in the affected area. This will provide a good grip for the filler material. To complete the repair, see the Section on filling and painting.

Repair of rust holes or gashes

8 Remove all paint from the affected area and from an inch or so of the surrounding metal using a sanding disk or wire brush mounted in a drill motor. If these are not available, a few sheets of sandpaper will do the job just as effectively.

9 With the paint removed, you will be able to determine the severity of the corrosion and decide whether to replace the whole panel, if possible, or repair the affected area. New body panels are not as expensive as most people think and it is often quicker to install a new panel than to repair large areas of rust.

10 Remove all trim pieces from the affected area except those which will act as a guide to the original shape of the damaged body, such as headlight shells, etc. Using metal snips or a hacksaw blade, remove all loose metal and any other metal that is badly affected by rust. Hammer the edges of the hole in to create a slight depression for the filler material.

11 Wire brush the affected area to remove the powdery rust from the surface of the metal. If the back of the rusted area is accessible, treat it with rust inhibiting paint.

12 Before filling is done, block the hole in some way. This can be done with sheet metal riveted or screwed into place, or by stuffing the hole with wire mesh.

13 Once the hole is blocked off, the affected area can be filled and painted. See the following subsection on filling and painting.

Filling and painting

14 Many types of body fillers are available, but generally speaking, body repair kits which contain filler paste and a tube of resin hardener are best for this type of repair work. A wide, flexible plastic or nylon applicator will be necessary for imparting a smooth and contoured finish to the surface of the filler material. Mix up a small amount of filler on a clean piece of wood or cardboard (use the hardener sparingly). Follow the manufacturer's instructions on the package, otherwise the filler will set incorrectly.

15 Using the applicator, apply the filler paste to the prepared area. Draw the applicator across the surface of the filler to achieve the desired contour and to level the filler surface. As soon as a contour that approximates the original one is achieved, stop working the paste. If you continue, the paste will begin to stick to the applicator. Continue to add thin layers of paste at 20-minute intervals until the level of the filler is just above the surrounding metal.

16 Once the filler has hardened, the excess can be removed with a body file. From then on, progressively finer grades of sandpaper should be used, starting with a 180-grit paper and finishing with 600-grit wet-or-dry paper. Always wrap the sandpaper around a flat rubber or wooden block, otherwise the surface of the filler will not be completely flat. During the sanding of the filler surface, the wet-or-dry paper should be periodically rinsed in water. This will ensure that a very smooth finish is produced in the final stage.

17 At this point, the repair area should be surrounded by a ring of bare metal, which in turn should be encircled by the finely feathered edge of good paint. Rinse the repair area with clean water until all of the dust produced by the sanding operation is gone.

18 Spray the entire area with a light coat of primer. This will reveal any imperfections in the surface of the filler. Repair the imperfections with fresh filler paste or glaze filler and once more smooth the surface with sandpaper. Repeat this spray-and-repair procedure until you are satisfied that the surface of the filler and the feathered edge of the paint are perfect. Rinse the area with clean water and allow it to dry completely.

19 The repair area is now ready for painting. Spray painting must be carried out in a warm, dry, windless and dust free atmosphere. These conditions can be created if you have access to a large indoor work area, but if you are forced to work in the open, you will have to pick the day very carefully. If you are working indoors, dousing the floor in the work area with water will help settle the dust which would otherwise be in the air. If the repair area is confined to one body panel, mask off the surrounding panels. This will help minimize the effects of a slight mismatch in paint color. Trim pieces such as chrome

strips, door handles, etc. will also need to be masked off or removed. Use masking tape and several thicknesses of newspaper for the masking operations.

20 Before spraying, shake the paint can thoroughly, then spray a test area until the spray painting technique is mastered. Cover the repair area with a thick coat of primer. The thickness should be built up using several thin layers of primer rather than one thick one. Using 600-grit wet-or-dry sandpaper, rub down the surface of the primer until it is very smooth. While doing this, the work area should be thoroughly rinsed with water and the wet-or-dry sandpaper periodically rinsed as well. Allow the primer to dry before spraying additional coats.

21 Spray on the top coat, again building up the thickness by using several layers of paint. Begin spraying in the center of the repair area and then, using a circular motion, work out until the whole repair area and about two inches of the surrounding original paint is covered. Remove all masking material 10 to 15 minutes after spraying on the final coat of paint. Allow the new paint at least two weeks to harden, then use a very fine rubbing compound to blend the edges of the new paint into the existing paint, Finally, apply a coat of wax.

6 Body repair - major damage

1 Major damage must be repaired by an auto body shop. These shops have the specialized equipment required to do the job properly.
2 If the damage is extensive, the frame must be checked for proper alignment or the vehicle's handling characteristics may be adversely affected and other components may wear at an accelerated rate.
3 Due to the fact that all of the major body components (hood, fenders, etc.) are separate and replaceable units, any seriously damaged components should be replaced rather than repaired. Sometimes the components can be found in a wrecking yard that specializes in used vehicle components, often at a considerable savings over the cost of new parts.

7 Hinges and locks - maintenance

Once every 3,000 miles, or every three months, the hinges and latch assemblies on the doors, hood and the tailgate should be given a few drops of light oil or lock lubricant. The door latch strikers should also be lubricated with a thin coat of grease to reduce wear and ensure free movement. Lubricate the door locks with spray-on graphite lubricant.

8 Hood - removal, installation and adjustment

Refer to illustrations 8.3 and 8.4
Warning: *Some models are equipped with airbags. Always disconnect the negative battery cable, then the positive battery cable and wait two minutes before working in the vicinity of the impact sensors, steering column or instrument panel to avoid the possibility of accidental deployment of the airbag, which could cause personal injury.*
Note: *The hood is heavy and somewhat awkward to remove and install - at least two people should perform this procedure.*

Removal and installation

1 Open the hood and support it in the open position with a long piece of wood.
2 Cover the fenders and cowl with blankets or heavy cloths to protect the paint.
3 Scribe or draw alignment marks around the hinge and bolt heads to ensure proper alignment on reinstallation **(see illustration)**.
4 Disconnect the underhood light connector on the driver's side and the ground strap on the passenger side of the hood **(see illustration)**.
5 Have an assistant hold onto the hood on one side while you hold the other side.
6 Remove the hood-to-hinge assembly bolts on your side of the hood, then hold your side of the hood while your assistant removes the hood-to-hinge bolts on the other side.
7 Lift the hood off.
8 Installation is the reverse of the removal

Steps with the following additions:
a) Align the hood and hinges using the alignment marks made in Step 3.
b) Be sure to tighten the bolts securely.

Adjustment

9 The hood can be adjusted to obtain a flush fit between the hood and fenders.
10 Loosen the hood retaining bolts.
11 Move the hood from side-to-side or front-to-rear until the hood is properly aligned with the fenders at the front. Tighten the bolts securely.
12 Loosen the bolts securing the hood latch assembly.
13 Move the latch until alignment is correct with the hood latch striker. Tighten the latch bolt securely.

9 Hood latch control cable - removal and installation

Refer to illustrations 9.2 and 9.3
Warning: *Some models are equipped with airbags. Always disconnect the negative battery cable, then the positive battery cable and wait two minutes before working in the vicinity of the impact sensors, steering column or instrument panel to avoid the possibility of accidental deployment of the airbag, which could cause personal injury.*

1 Open the hood and support it in the open position with a long piece of wood.
2 Scribe or draw alignment marks around the hood latch to ensure proper alignment on reinstallation **(see illustration)**.
3 Remove the bolts securing the hood latch and take the latch out **(see illustration)**.
4 Detach the cable bushing and lift the cable eye off the anchor post.
5 Remove the cable clips from the radiator support bracket and apron.
6 Remove the screws that secure the hood release handle to the instrument panel.
7 Remove the tie wrap that secures the cable to the steering column bracket.
8 Working in the passenger compartment, remove the cable grommet from the firewall, then pull the cable through.

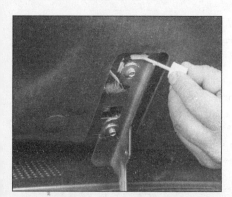

8.3 Scribe or draw alignment marks on the hood to ensure proper alignment on reinstallation

8.4 Detach the ground strap from the hood

9.2 Mark the position of the hood latch for proper installation

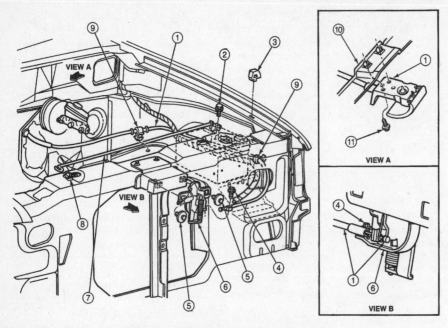

9.3 Hood release cable routing and details

1	Hood release cable	5	Bolt
2	Bolt	6	Hood latch
3	Hood bumper	7	Hood support rod retainer
4	Bolt		
8	Clip		
9	Hood cable clip		
10	Instrument panel		
11	Bolt		

9 Installation is the reverse of the removal steps with the following additions:
a) Insert the cable through the grommet in the firewall and make sure the grommet is properly seated within the firewall hole.
b) Before closing the hood, operate the control cable and make sure the latch control operates correctly.

10 Door - removal, installation and alignment

Removal and installation

1 With the door in the open position, place a jack under the door or have an assistant hold the door while the hinge bolts are removed. **Note:** *Place thick padding on top of the jack to protect the door's painted finish.*
2 On models so equipped, remove the upper and lower hinge access cover plates. Disconnect the electrical connectors to the door wiring harness.
3 Scribe or paint lines around the door hinges to ensure proper alignment on reinstallation.
4 Remove the hinge-to-door bolts and carefully lift off the door.
5 Installation is the reverse of the removal steps. Align the door.

Alignment

6 Check the alignment of the door all around the edge. The door should be evenly spaced in the opening and should be flush with the body.
7 Adjust if necessary as follows:
a) Note how the door is misaligned and determine which bolts need to be loosened to correct it.
b) Loosen the bolts just enough so the door can be moved with a padded prybar. Reposition the door as necessary, then tighten the bolts securely.
c) The door lock striker can also be adjusted both up-and-down and sideways to provide positive engagement with the lock mechanism (see Section 12).
8 Tighten the hinge-to-body bolts securely.

11 Door hinge - removal and installation

Note: *If both hinges are to be replaced, it is easier to leave the door in place and replace one hinge at a time.*

1 With the door in the open position, place a jack under the door or have an assistant hold the door while the door hinge bolts are removed. **Note:** *If a jack is used, place thick padding on top of the jack to protect the door's painted finish.*
2 On models so equipped, remove the upper and lower hinge access cover plates.
3 Scribe or draw lines around the door hinges to ensure proper alignment on reinstallation.
4 Remove the hinge-to-door bolts, then the hinge-to-body bolts. Carefully remove the hinge(s) or lift off the door.
5 Installation is the reverse of the removal Steps with the following additions:
a) Apply a sealant (Ford specification ESB-M2G150-A or equivalent) to the hinge before installing it onto the body panel.
b) Adjust the door if necessary (see Section 10).

12 Door latch striker - removal, installation and adjustment

Refer to illustration 12.1

Removal and installation

1 Use a Torx bit to unscrew the door latch striker stud from the door jamb **(see illustration)**.
2 Installation is the reverse of removal. Adjust if necessary.

Adjustment

3 The door latch striker can be adjusted vertically and laterally as well as fore-and-aft. **Note:** *Don't use the door latch striker to compensate for door misalignment.*
4 The door latch striker can be shimmed to obtain the correct clearance between the

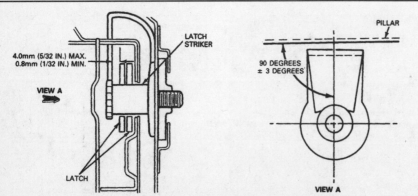

12.1 The door striker can be adjusted by repositioning it and by adding or removing a shim - don't use more than one shim per door

Chapter 11 Body

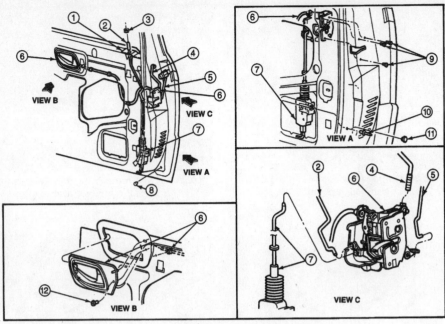

13.2 Front door latch and inside handle

1	Inner panel sound insulator	5	Lock cylinder rod
2	Lock button	6	Door latch
3	Bushing	7	Latch actuator
4	Latch actuating rod	8	Bumper
		9	Screw
		10	Rivet
		11	Plug
		12	Rivet

14.2 Details of the outer door handle assembly

1 Lock cylinder clip
2 Latch rod knob
3 Door handle
4 Rivet
5 Lock cylinder
6 Lock rod
7 Center body pillar
8 Door striker
9 Screw

latch and the striker. Don't use more than one shim per door.
5 To check the clearance between the latch jamb and the striker area, spread a layer of dark grease on the striker.
6 Open and close the door several times and note the pattern in the grease.
7 Move the door striker assembly laterally to provide a flush fit at the door and pillar or at the quarter panel.
8 Securely tighten the door latch striker after adjustment is complete.

13 Inner door handle and latch assembly - removal and installation

Refer to illustration 13.2

1 Remove the door trim panel and watershield (see Section 16).
2 Remove the screws securing the inner door handle assembly to the door **(see illustration)**.
3 Disconnect the handle link from the handle and lock cylinder.
4 Detach the rods from the latch, then remove the latch screws and take it out.
5 Installation is the reverse of the removal steps, plus the following additions:
   a) If a new latch is being installed, install the rod retaining clips in it.
   b) Attach the control rod and lock cylinder rod to the latch before installing the latch.
   c) Tighten the latch screws securely.
   d) Check operation of the handle and lock before installing the watershield and door panel.

14 Outer door handle - removal and installation

Refer to illustration 14.2

1 Remove the door trim panel and pull away the watershield (see Section 16).
2 Disconnect the door latch actuator rod from the latch **(see illustration)**.
3 Support the door handle in the open position.
4 Drill out both rivets that secure the handle to the door and remove the handle.
5 Installation is the reverse of the removal steps. Position the handle in the door and install the two pop-rivets.

15 Door lock cylinder - removal and installation

Refer to illustration 15.3

1 Raise the window all the way.
2 Remove the door trim panel and watershield (see Section 16).
3 Disconnect the door latch control-to-cylinder rod **(see illustration)**.
4 Use a pair of pliers and slide the cylinder retainer clip away from the lock cylinder and door.
5 Remove the lock cylinder from the door.
6 Install the lock cylinder into the door opening from the outside and push the retainer clip into place. Make sure it's seated correctly.
7 Reconnect the lock cylinder rod-to-door latch control.
8 Check for proper lock operation, then install the watershield and the door trim panel.

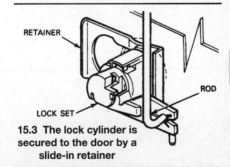

15.3 The lock cylinder is secured to the door by a slide-in retainer

16.1 Pull back the window crank trim and remove the screw (manual windows)

16.3a Remove the screw and the door trim finish panel . . .

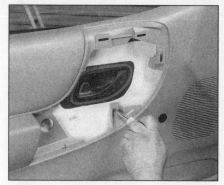

16.3b . . . and remove the panel screws from the armrest area

16.5 Remove one screw (if equipped) that secures the lower rear corner of the panel

16 Door trim panel and watershield - removal and installation

Refer to illustrations 16.1, 16.3a, 16.3b, 16.5 and 16.6

1 If the truck is equipped with manual windows, remove the window regulator handle **(see illustration)**.
2 If the truck is equipped with power windows or door locks, remove the switches (see Chapter 12).
3 Remove the screws from the armrest area of the door panel **(see illustrations)**.
4 Remove the inner door handle (see Section 13).
5 Remove the screw (if equipped) at the lower rear of the door panel **(see illustration)**.
6 Lift the panel at least 1-1/2 inches to disengage it from the slots along the bottom of the door, then remove it **(see illustration)**.
7 If necessary, carefully peel the watershield from the door. Be careful not to tear it.
8 Installation is the reverse of the removal Steps.

17 Front door window glass - replacement and adjustment

Refer to illustrations 17.4 and 17.5

Replacement

1 Remove the door trim panel and water-

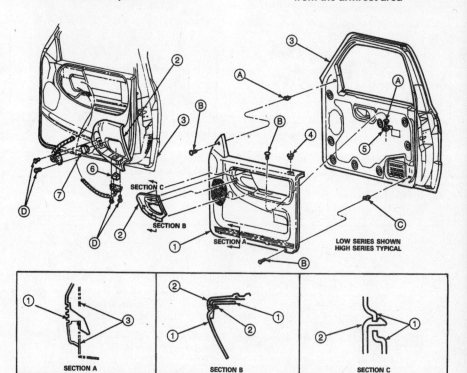

16.6 Door trim panel details (typical)

1 Door trim panel
2 Door trim finish panel
3 Door
4 Door lock rod grommet
5 Armrest support
6 Power door lock switch
7 Power window switch

shield (see Section 16).
2 Remove the door speakers (if equipped) (see Chapter 12).
3 Raise the window all the way.
4 Detach the rear glass run retainer from the door and take it out **(see illustration)**.
5 Lower the window enough to provide access to the glass bracket and retention rivets **(see illustration)**. Tap the center pin out of each rivet with a punch, then drill out the rivets with a 1/4-inch drill.
6 Lift the glass out through the top of the door.
7 Installation is the reverse of the removal Steps, with the following additions:

a) *Lubricate the regulator rollers, shafts and tracks with multi-purpose grease.*
b) *Raise and lower the window to check its operation before installing the speakers (if equipped), watershield and door trim panel.*

Adjustment

8 Lower the glass two-to-three inches from the full-up position.
9 Loosen the three guide assembly nut and washer assemblies **(see illustration 17.4)**.
10 Push the glass toward the rear until it bottoms out within the door frame.

Chapter 11 Body   11-9

17.5 Drill out the rivets to detach the glass

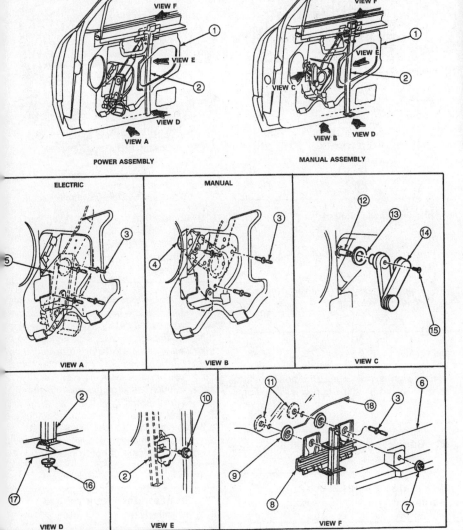

17.4 Front door window and mechanism

1 Door
2 Glass run
3 Rivet
4 Manual window regulator
5 Power window regulator drive
6 Door beltline
7 Nut and washer
8 Glass run retainer
9 Spacer
10 Screw
11 Door glass bracket retainer
12 Manual window regulator
13 Washer
14 Manual window crank
15 Screw
16 Nut and washer
17 Door
18 Window glass

3 Remove the window glass (see Section 17).
4 If you're working on a vehicle with power windows, disconnect the window motor electrical connector.
5 Carefully drill-out the rivets that secure the regulator base plate to the inner door panel **(see illustration 17.4)**.
6 Remove the nuts that secure the glass guide channel to the door.
7 Remove the regulator from the door, together with the power window motor (if equipped). If necessary, detach the window motor from the regulator. **Warning:** *Before removing the motor from the regulator, secure the regulator arm to the regulator body to prevent the arm from unloading violently when the motor is detached.*
8 Attach the regulator to the inner door panel. This can be done with rivets, or with suitable bolts, nuts and washers. If you use nuts and bolts, tighten them securely.
9 The remainder of installation is the reverse of the removal Steps.

19 Front bumper and valance - removal and installation

Refer to illustration 19.1

Warning: *If vehicle is equipped with airbags, refer to Chapter 12 to disarm the airbag system prior to performing any work described below.*

Note: *The bumper is heavy and somewhat awkward to remove and install - at least two people should perform this procedure.*

1 If you're working on a vehicle with fog lights, unbolt their brackets and remove them from the bumper **(see illustration)**.
2 Support the bumper with a jack and block of wood. Remove the mounting nuts on each side and detach the bumper from the vehicle.
3 Remove the screws and pushpins and detach the valance panel from the bumper.
4 Installation is the reverse of removal. Tighten the fasteners to the torques listed in this Chapter's Specifications.

11 Move the window guide post toward the rear within the retention slot in the door inner panel. Tighten the three guide assembly nuts securely.
12 Raise and lower the window several times and check for proper fit.
13 Install the watershield and door trim panel.

18 Front door window regulator - replacement

1 If you're working on a vehicle with power windows, disconnect the negative cable from the battery.
2 Remove the door trim panel and watershield (see Section 16).

Chapter 11 Body

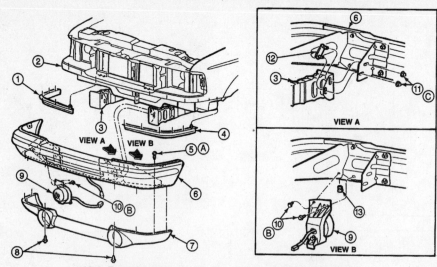

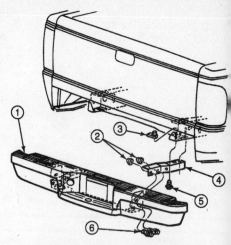

19.1 Front bumper details (Ford shown, Mazda similar)

1. Right stone deflector
2. Grille opening panel
3. Frame rail
4. Left stone deflector
5. Screw and washer
6. Front bumper
7. Valance panel
8. Push pin
9. Fog light bracket
10. Bolt
11. Nut and washer
12. Bolt and retainer
13. U-nut

20 Rear bumper - removal and installation

Refer to illustration 20.2
Note: *The bumper is heavy and somewhat awkward to remove and install - at least two people should perform this procedure.*
1. Support the rear bumper with a jack.
2. Unbolt the bumper brackets from the vehicle and lower it clear (see illustration).
3. Detach the brackets from the vehicle as necessary.
4. Installation is the reverse of the removal steps. Tighten the nuts and bolts to the torques listed in this Chapter's Specifications.

21 Radiator grille - removal and installation

Refer to illustrations 21.2a and 21.2b
Warning: *Some models are equipped with airbags. Always disconnect the negative battery cable, then the positive battery cable and wait two minutes before working in the vicinity of the impact sensors, steering column or instrument panel to avoid the possibility of accidental deployment of the airbag, which could cause personal injury (see Chapter 12 for more information on the airbag system).*
1. Open the hood.
2. Remove the screws and detach the grille from the vehicle (see illustrations).
3. Installation is the reverse of the removal procedure. Be careful not to overtighten the screws.

22 Front fender - removal and installation

Refer to illustration 22.3
Warning: *Some models are equipped with airbags. Always disconnect the negative bat-*

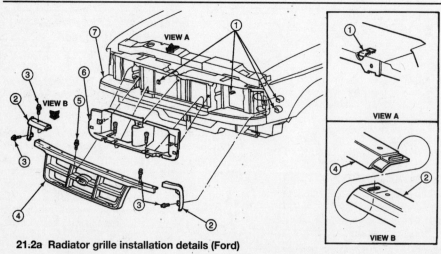

21.2a Radiator grille installation details (Ford)

1. Spring nut
2. Molding
3. Screw
4. Radiator grille
5. Screw
6. Radiator opening cover
7. Grille opening panel

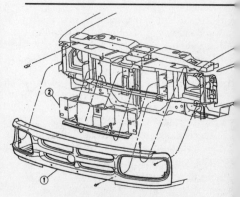

20.2 Rear bumper details (typical)

1. Bumper
2. Nut
3. U-nut
4. Isolator and bracket
5. Bolt
6. Bolts

21.2b Radiator grille installation details (Mazda)

1. Radiator grille
2. Radiator opening cover

Chapter 11 Body

tery cable, then the positive battery cable and wait two minutes before working in the vicinity of the impact sensors, steering column or instrument panel to avoid the possibility of accidental deployment of the airbag, which could cause personal injury (see Chapter 12 for more information on the airbag system).

1 Before you start, it's a good idea to check the fender fasteners for dirt, especially in the wheel well. The job will be easier if you wash off the dirt first with a high-pressure hose.
2 Remove the grille (see Section 21).
3 Unbolt the lower rear corner of the fender from the rocker panel **(see illustration)**.
4 Open the door and remove the single cowl-to-fender bolt.
5 Remove the screws and detach the inner fender panel.
6 Open the hood and remove the four bolts across the top of the fender.
7 Unbolt the fender brace from the radiator support. Carefully detach the fender from the vehicle.
8 Installation is the reverse of the removal procedure. Don't tighten any of the bolts until they are all installed.

23 Windshield and fixed glass - removal and installation

Replacement of the windshield and fixed glass requires the use of special fast-setting adhesive/caulk materials and some specialized tools and techniques. These operations should be left to a dealer service department or a shop specializing in automotive glass work.

24 Console - removal and installation

Refer to illustration 24.1
Warning: *If vehicle is equipped with airbags, refer to Chapter 12 to disarm the airbag system prior to performing any work described below.*
1 Carefully pry out the trim covers for access to the console rear mounting bolts **(see illustration)**, then remove the bolts.
2 Lift off the armrest and remove the single screw beneath it.
3 Remove the cup holder screws, lift off the cup holder and remove the two screws beneath it.
4 Lift the console away from the mounting brackets and remove it from the vehicle.
5 Installation is the reverse of the removal steps.

25 Steering column cover and shroud - removal and installation

Warning: *Some models are equipped with airbags. Always disconnect the negative bat-*

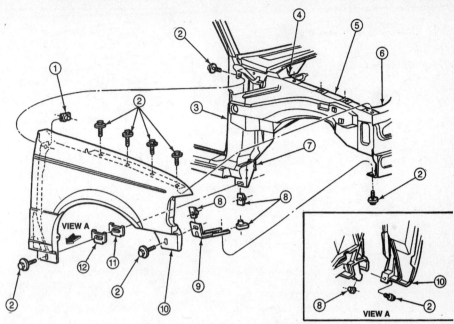

22.3 Front fender mounting details

1	U-nut	7	Rocker panel
2	Bolt and washer	8	U-nut
3	Body pillar	9	Support brace
4	Cowl top panel side reinforcement	10	Fender
5	Inner fender	11	Shim
6	Radiator support	12	Shim

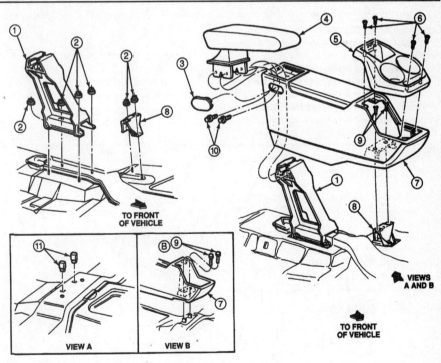

24.1 Console details

1	Rear mounting bracket	7	Console
2	Nut	8	Front mounting bracket
3	Armrest mount access cover	9	Screw
4	Armrest	10	Bolt
5	Cup holder	11	Nut
6	Screw		

Chapter 11 Body

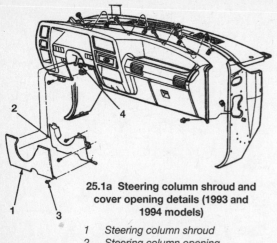

25.1a Steering column shroud and cover opening details (1993 and 1994 models)

1. Steering column shroud
2. Steering column opening reinforcement
3. Screw
4. U-nut

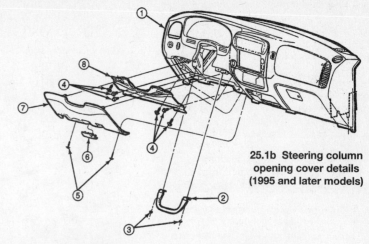

25.1b Steering column opening cover details (1995 and later models)

1. Instrument panel
2. Column opening center reinforcement
3. Nuts
4. Screw
5. Screw
6. Access cover
7. Steering column cover
8. Instrument panel reinforcement

tery cable, then the positive battery cable and wait two minutes before working in the vicinity of the impact sensors, steering column or instrument panel to avoid the possibility of accidental deployment of the airbag, which could cause personal injury (see Chapter 12 for more information on the airbag system). Refer to illustrations 25.1a, 25.1b, 25.3a, 25.3b and 25.3c

1 To remove the steering column cover and opening reinforcement, remove their screws and take them off the instrument panel **(see illustrations)**.
2 Remove the steering wheel (see Chapter 10).
3 If the vehicle is equipped with tilt steering, unscrew the tilt lever. Squeeze the top and bottom of the tilt lever collar and take it off the shrouds **(see illustration)**. Remove the screws from the underside of the lower column shroud and lower the shroud away from the column **(see illustrations)**. If you're working on a vehicle with automatic transmission, you may need to move the shift lever.
4 Lift the upper shroud half off the column.
5 Installation is the reverse of the removal steps.

26 Instrument cluster bezel - removal and installation

1993 and 1994 models

1 Pull rearward around the outer edge of the instrument cluster bezel to unsnap it from the cluster.
2 Pull the bezel out to clear the cluster.
3 Installation is the reverse of the removal steps.

1995 and later models

Refer to illustration 26.5

4 Remove the steering column cover and reinforcement (see Section 25).

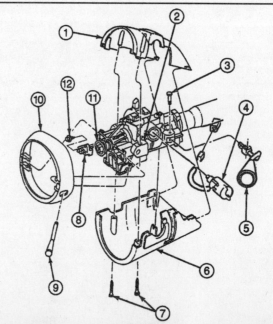

25.3a Steering column shroud details (1995 and later models shown, earlier models similar)

1. Upper shroud
2. Lock cylinder housing
3. Shift lever pin
4. Lock cylinder
5. Shift lever
6. Lower shroud
7. Screws
8. Horn brush
9. Tilt handle
10. Collar (1995 and later models only)
11. Flange
12. Horn brush screw

25.3b Remove the screws from the underside of the shroud . . .

25.3c . . . and lower it away from the steering column

Chapter 11 Body

26.5 Remove the cluster bezel screws and withdraw the bezel

5 Remove the cluster bezel screws and take the bezel out **(see illustration)**.
6 Installation is the reverse of the removal steps.

27 Instrument panel - removal and installation

Refer to illustration 27.8
Warning: *Some models are equipped with airbags. Always disconnect the negative battery cable, then the positive battery cable and wait two minutes before working in the vicinity of the impact sensors, steering column or instrument panel to avoid the possibility of accidental deployment of the airbag, which could cause personal injury (see Chapter 12 for more information on the airbag system).*

1 Disconnect the negative cable from the battery.
2 Open the hood. Locate the instrument cluster electrical connectors in the engine compartment and disconnect them.
3 Unbolt the hood release handle and parking brake control from the instrument panel.
4 If the vehicle is equipped with a console, remove it (see Section 24).
5 If you're working on a 4WD model, push the transfer case control all the way forward.
6 Remove the steering column cover and reinforcement from the instrument panel (see Section 25).
7 Unbolt the bracket at the right of the accelerator pedal.
8 Remove two nuts that secure the instrument panel yoke to the frame crossbeam **(see illustration)**.
9 Working inside the fuse box opening, remove two bolts that secure the instrument panel to the cowl.
10 Remove the single mounting bolt at the lower right of the instrument panel.
11 Remove the rivet and pushpins to detach the sound deadener from the lower right of the instrument panel.
12 Carefully pry up the edge of the defroster grille with a wooden tool and detach its clips, then lift it out.

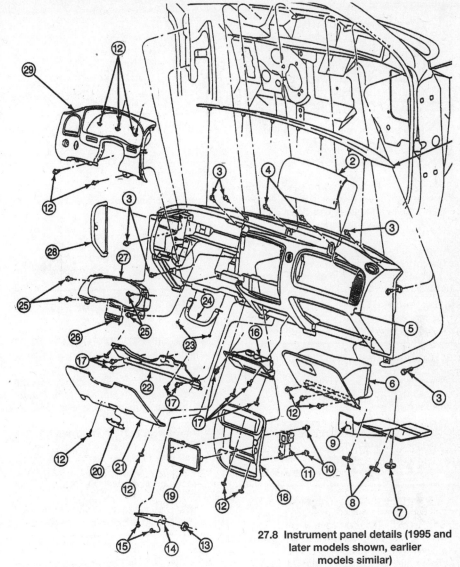

27.8 Instrument panel details (1995 and later models shown, earlier models similar)

1	Defroster grille	16	Ashtray
2	Airbag cover (if equipped)	17	Screw
3	Screw	18	Center finish panel
4	Screw	19	Radio opening panel
5	Instrument panel	20	Access cover
6	Glove box	21	Steering column cover
7	Push pin	22	Steering column cover reinforcement
8	Rivet	24	Center reinforcement
9	Sound deadener	25	Screw
10	Screw	26	Automatic transmission shift indicator
11	Center finish panel bracket	27	Instrument cluster
12	Screw	28	Fuse panel door
13	Nut	29	Cluster bezel
14	Brace		
15	Screw		

13 Remove the four mounting bolts at the top of the panel. Mark and unplug the remaining electrical connectors.
14 Check carefully to make sure the panel is completely detached from the vehicle, then lift it out.
15 Installation is the reverse of the removal steps.

28 Glove box - removal and installation

Refer to illustration 28.1
1 Open the glove box. Squeeze the sides of the glove box together to disengage the tabs and let the glove box hang down.

Chapter 11 Body

28.1 Swing the glove box down and remove the screws

Remove its mounting screws and take it out of the instrument panel **(see illustration)**.
2 Installation is the reverse of the removal steps.

29 Seats - removal and installation

Bucket and bench seats

Refer to illustrations 29.2a and 29.2b
1 Remove the seat track shield (if equipped).
2 Remove the four bolts securing the seat track to the floorpan and lift the seat from the vehicle **(see illustrations)**.
3 Installation is the reverse of the removal steps. Tighten the retaining bolts securely.

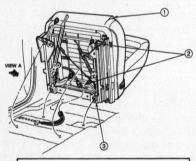

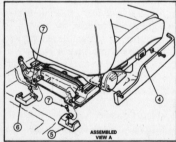

29.5a Tilt-slide seat details (1993 and 1994 Supercab models)

1	Seat	5	Seat track trim
2	Screws		
3	Seat track	6	Seat track trim
4	Seat cushion side shield		

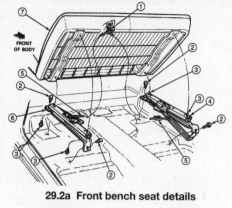

29.2a Front bench seat details

1	Seat track release cable hook	4	Seat track
		5	Seat latch tie-rod
2	Bolt	6	Seat track
3	Bolt	7	Seat

Tilt-slide seats (Supercab models)

Refer to illustrations 29.5a and 29.5b
4 Tilt the seat back and slide it all the way forward.
5 Pull the locktabs or pushpins on the seat track insulators clear of the tracks and unbolt the rear ends of the seat tracks from the floorpan **(see illustrations)**.
6 Tilt the seat upright and slide it to the rear.
7 Disconnect the seat electrical connector (if equipped). Unbolt the front ends of the seat tracks and remove the seat from the vehicle.
8 Installation is the reverse of the removal steps. Tighten the retaining bolts securely.

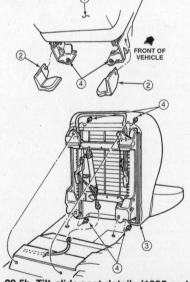

29.5b Tilt-slide seat details (1995 and later Supercab models)

1	Seat	3	Seat frame
2	Seat track insulators	4	Screw

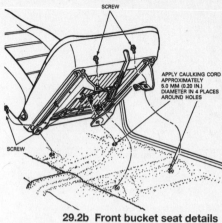

29.2b Front bucket seat details

30 Seat belt check

1 Check the seat belts, buckles, latch plates and guide loops for obvious damage and signs of wear.
2 Check that the seat belt reminder light comes on when the ignition key is turned to the Run or Start position.
3 The seat belts are designed to lock up during a sudden stop or impact, yet allow free movement during normal driving. Check that the retractors return the belt against your chest while driving and rewind the belt fully when the buckle is unlatched.
4 If any of the above checks reveal problems with the seat belt system, replace parts as necessary.

31 Mirrors (exterior) - removal and installation

Refer to illustration 31.4
1 Remove the door trim panel (see Section 16).
2 Disconnect the electrical connector for the power mirror (if equipped).
3 Where necessary, lower the window for access to the mirror mounting nuts.
4 Remove the mounting nuts and take the mirror out **(see illustration)**.
5 Installation is the reverse of the removal steps.

31.4 Remove the screws (arrows) to detach the mirror

Chapter 12
Chassis electrical system

Contents

Section		Section
Airbag - general information	28	Instrument cluster - removal and installation	21
Bulb replacement	17	Power door lock system - description and check	26
Circuit breakers - general information	6	Power mirrors - check, removal and installation	25
Connectors - general information	3	Power window system - description and check	24
Cruise control system - description and check	27	Radio - removal and installation	18
Electrical troubleshooting - general information	2	Radio antenna - removal and installation	19
Fuses - general information	4	Speakers - removal and installation	20
Fusible links - general information	5	Speedometer cable - replacement	23
Gauges - replacement	22	Steering column switches - replacement	7
General information	1	Turn signal/hazard flasher relay - replacement	8
Headlight bulb housing - replacement	15	Windshield washer reservoir and pump assembly - removal and installation	13
Headlight switch - replacement	14	Windshield wiper arm - removal and installation	12
Headlights - adjusting	16	Windshield wiper motor - removal and installation	11
Horn - check and replacement	10	Wiring diagrams - general information	29
Ignition switch/key lock cylinder - check and replacement	9		

1 General information

Warning: *To prevent electrical shorts, fires and injury, always disconnect the cable from the negative terminal of the battery before checking, repairing or replacing electrical components.*
Note: *Whenever the battery is disconnected, the powertrain control module must reset itself. This may cause driveability problems until the vehicle has been driven 10 miles or more.*

The chassis electrical system of this vehicle is a 12-volt, negative ground type. Power for the lights and all electrical accessories is supplied by a lead/acid-type battery which is charged by the alternator.

This chapter covers repair and service procedures for various chassis (non-engine related) electrical components. For information regarding the engine electrical system components (battery, alternator, distributor and starter motor), see Chapter 5.

2 Electrical troubleshooting - general information

A typical electrical circuit consists of an electrical component, any switches, relays, motors, fuses, fusible links or circuit breakers, etc. related to that component and the wiring and connectors that link the components to both the battery and the chassis. To help you pinpoint an electrical circuit problem, wiring diagrams are included at the end of this book.

Before tackling any troublesome electrical circuit, first study the appropriate wiring diagrams to get a complete understanding of what makes up that individual circuit. Trouble spots, for instance, can often be isolated by noting if other components related to that circuit are often routed through the same fuse and ground connections.

Electrical problems usually stem from simple causes such as loose or corroded connectors, a blown fuse, a melted fusible link or a bad relay. Visually inspect the condition of all fuses, wires and connectors in a problem circuit before troubleshooting it.

The basic tools needed for electrical troubleshooting include a circuit tester, a high impedance (10 K-ohm) digital voltmeter, a continuity tester, and a jumper wire with an inline circuit breaker for bypassing electrical components. Before attempting to locate or define a problem with electrical test instruments, use the wiring diagrams to decide where to make the necessary connections.

Voltage checks

Perform a voltage check first when a circuit is not functioning properly. Connect one lead of a circuit tester to either the negative battery terminal or a known good ground. Connect the other lead to a connector in the circuit being tested, preferably nearest to the battery or fuse. If the bulb of the tester lights up, voltage is present, which means that the part of the circuit between the connector and the battery is problem free. Continue checking the rest of the circuit in the same fashion.

When you reach a point at which no voltage is present, the problem lies between that point and the last test point with voltage. Most of the time the problem can be traced to a loose connection. **Note:** *Keep in mind that some circuits receive voltage only when the ignition key is in the Accessory or Run position.*

Finding a short circuit

One method of finding shorts in a circuit is to remove the fuse and connect a test light or voltmeter in its place. There should be no voltage present in the circuit. Move the electrical connectors from side-to-side while watching the test light. If the bulb goes on, there is a short to ground somewhere in that area, probably where the insulation has been rubbed through. The same test can be performed on each component in a circuit, even a switch.

Ground check

Perform a ground test to check whether a component is properly grounded. Disconnect the battery and connect one lead of a self-powered test light, known as a continuity tester, to a known good ground. Connect the other lead to the wire or ground connection being tested. If the bulb goes on, the ground is good.

If the bulb does not go on, the ground is not good.

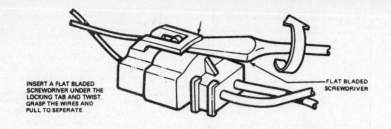
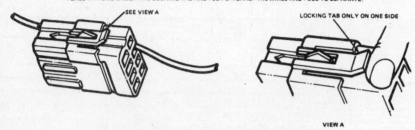
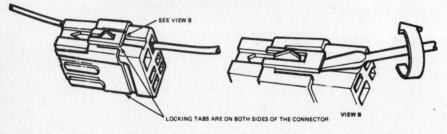
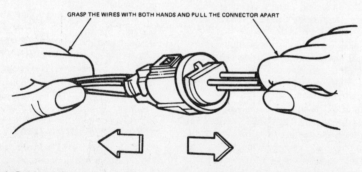

3.1 Some of the various types of inline connectors used on these models

3 Connectors - general information

Refer to illustration 3.1

Always release the lock lever(s) before attempting to unplug inline type connectors. There are a variety of lock lever configurations **(see illustration)**. Although nothing more than a finger is usually necessary to pry lock levers open, a small pocket screwdriver is effective for hard-to-release levers. Once the lock levers are released, try to pull on the connectors themselves when unplugging two connector halves (there are times, however, when this is not possible - use good judgment).

It is usually necessary to know which side, male or female, of the connector you're checking. Male connectors are easily distinguished from females by the shape of their internal pins.

When checking continuity or voltage with a circuit tester, insertion of the test probe into the receptacle may open the fitting to the connector and result in poor contact. Instead, insert the test probe from the wire harness side of the connector (known as "backprobing").

4 Fuses - general information

Refer to illustrations 4.1a through 4.1h, 4.5 and 4.7

The electrical circuits are protected by a combination of fuses, fusible links and circuit breakers. There are two fuse panels; one on the power distribution box in the engine compartment and one above the driver's footwell **(see illustrations)**.

The fuse panels are equipped with miniaturized fuses because their compact dimensions and convenient blade-type terminal design allow fingertip removal and installation.

Each fuse protects one or more circuits. A fuse guide is included here but consult your owner's manual - it will have the most accurate guide for your vehicle.

The cover of the footwell fuse panel contains spare fuses. A fuse puller tool is mounted in the cover on 1993 and 1994 models. On 1995 and later models, the fuse puller is mounted in the fuse box.

If an electrical component fails, always check the fuse first. A blown fuse, which is nothing more than a broken element, is easily identified through the clear plastic body or by checking it with a test light **(see illustration)**.

Use the fuse puller tool to remove and insert fuses. Don't twist the fuses during removal or installation. Twisting could force the terminal open too far, resulting in a bad connection.

Be sure to replace blown fuses with the correct type and amp rating. Fuses of different ratings are physically interchangeable, but replacing a fuse with one of a higher or lower value than specified is not recommended. Each electrical circuit needs a spe-

Continuity check

A continuity check determines if there are any breaks in a circuit - if it is conducting electricity properly. With the circuit off (no power in the circuit), a self-powered continuity tester can be used to check the circuit. Connect the test leads to both ends of the circuit, and if the test light comes on the circuit is passing current properly. If the light doesn't come on, there is a break somewhere in the circuit. The same procedure can be used to test a switch, by connecting the continuity tester to the power in and power out sides of the switch. With the switch turned on, the test light should come on.

Finding an open circuit

When diagnosing for possible open circuits it is often difficult to locate them by sight because oxidation or terminal misalignment are hidden by the connectors. Merely wiggling a connector on a sensor or in the electrical connector may correct the open circuit condition. Remember this if an open circuit is indicated when troubleshooting a circuit. Intermittent problems may also be caused by oxidized or loose connections.

Electrical troubleshooting is simple if you keep in mind that all electrical circuits are basically electricity running from the battery, through the wires, switches, relays, fuses and fusible links to each electrical component (light bulb, motor, etc.) and then to ground, from which it is passed back to the battery. Any electrical problem is an interruption in the flow of electricity to and from the battery.

Chapter 12 Chassis electrical system

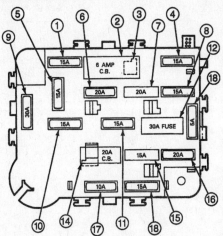

Fuse Position	Amps	Circuits Protected
1	15 Amp. Fuse	Hazard Flasher, Stoplamps, Indicator Lamps, Speed Control, Brake Shift Interlock
2	6.0 Amp C.B.	Windshield Wiper/Washer
3		Not Used
4	15 Amp Fuse	Side Marker Lamps, Headlamp On Warning Chime, Underhood Lamp, Daytime Running Lamps, Parking Lamps, License Lamps, Instrument Panel Illumination
5	15 Amp Fuse	Turn Signal Flasher, Day/Night Mirror, Airbag Module, Back-up Lamps, DRL Module, Rear Defroster Switch, HEGO Sensors, Turn Signal Indicator, Brake Shift Interlock
6	20 Amp Fuse	Speed Control Amplifier, Warning Chime Module, A/C Clutch, Door Ajar Relay, Autolamp Relay
7	20 Amp. Fuse	Rear Anti-Lock Brake System (RABS) Module
8	15 Amp. Fuse	Illuminated Entry, RABS "Keep Alive", Stepwell Lamp, Cargo Lamp, Power Mirrors, Courtesy Lamp, PSOM, Clock, Glove Box Lamp, Visor Vanity Lamps, I/P Courtesy Lamps, EIC (Electronic Instrument Cluster), Dome Lamp, Reading Lamps
9	30 Amp. Fuse	Aux. A/C, Front A/C Blower Motor
10	15 Amp. Fuse	Airbag Module
11	15 Amp. Fuse	Graphic Equalizer, Radio, Amplifier
12	30 Amp. Fuse	Power Lumbar, Flash to Pass. High Beam Indicator, Horns
13	5 Amp. Fuse	Instrument and Control Illumination
14	20 Amp. C.B.	Power Window Motors
15	15 Amp. Fuse	A.W.D. Module and Indicator
16	20 Amp. Fuse	Front Cigar Lighter, Rear Cigar Lighter
17	10 Amp. Fuse	EIC, PSOM
18	15 Amp. Fuse	Warning Chime, Indicator Lamps, Rear Wiper Motor, Rear Washer Pump
Fuse Block Bracket	4.5 Amp. C.B.	Rear Wiper Motor, Rear Washer Pump

4.1a Passenger compartment fuses - 1993 models

cific amount of protection. The amperage value of each fuse is usually molded into the fuse body. Different colors are also used to denote fuses of different amperage types. The accompanying color code (see illustration) shows common amperage values and their corresponding colors. **Caution:** *Always turn off all electrical components and the ignition switch before replacing a fuse. Never bypass a fuse with pieces of metal or foil. Serious damage to the electrical system, or even a fire, could result.*

If the replacement fuse immediately fails, do not replace it again until the cause of the problem is isolated and corrected. In most cases, this will be a short circuit in the wiring caused by a broken or deteriorated wire.

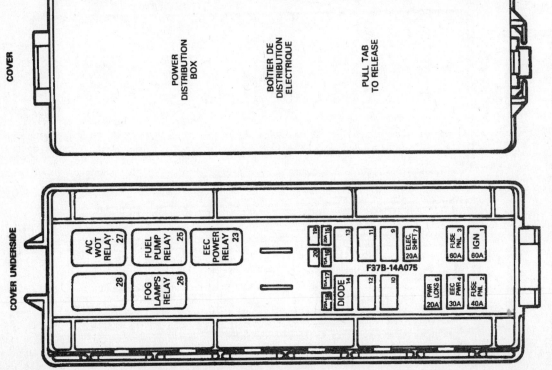

4.1b Engine compartment fuses - 1993 models

1. Ignition - 60 amps
2. Fuse panel - 40 amps
3. Fuse panel - 60 amps
4. PCM power - 30 amps
5. Not used
6. Power locks - 20 amps
7. Transfer case shift - 20 amps
8. Not used
9. Not used
10. Not used
11. Not used
12. Not used
13. Not used
14. EEC system - Diode
15. Fog lights - 20 amps
16. Alternator - 15 amps
17. Hood, daytime running lights
18. Fuel pump - 20 amps
19. Not used
20. Not used
21. Not used
22. Not used

Chapter 12 Chassis electrical system

Fuse Position	Rating	Circuit Protected
1	15 Amp Fuse	Four-Way Flash
2	20 Amp Fuse	Horns
3	20 Amp C.B.	Cigar Lighter, Flash to Pass/Power Lumbar
4	10 Amp Fuse	Instrument Panel Illumination
5	20 Amp Fuse	Premium Radio Amplifier
6	30 Amp C.B.	Power Windows
7	20 Amp Fuse	RABS Module
8	15 Amp Fuse	HO2S Heater
9	10 Amp Fuse	A/C Switches, Clutch Coil
10	15 Amp Fuse	Radio
11	15 Amp Fuse	Park/License Lamps/Stoplamps/RABS "Keep Alive"
12	30 Amp Fuse	Blower Motor
13	15 Amp Fuse	Turn Lamps, B/U Lamps, Turn Indicator, Daytime Running Lamps
14	15 Amp Fuse	Dome/Courtesy Lamp, Radio Memory, Power Mirror

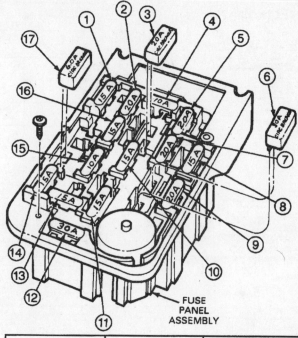

Fuse Position	Rating	Circuit Protected
15	10 Amp Fuse	Speed Control
16	15 Amp Fuse	Warning Gauges/Lamps, Electric Shift 4x4, Chime Module
17	6 Amp C.B.	Front Wash/Wipe

4.1c Passenger compartment fuses - 1994 models

4.1d Engine compartment fuses - 1994 models

1 Ignition - 60 amps
2 Fuse panel/headlights - 40 amps
3 Fuse panel - 60 amps
4 PCM power - 30 amps
5 Blower motor - 50 maps
6 Power door locks - 30 amps
7 Transfer case shift - 20 amps
8 Power seat - 60 amps
9 Anti-lock Brake System - 30 amps
10 Anti-lock Brake system - 30 amps
11 Heated rear window - 40 amps
12 Not used
13 ABS diode
14 EEC diode
15 Fog lights - 20 amps
16 Alternator - 15 amps
17 Underhood light/daytime running lights - 15
18 Fuel pump - 20 amps
19 Trailer tow - 20 amps
20 Not used
21 Not used
22 Not used

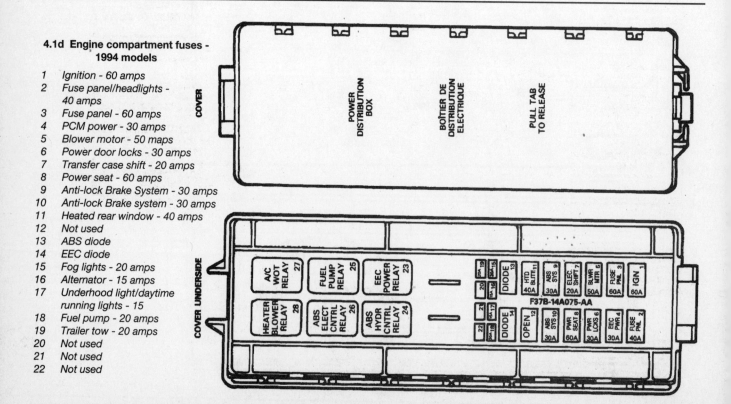

Chapter 12 Chassis electrical system

Fuse Panel, Ranger

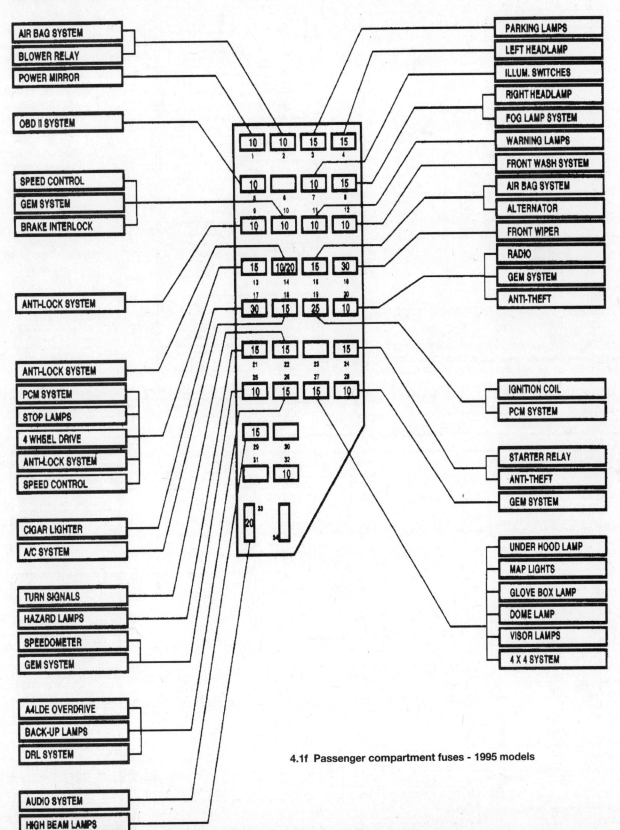

4.1f Passenger compartment fuses - 1995 models

Chapter 12 Chassis electrical system

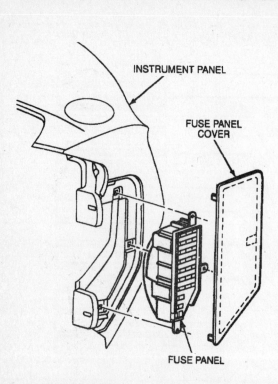

4.1e The passenger compartment fuses on 1995 and later models are located on the left (driver's) end of the instrument panel - remove the cover to expose the fuses

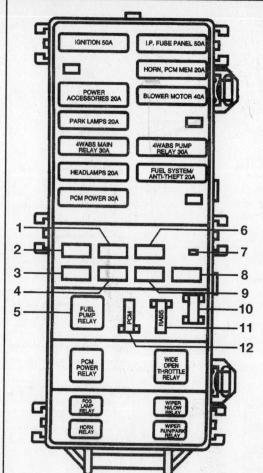

4.1g Engine compartment fuses - 1995 models

1. 4WD system - 20 amps
2. Daytime running lights (if equipped), fog lights - 15 amps
3. Airbag - 10 amps
4. Powertrain control module oxygen sensor system - 15 amps
5. Fuel pump relay
6. Alternator - 15 amps
7. Not used
8. JBL system - 30 amps
9. Power point - 30 amps
10. RABS/low brake fluid resistor
11. RABS diode
12. Powertrain control module diode

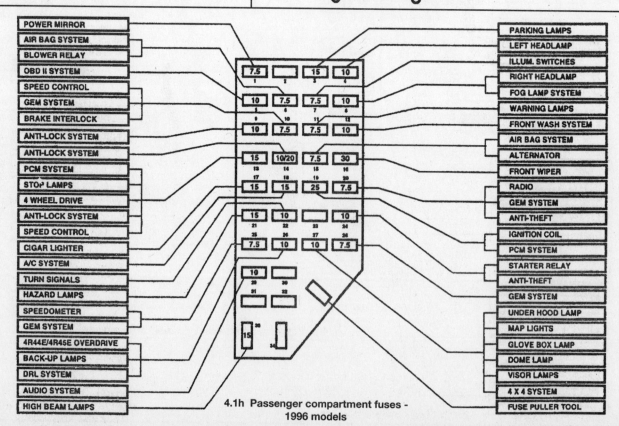

4.1h Passenger compartment fuses - 1996 models

Chapter 12 Chassis electrical system

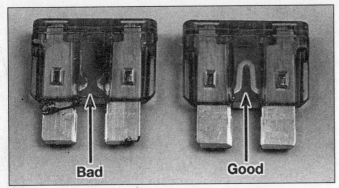

Fuse Value Amps	Color Code
4	Pink
5	Tan
10	Red
15	Light Blue
20	Yellow
25	Natural
30	Light Green

4.5 To test for a blown fuse you can pull it out and inspect it for a broken element (left)

4.7 Each fuse amp value has a corresponding color code

5 Fusible links - general information

Refer to illustration 5.1

Some circuits are protected by fusible links. These links are used in circuits which are not ordinarily fused, such as the ignition circuit **(see illustration)**.

Although fusible links appear to be of heavier gauge than the wire they are protecting, their appearance is due to thicker insulation. All fusible links are four wire gauges smaller than the wire they are designed to protect.

Fusible links cannot be repaired, but a new link of the same size wire can be put in its place. The procedure is as follows:

a) Disconnect the negative cable at the battery.
b) Disconnect the fusible link from the electrical connectors.
c) Cut the damaged fusible link out of the wiring just behind the connector.
d) Strip the insulation back approximately 1/2-inch.
e) Position the connector on the new fusible link and crimp it into place.
f) Splice and solder the new fusible link to the wires from which the old link was cut. Use rosin core solder at each end of the new link to obtain a good solder joint.
g) Wrap the splices completely with vinyl electrical tape around the soldered joint. No wires should be exposed.
h) Install the repaired wiring as before, using existing clips, if provided.
i) Connect the battery ground cable.
j) Test the circuit for proper operation.

6 Circuit breakers - general information

Circuit breakers protect accessories such as the windshield wipers, windshield washer pump, interval wiper, low washer fluid, etc. Circuit breakers are located in the fuse box. Refer to the fuse panel guide in Section 4 and the fuse panel guide in your owner's manual for the location of the circuit breakers used in your vehicle.

Because a circuit breaker resets itself automatically, an electrical overload in a circuit breaker protected system will cause the circuit to fail momentarily, then come back on. If the circuit does not come back on, check it immediately.

a) Remove the circuit breaker from the fuse panel **(see illustrations 4.1a, 4.1c and 4.1g)**.
b) Using an ohmmeter, verify that there is continuity between both terminals of the circuit breaker. If there is no continuity, replace the circuit breaker.

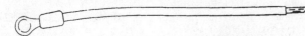

5.1 Fusible link service procedures

12-8 Chapter 12 Chassis electrical system

7.4a Remove the lower screw . . .

7.4b . . . and the upper screw

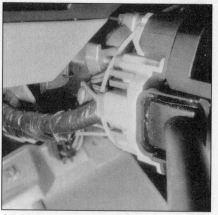

7.5 Use a small screwdriver and release the tangs to allow connector removal

c) *Install the old or new circuit breaker. If it continues to cut out, a short circuit is indicated. Troubleshoot the appropriate circuit (see the Wiring Diagrams at the back of this book) or have the system checked by a professional mechanic.*

7 Steering column switches - replacement

Refer to illustrations 7.4a, 7.4b and 7.5
Warning: *Some models are equipped with airbags. Always disconnect the negative battery cable, then the positive battery cable and wait two minutes before working in the vicinity of the impact sensors, steering column or instrument panel to avoid the possibility of accidental deployment of the airbag, which could cause personal injury.*

1 Disconnect the negative cable from the battery (and if equipped with an airbag, the positive cable, too).
2 If you're working on a 1993 or 1994 model, remove the steering wheel (see Chapter 10).
3 Remove the steering column lower cover and shrouds (see Chapter 11).

4 Remove the two screws securing the switch assembly to the steering column **(see illustrations)**.
5 Use a small screwdriver and release the tangs, then pull the electrical connector from the switch **(see illustration)**.
6 Remove the switch assembly.
7 Installation is the reverse of the removal steps. Be sure that the horn harness grommet (if equipped) is positioned correctly **(see illustration 7.4a)**. Check the steering column and switch for proper operation.

8 Turn signal/hazard flasher relay - replacement

1 Disconnect the negative cable from the battery.
2 The flasher relay on 1993 and 1994 models is located on the interior fuse panel; on 1995 and later models it's under the instrument panel to the right of the steering column. Remove the flasher by pulling straight out.
3 Install the new flasher unit. Be sure to line up the metal contacts with the slots in the fuse panel. Press the flasher firmly into place.

9 Ignition switch/key lock cylinder - check and replacement

Warning: *Some models are equipped with airbags. Always disconnect the negative battery cable, then the positive battery cable and wait two minutes before working in the vicinity of the impact sensors, steering column or instrument panel to avoid the possibility of accidental deployment of the airbag, which could cause personal injury.*

1 The ignition switch on 1993 and 1994 models is mounted on top of the steering column and secured by two nuts. On 1995 and later models, the ignition switch is secured to the underside of the steering column by two screws.

Ignition switch check

Refer to illustrations 9.4a, 9.4b, 9.5a and 9.5b
2 The ignition switch can be tested with an ohmmeter.
3 Disconnect the negative cable from the battery (and if equipped with an airbag, the positive cable, too).
4 Remove the steering column cover and shrouds (see Chapter 11). Disconnect the

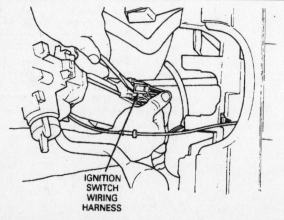

9.4a Pry open the tabs and unplug the electrical connector

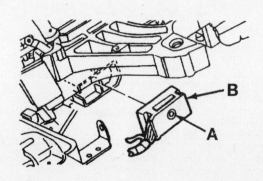

9.4b Detach the connector from the switch

A Screw   B Ignition switch connector

Chapter 12 Chassis electrical system

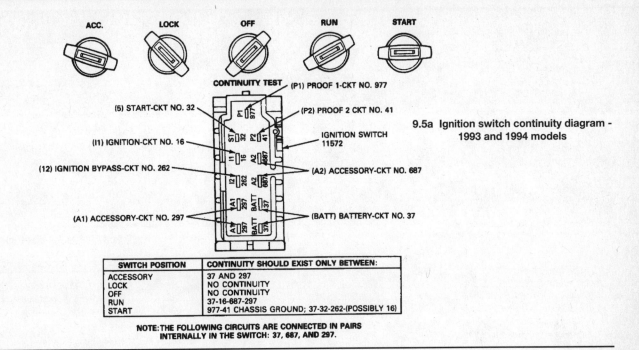

9.5a Ignition switch continuity diagram - 1993 and 1994 models

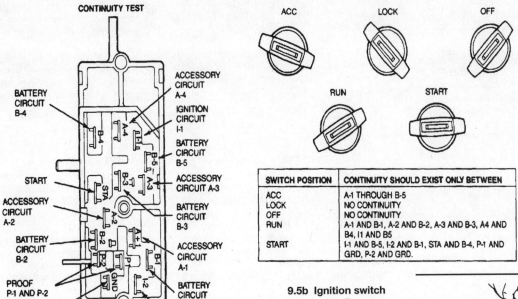

9.5b Ignition switch continuity diagram, 1995 and later models

electrical connector from the ignition switch **(see illustrations)**.

5   Place the key in the lock cylinder. Turn it to the various switch positions and connect the ohmmeter between the terminals specified in the accompanying illustrations **(see illustrations)**. The ohmmeter should indicate continuity or no continuity as indicated in the illustrations. If it doesn't, replace the switch.

Ignition switch replacement

1993 and 1994 models

Refer to illustration 9.7

6   Disconnect the negative cable from the battery.

7   Place the ignition key in the Lock position. Remove the switch mounting nuts **(see illustration)**. Lift the switch off the studs and disengage it from the actuator rod.

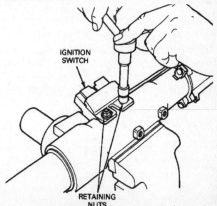

9.7 Remove the nuts and lift the switch off the studs

Chapter 12 Chassis electrical system

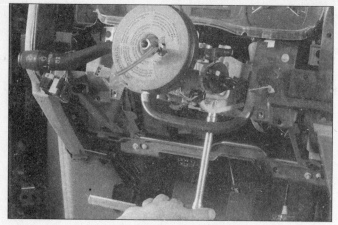

9.11a Remove the nuts . . .

9.11b . . . lower the switch and unplug the electrical connector

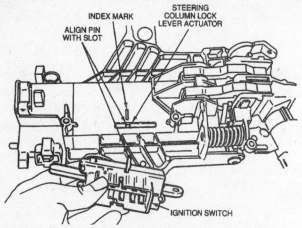

9.12 Align the ignition switch pin with the slot in the steering column

9.18 Push in on the actuator pin and pull the lock cylinder out

8 Installation is the reverse of removal. Make sure the new switch is in the Lock position, then engage it with the actuator rod and position it on the studs. Tighten the nuts securely. Check to make sure the switch operates correctly in all of the key positions.

1995 and later models

Refer to illustrations 9.11a, 9.11b and 9.12

9 Disconnect the negative cable from the battery (and if equipped with an airbag, the positive cable, too).
10 Remove the steering column cover and reinforcement (see Chapter 11).
11 Remove the ignition switch screws and take it off the column **(see illustrations)**. Disconnect the ignition switch electrical connector.
12 Installation is the reverse of the removal steps, with the following additions:
  a) Place the ignition key and the ignition switch in the Run positions.
  b) Align the ignition switch pin with the slot in the steering column, then install the switch and tighten the screws **(see illustration)**.
  c) Check to make sure the switch operates correctly in all of the key positions.

Key lock cylinder replacement

13 Disconnect the negative cable from the battery. If you're working on a 1995 or later model, disconnect the positive cable as well.

1993 and 1994 models

14 Remove the steering wheel, steering column lower cover and steering column shrouds (see Chapter 11).
15 Turn the steering lock to the On position.

1995 and later models

16 Remove the steering column lower cover and shroud (see Chapter 11).
17 Turn the steering lock to the Run position.

All models

Refer to illustration 9.18

18 Insert a 1/8-inch diameter wire, punch or small screwdriver into the lock cylinder housing and release the actuator pin while pulling

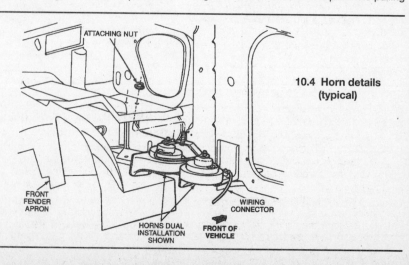

10.4 Horn details (typical)

Chapter 12 Chassis electrical system

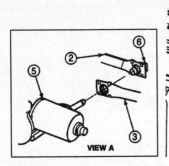
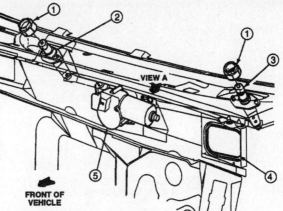

11.5 Windshield wiper linkage details

1. Pivot nuts
2. Arm and pivot shaft
3. Arm and pivot shaft
4. Linkage access cover
5. Wiper motor
6. Clip

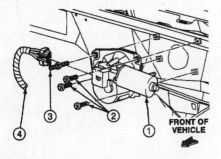

11.10 Windshield wiper motor

1. Wiper motor
2. Bolts
3. Stud
4. Wiring harness

out on the cylinder **(see illustration)**.
19 Installation is the reverse of the removal steps.

10 Horn - check and replacement

Refer to illustration 10.4
Note: *Check the fuse and relay before beginning electrical diagnosis.*
1 Disconnect the electrical connector from the horn.
2 Test the horn by connecting battery voltage to the terminal with a jumper wire.
3 If the horn doesn't sound, replace it. It does sound, the problem lies in the switch, relay or the wiring between components.
4 Disconnect the electrical connector(s) and remove the mounting bolt **(see illustration)**.
5 Installation is the reverse of removal.

11 Windshield wiper motor - removal and installation

Refer to illustrations 11.5 and 11.10
1 Turn the windshield wiper switch On.
2 Turn the ignition switch on and keep your hand on the key. When the wiper arms move to the straight up position, turn the ignition switch Off.
3 Disconnect the negative cable from the battery.
4 Remove the right side wiper arm and blade assembly (see Section 12).

5 Remove the pivot nut from the right side linkage post **(see illustration)**. Allow the linkage to drop down into the cowl.
6 Remove the wiper linkage access cover. Reach through the access opening and unsnap the wiper motor clip.
7 Push the clip away from the linkage until it clears the crank pin nib, then push the clip off the linkage.
8 Remove the wiper linkage from the motor crankpin.
9 Disconnect the electrical connector from the motor.
10 Remove the screws securing the motor and remove the motor **(see illustration)**.
11 Installation is the reverse of the removal steps. Make sure the wiper blades are in the parked position before attaching the linkage to the motor.
12 Installation is the reverse of the removal steps.

12 Windshield wiper arm - removal and installation

Refer to illustrations 12.2a and 12.2b
1 **Note:** *To prevent damage to the windshield and also to make sure the wipers are operating under normal conditions, keep the windshield wet during this step. Prior to removing the wiper arm, turn the windshield wipers switch On, allow the wipers to travel through several cycles, then turn it Off. This will ensure that the wiper arm is in the parked position parallel to the base of the windshield.*
2 Lift the wiper arm and blade assembly away from the windshield and pull-out the small latch on the wiper arm near the spindle **(see illustrations)**. Lower the arm and allow it to rest on the latch. The blade will rest several inches away from the windshield. Grasp the arm, wiggle it back-and-forth and remove it from the spindle.
3 Installation is the reverse of the removal steps with the following additions.
 a) Position the arm back onto the post in the parked position. Do not position the arm too low as it will hit the base of the windshield during normal wiper operation.

12.2a Lift the wiper blade away from the windshield, pull the latch out and allow the arm to rest on the latch - this prevents the arm from binding on the spindle

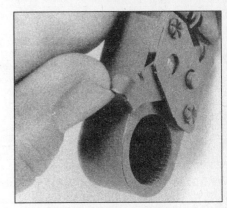

12.2b Be sure to pull the latch back far enough to clear the spindle

 b) Make sure the latch is correctly seated in the groove in the wiper spindle.

13 Windshield washer reservoir and pump assembly - removal and installation

Refer to illustrations 13.1, 13.6 and 13.9
1 Use a small screwdriver to unlock the

13.1 Disconnect the electrical connector for the pump motor (arrow)

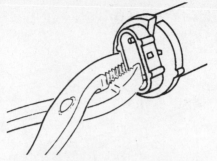

13.6 Grip the wall around the electrical terminal and pull the motor/pump assembly from the reservoir

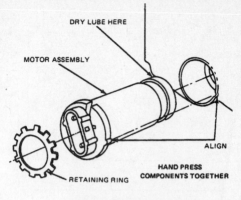

13.9 Windshield washer motor/pump installation

electrical connector tabs, then unplug the washer pump's electrical connector **(see illustration)**.

2   Remove the two screws securing the washer reservoir and pump motor assembly to the fenderwell.

3   Lift the washer reservoir up and disconnect the small hose from the base of the reservoir. Place your finger over the end of the small hose fitting on the reservoir to prevent spilling the washer fluid in the engine compartment. Remove the reservoir.

4   Drain the washer fluid from the reservoir into a clean container. If the fluid is kept clean it can be reused.

5   Use a small screwdriver and carefully pry out the retaining ring securing the pump motor in the reservoir receptacle.

6   Grasp one wall surrounding the electrical terminal with a pair of pliers, then pull the motor, seal and impeller assembly out of the reservoir **(see illustration)**. **Note:** *If the impeller and seal separate from the motor they can be reassembled after removal.*

7   Flush out the reservoir with clean water to remove any residue. Inspect it for any foreign matter.

8   Inspect the reservoir pump chamber prior to installing an old motor into a new reservoir. Clean it if necessary.

9   Lubricate the outer surface of the seal with powdered graphite to make installation easier **(see illustration)**.

10   Align the small projection on the motor end cap with the slot in the reservoir and push it in until the seal seats against the bottom of the motor receptacle in the reservoir.

11   Use a 1-inch 12-point socket and hand press the retaining ring securely against the motor and plate.

12   Connect the hose to the fitting on the base of the reservoir.

13   Install the reservoir in the engine compartment and secure with the two screws.

14   Connect the electrical connector.

15   **Caution:** *Do not operate the pump without fluid in the reservoir, as it would be damaged. Fill the reservoir with fluid, operate the pump and check for leaks.*

14   Headlight switch - replacement

1   Disconnect the negative battery cable from the battery.

1993 and 1994 models

Refer to illustrations 14.4 and 14.5

2   Remove the ashtray and two instrument panel trim panel screws. Snap off the trim

14.4 Depress the shaft release button on the switch assembly and pull the shaft out (switch removed for clarity)

panel and the storage bin.

3   Pull the headlight switch out as far as it will go.

4   Reach up along the side of the switch (on the side nearest the driver's door) and push the knob release button **(see illustration)**. While holding the button down, pull the knob and shaft out of the switch.

5   Unscrew the switch retaining nut with your fingers **(see illustration)**. **Caution:** *The nut is plastic. Don't use tools on it.*

6   Disconnect the electrical connector from the switch. Take the switch out.

7   Installation is the reverse of the removal Steps. Align the switch tab with the notch in the instrument panel.

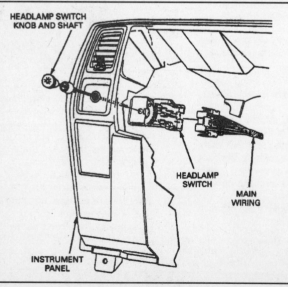

14.5 Unscrew the plastic retaining nut with your fingers only, then pry open the tabs and unplug the electrical connector

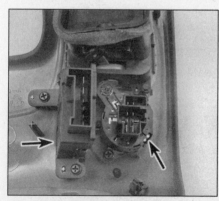

14.9 The light switch (right arrow) and instrument panel dimmer (left arrow) are mounted on the back of the instrument cluster bezel

Chapter 12 Chassis electrical system

15.2 Unscrew the retainer from the headlight

15.3 Pull the socket out of the housing, then pull the bulb straight out of the socket and push the new one in - don't touch the bulb glass with your bare fingers

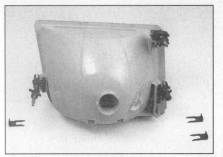

15.11 The headlight housing is secured by three clips

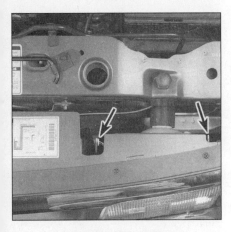

16.1 The headlight adjusting screws (arrows) are accessible from the rear

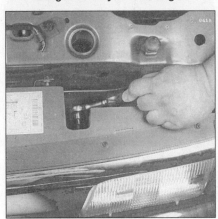

16.6 Turn the adjusting screws with a socket

1995 and later models

Refer to illustration 14.9

8 Remove the instrument cluster bezel (see Chapter 11).
9 Remove the switch screws and detach the switch from the bezel **(see illustration)**.
10 Installation is the reverse of the removal steps.

15 Headlight bulb and housing - replacement

Bulb replacement

Refer to illustrations 15.2 and 15.3

Warning 1: *Halogen bulbs get hot enough to cause burns. If the bulb has just burned out, allow it to cool before replacing it.*

Warning 2: *Halogen gas filled bulbs are under pressure and may shatter if the surface is scratched or the bulb is dropped. Wear eye protection and handle the bulbs carefully, grasping only the base whenever possible. Do not touch the surface of the bulb with your fingers because the oil from your skin could cause the bulb to overheat and fail prematurely. If you do touch the bulb surface, clean it with rubbing alcohol.*

1 Open the hood.
2 Unscrew the retainer from the back of the headlight **(see illustration)**.
3 Pull the bulb socket out of the headlight assembly **(see illustration)**.
4 Carefully pull the bulb straight out of the socket **(see illustration 15.3)**. Do not rotate the bulb during removal.
5 Without touching the glass with bare fingers, insert the bulb into the socket. Position the flat on the plastic base up.
6 Align the socket locating tabs with the notches in the headlight housing, then push the socket firmly into the housing. Make sure the socket mounting flange contacts the housing.
7 Slide the retainer on. Turn it clockwise until it hits the stop.
8 Test headlight operation, then close the hood.
9 Have the headlight adjustment checked and, if necessary, adjusted by a dealer service department or service station at the earliest opportunity.

Housing removal and installation

Refer to illustration 15.11

10 Perform Steps 1 through 3 to detach the bulb and wiring from the housing.
11 Pull out the housing clips and remove the housing from the grille opening panel **(see illustration)**.

16 Headlights - adjusting

Refer to illustrations 16.1 and 16.6

Note: *The headlights must be aimed correctly. If adjusted incorrectly they could momentarily blind the driver of an oncoming vehicle and cause a serious accident or seriously reduce your ability to see the road. The headlights should be checked for proper aim every 12 months and any time a new headlight is installed or front end body work is performed. It should be emphasized that the following procedure is only an interim step which will provide temporary adjustment until the headlights can be adjusted by a properly equipped shop.*

1 Headlights have two hex-head adjusting screws, one controlling up-and-down movement and one controlling left-and-right movement **(see illustration)**.
2 There are several methods of adjusting the headlights. The simplest method requires a blank wall 25 feet in front of the vehicle and a level floor.
3 Position masking tape vertically on the wall in reference to the vehicle centerline and the centerline of both headlights.
4 Position a horizontal line in reference to the centerline of all headlights. **Note:** *It may be easier to position the tape on the wall with the vehicle parked only a few inches away.*
5 Adjustment should be made with the vehicle sitting level, the gas tank half-full and no unusually heavy load in the vehicle.
6 Starting with the low beam adjustment, position the high intensity zone so it is two inches below the horizontal line and two inches to the right of the headlight vertical line. Turn the vertical adjusting screw as necessary to raise or lower the beam. The other adjusting screw should be used in the same manner to move the beam left or right **(see illustration)**.

Chapter 12 Chassis electrical system

17.1a Remove the screw that secures the top of the lamp housing . . .

17.1b . . . and the nut from the back

17.2 Pull the housing out of the body and detach the bulb socket

17.3 Pull the bulb out of the socket

17.5 Remove the lamp housing screws

17.7 Rotate the bulb socket and remove it from the housing

7   With the high beams on, the high intensity zone should be vertically centered with the exact center just below the horizontal line. **Note:** *It may not be possible to position the headlight aim exactly for both high and low beams. If a compromise must be made, keep in mind that the low beams are the most used and have the greatest effect on driver safety.*

8   Have the headlights adjusted by a dealer service department or service station at the earliest opportunity.

17.8 Pry the retainer off and pull the bulb out of the socket

17   Bulb replacement

Parking/turn signal/front side marker lamp

Refer to illustrations 17.1a, 17.1b, 17.2 and 17.3

1   Open the hood. Remove the screw from the top of the lamp assembly and the nut from the back and remove the lamp assembly from the grille panel **(see illustrations)**.
2   Pull the assembly straight out and detach the bulb socket **(see illustration)**.
3   Pull the bulb out of the socket **(see illustration)**.
4   Installation is the reverse of the removal Steps.

Rear combination lights (stop, tail, turn and backup)

Refer to illustrations 17.5, 17.7 and 17.8

5   Remove the light assembly screws **(see illustration)**.
6   Take the assembly out of the fender **(see illustration 17.5)**.
7   Rotate the bulb socket to align the lugs and take the socket out of the housing **(see illustration)**.

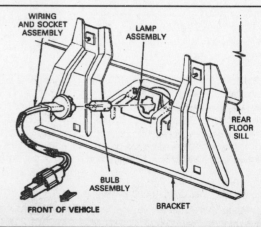

17.10 Rotate the bulb socket and remove it from the bracket

Chapter 12 Chassis electrical system

17.11a Carefully pry one side of the light out - if it won't go, try the other side - don't force it

17.11b Pull the socket out and pull out the bulb

17.13 Remove the light housing screws

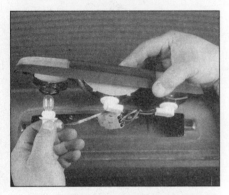
17.14 Lift the housing away from the body, detach the socket and pull out the bulb

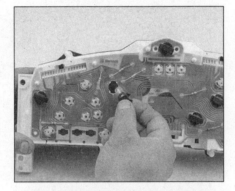
17.17 Twist the socket and pull it out of the cluster, then pull the bulb straight out of the socket

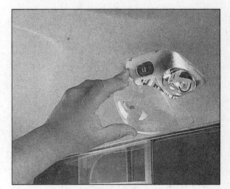
17.18 Squeeze the retainer tabs on the light assembly and pull it out

8 Carefully pry off the retaining clip and pull the bulb out of the socket (see illustration).
9 Installation is the reverse of the removal Steps.

License plate bulb

Refer to illustrations 17.10, 17.11a and 17.11b
10 If the vehicle doesn't have a rear bumper, twist the socket counterclockwise and remove it from the housing (see illustration). Pull the bulb out of the socket.
11 If the vehicle has a rear bumper, carefully pry the bulb socket out of the bumper, then pull the bulb out of the socket (see illustrations).
12 Installation is the reverse of the removal Steps.

Cargo light and high-mount brake light

Refer to illustrations 17.13 and 17.14
13 Remove the light assembly screws (see illustration).
14 Take the light assembly out and pull the bulb out of the socket (see illustration).
15 Installation is the reverse of the removal Steps.

Instrument cluster bulbs

Refer to illustration 17.17
16 Caution: The instrument cluster printed circuit board is fragile. Be careful when removing the bulb assemblies. Reach up under the instrument panel and turn the bulb/socket assembly 1/4-turn counterclockwise.
17 Remove the bulb from the socket and install a new bulb (see illustration). Note: If you cannot reach some of the bulb/sockets, remove the instrument cluster to gain access to the bulbs (see Section 21).

Dome light

Refer to illustration 17.18
18 Snap the lens free from the dome light assembly and remove the bulb (see illustration).
19 Installation is the reverse of the removal steps.

18 Radio - removal and installation

Refer to illustration 18.2
1 Remove the instrument cluster trim panel (see Section 21).
2 Insert a radio removal tool, available at most auto parts stores, or equivalent, into the

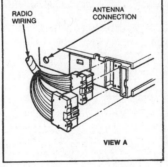
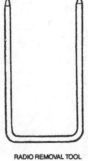
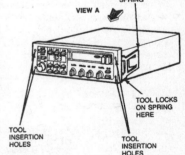

18.2 Special tool used for radio removal

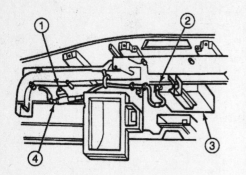

19.1 Antenna cable routing details

1. Screw
2. Antenna lead-in cable
3. Radio
4. Antenna base cable

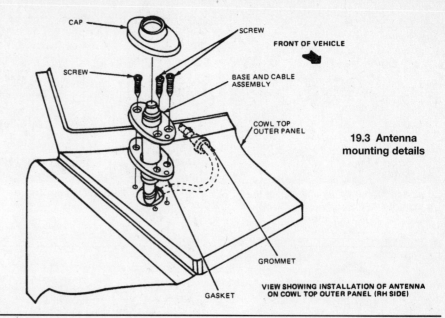

19.3 Antenna mounting details

radio face plate **(see illustration)**. Push the tool in about 1-inch to release the retaining clips on each side.
3 Using the special tool, pull the radio partially out of the instrument panel, then disconnect the power, speaker and antenna wires from the radio.
4 Pull the radio assembly out of the instrument panel.
5 Installation is the reverse of the removal steps.

19 Radio antenna - removal and installation

Refer to illustrations 19.1 and 19.3

1 Reach under the instrument panel and disconnect the antenna cable at the lead-in connector **(see illustration)**.
2 Detach the antenna cable from the plastic clips under the instrument panel **(see illustration 19.1)**.
3 Unscrew the antenna from the base. Use a small screwdriver and carefully pry the antenna cap from the base and remove the cap **(see illustration)**.
4 Tie a piece of string or wire onto the radio end of the antenna cable. This will be used to pull the new antenna cable back through the same path as the old one.
5 Remove the screws securing the antenna base to the body and slowly pull the antenna cable out through the body opening. Remove the antenna base and gasket.
6 If necessary, disconnect the antenna lead-in cable from the radio and detach it from its clips **(see illustration 19.1)**.
7 Installation is the reverse of the removal steps with the following additions:

a) Be sure to install a new gasket.

b) Untie the string or wire and attach it to the new antenna cable. Slowly pull the string and antenna cable back through the body opening and into place. Remove the string or wire.

20 Speakers - removal and installation

Note: The models covered by this manual come equipped from the factory with speakers in the doors, with additional rear speakers on vehicles equipped with AM/FM sound systems. Your particular model may vary from the following procedure due to the size and configuration of aftermarket speakers installed in some models.

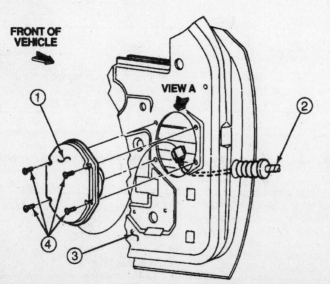

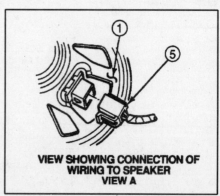

20.2 Door speaker installation details (typical)

1. Speaker
2. Wiring harness
3. Door
4. Screws
5. Connector

Chapter 12 Chassis electrical system

12-17

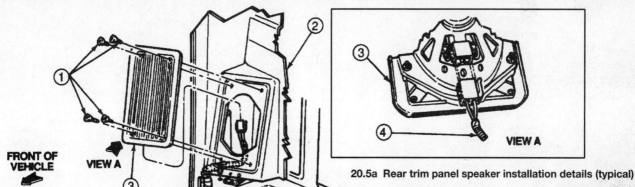

20.5a Rear trim panel speaker installation details (typical)

1 Screws
2 Trim panel
3 Speaker grille
4 Wiring harness

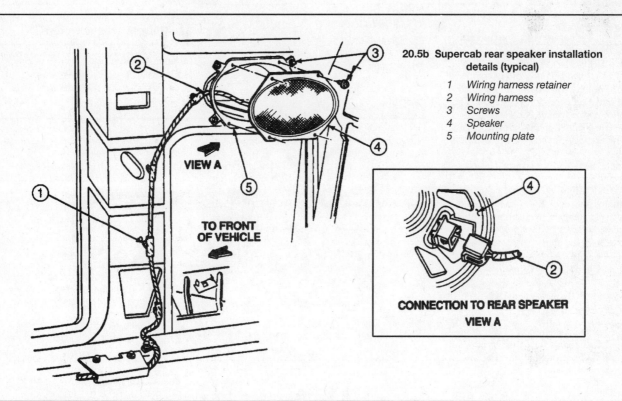

20.5b Supercab rear speaker installation details (typical)

1 Wiring harness retainer
2 Wiring harness
3 Screws
4 Speaker
5 Mounting plate

Door speakers

Refer to illustration 20.2

1 Remove the door trim panel (see Chapter 11).
2 Remove the screws securing the speaker to the door **(see illustration)**.
3 Lift out the speaker and disconnect the electrical connector at the speaker. **Caution:** *Do not operate the radio with the speakers disconnected.*
4 Installation is the reverse of the removal steps.

Rear speakers

Refer to illustrations 20.5a and 20.5b

5 Remove the screws and lift the speaker out of the trim panel **(see illustrations)**.
6 Disconnect the electrical connector at the speaker. **Caution:** *Do not operate the radio with the speakers disconnected.*
7 Installation is the reverse of the removal steps.

21 Instrument cluster - removal and installation

Refer to illustrations 21.6a, 21.6b, 21.9 and 21.10

Warning: *Some models are equipped with airbags. Always disconnect the negative battery cable, then the positive battery cable and wait two minutes before working in the vicinity of the impact sensors, steering column or instrument panel to avoid the possibility of accidental deployment of the airbag, which could cause personal injury.*

Removal

1 Take note of the preceding **Warning**, then disconnect the battery cables.

1993 and 1994 models

2 Pull out the ashtray, remove the four ashtray retaining bracket screws and remove the bracket.
3 Remove the left and right air conditioning vents and the instrument cluster bezel (see Chapter 11).

1995 and later models

4 Remove the radio (see Section 18).
5 Remove the center and lower trim panels, the knee bolster and the instrument cluster bezel (see Chapter 11).

Chapter 12 Chassis electrical system

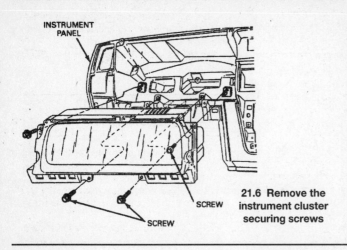

21.6 Remove the instrument cluster securing screws

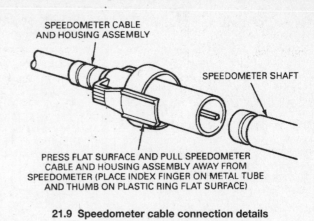

21.9 Speedometer cable connection details

All models

6 Remove the screws securing the instrument cluster **(see illustrations)**.

7 On automatic transmission models, remove the screws securing the gear indicator to the instrument cluster and slide the indicator down and out of the cluster. It is not necessary to disconnect the indicator.

8 Carefully pull the instrument cluster assembly partially out from the instrument panel.

9 If the vehicle has a speedometer cable, reach under the instrument panel, press on the flat portion of the quick-disconnect plastic connector and detach the cable from the instrument cluster **(see illustration)**. **Note:** *If there isn't enough slack in the speedometer cable to pull the cluster out far enough, disconnect the cable at the transmission or transfer case under the vehicle, then push it inside the vehicle while an assistant tugs on the cluster. Once there is enough slack in the cable, reach behind the cluster and disconnect it.*

10 Squeeze the locking tabs on the printed circuit electrical connectors, then disconnect them from the printed circuit board **(see illustration)**.

11 Remove the instrument cluster assembly.

Installation

12 Installation is the reverse of the removal steps, with the following addition: If the vehicle has a speedometer cable, apply a 3/16-inch diameter ball of silicone dielectric compound to the drive hole of the speedometer head prior to installing the cable.

22 Gauges - replacement

Refer to illustration 22.2

Note: *US Federal law requires that the odometer in any replacement speedometer must register the same mileage as that on the*

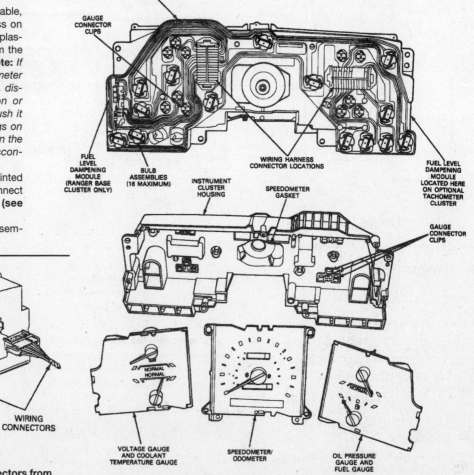

21.10 Disconnect the wiring connectors from the instrument cluster

22.2 Instrument cluster details (typical)

Chapter 12 Chassis electrical system

removed speedometer.

1 Remove the instrument cluster (see Section 21).
2 Remove the lens and mask from the instrument cluster **(see illustration)**.
3 Gently push on the back of the speedometer to loosen it from the retainer clips in the cluster housing.
4 Lift the gauges with a gentle rocking motion to separate them from the retainer prongs in the cluster housing.
5 Installation is the reverse of the removal steps. Apply a 3/16-inch diameter ball of silicone dielectric compound to the drive hole of the speedometer head prior to installing the cable.

23 Speedometer cable - replacement

Refer to illustration 23.3

1 Perform Steps 1 through 9 of Section 21.
2 To remove the cable only, pull out the cable core from the casing and install a new one by reversing the removal steps.
3 Remove the casing as follows:
  a) *Remove the screw and clamp securing the speedometer cable to the firewall.*
  b) *Working under the vehicle, remove the clamp screw and remove the cable and casing from the transmission or transfer case (see illustration).*
4 Connect the speedometer casing to the head and (if removed) install the casing to the transmission using a new O-ring.
5 Install the instrument cluster (see Section 21).

24 Power window system - description and check

1 The power window system operates the electric motors mounted in the doors which lower and raise the windows. The system consists of the control switches, the motors (regulators), glass mechanisms and associated wiring.
2 The windows can be lowered and raised

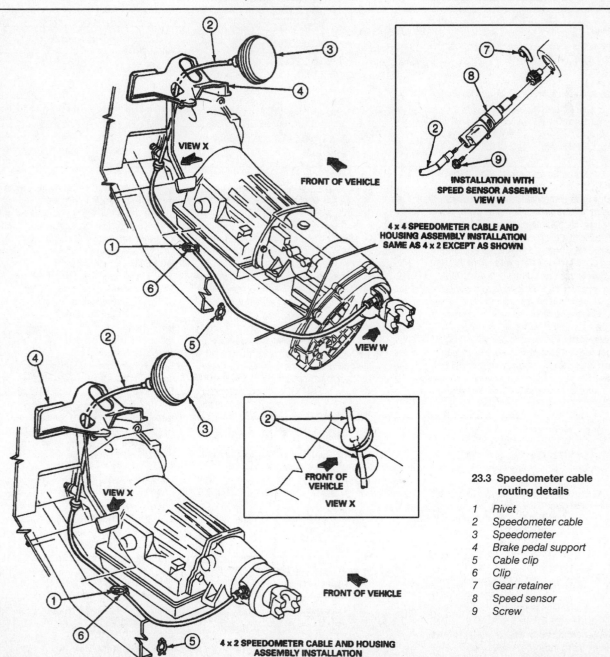

23.3 Speedometer cable routing details

1 Rivet
2 Speedometer cable
3 Speedometer
4 Brake pedal support
5 Cable clip
6 Clip
7 Gear retainer
8 Speed sensor
9 Screw

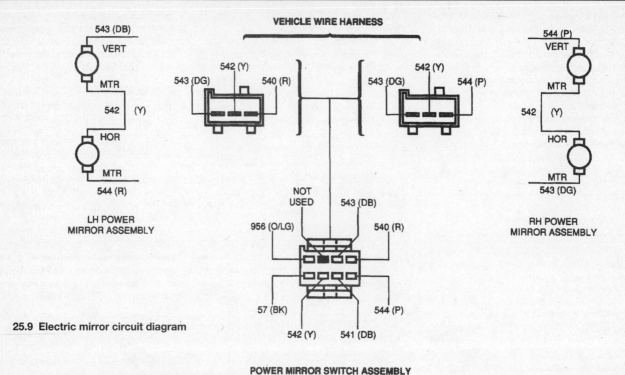

25.9 Electric mirror circuit diagram

from the master control switch by the driver or by a remote switch located at the passenger window. Each window has a separate motor which is reversible. The position of the control switch determines the polarity and therefore the direction of operation. Some systems are equipped with relays that control current flow to the motors.

3   Each motor is equipped with a separate circuit breaker in addition to the fuse or circuit breaker protecting the whole circuit. This prevents one stuck window from disabling the whole system.

4   The power window system will only operate when the ignition switch is ON. In addition, when activated the window lockout switch at the master control switch disables the switches at the passenger's window also. Always check these items before troubleshooting a window problem.

5   These procedures are general in nature, so if you can't find the problem using them, take the vehicle to a dealer service department.

6   If the power windows don't work at all, check the fuse or circuit breaker.

7   If only the rear windows are inoperative, or if the windows only operate from the master control switch, check the rear window lockout switch for continuity in the unlocked position. Replace it if it doesn't have continuity.

8   Check the wiring between the switches and fuse panel for continuity. Repair the wiring, if necessary.

9   If the passenger window is inoperative from the master control switch, try the other control switch at the window. **Note:** *This doesn't apply to the driver's door window.*

10 If the same window works from one switch, but not the other, check the switch for continuity.

11 If the switch tests OK, check for a short or open in the wiring between the affected switch and the window motor.

12 If the passenger window is inoperative from both switches, remove the trim panel from the door and check for voltage at the motor while the switch is operated.

13 If voltage is reaching the motor, disconnect the glass from the regulator (see Chapter 11). Move the window up and down by hand while checking for binding and damage. Also check for binding and damage to the regulator. If the regulator is not damaged and the window moves up and down smoothly, replace the motor. If there's binding or damage, lubricate, repair or replace parts, as necessary.

14 If voltage isn't reaching the motor, check the wiring in the circuit for continuity between the switches and motors. You'll need to consult the wiring diagram for the vehicle.

15 Test the windows after you are done to confirm proper repairs.

25 Power mirrors - check, removal and installation

Refer to illustration 25.9

1   The electric rear view mirrors use two motors to move the glass; one for up and down adjustments and one for left-right adjustments.

2   The control switch has a selector portion which sends voltage to the left or right side mirror. With the ignition switch in the ACC position, roll down the windows and operate the mirror control switch through all functions (left-right and up-down) for both the left and right side mirrors.

3   Listen carefully for the sound of the electric motors running in the mirrors.

4   If the motors can't be heard or if the mirror glass doesn't move, there's probably a problem with the drive mechanism inside the mirror. Remove and disassemble the mirror to locate the problem.

5   If the mirrors don't operate and no sound comes from them, check the mirror fuse in the fuse block (see Section 4).

6   If the fuse is OK, remove the switch bezel (see Chapter 11) for access to the back of the mirror control switch without disconnecting the wire attached to it. Turn the ignition On and check for voltage at the switch. There should be voltage at one terminal. If there's no voltage at the switch, check for an open or short in the wiring between the fuse panel and switch.

7   If there's voltage at the switch, disconnect it. Check the switch for continuity in all its operating positions. If the switch does not have continuity, replace it.

8   Re-connect the switch. Locate the wire going from the switch to ground. Leaving the switch connected, connect a jumper wire between this wire and ground. If the mirror works normally with this wire in place, repair the faulty ground connection.

9   If the mirror still doesn't work, remove the mirror (see Chapter 11) and check the

Chapter 12 Chassis electrical system

wires at the mirror for voltage **(see illustration)**. Check with the ignition On and the mirror selector switch on the appropriate side. Operate the mirror switch in all its positions. There should be voltage at one of the switch-to-mirror wires in each switch position (except the neutral "off" position).

10 If there's not voltage in each switch position, check the wiring between the mirror and switch for opens and shorts.

11 If there's voltage, remove the mirror and test it off the vehicle with jumper wires. Replace the mirror if it fails this test.

26 Power door lock system - description and check

The power door lock system operates the door lock actuators mounted in each door. The system consists of the switches, actuators and associated electrical wiring.

Diagnosis can usually be limited to simple checks of the wiring connectors and actuators for minor faults which can be easily repaired. These include:

a) *Check the system fuse and circuit breaker.*
b) *Check the switch wiring for damage or loose connections.*
c) *Check the switch for continuity in the closed position. Also check the switches for sticking. If a switch sticks closed the actuator internal circuit breaker will trip and stay tripped until voltage is removed.*
d) *Remove the door panel(s) and check the actuator electrical connections for looseness or damage. Inspect the actuator rods and door lock linkage to make sure they are not bent, damaged or binding.*
e) *Remove the electrical connector at the actuator and with a test light or voltmeter, check for available voltage to the actuator as you cycle the switch. If voltage is present, but the actuator doesn't operate when connected, the actuator is probably defective. If no voltage is present at the connector, the switch, relay (if equipped) or wiring is probably defective.*

27 Cruise control system - description and check

1 The cruise control system on 1993 and 1994 models maintains vehicle speed with a vacuum actuated servo motor mounted in the engine compartment, which is connected to the throttle linkage by a cable. The system consists of the servo motor, brake switch, clutch switch (manual transmission models), control switches, a relay and associated vacuum hoses.

2 The electronic cruise control system used on 1995 and later models is generally similar to the earlier design, but the system doesn't need engine vacuum to operate, so there are no vacuum hoses.

3 Cruise controls all work by the same basic principles. However, the hardware used varies considerably depending on model and year or manufacture. Some later systems require special testers and diagnostic procedures which are beyond the scope of the home mechanic. Listed below are some general procedures that may be used to locate common problems.

4 Locate and check the fuse (see Section 4).

5 Have an assistant operate the brake lights while you check their operation (voltage from the brake switch deactivates the cruise control).

6 If the brake lights don't come on or don't shut off, correct the problem and retest the cruise control.

7 Inspect the actuator cable between the cruise control servo and the throttle linkage. The cruise control servo is usually fist-sized or slightly larger and is located in the left rear corner of the engine compartment (1993 and 1994) or on the right front fender (1995 and later).

8 Visually inspect the vacuum hoses (1993 and 1994) and wires connected to the cruise control servo and replace as necessary.

9 Cruise controls use a variety of speed sensing devices. On these models the speed sensor pick-up is located at the transmission.

10 Test drive the vehicle to determine if the cruise control is now working. If it isn't, take it to a dealer service department or an automotive electrical specialist for further diagnosis and repair.

28 Airbag - general information

Later models are equipped with a Supplemental Inflatable Restraint (SIR) System, more commonly known as an airbag. This system is designed to protect the driver (and front seat passenger on some models) from serious injury in the event of a head-on or frontal collision. It consists of an airbag module in the center of the steering wheel (driver) and the right side of the instrument panel (passenger), two crash sensors mounted at the front of the vehicle and a diagnostic monitor which also contains a backup power supply located in the passenger compartment.

Airbag module

Steering wheel-mounted

The airbag inflator module contains a housing incorporating the cushion (airbag) and inflator unit, mounted in the center of the steering wheel. The inflator assembly is mounted on the back of the housing over a hole through which gas is expelled, inflating the bag almost instantaneously when an electrical signal is sent from the system. A coil assembly on the steering column under the module carries this signal to the module.

This coil assembly can transmit an electrical signal regardless of steering wheel position. The igniter in the airbag converts the electrical signal to heat and ignites the sodium azide/copper oxide powder, producing nitrogen gas, which inflates the bag.

Instrument panel-mounted

The passenger airbag (if equipped) is mounted above the glove compartment and designated by the letters SRS (Supplemental Restraint System). It consists of an inflator containing an igniter, a bag assembly, a reaction housing and a trim cover.

This airbag is considerably larger than the steering wheel-mounted unit and is supported by the steel reaction housing. The trim cover is textured and painted to match the instrument panel and has a molded seam which splits when the bag inflates. As with the steering-wheel-mounted airbag, the igniter electrical signal converts to heat, converting sodium azide/iron oxide powder to nitrogen gas, inflating the bag.

Sensors

The system has three sensors: two forward sensors at the front of the vehicle and a safing sensor inside the electronic diagnostic monitor.

The forward and passenger compartment sensors are basically pressure sensitive switches that complete an electrical circuit during an impact of sufficient G force. The electrical signal from these sensors is sent to the electronic diagnostic monitor which then completes the circuit and inflates the airbag(s).

Electronic diagnostic monitor

The electronic diagnostic monitor supplies the current to the airbag system in the event of the collision, even if battery power is cut off. It checks this system every time the vehicle is started, causing the "AIR BAG" light to go on then off, if the system is operating properly. If there is a fault in the system, the light will go on and stay on, flash, or the dash will make a beeping sound. If this happens, the vehicle should be taken to your dealer immediately for service.

Disabling the system

Whenever working in the vicinity of the steering wheel, steering column or near other components of the airbag system, the system should be disarmed. To do this, perform the following steps:

a) *Turn the ignition switch to Off.*
b) *Detach the cable from the negative battery terminal, then detach the positive cable. Wait two minutes for the electronic module backup power supply to be depleted.*

Enabling the system

a) *Turn the ignition switch to the Off position.*
b) *Connect the positive battery cable first, then connect the negative cable.*

On models equipped with a passenger side airbag, a switch is included that can be used to disable the airbag when an infant or small child is being transported in a rear-facing child safety seat. The switch is located in the center of the dash and can be operated by inserting any Ford ignition key.

29 Wiring diagrams - general information

Refer to illustration 29.3

Since it isn't possible to include all wiring diagrams for every model year covered by this manual, the following diagrams are those that are typical and most commonly needed.

Prior to troubleshooting any circuit, check the fuses and circuit breakers to make sure they're in good condition. Make sure the battery is fully charged and check the cable connections (see Chapter 1). Make sure all connectors are clean, with no broken or loose terminals.

Refer to the accompanying table for the wire color codes applicable to your vehicle **(see illustration)**.

B	Black			P	Purple
BR	Brown			PK	Pink
DB	Dark blue			R	Red
DG	Dark green			T	Tan
GY	Gray			W	White
LB	Light blue			Y	Yellow
LG	Light green			(H)	Hash
N	Natural			(D)	Dot
O	Orange				

29.3 Wiring diagram color codes

Chapter 12 Chassis electrical system 12-23

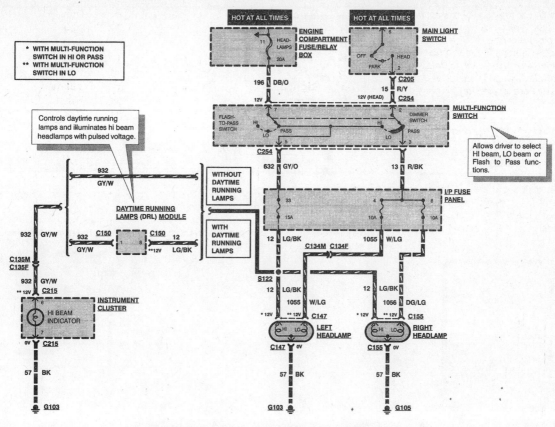

Headlight circuit (1996 model shown; others similar)

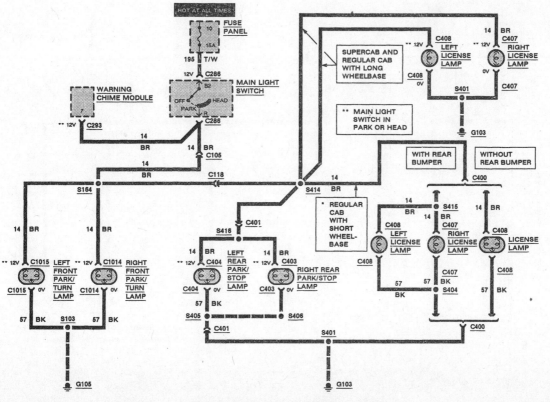

Exterior lighting circuit (1993 and 1994)

12-24 Chapter 12 Chassis electrical system

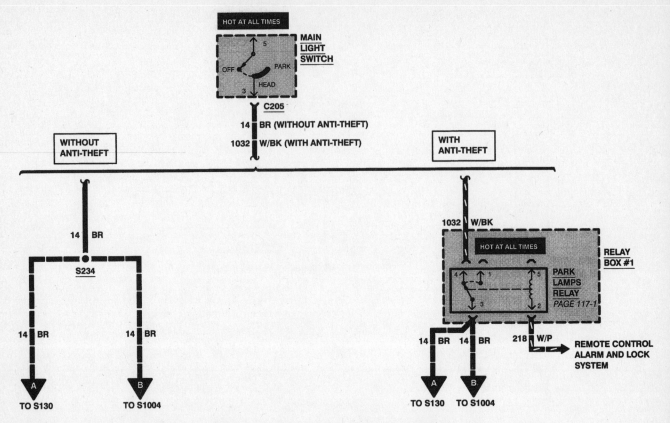

Exterior lighting circuit (1995-on; part 1 of 2)

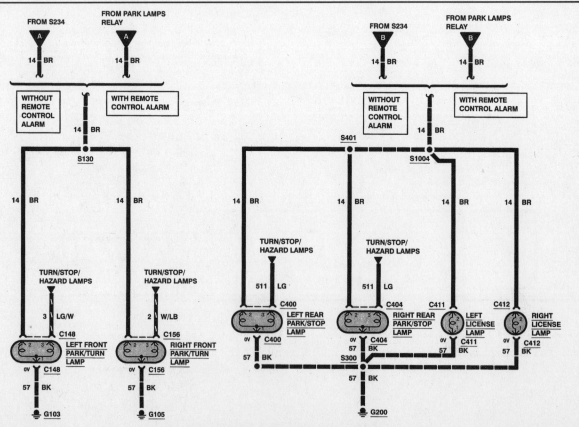

Exterior lighting circuit (1995-on; part 2 of 2)

Chapter 12 Chassis electrical system   12-25

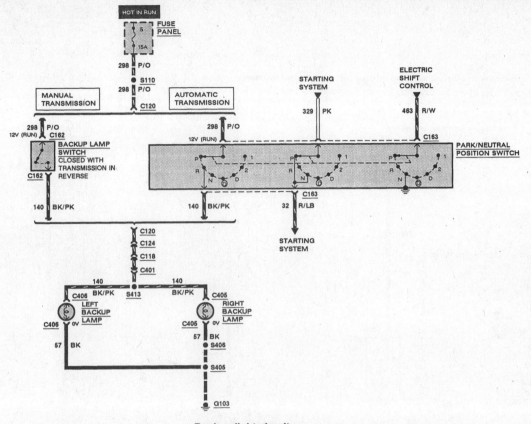

Backup light circuit

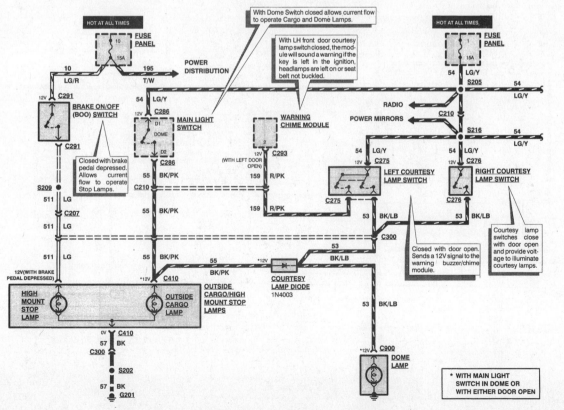

Courtesy light circuit (1993 and 1994 models; part 1 of 2)

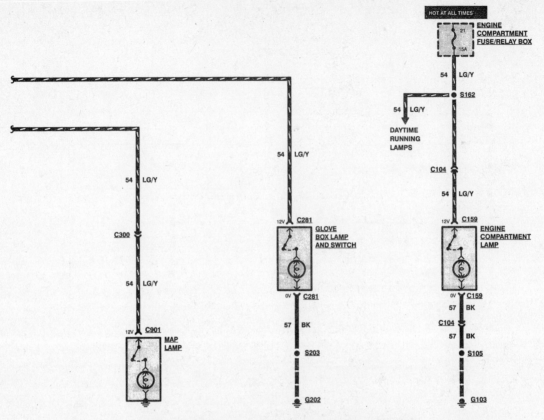

Courtesy light circuit (1993 and 1994 models; part 2 of 2)

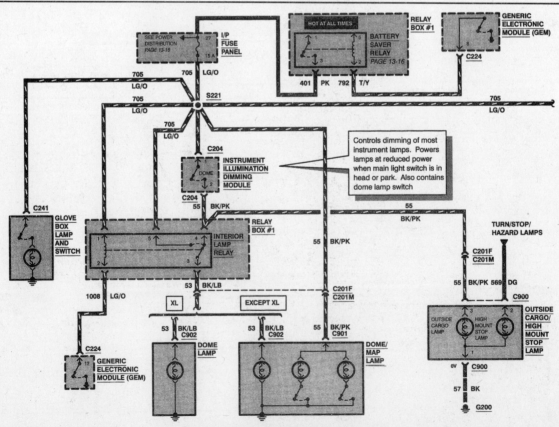

Courtesy light circuit (1995-on; part 1 of 2)

Chapter 12 Chassis electrical system

12-27

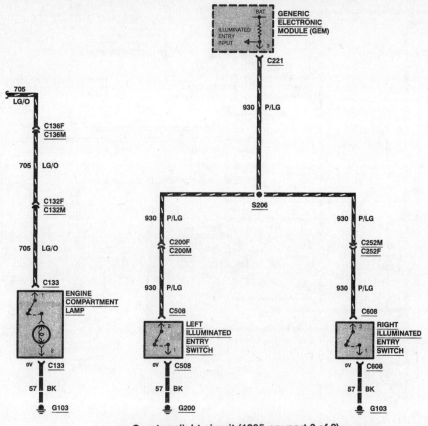

Courtesy light circuit (1995-on; part 2 of 2)

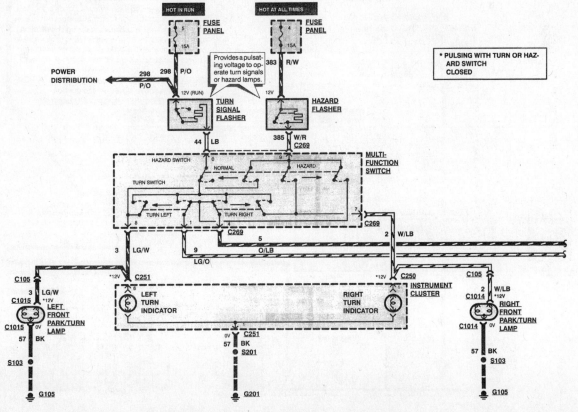

Turn/stop/hazard lighting circuit (1993 and 1994; part 1 of 2)

12-28 Chapter 12 Chassis electrical system

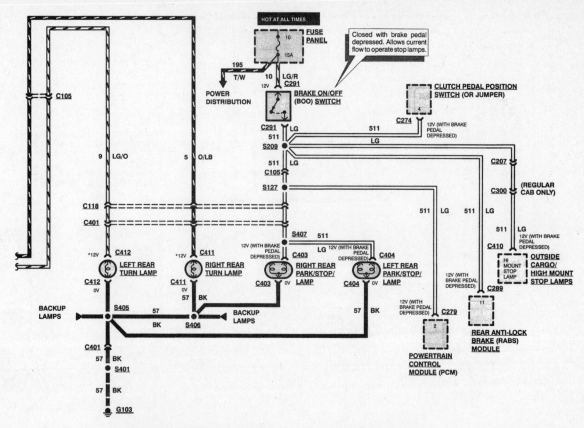

Turn/stop/hazard lighting circuit (1993 and 1994; part 2 of 2)

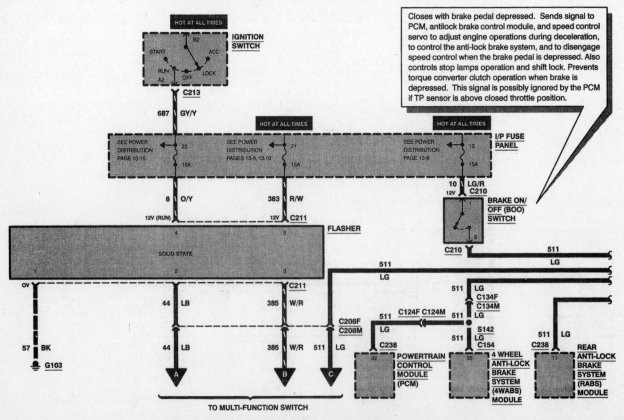

Turn/stop/hazard lighting circuit (1995-on; part 1 of 3)

Chapter 12 Chassis electrical system 12-29

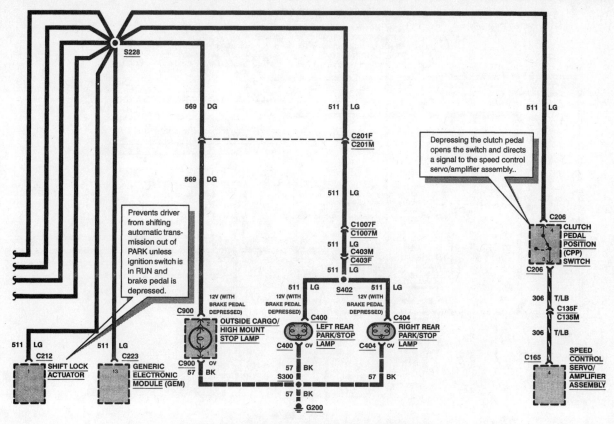

Turn/stop/hazard lighting circuit (1995-on; part 2 of 3)

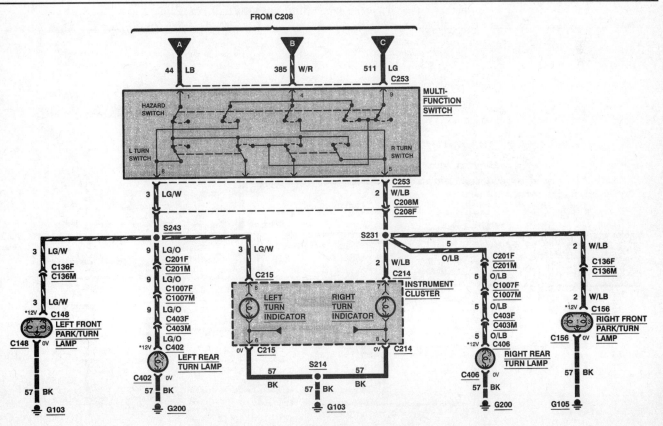

Turn/stop/hazard lighting circuit (1995-on; part 3 of 3)

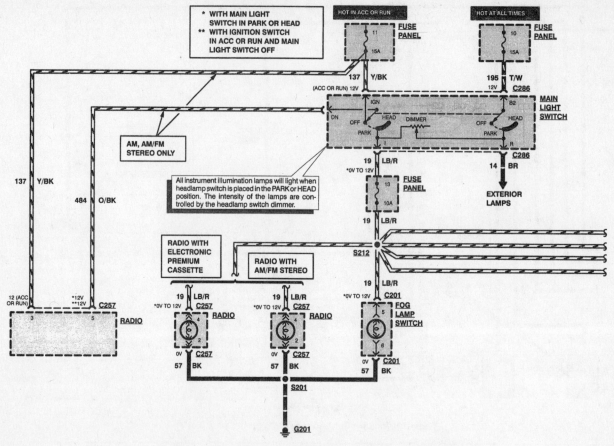

Instrument panel lighting circuit (1993 and 1994; part 1 of 2)

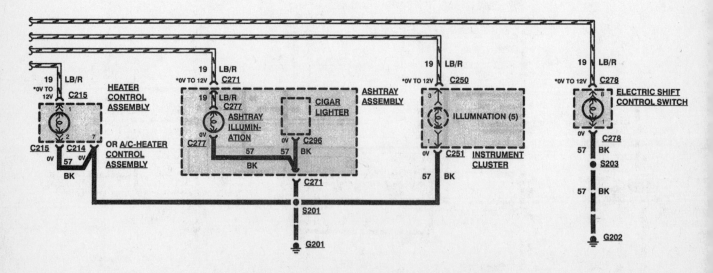

Instrument panel lighting circuit (1993 and 1994; part 2 of 2)

Chapter 12 Chassis electrical system

12-31

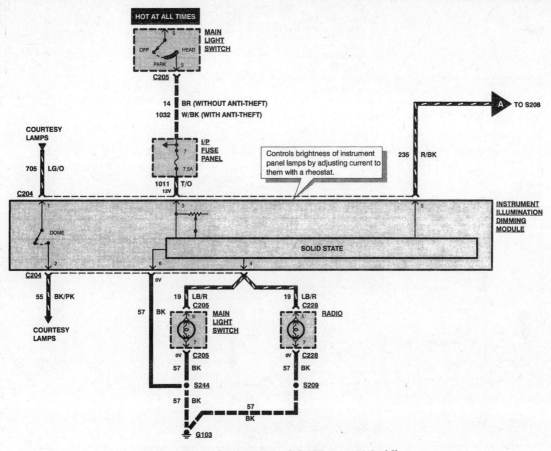

Instrument panel lighting circuit (1995-on; part 1 of 2)

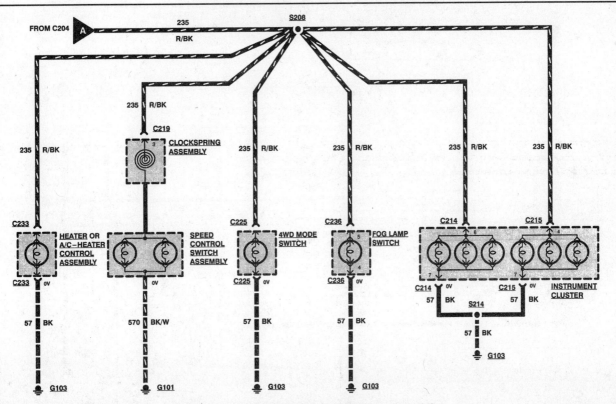

Instrument panel lighting circuit (1995-on; part 2 of 2)

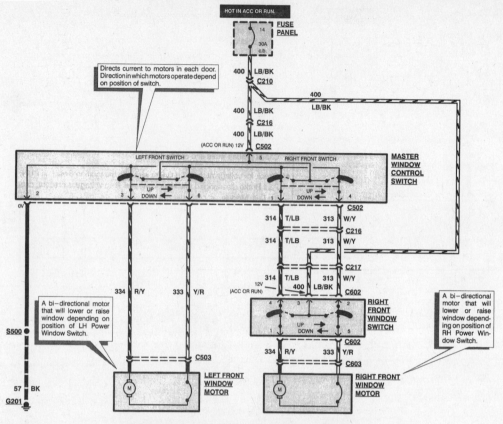

Power window circuit (1993 and 1994 models)

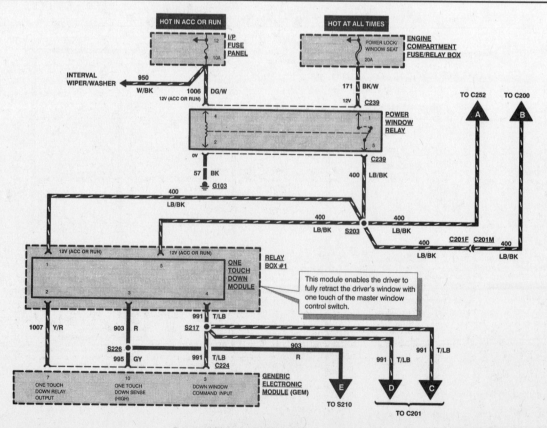

Power window circuit (1995-on; part 1 of 2)

Chapter 12 Chassis electrical system   12-33

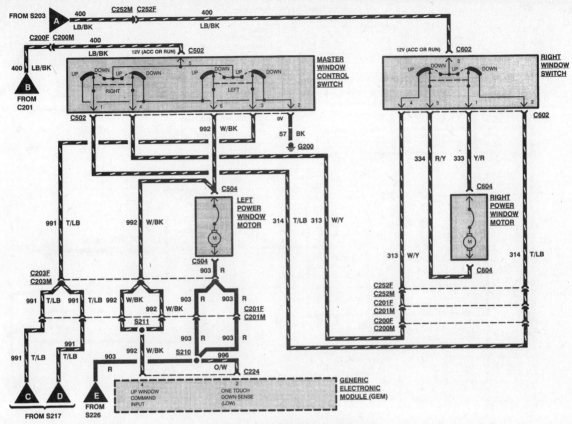

Power window circuit (1995-on; part 2 of 2)

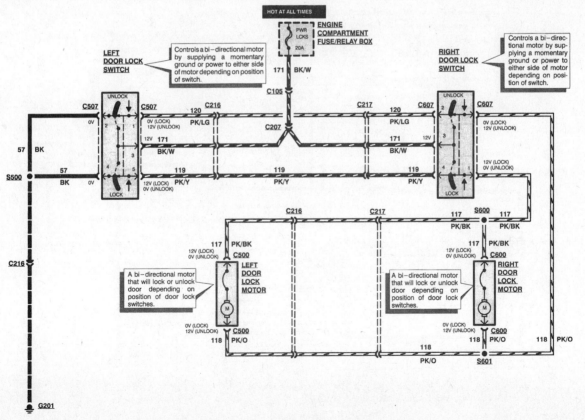

Power door lock circuit (1993 and 1994 models)

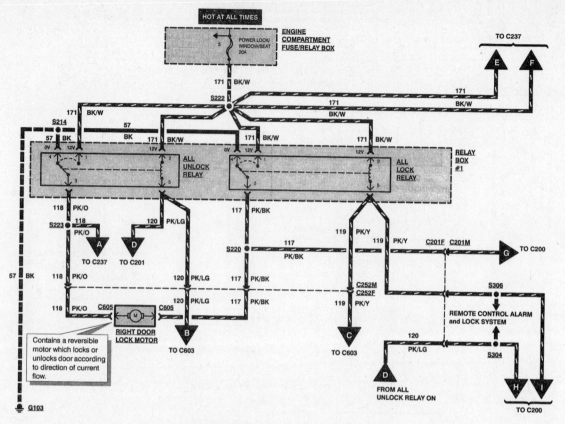

Power door lock circuit (1995-on; part 1 of 2)

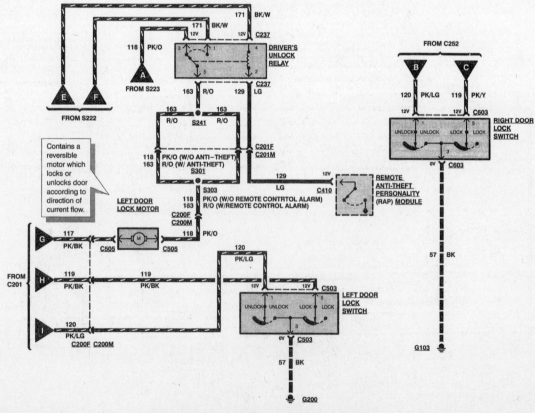

Power door lock circuit (1995-on; part 2 of 2)

Chapter 12 Chassis electrical system

12-35

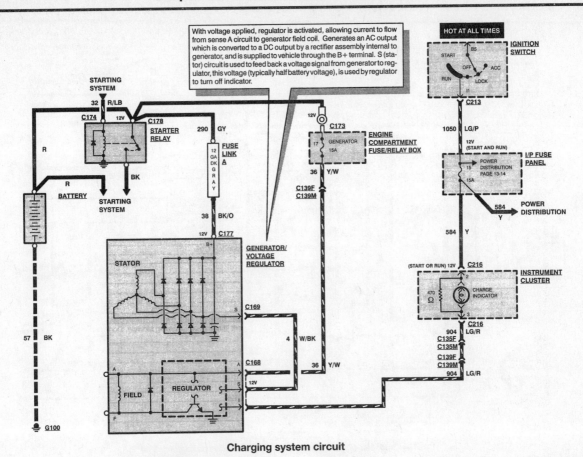

Charging system circuit

Starting system circuit (1996 3.0L and 4.0L models shown; others similar)

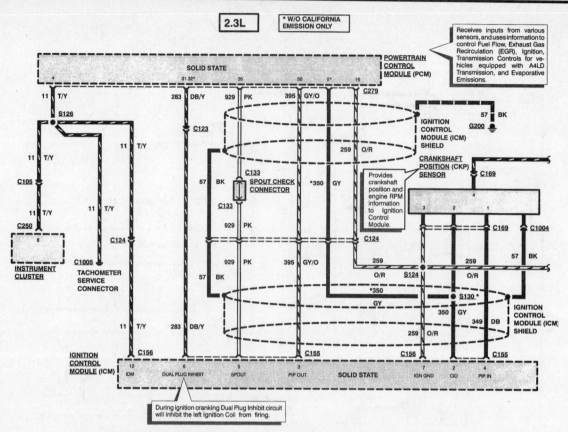

Ignition system circuit (1993 and 1994 2.3L models; part 1 of 2)

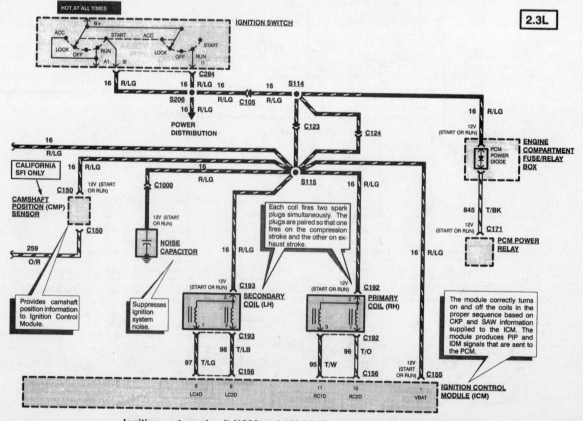

Ignition system circuit (1993 and 1994 2.3L models; part 2 of 2)

Chapter 12 Chassis electrical system

12-37

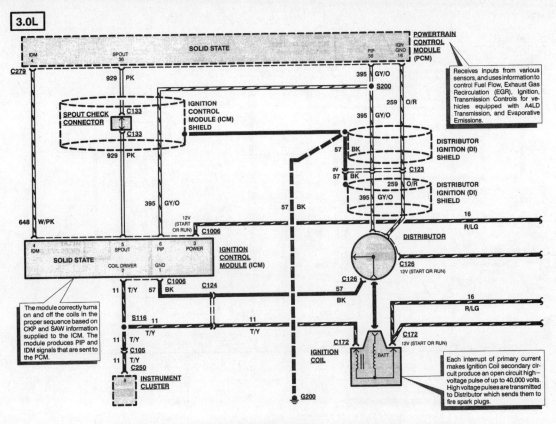

Ignition system circuit (1993 and 1994 3.0L models; part 1 of 2)

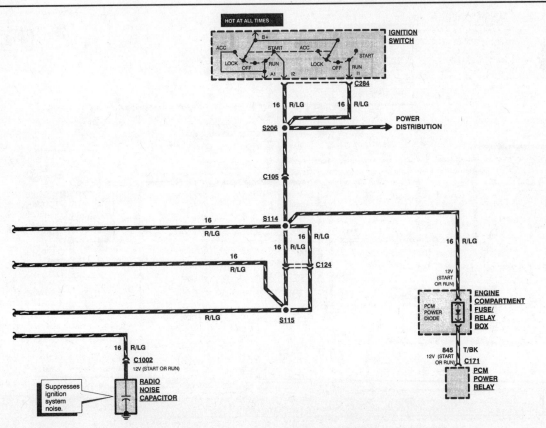

Ignition system circuit (1993 and 1994 3.0L models; part 2 of 2)

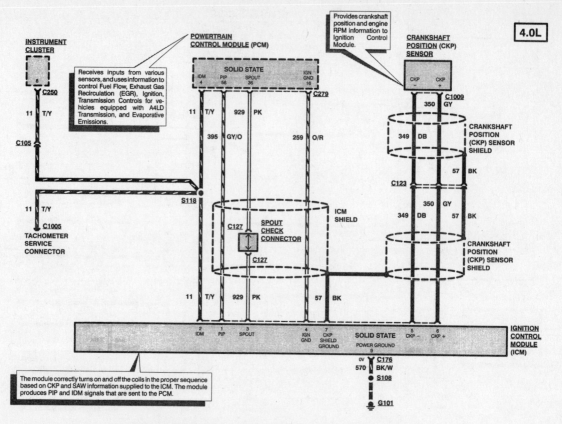

Ignition system circuit (1993 and 1994 4.0L models; part 1 of 2)

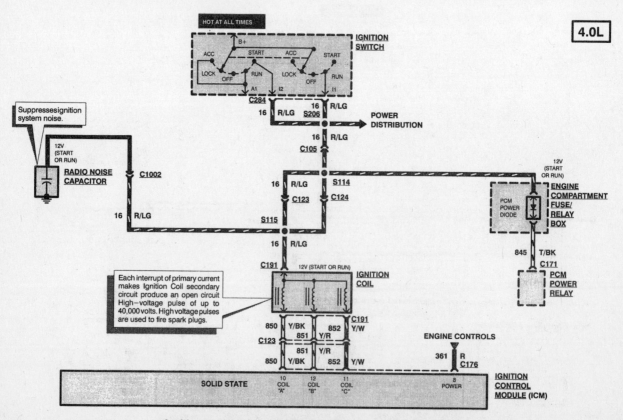

Ignition system circuit (1993 and 1994 4.0L models; part 2 of 2)

Chapter 12 Chassis electrical system

12-39

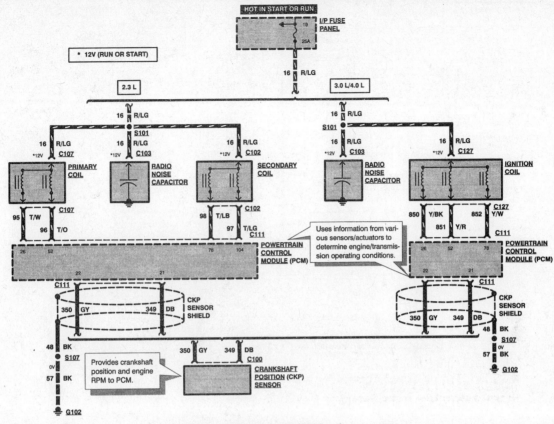

Ignition system circuit (1995-on)

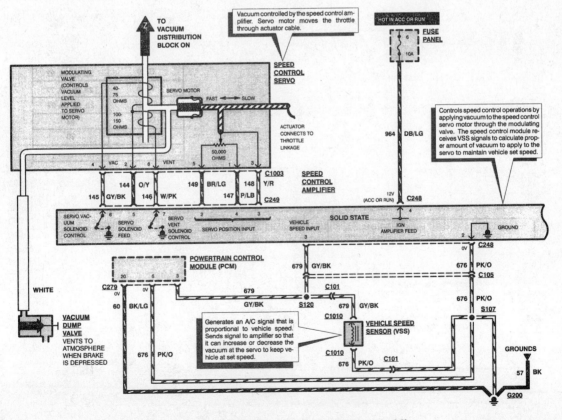

Cruise control circuit (1993 and 1994 models; part 1 of 2)

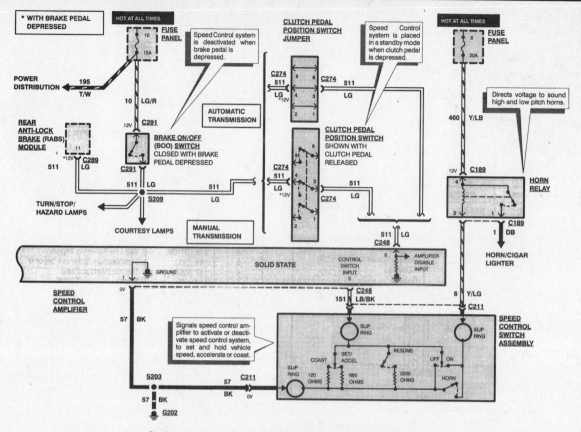

Cruise control circuit (1993 and 1994 models; part 2 of 2)

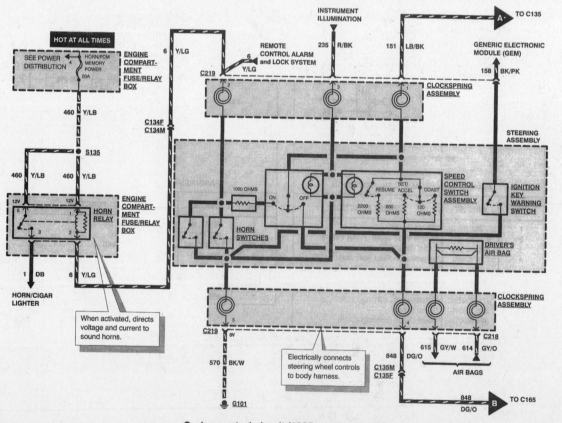

Cruise control circuit (1995-on; part 1 of 2)

Chapter 12 Chassis electrical system 12-41

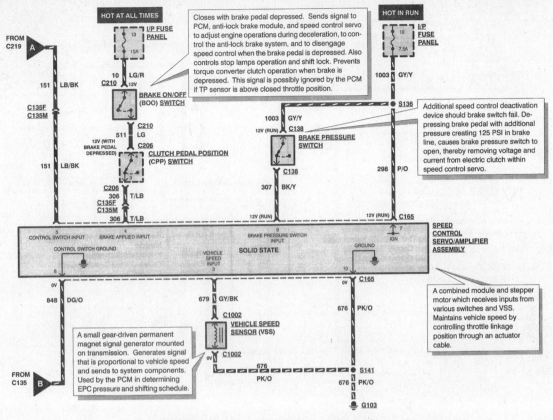

Cruise control circuit (1995-on; part 2 of 2)

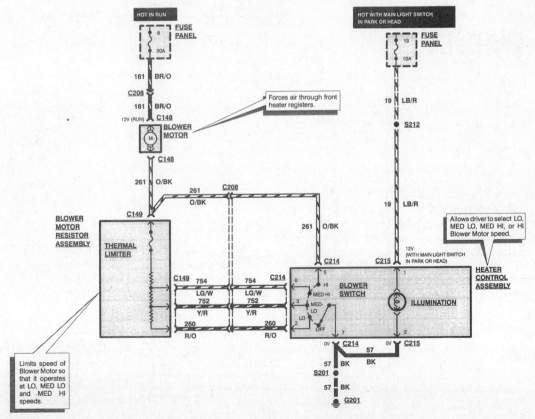

Heater circuit (without air conditioner) (1993 and 1994 models)

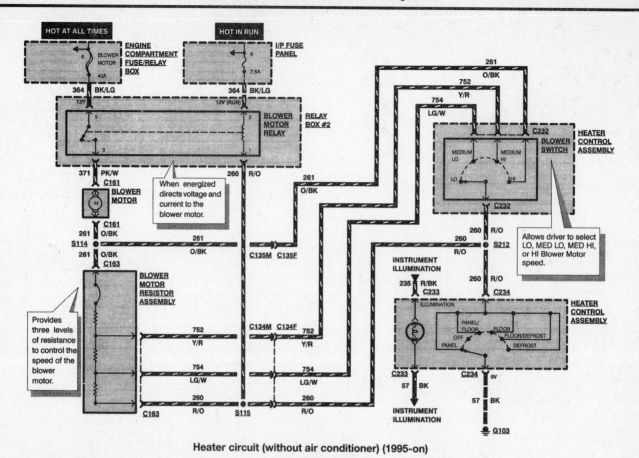

Heater circuit (without air conditioner) (1995-on)

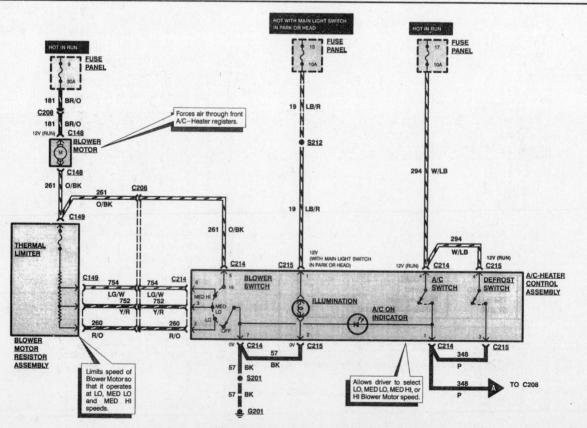

Heater/air conditioner circuit (1993 and 1994 models; part 1 of 2)

Chapter 12 Chassis electrical system

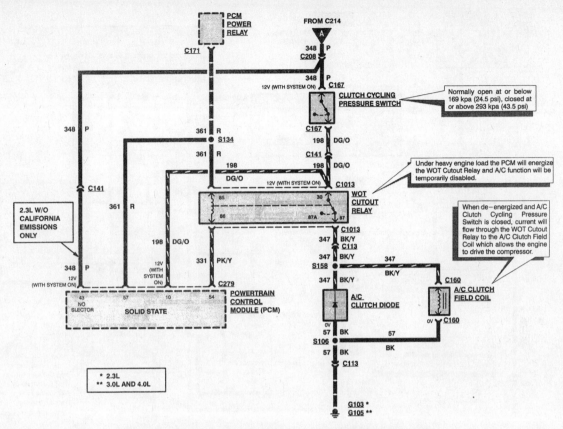

Heater/air conditioner circuit (1993 and 1994 models; part 2 of 2)

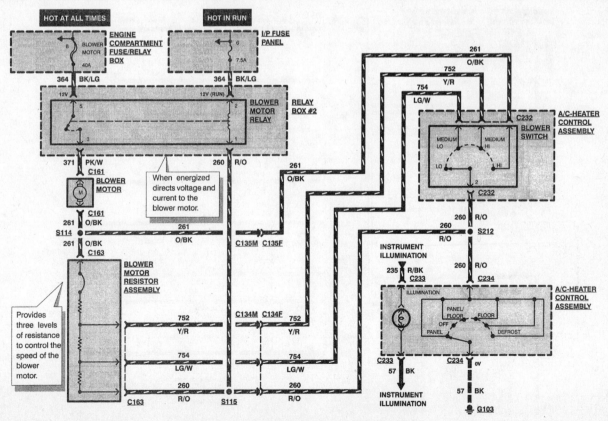

Heater/air conditioner circuit (1995-on; part 1 of 2)

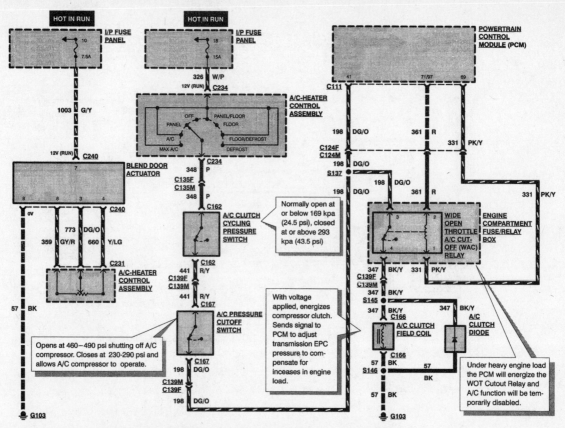

Heater/air conditioner circuit (1995-on; part 2 of 2)

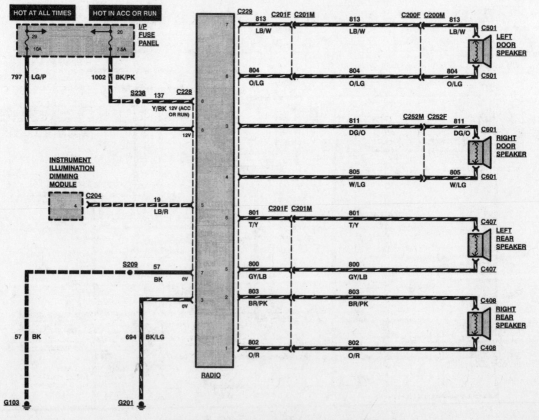

Radio and speaker wiring diagram (typical)

Chapter 12 Chassis electrical system 12-45

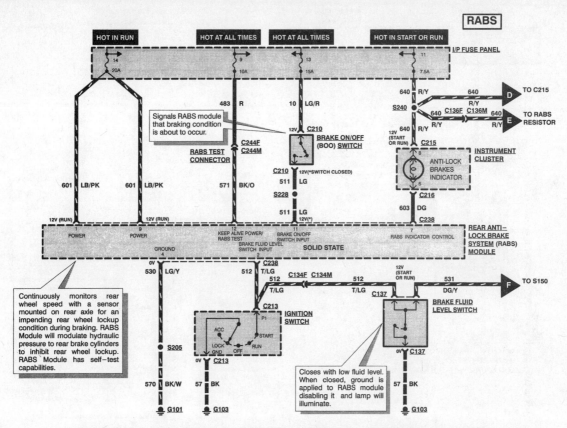

Rear Anti-lock Brake System (RABS) (1996 model shown; others similar; part 1 of 2)

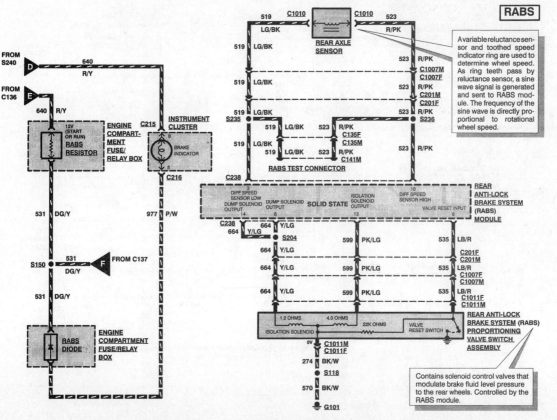

Rear Anti-lock Brake System (RABS) (1996 model shown; others similar; part 2 of 2)

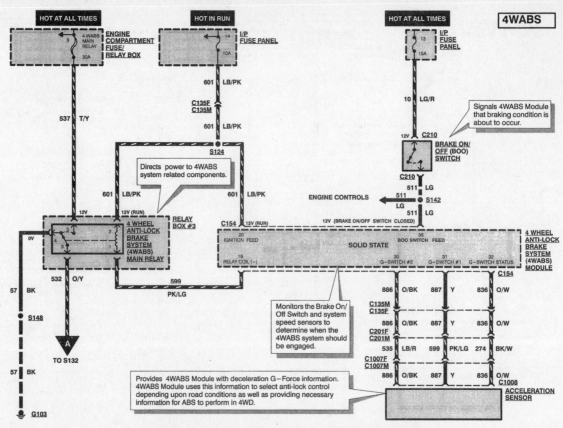

4-Wheel Anti-lock Brake System (4WABS) (part 1 of 3)

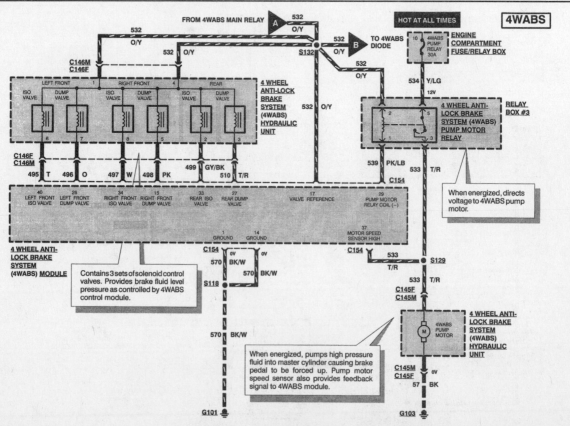

4-Wheel Anti-lock Brake System (4WABS) (part 2 of 3)

Chapter 12 Chassis electrical system 12-47

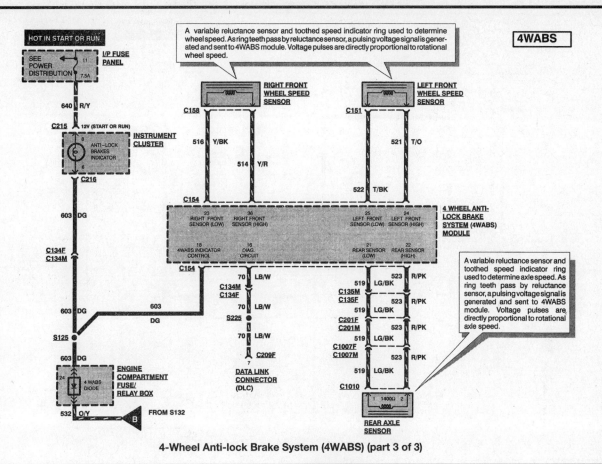

4-Wheel Anti-lock Brake System (4WABS) (part 3 of 3)

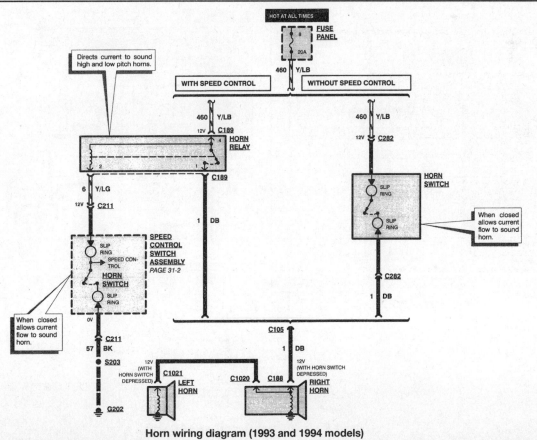

Horn wiring diagram (1993 and 1994 models)

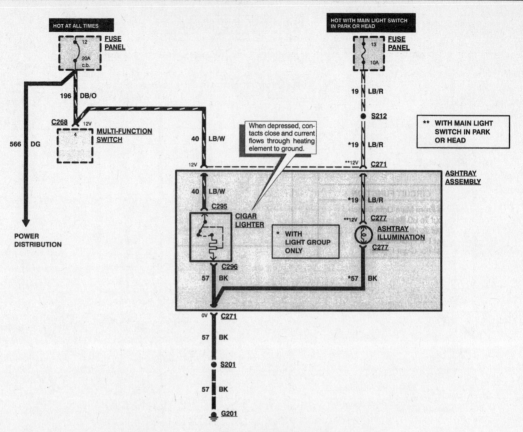

Cigarette lighter wiring diagram (1993 and 1994 models)

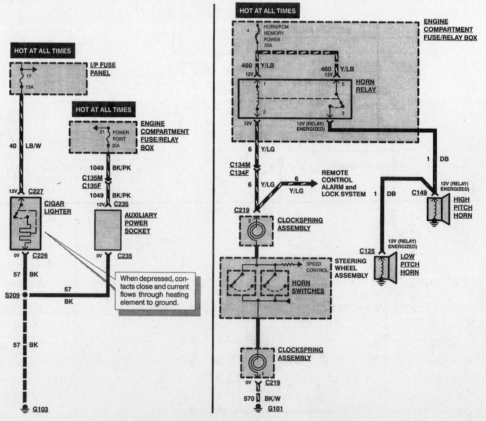

Horn and cigarette lighter wiring diagram (1995-on)

Chapter 12 Chassis electrical system

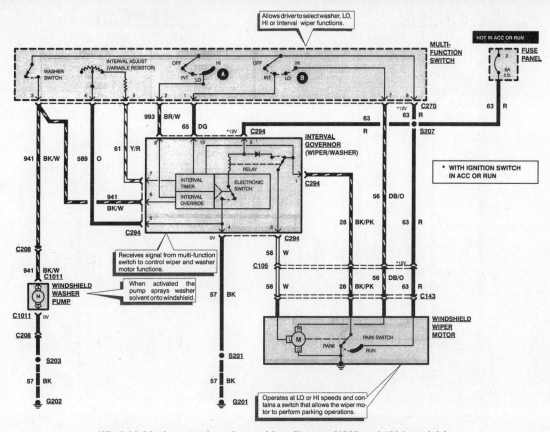

Windshield wipers and washers wiring diagram (1993 and 1994 models)

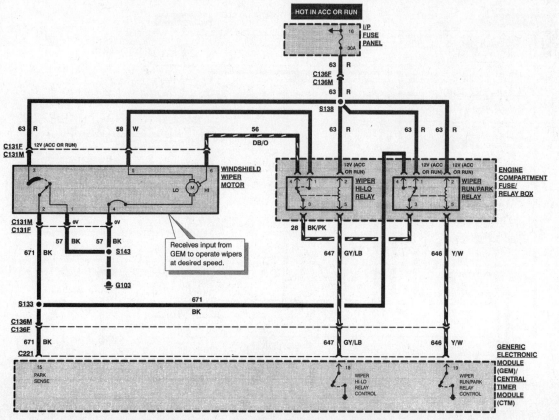

Windshield wipers and washers wiring diagram (1995-on; part 1 of 2)

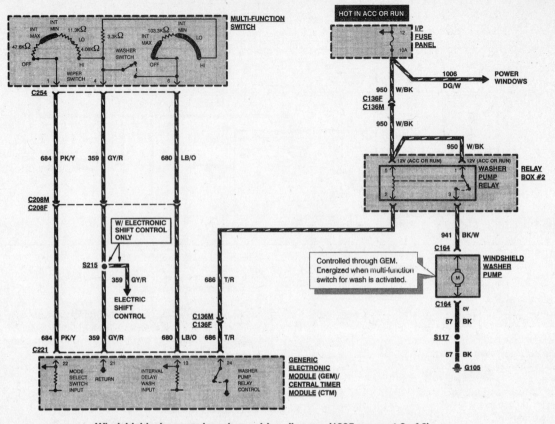

Windshield wipers and washers wiring diagram (1995-on; part 2 of 2)

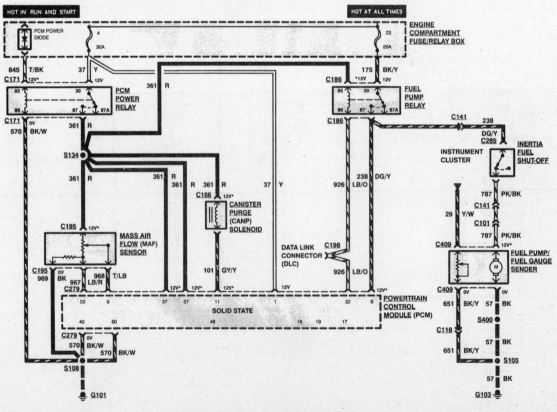

Typical fuel pump wiring diagram

Index

A

About this manual, 0-2
Acknowledgements, 0-2
Air cleaner housing, removal and installation, 4-6
Air conditioning system
    accumulator, removal and installation, 3-11
    check and maintenance, 3-9
    compressor, removal and installation, 3-10
    condenser, removal and installation, 3-11
    evaporator, replacement, 3-12
Air filter replacement, 1-35
Airbag, general information, 12-21
Alternator, removal and installation, 5-10
Antifreeze/coolant, general information, 3-2
Anti-lock Brake System (ABS)
    brake control module, 9-6
    diagnosis and repair, 9-6
    exciter ring, 9-5
    Four-wheel Anti-lock Brake System (4WABS), 9-5
    general information, 9-3
    Rear-wheel Anti-lock Brake System (RABS), 9-3
    relays, 9-6
    speed sensor, 9-5
    wheel sensors, 9-5
Automatic transmission, 7B-1 through 7B-6
    diagnosis, general, 7B-1
    extension housing oil seal, replacement, 7B-3
    fluid
        and filter change, 1-12
        level check/change, 1-12
    general information, 7B-1
    kickdown cable
        adjustment, 7B-3
        replacement, 7B-3
    mount, check and replacement, 7B-6
    neutral start switch, removal, installation and adjustment, 7B-5
    removal and installation, 7B-4
    shift cable
        adjustment, 7B-2
        replacement, 7B-2
    shift linkage lubrication, 1-21
    vacuum diaphragm, removal and installation, 7B-5
Automotive chemicals and lubricants, 0-14
Axle
    arm, front, removal and installation, 10-8
    pivot bracket, removal and installation, 10-9
    pivot bushing, removal and installation, 10-8
    rear, description and check, 8-9

Axle assembly, removal and installation
    front (4WD models), 8-12
    rear, 8-9
Axleshaft and joint assemblies (front), removal, component replacement and installation, 8-14
Axleshafts, bearings and oil seals (rear), removal and installation, 8-11

B

Balljoints, removal and installation, 10-5
Battery
    cables, check and replacement, 5-2
    check, maintenance and charging, 1-16
    removal and installation, 5-2
Body, 11-1 through 11-14
    general information, 11-1
    maintenance, 11-1
    repair
        major damage, 11-5
        minor damage, 11-4
Booster battery (jump) starting, 0-13
Brake light switch, removal and installation, 9-18
Brake On/Off (BOO) switch, 6-13
Brakes, 9-1 through 9-18
    Anti-lock Brake System (ABS), general information, 9-3
    brake control module, 9-6
    brake disc, inspection, removal and installation, 9-8
    caliper, removal, overhaul and installation, 9-6
    general information, 9-2
    hoses and lines, inspection and replacement, 9-15
    hydraulic system, bleeding, 9-16
    master cylinder, removal, overhaul and installation, 9-14
    pads, replacement, 9-5
    parking
        adjustment, 9-17
        cables, replacement, 9-17
    power brake booster, check, removal, installation and adjustment, 9-18
    shoes (rear), replacement, 9-9
    system check, 1-28
    wheel cylinder, removal, overhaul and installation, 9-13
Bulb replacement, 12-14
Bumper
    and valance, front, removal and installation, 11-9
    rear, removal and installation, 11-10
Buying parts, 0-6

C

Camshaft
  position sensor, 6-11
  lifters and bearings, inspection and replacement, 2D-18
Catalytic converter, 6-18
Center bearing, replacement, 8-8
Charging system
  check, 5-9
  general information and precautions, 5-8
Chassis electrical system, 12-1 through 12-50
Chassis lubrication, 1-22
Circuit breakers, general information, 12-7
Clutch and drivetrain, 8-1 through 8-18
Clutch
  clutch/starter interlock switch, removal and installation, 8-6
  components, removal, inspection and installation, 8-2
  description and check, 8-2
  hydraulic line quick-disconnect fittings, general information, 8-4
  hydraulic system, bleeding, 8-5
  master cylinder and reservoir, removal and installation, 8-5
  pedal, removal and installation, 8-6
  release bearing, removal, inspection and installation, 8-4
  release cylinder, removal and installation, 8-5
Coil spring, removal and installation, 10-6
Connecting rod, steering, removal and installation, 10-15
Connectors, general information, 12-2
Console, removal and installation, 11-11
Conversion factors, 0-15
Coolant
  reservoir, removal and installation, 3-5
  temperature sending unit, check and replacement, 3-5
Cooling and heating systems
  antifreeze/coolant, general information, 3-2
  coolant reservoir, removal and installation, 3-5
  coolant temperature sending unit, check and replacement, 3-5
  cooling fan and viscous clutch, inspection, removal and installation, 3-4
  heater
    and air conditioning control assembly, removal and installation, 3-7
    blower motor and circuit, check and replacement, 3-8
    control cables, check and adjustment, 3-7
    core, replacement, 3-8
    general information, 3-6
  radiator, removal and installation, 3-3
  system
    check, 1-26
    servicing, 1-34
  thermostat, check and replacement, 3-2
  water pump, check and replacement, 3-6
Cooling, heating and air conditioning systems, 3-1 through 3-12
Crankshaft
  position sensor, 5-4, 6-13
  inspection, 2D-16
  installation and main bearing oil clearance check, 2D-19
  removal, 2D-13
Cruise control system, description and check, 12-21
Cylinder
  compression check, 2D-7
  head
    cleaning and inspection, 2D-10
    disassembly, 2D-9
    reassembly, 2D-12
    removal and installation
      2.3L four-cylinder engine, 2A-9
      3.0L V6 engine, 2B-7
      4.0L V6 engine, 2C-9
    honing, 2D-15

D

Diagnostic systems, general information, 6-2
Diagnostic Trouble Codes (EEC-IV, 1993 and 1994 models), 6-5
Differential lubricant level check and change, 1-23
Distributor
  removal and installation, 5-4
  stator assembly, check and replacement, 5-5
Door
  hinge, removal and installation, 11-6
  latch striker, removal, installation and adjustment, 11-6
  lock cylinder, removal and installation, 11-7
  lock system, power, description and check, 12-21
  trim panel and watershield, removal and installation, 11-8
  removal, installation and alignment, 11-6
  window glass, front, replacement and adjustment, 11-8
  window regulator, front, replacement, 11-9
Drag link, removal and installation, 10-15
Driveaxle boot replacement and CV joint overhaul, 8-17
Drivebelt check, adjustment and replacement, 1-14
Driveshaft
  and driveaxle yoke lubrication (4WD models), 1-36
  and universal joints, description and check, 8-7
  removal and installation, 8-7

E

Electrical troubleshooting, general information, 12-1
Electronic Engine Control (EEC-IV, EEC-V) system and trouble codes, 6-2
Electronic shift control
  module, removal and installation, 7C-3
  removal and installation, 7C-2
Emissions control systems, 6-1 through 6-18
  catalytic converter, 6-18
  diagnostic systems, general information, 6-2
  Electronic Engine Control (EEC-IV, EEC-V) system and trouble codes, 6-2
  Evaporative Emissions Control System (EECS), 6-15
  Exhaust Gas Recirculation (EGR) system, 6-14
  general information, 6-1
  information sensors, 6-7
  inlet air temperature control system (1993 and 1994 models), 6-17
  Positive Crankcase Ventilation (PCV) system, 6-16
  Powertrain Control Module (PCM), 6-7
Engine
  four-cylinder engines
    auxiliary shaft (1993 and 1994 models), removal, inspection and installation, 2A-7
    camshaft, removal, inspection and installation, 2A-6
    cylinder head, removal and installation, 2A-9
    exhaust manifold, removal and installation, 2A-8
    flywheel/driveplate, removal and installation, 2A-11
    front oil seals, replacement, 2A-7
    general information, 2A-2
    intake manifold, removal and installation, 2A-8
    mounts, removal and installation, 2A-12
    oil pan, removal and installation, 2A-10
    oil pump, removal and installation, 2A-10
    rear main oil seal, replacement, 2A-12
    repair operations possible with the engine in the vehicle, 2A-2
    timing belt, removal and installation, 2A-4
    Top Dead Center (TDC) for number one piston, locating, 2A-2
    valve cover, removal and installation, 2A-3
    valve train components, removal and installation, 2A-3
  3.0L V6 engine, 2B-1 through 2B-12
    camshaft, removal, inspection and installation, 2B-11
    crankshaft front oil seal, replacement, 2B-8
    cylinder heads, removal and installation, 2B-7

exhaust manifolds, removal and installation, 2B-6
general information, 2B-2
intake manifold, removal and installation, 2B-5
mounts, removal and installation, 2B-12
oil pan, removal and installation, 2B-11
oil pump and pick-up, removal and installation, 2B-12
rear main oil seal, replacement, 2B-12
repair operations possible with the engine in the vehicle, 2B-2
rocker arms and pushrods, removal, inspection and installation, 2B-3
timing chain and sprockets, check, removal and installation, 2B-9
timing chain cover, removal and installation, 2B-8
Top Dead Center (TDC) for number one piston, locating, 2B-2
valve covers, removal and installation, 2B-3
valve lifters, removal, inspection and installation, 2B-10
valve springs, retainers and seals, replacement, 2B-4
4.0L V6 engine, 2C-1 through 2C-18
camshaft, removal, inspection and installation, 2C-14
crankshaft
    oil seals replacement, 2C-17
    pulley and front oil seal, removal and installation, 2C-10
cylinder heads, removal and installation, 2C-9
exhaust manifolds, removal and installation, 2C-8
flywheel/driveplate, removal and installation, 2C-17
general information, 2C-2
intake manifold, removal and installation, 2C-7
mounts, check and replacement, 2C-18
oil pan and baffle, removal and installation, 2C-15
oil pump, removal and installation, 2C-16
repair operations possible with the engine in the vehicle, 2C-2
rocker arms and pushrods, removal, inspection and installation, 2C-5
timing chain and sprockets, inspection, removal and installation, 2C-11
timing chain cover, removal and installation, 2C-11
Top Dead Center (TDC) for number one piston, locating, 2C-2
valve covers, removal and installation, 2C-3
valve lifters, removal, inspection and installation, 2C-13
valve springs, retainers and seals, replacement, 2C-6
general overhaul procedures
    block
        cleaning, 2D-13
        inspection, 2D-14
    general information, 2D-6
oil and filter change, 1-10
overhaul
    disassembly sequence, 2D-9
    general information, 2D-6
    reassembly sequence, 2D-18
rebuilding alternatives, 2D-7
removal, methods and precautions, 2D-7
removal and installation, 2D-8

Engine Coolant Temperature (ECT) sensor, 6-7
Engine electrical systems, 5-1 through 5-12
Evaporative Emissions Control System (EECS), 6-15
Exciter ring, 9-5
Exhaust Gas Recirculation (EGR) system, 6-14
Exhaust system
air cleaner housing, removal and installation, 4-6
check, 1-26
service, general information, 4-12
Extension housing oil seal, replacement
2WD models, 7A-4
4WD models, 7B-3

F

Fender, front, removal and installation, 11-11

Fluid level checks, 1-7
brake and clutch fluid, 1-8
engine coolant, 1-8
engine oil, 1-7
windshield washer fluid, 1-9
Four wheel Anti-lock Brake System (4WABS), 9-5
Front end alignment, general information, 10-22
Front hub lock, spindle bearing and wheel bearing maintenance (4WD models), 1-31
Fuel and exhaust systems, 4-1 through 4-12
Fuel filter replacement, 1-36
Fuel system
accelerator cable, removal and installation, 4-7
check, 1-25
electronic fuel injection system, component check and replacement, 4-8
electronic fuel injection system, general checks, 4-8
fuel
    level sending unit, check and replacement, 4-6
    lines and fittings, general information, 4-2
    pressure relief procedure, 4-2
    pump, removal and installation, 4-5
    pump/fuel pressure, check, 4-4
    tank
        cleaning and repair, 4-5
        removal and installation, 4-4
Fuses, general information, 12-2
Fusible links, general information, 12-7

G

Gauges, replacement, 12-18
General engine overhaul procedures, 2D-1 through 2D-22
Glove box, removal and installation, 11-14

H

Headlight
adjusting, 12-13
bulb and housing, replacement, 12-13
switch, replacement, 12-12
Heater
and air conditioning control assembly, removal and installation, 3-7
blower motor and circuit, check and replacement, 3-8
control cables, check and adjustment, 3-7
core, replacement, 3-8
general information, 3-6
Hinges and locks, maintenance, 11-5
Hood
latch control cable, removal and installation, 11-5
removal, installation and adjustment, 11-5
Horn, check and replacement, 12-11
Hub and bearing assembly (1998 and later 4WD models), removal and installation, 10-10

I

Idle Air Control (IAC) solenoid, 6-11
Ignition
coil, check, removal and installation, 5-3
module, check, removal and installation, 5-6
switch/key lock cylinder, check and replacement, 12-8
system
    check, 5-2
    general information, 5-2
timing procedure, 5-8

Information sensors, 6-7
    Brake On/Off (BOO) switch, 6-13
    camshaft position sensor, 6-11
    crankshaft position sensor, 6-13
    Engine Coolant Temperature (ECT) sensor, 6-7
    Idle Air Control (IAC) solenoid, 6-11
    Intake Air Temperature (IAT) Sensor, 6-10
    Mass Airflow Sensor (MAF), 6-9
    oxygen sensor, 6-7
    power steering pressure switch, 6-11
    Throttle Position Sensor (TPS), 6-8
    Vehicle Speed Sensor (VSS), 6-9
Initial start-up and break-in after overhaul, 2D-22
Inlet air temperature control system (1993 and 1994 models), 6-17
Inner door handle and latch assembly, removal and installation, 11-7
Instrument cluster
    bezel, removal and installation, 11-12
    cluster, removal and installation, 12-17
Instrument panel, removal and installation, 11-13
Intake Air Temperature (IAT) Sensor, 6-10
Introduction to the Ford Ranger/Mazda B-series pick-ups, 1-7

J

Jacking and towing, 0-12

K

Kickdown cable
    adjustment, 7B-3
    replacement, 7B-3

L

Leaf spring, removal and installation, 10-12
Linkage adjustment, manual shift transfer case, 7C-1

M

Main and connecting rod bearings, inspection, 2D-17
Maintenance schedule, 1-6
Maintenance techniques, tools and working facilities, 0-6
Manual transmission, 7A-1 through 7A-4
    extension housing oil seal (2WD models), replacement, 7A-4
    general information, 7A-2
    lubricant level check and change, 1-23
    removal and installation, 7A-3
    shift lever, removal and installation, 7A-2
    speedometer pinion gear and seal, removal and installation, 7A-3
    transmission overhaul, general information, 7A-4
Mass Airflow Sensor (MAF), 6-9
Master cylinder, removal, overhaul and installation, 9-14
Mirrors
    exterior, removal and installation, 11-14
    power, check, removal and installation, 12-20

N

Neutral start switch, removal, installation and adjustment, 7B-5

O

Outer door handle, removal and installation, 11-7
Output shaft oil seal, front, removal and installation, 7C-3
Oxygen sensor, 6-7

P

Parking brake
    adjustment, 9-17
    cables, replacement, 9-17
    check, 1-29
Pilot bearing, inspection and replacement, 8-3
Piston rings, installation, 2D-18
Pistons/connecting rods
    inspection, 2D-15
    installation and rod bearing oil clearance check, 2D-21
    removal, 2D-12
Positive Crankcase Ventilation (PCV) system, 6-16
Power brake booster, check, removal, installation and adjustment, 9-18
Power steering
    fluid level check, 1-11
    line quick-disconnect fittings, 10-18
    pressure switch, 6-11
    pump, removal and installation, 10-17
    system, bleeding, 10-19
Powertrain Control Module (PCM), 6-7

R

Radiator
    grille, removal and installation, 11-10
    removal and installation, 3-3
Radio
    antenna, removal and installation, 12-16
    removal and installation, 12-15
Radius arm
    insulators, replacement, 10-8
    removal and installation, 10-7
Rear Wheel Anti-lock Brake System (RABS), 9-3

S

Safety checks, 1-10
Safety first!, 0-16
Seat belt check, 11-14
Seats, removal and installation, 11-14
Shift cable
    adjustment, 7B-2
    replacement, 7B-2
Shift lever, removal and installation, 7A-2
Shock absorbers, inspection, removal and installation
    front, 10-3
    rear, 10-12
Slip yoke (right) and stub shaft assembly, carrier, carrier oil seals and bearings, removal and installation, 8-15
Spark plugs, wires, distributor cap and rotor check and replacement, 1-17
Speakers, removal and installation, 12-16
Speed sensor, 9-5
Speedometer
    cable, replacement, 12-19
    pinion gear and seal, removal and installation, 7A-3

Index

Stabilizer bar
    front, removal and installation, 10-10
    bar, rear, removal and installation, 10-12
Starter
    motor
        and circuit, in-vehicle check, 5-11
        removal and installation, 5-12
    relay, removal and installation, 5-12
Starting system, general information, 5-11
Steering system
    and suspension check, 1-27
    column cover and shroud, removal and installation, 11-12
    column switches, replacement, 12-8
    connecting rod, removal and installation, 10-15
    drag link, removal and installation, 10-15
    front end alignment, general information, 10-22
    gear, removal and installation, 10-16
    general information, 10-16
    power steering
        line quick disconnect fittings, 10-21
        pump, removal and installation, 10-20
        system, bleeding, 10-19
    steering gear, removal and installation, 10-19
    steering wheel, removal and installation, 10-16
    tie-rod ends, removal and installation, 10-16
    wheel, removal and installation, 10-16
    wheels and tires, general information, 10-22
Suspension and steering systems, 10-1 through 10-22
Suspension system
    axle
        arm, front, removal and installation, 10-8
        pivot bracket, removal and installation, 10-9
        pivot bushing, removal and installation, 10-8
    balljoints, removal and installation, 10-5
    coil spring, removal and installation, 10-6
    control arm (1998 and later models), removal, inspection and installation
        lower, 10-13
        upper, 10-14
    leaf spring, removal and installation, 10-15
    radius arm
        insulators, replacement, 10-8
        removal and installation, 10-7
    shock absorbers
        front, inspection, removal and installation, 10-4
        rear, inspection, removal and installation, 10-14
    stabilizer bar
        front, removal and installation, 10-10, 10-13
        rear, removal and installation, 10-15
    wheel spindle, front, removal and installation, 10-5

T

Thermostat, check and replacement, 3-2
Throttle Position Sensor (TPS), 6-8
Tie-rod ends, removal and installation, 10-16
Tire and tire pressure checks, 1-9
Tire rotation, 1-13
Tools, 0-8
Torsion bar (1998 and later 4WD models), removal and installation, 10-13

Transfer case, 7C-1 through 7C-6
    electronic shift control
        module, removal and installation, 7C-3
        removal and installation, 7C-2
    electronic shift motor, general information, diagnosis and replacement, 7C-5
    front output shaft oil seal, removal and installation, 7C-3
    general information, 7C-1
    linkage adjustment, manual shift transfer case, 7C-1
    lubricant level check and change, 1-24
    overhaul, general information, 7C-5
    rear output shaft oil seal, removal and installation, 7C-3
    removal and installation, 7C-4
    shift lever, removal and installation, 7C-2
Transmission mount, check and replacement, 7A-2
Transmission overhaul, general information, 7A-4
Troubleshooting, 0-17
Tune-up
    and routine maintenance, 1-1 through 1-36
    general information, 1-7
Turn signal/hazard flasher relay, replacement, 12-8

U

Underhood hose check and replacement, 1-13
Universal joints, replacement, 8-7
Upholstery and carpets, maintenance, 11-4

V

Vacuum diaphragm, removal and installation, 7B-5
Valves, servicing, 2D-11
Vehicle identification numbers, 0-5
Vehicle Speed Sensor (VSS), 6-9
Vinyl trim, maintenance, 11-1
Voltage regulator/alternator brushes, replacement, 5-11

W

Water pump, check and replacement, 3-6
Wheel
    bearing (front) check, repack and adjustment (2WD models), 1-29
    cylinder, removal, overhaul and installation, 9-14
    sensors, 9-5
    spindle, front, removal and installation
        1993 through 1997 models, 10-5
        1998 and later models, 10-12
Wheels and tires, general information, 10-22
Window, power system, description and check, 12-19
Windshield
    and fixed glass, removal and installation, 11-11
    wiper
        arm, removal and installation, 12-11
        blade check and replacement, 1-33
        motor, removal and installation, 12-11
Wiring diagrams, general information, 12-22
Working facilities, 0-11

Haynes Automotive Manuals

NOTE: New manuals are added to this list on a periodic basis. If you do not see a listing for your vehicle, consult your local Haynes dealer for the latest product information.

ACURA
- *12020 Integra '86 thru '89 & Legend '86 thru '90

AMC
- Jeep CJ - see JEEP (50020)
- 14020 Concord/Hornet/Gremlin/Spirit '70 thru '83
- 14025 (Renault) Alliance & Encore '83 thru '87

AUDI
- 15020 4000 all models '80 thru '87
- 15025 5000 all models '77 thru '83
- 15026 5000 all models '84 thru '88

AUSTIN
- Healey Sprite - see MG Midget (66015)

BMW
- *18020 3/5 Series '82 thru '92
- *18021 3 Series except 325iX models '92 thru '97
- 18025 320i all 4 cyl models '75 thru '83
- 18035 528i & 530i all models '75 thru '80
- 18050 1500 thru 2002 except Turbo '59 thru '77

BUICK
- Century (FWD) - see GM (38005)
- *19020 Buick, Oldsmobile & Pontiac Full-size (Front wheel drive) '85 thru '98
  - Buick Electra, LeSabre and Park Avenue; Oldsmobile Delta 88 Royale, Ninety Eight and Regency; Pontiac Bonneville
- 19025 Buick Oldsmobile & Pontiac Full-size (Rear wheel drive)
  - Buick Estate '70 thru '90, Electra '70 thru '84, LeSabre '70 thru '85, Limited '74 thru '79 Oldsmobile Custom Cruiser '70 thru '90, Delta 88 '70 thru '85, Ninety-eight '70 thru '84 Pontiac Bonneville '70 thru '81, Catalina '70 thru '81, Grandville '70 thru '75, Parisienne '83 thru '86
- 19030 Mid-size Regal & Century '74 thru '87
  - Regal - see GENERAL MOTORS (38010)
  - Skyhawk - see GM (38030)
  - Skylark - see GM (38020, 38025)
  - Somerset - see GENERAL MOTORS (38025)

CADILLAC
- *21030 Cadillac Rear Wheel Drive '70 thru '93
  - Cimarron, Eldorado & Seville - see GM (38015, 38030)

CHEVROLET
- 10305 Chevrolet Engine Overhaul Manual
- *24010 Astro & GMC Safari Mini-vans '85 thru '93
- 24015 Camaro V8 all models '70 thru '81
- 24016 Camaro all models '82 thru '92
  - Cavalier - see GM (38015)
  - Celebrity - see GM (38005)
- 24017 Camaro & Firebird '93 thru '97
- 24020 Chevelle, Malibu, El Camino '69 thru '87
- 24024 Chevette & Pontiac T1000 '76 thru '87
  - Citation - see GENERAL MOTORS (38020)
- *24032 Corsica/Beretta all models '87 thru '96
- 24040 Corvette all V8 models '68 thru '82
- *24041 Corvette all models '84 thru '96
- 24045 Full-size Sedans Caprice, Impala, Biscayne, Bel Air & Wagons '69 thru '90
- 24046 Impala SS & Caprice and Buick Roadmaster '91 thru '96
  - Lumina '90 thru '94 - see GM (38010)
- 24048 Lumina & Monte Carlo '95 thru '98
  - Lumina APV - see GM (38038)
- 24050 Luv Pick-up 2WD & 4WD '72 thru '82
- 24055 Monte Carlo all models '70 thru '88
  - Monte Carlo '95 thru '98 - see LUMINA
- 24059 Nova all V8 models '69 thru '79
- *24060 Nova/Geo Prizm '85 thru '92
- 24064 Pick-ups '67 thru '87 - Chevrolet & GMC, all V8 & in-line 6 cyl, 2WD & 4WD '67 thru '87; Suburbans, Blazers & Jimmys '67 thru '91
- *24065 Pick-ups '88 thru '98 - Chevrolet & GMC, all full-size pick-ups '88 thru '98; Blazer & Jimmy '92 thru '94; Suburban '92 thru '98; Tahoe & Yukon '95 thru '98
- *24070 S-10 & GMC S-15 Pick-ups '82 thru '93
- 24071 S-10, Gmc S-15 & Jimmy '94 thru '96
- *24075 Sprint & Geo Metro '85 thru '94
- *24080 Vans - Chevrolet & GMC '68 thru '96

CHRYSLER
- 10310 Chrysler Engine Overhaul Manual
- *25015 Chrysler Cirrus, Dodge Stratus, Plymouth Breeze, '95 thru '98
- *25020 Full-size Front-Wheel Drive '88 thru '93
  - K-Cars - see DODGE Aries (30008)
  - Laser - see DODGE Daytona (30030)
- 25025 Chrysler LHS, Concorde & New Yorker, Dodge Intrepid, Eagle Vision, '93 thru '97
- *25030 Chrysler/Plym. Mid-size '82 thru '95
  - Rear-wheel Drive - see DODGE (30050)

DATSUN
- 28005 200SX all models '80 thru '83
- 28007 B-210 all models '73 thru '78
- 28009 210 all models '78 thru '82
- 28012 240Z, 260Z & 280Z Coupe '70 thru '78
- 28014 280ZX Coupe & 2+2 '79 thru '83
  - 300ZX - see NISSAN (72010)
- 28016 310 all models '78 thru '82
- 28018 510 & PL521 Pick-up '68 thru '73
- 28020 510 all models '78 thru '81
- 28022 620 Series Pick-up all models '73 thru '79
  - 720 Series Pick-up - NISSAN (72030)
- 28025 810/Maxima all gas models, '77 thru '84

DODGE
- 400 & 600 - see CHRYSLER (25030)
- *30008 Aries & Plymouth Reliant '81 thru '89
- 30010 Caravan & Ply. Voyager '84 thru '95
- *30011 Caravan & Ply. Voyager '96 thru '98
- 30012 Challenger/Plymouth Saporro '78 thru '83
  - Challenger '67-'76 - see DART (30025)
- 30016 Colt/Plymouth Champ '78 thru '87
- *30020 Dakota Pick-ups all models '87 thru '96
- 30025 Dart, Challenger/Plymouth Barracuda & Valiant 6 cyl all models '67 thru '76
- *30030 Daytona & Chrysler Laser '84 thru '89
  - Intrepid - see Chrysler (25025)
- *30034 Dodge & Plymouth Neon '95 thru '97
- *30035 Omni & Plymouth Horizon '78 thru '90
- 30040 Pick-ups all full-size models '74 thru '93
- *30041 Pick-ups all full-size models '94 thru '96
- *30045 Ram 50/D50 Pick-ups & Raider and Plymouth Arrow Pick-ups '79 thru '93
- 30050 Dodge/Ply./Chrysler RWD '71 thru '89
- *30055 Shadow/Plymouth Sundance '87 thru '94
- *30060 Spirit & Plymouth Acclaim '89 thru '95
- *30065 Vans - Dodge & Plymouth '71 thru '96

EAGLE
- Talon - see MITSUBISHI Eclipse (68030)
- Vision - see CHRYSLER (25025)

FIAT
- 34010 124 Sport Coupe & Spider '68 thru '78
- 34025 X1/9 all models '74 thru '80

FORD
- 10355 Ford Automatic Transmission Overhaul
- 10320 Ford Engine Overhaul Manual
- *36004 Aerostar Mini-vans '86 thru '96
  - Aspire - see FORD Festiva (36030)
- *36006 Contour/Mercury Mystique '95 thru '98
- 36008 Courier Pick-up all models '72 thru '82
- 36012 Crown Victoria & Mercury Grand Marquis '88 thru '96
- 36016 Escort/Mercury Lynx '81 thru '90
- *36020 Escort/Mercury Tracer '91 thru '96
  - Expedition - see FORD Pick-up (36059)
- *36024 Explorer & Mazda Navajo '91 thru '95
- 36028 Fairmont & Mercury Zephyr '78 thru '83
- 36030 Festiva & Aspire '88 thru '97
- 36032 Fiesta all models '77 thru '80
- 36036 Ford & Mercury Full-size,
  - Ford LTD & Mercury Marquis ('75 thru '82); Ford Custom 500,Country Squire, Crown Victoria & Mercury Colony Park ('75 thru '87); Ford LTD Crown Victoria & Mercury Gran Marquis ('83 thru '87)
- 36040 Granada & Mercury Monarch '75 thru '80
- 36044 Ford & Mercury Mid-size,
  - Ford Thunderbird & Mercury Cougar ('75 thru '82); Ford LTD & Mercury Marquis ('83 thru '86); Ford Torino,Gran Torino, Elite, Ranchero pick-up, LTD II, Mercury Montego, Comet, XR-7 & Lincoln Versailles ('75 thru '86)
- 36048 Mustang V8 all models '64-1/2 thru '73
- 36049 Mustang II 4 cyl, V6 & V8 '74 thru '78
- 36050 Mustang & Mercury Capri incl. Turbo Mustang, '79 thru '93; Capri, '79 thru '86
- *36051 Mustang all models '94 thru '97
- *36054 Pick-ups and Bronco '73 thru '79
- *36058 Pick-ups and Bronco '80 thru '96
- *36059 Pick-ups, Expedition & Lincoln Navigator '97 thru '98
- 36062 Pinto & Mercury Bobcat '75 thru '80
- 36066 Probe all models '89 thru '92
- *36070 Ranger/Bronco II gas models '83 thru '92
- *36071 Ford Ranger '93 thru '97 & Mazda Pick-ups '94 thru '97
- *36074 Taurus & Mercury Sable '86 thru '95
- *36075 Taurus & Mercury Sable '96 thru '98
- *36078 Tempo & Mercury Topaz '84 thru '94
- 36082 Thunderbird/Mercury Cougar '83 thru '88
- *36086 Thunderbird/Mercury Cougar '89 and '97
- 36090 Vans all V8 Econoline models '69 thru '91
- *36094 Vans full size '92 thru '95
- *36097 Windstar Mini-van '95 thru '98

GENERAL MOTORS
- *10360 GM Automatic Transmission Overhaul
- *38005 Buick Century, Chevrolet Celebrity, Olds Cutlass Ciera & Pontiac 6000 all models '82 thru '96
- *38010 Buick Regal, Chevrolet Lumina, Oldsmobile Cutlass Supreme & Pontiac Grand Prix front wheel drive '88 thru '93
- *38015 Buick Skyhawk, Cadillac Cimarron, Chevrolet Cavalier, Oldsmobile Firenza Pontiac J-2000 & Sunbird '82 thru '94
- *38016 Chevrolet Cavalier & Pontiac Sunfire '95 thru '98
- 38020 Buick Skylark, Chevrolet Citation, Olds Omega, Pontiac Phoenix '80 thru '85
- 38025 Buick Skylark & Somerset, Olds Achieva, Calais & Pontiac Grand Am '85 thru '95
- 38030 Cadillac Eldorado & Oldsmobile Toronado '71 thru '85, Seville '80 thru '85, Buick Riviera '79 thru '85
- *38035 Chevrolet Lumina APV, Oldsmobile Silhouette & Pontiac Trans Sport '90 thru '95
  - General Motors Full-size Rear-wheel Drive - see BUICK (19020)

GEO
- Metro - see CHEVROLET Sprint (24075)
- Prizm - see CHEVROLET (24060) or TOYOTA (92036)
- *40030 Storm all models '90 thru '93
  - Tracker - see SUZUKI Samurai (90010)

GMC
- Safari - see CHEVROLET ASTRO (24010)
- Vans & Pick-ups - see CHEVROLET

HONDA
- 42010 Accord CVCC all models '76 thru '83
- 42011 Accord all models '84 thru '89
- 42012 Accord all models '90 thru '93
- *42013 Accord all models '94 thru '95
- 42020 Civic 1200 all models '73 thru '79
- 42021 Civic 1300 & 1500 CVCC '80 thru '83
- 42022 Civic 1500 CVCC all models '75 thru '79
- 42023 Civic all models '84 thru '91
- 42024 Civic & del Sol '92 thru '95
  - Passport - see ISUZU Rodeo (47017)
- *42040 Prelude CVCC all models '79 thru '89

HYUNDAI
- *43015 Excel all models '86 thru '94

ISUZU
- Hombre - see CHEVROLET S-10 (24071)
- *47017 Rodeo '91 thru '97, Amigo '89 thru '94, Honda Passport '95 thru '97
- *47020 Trooper '84 thru '91, Pick-up '81 thru '93

JAGUAR
- *49010 XJ6 all 6 cyl models '68 thru '86
- *49011 XJ6 all models '88 thru '94
- *49015 XJ12 & XJS all 12 cyl models '72 thru '85

JEEP
- *50010 Cherokee, Comanche & Wagoneer Limited all models '84 thru '96
- 50020 CJ all models '49 thru '86
- *50025 Grand Cherokee all models '93 thru '98
- *50029 Grand Wagoneer & Pick-up '72 thru '91
- *50030 Wrangler all models '87 thru '95

LINCOLN
- Navigator - see FORD Pick-up (36059)
- 59010 Rear Wheel Drive all models '70 thru '96

MAZDA
- 61010 GLC (rear wheel drive) '77 thru '83
- 61011 GLC (front wheel drive) '81 thru '85
- *61015 323 & Protegé '90 thru '97
- *61016 MX-5 Miata '90 thru '97
- *61020 MPV all models '89 thru '94
  - Navajo - see FORD Explorer (36024)
- 61030 Pick-ups '72 thru '93
  - Pick-ups '94 on - see Ford (36071)
- 61035 RX-7 all models '79 thru '85
- *61036 RX-7 all models '86 thru '91
- *61040 626 (rear wheel drive) '79 thru '82
- *61041 626 & MX-6 (front wheel drive) '83 thru '91

MERCEDES-BENZ
- 63012 123 Series Diesel '76 thru '85
- *63015 190 Series 4-cyl gas models, '84 thru '88
- 63020 230, 250 & 280 6 cyl sohc '68 thru '72
- 63025 280 123 Series gas models '77 thru '81
- 63030 350 & 450 all models '71 thru '80

MERCURY
- See FORD Listing

MG
- 66010 MGB Roadster & GT Coupe '62 thru '80
- 66015 MG Midget & Austin Healey Sprite Roadster '58 thru '80

MITSUBISHI
- *68020 Cordia, Tredia, Galant, Precis & Mirage '83 thru '93
- *68030 Eclipse, Eagle Talon & Plymouth Laser '90 thru '94
- *68040 Pick-up '83 thru '96, Montero '83 thru '93

NISSAN
- 72010 300ZX all models incl. Turbo '84 thru '89
- *72015 Altima all models '93 thru '97
- *72020 Maxima all models '85 thru '91
- *72030 Pick-ups '80 thru '96, Pathfinder '87 thru '95
- 72040 Pulsar all models '83 thru '86
- 72050 Sentra all models '82 thru '94
- *72051 Sentra & 200SX all models '95 thru '98
- *72060 Stanza all models '82 thru '90

OLDSMOBILE
- *73015 Cutlass '74 thru '88
  - For other OLDSMOBILE titles, see BUICK, CHEVROLET or GENERAL MOTORS listing.

PLYMOUTH
- For PLYMOUTH titles, see DODGE.

PONTIAC
- 79008 Fiero all models '84 thru '88
- 79018 Firebird V8 all models except Turbo '70 thru '81
- 79019 Firebird all models '82 thru '92
  - For other PONTIAC titles, see BUICK, CHEVROLET or GENERAL MOTORS listing.

PORSCHE
- *80020 911 Coupe & Targa models '65 thru '89
- 80025 914 all 4 cyl models '69 thru '76
- 80030 924 all models incl. Turbo '76 thru '82
- *80035 944 all models incl. Turbo '83 thru '89

RENAULT
- Alliance, Encore - see AMC (14020)

SAAB
- *84010 900 including Turbo '79 thru '88

SATURN
- *87010 Saturn all models '91 thru '96

SUBARU
- 89002 1100, 1300, 1400 & 1600 '71 thru '79
- *89003 1600 & 1800 2WD & 4WD '80 thru '94

SUZUKI
- *90010 Samurai/Sidekick/Geo Tracker '86 thru '96

TOYOTA
- 92005 Camry all models '83 thru '91
- *92006 Camry all models '92 thru '96
- 92015 Celica Rear Wheel Drive '71 thru '85
- *92020 Celica Front Wheel Drive '86 thru '93
- 92025 Celica Supra all models '79 thru '92
- 92030 Corolla all models '75 thru '79
- *92032 Corolla rear wheel drive models '80 thru '87
- *92035 Corolla front wheel drive models '84 thru '92
- *92036 Corolla & Geo Prizm '93 thru '97
- 92040 Corolla Tercel all models '80 thru '82
- 92045 Corona all models '74 thru '82
- 92050 Cressida all models '78 thru '82
- 92055 Land Cruiser Series FJ40, 43, 45 & 55 '68 thru '82
- *92056 Land Cruiser Series FJ60, 62, 80 & FZJ80 '68 thru '82
- *92065 MR2 all models '85 thru '87
- *92070 Pick-up '69 thru '78
- *92075 Pick-up all models '79 thru '95
- *92076 Tacoma '95 thru '98, 4Runner '96 thru '98, T100 '93 thru '98
- *92080 Previa all models '91 thru '95
- *92085 Tercel all models '87 thru '94

TRIUMPH
- 94007 Spitfire all models '62 thru '81
- 94010 TR7 all models '75 thru '81

VW
- 96008 Beetle & Karmann Ghia '54 thru '79
- 96012 Dasher all gasoline models '74 thru '81
- *96016 Rabbit, Jetta, Scirocco, & Pick-up gas models '74 thru '91 & Convertible '80 thru '92
- *96017 Golf & Jetta '93 thru '97
- *96020 Rabbit, Jetta, Pick-up diesel '77 thru '84
- 96030 Transporter 1600 all models '68 thru '79
- 96035 Transporter 1700, 1800, 2000 '72 thru '79
- 96040 Type 3 1500 & 1600 '63 thru '73
- 96045 Vanagon air-cooled models '80 thru '83

VOLVO
- 97010 120, 130 Series & 1800 Sports '61 thru '73
- 97015 140 Series all models '66 thru '74
- *97020 240 Series all models '76 thru '93
- 97025 260 Series all models '75 thru '82
- *97040 740 & 760 Series all models '82 thru '88

TECHBOOK MANUALS
- 10205 Automotive Computer Codes
- 10210 Automotive Emissions Control Manual
- 10215 Fuel Injection Manual, 1978 thru 1985
- 10220 Fuel Injection Manual, 1986 thru 1996
- 10225 Holley Carburetor Manual
- 10230 Rochester Carburetor Manual
- 10240 Weber/Zenith/Stromberg/SU Carburetor
- 10305 Chevrolet Engine Overhaul Manual
- 10310 Chrysler Engine Overhaul Manual
- 10320 Ford Engine Overhaul Manual
- 10330 GM and Ford Diesel Engine Repair
- 10340 Small Engine Repair Manual
- 10345 Suspension, Steering & Driveline
- 10355 Ford Automatic Transmission Overhaul
- 10360 GM Automatic Transmission Overhaul
- 10405 Automotive Body Repair & Painting
- 10410 Automotive Brake Manual
- 10415 Automotive Detailing Manual
- 10420 Automotive Eelectrical Manual
- 10425 Automotive Heating & Air Conditioning
- 10430 Automotive Reference Dictionary
- 10435 Automotive Tools Manual
- 10440 Used Car Buying Guide
- 10445 Welding Manual
- 10450 ATV Basics

SPANISH MANUALS
- 98903 Reparación de Carrocería & Pintura
- 98905 Códigos Automotrices de la Computadora
- 98910 Frenos Automotriz
- 98915 Inyección de Combustible 1986 al 1994
- 99040 Chevrolet & GMC Camionetas '67 al '87
- 99041 Chevrolet & GMC Camionetas '88 al '95
- 99042 Chevrolet & GMC Camionetas Cerradas '68 al '95
- 99055 Dodge Caravan/Ply. Voyager '84 al '95
- 99075 Ford Camionetas y Bronco '80 al '94
- 99077 Ford Camionetas Cerradas '69 al '91
- 99088 Ford Modelos de Tamaño Grande '75 al '87
- 99089 Ford Modelos de Tamaño Mediano '75 al '86
- 99091 Ford Taurus & Mercury Sable '75 al '86
- 99095 GM Modelos de Tamaño Grande '70 al '90
- 99100 GM Modelos de Tamaño Mediano '70 al '88
- 99110 Nissan Camionetas '80 al '96, Pathfinder '87 al '95
- 99118 Nissan Sentra '82 al '94
- 99125 Toyota Camionetas y 4-Runner '79 al '95

* *Listings shown with an asterisk (*) indicate model coverage as of this printing. These titles will be periodically updated to include later model years - consult your Haynes dealer for more information.*

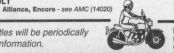

Nearly 100 Haynes motorcycle manuals also available

5-98

Haynes North America, Inc., 861 Lawrence Drive, Newbury Park, CA 91320 • (805) 498-6703